Mountain Biodiversity, Ecosystem Functioning and Services

Mountain Biodiversity, Ecosystem Functioning and Services

Editors

Lin Zhang
Jinniu Wang

MDPI • Basel • Beijing • Wuhan • Barcelona • Belgrade • Manchester • Tokyo • Cluj • Tianjin

Editors
Lin Zhang
Chinese Academy of Sciences
Beijing, China

Jinniu Wang
Chinese Academy of Sciences
Chengdu, China

Editorial Office
MDPI
St. Alban-Anlage 66
4052 Basel, Switzerland

This is a reprint of articles from the Special Issue published online in the open access journal *Diversity* (ISSN 1424-2818) (available at: https://www.mdpi.com/journal/diversity/special_issues/mountain_biodiversity).

For citation purposes, cite each article independently as indicated on the article page online and as indicated below:

LastName, A.A.; LastName, B.B.; LastName, C.C. Article Title. *Journal Name* **Year**, *Volume Number*, Page Range.

ISBN 978-3-0365-8306-8 (Hbk)
ISBN 978-3-0365-8307-5 (PDF)

Cover image courtesy of Jinniu Wang

Contents

About the Editors

Lin Zhang

Lin Zhang, Ph.D., is a Professor in the Institute of Tibetan Plateau Research, Chinese Academy of Sciences (CAS). He is currently the Deputy Director of Motuo Observation Research Center for Earth Landscape and Earth System, CAS. As a member of the Alpine Ecology Committee of the Ecological Society of China, Plant ex situ Conservation Committee of the China Wild Plant Conservation Association, etc., he is active in many fields of ecology, with current research interests in the adaption and responses of alpine plants to climate change, plant geography and functional traits. He has co-authored more than 80 papers worldwide, and participated in more than 20 projects, including the National Natural Science Foundation of China, the S&T Basic Work of Sciences and Technology, and the Second Tibetan Plateau Scientific Expedition and Research (STEP) program.

Jinniu Wang

Jinniu Wang, Ph.D., is an Associate Professor, Chengdu Institute of Biology, Chinese Academy of Sciences. His main research interests including alpine plant functional ecology, vegetation patterns and their drivers in forest and grassland ecotones, and sustainable development in mountainous areas. He is also acting as the Deputy Secretary of the specialized committee in Alpine Ecology, Ecological Society of China. Recently, he was appointed Assistant Head of Mangkang Ecological Station, Tibet Ecological Safety Monitor Network. Dr. Wang has published more than 60 peer reviewed academic papers sponsored by the National Natural Science Foundation of China. Besides acting as guest editor of Forests, Diversity, and Frontiers in Ecology and Evolution—Dryland section, he is an invited Editorial Board Member of the Chinese Journal of Applied and Environmental Biology, and Pratacultural Science. He has acted as a reviewer for more than a dozen international and domestic peer-reviewed academic journals.

diversity

MDPI

Editorial

Mountain Biodiversity, Species Distribution and Ecosystem Functioning in a Changing World

Lin Zhang [1,2,*] and Jinniu Wang [3]

1 State Key Laboratory of Tibetan Plateau Earth System, Resources and Environment (TPESRE), Institute of Tibetan Plateau Research, Chinese Academy of Sciences, Beijing 100101, China
2 Institute of Science and Technology Information Research of Tibet Autonomous Region, Lhasa 850000, China
3 Chengdu Institute of Biology, Chinese Academy of Sciences, Chengdu 610041, China; wangjn@cib.ac.cn
* Correspondence: zhanglin@itpcas.ac.cn; Tel.: +86-010-8409-7055

Citation: Zhang, L.; Wang, J. Mountain Biodiversity, Species Distribution and Ecosystem Functioning in a Changing World. *Diversity* **2023**, *15*, 799. https://doi.org/10.3390/d15070799

Received: 4 June 2023
Revised: 14 June 2023
Accepted: 21 June 2023
Published: 22 June 2023

Mountains encompass more than 30% of all land and 23% of the Earth's forests, with high levels of biodiversity and endemism, and they support diverse habitats and refuges for approximately 85% of amphibian, bird, and mammal species. More than 1/4 of the global human population inhabits mountain environments, many of whom are among the world's poorest people. However, mountains have had a tremendous impact on biodiversity and ecosystems around the world. Around 4000 m above the sea level on the Qinghai–Xizang Plateau, "the roof of the world" is characterized by a low temperature, short growing season, and diverse ecosystems, with "colder soils in a warmer world" and "escalating woody-plants encroachment into grasslands", which might reduce biodiversity and alter ecosystem functions. Biodiversity and people benefitting from nature are essential to human development and the success of the new Sustainable Development Goals [1]. According to the 2022 UN biodiversity Conference COP15, we must fulfill goals related to biodiversity conservation, the construction of ecological civilization, and harmony between humans and nature in the future. This Special Issue, therefore, focuses on studies dealing with many organisms and ecosystems in mountain areas and the responses to climate change and human activities from genes to ecosystems and from valleys to remote alpine regions.

The study of geographic diversity is attractive throughout the world. In northern China, Liang et al. investigated plant diversity along an altitudinal gradient in the Taihang Mountains and found that the plant diversity showed a hump-shaped pattern and anthropogenic factors play an important role [2]. Similarly, arbuscular mycorrhizal fungi (AMF) diversity associated with roots reflected a positive quadratic function with increasing altitude in the Taibai Mountains, which were mostly associated with soil and root nutrients [3]. In the pan-Himalaya region, the species richness in Meconopsis also showed a hump-shaped pattern along environmental gradients, which provides a theoretical background for the conservation and sustainable exploitation of Meconopsis species [4]. Concerning community assembly in alpine grasslands, based on an eleven-year-long fencing experiment along an altitudinal gradient (4400–5200 m) in Central Tibet, Deng et al. proved that stochastic processes drove plant community assembly during the restoration period [5]. Plant diversity is closely related to plant distribution and functions. In this issue, Li et al. found that the species richness level was higher on the north-facing slopes than it was on the south-facing slopes in the alpine meadows of the eastern Tibetan Plateau, and it was negatively correlated with leaf N concentration, providing evidence for interpreting the functional significance of leaf N:P stoichiometry with species composition [6].

Species distribution has frequently been studied under scenarios of future climate change. Xia et al. reviewed the progress on the geographical distribution and driving factors of Nepalese alder (*Alnus nepalensis*) and assumed that temperature would be the main environmental limiting factor since this species presented a clear northern limit of distribution and an upper limit of elevation [7]. Climate warming would, therefore, benefit its expansion. However, the responses to climate change differ among different species. For example, the suitable habitat for *Sabina przewalskii* and *Potentilla parvifolia* will enlarge in the 2070s, while that of *Picea crassifolia* will shrink [8]. *Capra sibirica*, a Siberian ibex in eastern Pamir, may respond to climate change by shifting to higher latitudes due to habitat reduction under future climate scenarios [9]. It seems that animals and birds are more acclimated to the changing climate than plants are. As an example, riparian birds in the Colorado mineral belt appears to be resilient to climate-change-driven increases in metals and rare earth element contents in water and aquatic macroinvertebrates [10].

At larger scales, how does vegetation vary in response to changes in climate and anthropogenic activity? Two papers in this issue analyzed the variation in vegetation using remote sensing data (MODIS NDVI) in a dry mountainous area over the last twenty years. Both cases reported an increased NDVI along the time series, which was associated with precipitation rather than temperature [11,12]. However, the NDVI was controlled by both precipitation and grazing regimes in some parts of the Mt. Qomolangma National Nature Reserve, which is probably because of the high grazing intensity due to tourism and the corresponding high population density [12]. In terms of scenery, according to the difference in plant height, ecotones between abandoned farmland and forest ecosystems can be accurately delineated via high spatial resolution imagery obtained using an aircraft system, which may allow researchers to gain a comprehensive understanding of agricultural abandoned land restoration and support the "Returning Farmland to Forest Program" in China [13]. Luo et al. applied ecological intactness scores (EIS) to indicate the status of the Giant Panda National Park (GPNP). They pointed out that the most of the GPNP's ecological intactness has remained stable during the past 40 years, while only 14% of the GPNP was degraded mainly due to LUCC and road construction [14].

In a practical application, Yunnan pine (*Pinus yunnanensis*) plantation mixed with broadleaved trees, like Nepalese alder, could improve soil quality and enzymes activity [15], which may improve the economy of forest land and enhance the ecological value. In addition, mixed forests have acquired remarkable insect resistance abilities and contributed to the protection of mountain forest ecosystem in southwest China. Li et al. identified the detoxification genes of *Tomicus yunnanensis*, a trunk borer that causes a large number of tree deaths, which provides a theoretical basis for insect resistance in mixed forests [16].

Concerning the relationship between biodiversity and ecosystem functioning, Hou et al. summarized the factors that affect biodiversity and ecosystem functioning (BEF). They further constructed a new conceptual model to provide a systematic means of understanding how different factors affect BEF. The model shows that the correlation between biodiversity and EFs involves a cascade process, while the separation of biodiversity and EFs from ecosystems without considering integrated features is not appropriate for BEF-related research [17].

This Special Issue features research articles with a focus on the Tibetan Plateau and the south-west mountain areas in China. The topic deals with the application of multiscale methods and approaches in the context of biodiversity from the genes and individuals and ecosystems to landscapes, including the adaptations of organisms, like plants, birds, and micro-organisms, spatial–temporal variations in vegetation patterns and land use statuses, and the possible driving factors for the functional adaptation of organisms to climate and anthropogenic changes. It is a shared mission to guarantee the harmonious development of humans and nature around the world and create a greener future. One indispensable component is to enable sustainable and resilient mountain development, which can enhance mountain people's ability to adapt to climate, environmental, and socioeconomic changes and enjoy the benefits and opportunities afforded by natural endowment.

Author Contributions: Writing—review and editing, L.Z. and J.W.; funding acquisition, L.Z. and J.W. All authors have read and agreed to the published version of the manuscript.

Funding: The Natural Science Foundation of Tibet Autonomous Region (XZ202301ZR0027G), the National Natural Science Foundation of China (31971436), and the Second Tibetan Plateau Scientific Expedition and Research (STEP) program (2019QZKK0301-1).

Institutional Review Board Statement: Not applicable.

Data Availability Statement: Not applicable.

Acknowledgments: We are grateful to all authors who contributed to this Special Issue, as well as the reviewers and editors for contributing their time and expertise.

Conflicts of Interest: The authors declare no conflict of interest.

References

1. IPBES. *Global Assessment Report on Biodiversity and Ecosystem Services of the Intergovernmental Science-Policy Platform on Biodiversity and Ecosystem Services;* Brondizio, E.S., Settele, J., Díaz, S., Ngo, H.T., Eds.; IPBES Secretariat: Bonn, Germany, 2019; 1148p. [CrossRef]
2. Liang, H.; Fu, T.; Gao, H.; Li, M.; Liu, J. Climatic and Non-Climatic Drivers of Plant Diversity along an Altitudinal Gradient in the Taihang Mountains of Northern China. *Diversity* **2023**, *15*, 66. [CrossRef]
3. Zhang, M.; Yang, M.; Shi, Z.; Gao, J.; Wang, X. Biodiversity and Variations of Arbuscular Mycorrhizal Fungi Associated with Roots along Elevations in Mt. *Taibai of China. Diversity* **2022**, *14*, 626. [CrossRef]
4. Shi, N.; Wang, C.; Wang, J.; Wu, N.; Naudiyal, N.; Zhang, L.; Wang, L.; Sun, J.; Du, W.; Wei, Y.; et al. Biogeographic Patterns and Richness of the Meconopsis Species and Their Influence Factors across the Pan-Himalaya and Adjacent Regions. *Diversity* **2022**, *14*, 661. [CrossRef]
5. Deng, Z.; Zhao, J.; Wang, Z.; Li, R.; Guo, Y.; Luo, T.; Zhang, L. Stochastic Processes Drive Plant Community Assembly in Alpine Grassland during the Restoration Period. *Diversity* **2022**, *14*, 832. [CrossRef]
6. Li, X.; Hu, Y.; Zhang, R.; Zhao, X.; Qian, C. Linking Leaf N:P Stoichiometry to Species Richness and Composition along a Slope Aspect Gradient in the Eastern Tibetan Meadows. *Diversity* **2022**, *14*, 245. [CrossRef]
7. Xia, C.; Zhao, W.; Wang, J.; Sun, J.; Cui, G.; Zhang, L. Progress on Geographical Distribution, Driving Factors and Ecological Functions of Nepalese Alder. *Diversity* **2023**, *15*, 59. [CrossRef]
8. Hu, H.; Wei, Y.; Wang, W.; Wang, C. The Influence of Climate Change on Three Dominant Alpine Species under Different Scenarios on the Qinghai–Tibetan Plateau. *Diversity* **2021**, *13*, 682. [CrossRef]
9. Zhuo, Y.; Wang, M.; Zhang, B.; Ruckstuhl, K.E.; Alves da Silva, A.; Yang, W.; Alves, J. Siberian Ibex Capra sibirica Respond to Climate Change by Shifting to Higher Latitudes in Eastern Pamir. *Diversity* **2022**, *14*, 750. [CrossRef]
10. Watson, K.E.; McKnight, D.M. Riparian Bird Occupancy in a Mountain Watershed in the Colorado Mineral Belt Appears Resilient to Climate-Change-Driven Increases in Metals and Rare Earth Elements in Water and Aquatic Macroinvertebrates. *Diversity* **2023**, *15*, 712. [CrossRef]
11. Jiang, Y.; Du, W.; Chen, J.; Wang, C.; Wang, J.; Sun, W.; Chai, X.; Ma, L.; Xu, Z. Climatic and Topographical Effects on the Spatiotemporal Variations of Vegetation in Hexi Corridor, Northwestern China. *Diversity* **2022**, *14*, 370. [CrossRef]
12. Zhao, W.; Luo, T.; Wei, H.; Alamu; Zhang, L. Relative Impact of Climate Change and Grazing on NDVI Changes in Grassland in the Mt. Qomolangma Nature Reserve and Adjacent Regions during 2000–2018. *Diversity* **2022**, *14*, 171. [CrossRef]
13. Wang, B.; Sun, H.; Cracknell, A.P.; Deng, Y.; Li, Q.; Lin, L.; Xu, Q.; Ma, Y.; Wang, W.; Zhang, Z. Detection and Quantification of Forest-Agriculture Ecotones Caused by Returning Farmland to Forest Program Using Unmanned Aircraft Imagery. *Diversity* **2022**, *14*, 406. [CrossRef]
14. Luo, C.; Yang, H.; Luo, P.; Liu, S.; Wang, J.; Wang, X.; Li, H.; Mou, C.; Mo, L.; Jia, H.; et al. Spatial-Temporal Change for Ecological Intactness of Giant Panda National Park and Its Adjacent Areas in Sichuan Province, China. *Diversity* **2022**, *14*, 485. [CrossRef]
15. Liang, C.; Liu, L.; Zhang, Z.; Ze, S.; Ji, M.; Li, Z.; Yu, J.; Yang, B.; Zhao, N. Do Mixed Pinus yunnanensis Plantations Improve Soil's Physicochemical Properties and Enzyme Activities? *Diversity* **2022**, *14*, 214. [CrossRef]
16. Li, W.; Yang, B.; Liu, N.; Zhu, J.; Li, Z.; Ze, S.; Yu, J.; Zhao, N. Identification and Characterization of the Detoxification Genes Based on the Transcriptome of Tomicus yunnanensis. *Diversity* **2022**, *14*, 23. [CrossRef]
17. Hou, J.; Feng, H.; Wu, M. Incorporating Effect Factors into the Relationship between Biodiversity and Ecosystem Functioning (BEF). *Diversity* **2022**, *14*, 274. [CrossRef]

diversity

MDPI

Article

Riparian Bird Occupancy in a Mountain Watershed in the Colorado Mineral Belt Appears Resilient to Climate-Change-Driven Increases in Metals and Rare Earth Elements in Water and Aquatic Macroinvertebrates

Kelly E. Watson and Diane M. McKnight *

Institute of Arctic and Alpine Research, University of Colorado, Boulder, CO 80303, USA
* Correspondence: diane.mcknight@colorado.edu

Abstract: Acid rock drainage (ARD) impacts species composition in mountain streams. The potential impact for riparian birds experiencing elevated metal uptake by consumption of benthic invertebrates is concerning but not well studied. We investigated the influence of metal and rare earth element (REE) content in benthic invertebrates on the presence of breeding birds in an ARD-impacted watershed in Colorado, USA, where tree swallows in nest boxes had previously been found to have elevated metal concentrations at some sites. The concentrations of particular REEs in invertebrates were higher than those for cadmium or lead. Avian point counts indicated that most bird species detected were present at most sites, and that tree swallows were rarely found. Occupancy models showed that the availability of shrub or forest habitat was a good predictor for a few habitat-specialists, but metal and REE concentrations in water and invertebrates were not good predictors of avian presence. For other species, neither habitat type nor water quality were important predictors. Overall, this study indicates that the climate-change-driven increases in metals and REEs may not influence the presence of riparian birds in ARD-impacted streams.

Keywords: acid mine drainage; riparian birds; rare earth elements; benthic invertebrates

Citation: Watson, K.E.; McKnight, D.M. Riparian Bird Occupancy in a Mountain Watershed in the Colorado Mineral Belt Appears Resilient to Climate-Change-Driven Increases in Metals and Rare Earth Elements in Water and Aquatic Macroinvertebrates. *Diversity* **2023**, *15*, 712. https://doi.org/10.3390/d15060712

Academic Editors: Lin Zhang, Jinniu Wang and Fabio Stoch

Received: 5 March 2023
Revised: 23 May 2023
Accepted: 24 May 2023
Published: 27 May 2023

1. Introduction

Acidic drainage can occur naturally wherever geology puts sulfidic minerals in contact with air and water and is referred to as acid rock drainage (ARD). Major water quality concerns arise in regions where mining activities greatly accelerate this weathering process, which is referred to as acid mine drainage (AMD) [1]. In the late 1800s, during the Colorado Gold Rush, extensive hard-rock mining took place throughout the Colorado Mineral Belt in the Rocky Mountains (USA). Many mining claims were abandoned just as quickly as they were excavated and piles of waste rock or mine tailings are common throughout the Colorado Rocky Mountains to this day. The US Government Accountability Office has estimated that there are hundreds of thousands of abandoned mines scattered throughout the Western United States [2]. These abandoned mines continue to contribute acidic, metal-enriched water to mountain streams, impairing the aquatic ecosystems and acting as a source of trace metals to drinking-water supplies [3]. Recent studies have found that rare earth elements (REEs) are also elevated in ARD-impacted streams in the Colorado Mineral Belt and that summertime concentrations of both metals and REEs are increasing due to climate change [4].

A pervasive impact of ARD streams is the reduction in species richness, occurring across microbial, algal, and invertebrate communities [5,6]. Genomics methods have shown that microbial taxa in ARD streams are dominated by specialized acidophile organisms [7,8]. Fish and amphibians are often absent due to metal or acid toxicity [9,10]. The drivers for ARD impacts are not only the acidity itself and metal toxicity but also the precipitation of iron and aluminum oxides [11]. In streams with high amounts of metal oxide precipitates covering the streambed, benthic organisms such as primary producers, periphyton, and invertebrates can be smothered completely [12,13]. Ecosystem processes, such as litter decomposition, are also suppressed [14,15]. However, in streams with a low pH and high dissolved metal concentrations, but without precipitates, a few tolerant species of benthic invertebrates may be as abundant as in pristine streams. For example, McKnight and Feder [12] found no difference in the benthic invertebrate abundance in an ARD stream compared to a nearby pristine stream, and the community was dominated by the stonefly *Zapada haysii* which is common in ARD streams.

Benthic macroinvertebrates are often used as bioindicators of stream health, and studies of acidic drainages have shown that invertebrates accumulate metals in their bodies [16,17]. There are few studies of REE content in benthic invertebrates and no studies of REEs' accumulation in benthic invertebrates in ARD streams. In a study of temperate lakes in Canada, Amyot et al. [18] found that REE concentrations ranged from 0.1 to 1 ng/g and that REEs were subject to trophic dilution, with lower concentrations at higher trophic levels. The accumulation of metals and REEs in benthic invertebrates is of concern from an ecosystem perspective because of the potential for transport up the food web through bioaccumulation to fish and other predators [19,20].

There has been little investigation into whether elevated metal or REE concentrations in streams affect nearby terrestrial organisms. One potential pathway for these elements to enter terrestrial food chains is the emergence of aquatic insects that spend their larval stages in streams and can be consumed by birds as adult flying insects. Studies have shown that birds acquire elevated metals from prey in a variety of environments, including from aquatic sources [21–23]. Environmental acidification and the resulting increased mobility of metals, especially cadmium, mercury, and lead, has been a long-term concern for birds [24,25].

The few studies of ARD effects on terrestrial organisms have addressed concerns related to ingesting elevated levels of toxic metals such as lead (Pb) and cadmium (Cd), which may lead to a lower survival in breeding birds. There is some evidence for reduced density, delayed breeding, and fewer offspring produced by Eurasian dippers (*Cinclus cinclus*) and Louisiana waterthrushes (*Seiurus motacilla*) breeding along acidified streams in Wales and Pennsylvania, respectively [26,27]. The most acidic streams in the dipper and waterthrush studies had pH values of 4.5, which is higher than many ARD streams. In a study of tree swallows (*Tachycineta bicolor*) occupying nest boxes adjacent to ARD streams in the Snake River watershed in Colorado (USA), Custer et al. [28] found that at sites where invertebrates had elevated metal concentrations, nestlings also had elevated, but not toxic, Pb concentrations in liver tissue. It has not yet been determined whether ARD affects the behavior, survival, reproduction, or population dynamics of free-living avian species living adjacent to ARD streams. In a recent study of 35 first- and second-order stream ecosystems impacted by ARD in the Colorado Mineral Belt, Kraus et al. [29] found that adult aquatic invertebrates had lower contents of Cd and Pb than the larval stages living in the streams, reducing the impact of metal transfer to insectivorous riparian spiders. These lower metal contents could potentially also limit the impact of metal transfer to insectivorous riparian birds.

The present study was conducted in the same watershed as the study by Custer et al. [28] and assessed the potential impacts to all riparian birds in a large montane watershed with both pristine and metal-impacted sub-watersheds. The two objectives were to determine (i) both the trace metal and REE contents of benthic invertebrates in the watershed to obtain more complete information on ARD impacts, and (ii) whether birds breeding in riparian forests and wetlands in the watershed are less likely to occupy sites adjacent to the ARD-impacted streams compared to the pristine streams. Given that climate change is expected to continue to exacerbate the impacts of ARD [30], the results of this study help to establish the degree of impact of ARD on both aquatic and terrestrial wildlife.

2. Methods

2.1. Study Area

This study builds on the extensive and ongoing record of water-quality monitoring and hydrologic studies in the Snake River watershed (e.g., [4,12,31–33]). For this study, the lowest point of the watershed was defined as the confluence of the Snake River and Peru Creek, forming a watershed with an area of 80 km^2 (Figure 1). Within each subwatershed, valleys are dominated by spruce-fir forest, with riparian willow shrublands near the streams. At higher elevations, the dominant land-cover types are alpine turf and exposed bedrock or scree (based on personal observation and the USGS GAP/LANDFIRE National Terrestrial Ecosystems dataset). Runoff in the watershed is dominated by spring snowmelt, as recorded by a United States Geological Survey (USGS) stream gage (09047500) seven kilometers downstream of the study area (Figure S1). Flow in the Upper Snake River is well correlated with flow at the USGS gage [3].

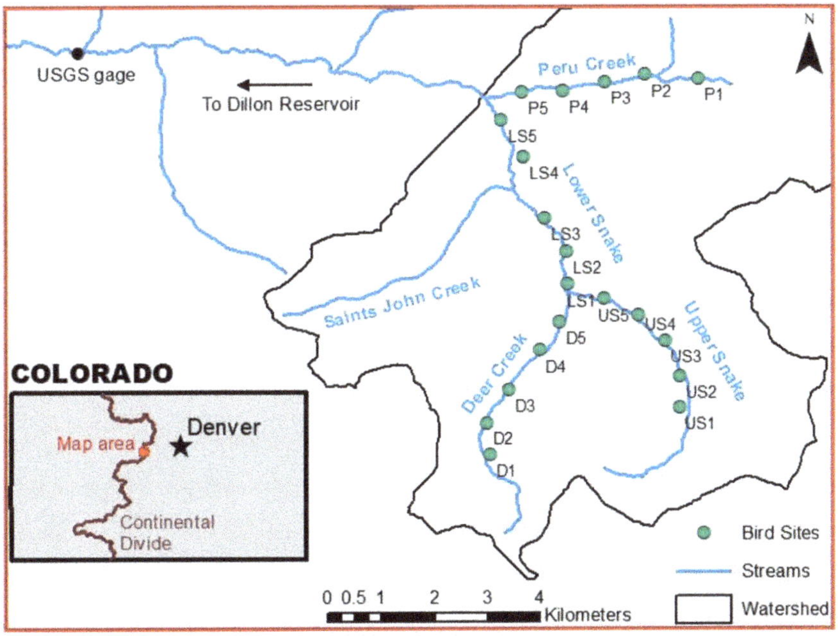

Figure 1. Map of the Snake River watershed in Summit County, Colorado, showing the sites where water-quality samples and benthic invertebrates were sampled and where avian point counts were conducted.

Snake River and Peru Creek are identified on Colorado's 2022 303(d) list of impaired water bodies due to high concentrations of zinc, cadmium, copper, and lead [34]. The Upper Snake River contains naturally-occurring ARD due to underlying acid-producing schists and gneisses [35]. The pH has ranged between 3.4 and 5.3 since 1980 [3,36]. Monitoring records since 1980 for the Upper Snake indicate that warmer summer temperatures have driven increasing concentrations of sulfate, zinc, manganese, other trace metals, and REEs [3,4]. A "tributary of interest" contributes significantly to the trace metal and REE load in the Upper Snake [4,33]. The elevated concentrations of REEs have made REEs an emerging concern for the watershed [4,37].

In contrast to the Upper Snake, the water quality in Deer Creek is good, despite the presence of abandoned mines [38]. The pH values since 1980 have ranged between 6.2 and 7.6.

At the confluence of the Upper Snake with Deer Creek, the pH increases rapidly and there is subsequent precipitation of aluminum and iron oxides. The pH below the confluence has ranged between 4.8 and 6.8 since 1980 [39–41]. Peru Creek, which joins the Lower Snake 4.3 km downstream of the confluence of the Upper Snake and Deer Creek, is heavily impacted by AMD from abandoned mines. The main AMD source in Peru Creek is the Pennsylvania Mine, which was recently remediated to improve water quality.

The Rocky Mountains of Colorado include a wide array of avifauna. In riparian habitats above 3000 m (10,000 feet), common species are the American robin (*Turdus migratorius*), Wilson's warbler (*Cardellina pusilla*), Lincoln's sparrow (*Melospiza lincolnii*), and the white-crowned sparrow (*Zonotrichia leucophrys*) [42]. In spruce-fir forest above 3000 m, common species again include the American robin, as well as the mountain chickadee (*Poecile gambeli*), golden-crowned kinglet (*Regulus satrapa*), ruby-crowned kinglet (*Regulus calendula*), yellow-rumped warbler (*Setophaga coronata*), dark-eyed junco (*Junco hyemalis*), and Cassin's finch (*Haemorhous cassinii*) [42].

Twenty study sites were established with five sites along each of four main streams in the Snake River watershed: Deer Creek, the Upper Snake, the Lower Snake, and Peru Creek (Figure 1). Sites were numbered from upstream to downstream (i.e., D1 was the highest site on Deer Creek, LS5 was the lowest site along the Lower Snake). Six of the sites corresponded to sites used by Custer et al. [28] and Yang [36]: D1 = Upper Deer (Custer et al. site), D3 = Middle Deer, D4 = Lower Deer, US1 = Upper Snake, US4 = Middle Snake, and LS4 = Montezuma. Using satellite imagery, additional sites were selected to ensure at least 600 m between sites. Given that the home ranges of the birds commonly found in the area are much less than 600 m, this buffer meant that the double-counting of birds was unlikely. A GPS unit (Garmin GPSmap 64) was used for navigation in the field. Final sites were determined based on access and proximity to the stream. The elevations of the sites ranged from 3047 to 3465 m (10,000 to 11,370 feet) above sea level. Photos of select sites are provided in Figure S1. For each site, bird counts were conducted between 23 and 140 m away from the stream. The bird counts could not be conducted directly adjacent to the stream because the sound of running water negatively impacted the observer's ability to detect birds.

2.2. Sample Collection, Observations, and Analytical Methods

In order to address the first objective of this study, water and benthic macroinvertebrate samples were collected from both acidic and relatively pristine sites and analyzed for concentrations of trace metals and rare earth elements (REEs). Water-quality data in the Snake River, Deer Creek, and Peru Creek were collected at 14 sites (D1, D2, D3, D4, D5, US1, US3, US4, LS1, LS3, LS5, P1, P2, P4; see Figure 1) between 17 July and 26 July 2017. The mean daily discharge at USGS gage 09047500 was between 96 and 124 cfs on the four sampling days, with higher flows during the first week (Figure S1). Ten sites were visited twice and four sites were only visited once due to access constraints. Specific conductance, water temperature, and pH were measured with YSI meters. Water samples '(150 mL volume) were collected in acid-washed HDPE bottles, filtered on the day of collection through 0.45 μm glass-fiber filters, and acidified to below pH 2 with Fisher Scientific Trace Metal Grade nitric acid.

Benthic macroinvertebrates were collected with a Surber sampler at all sites except P2, where the stream was too deep and fast for the sampler [43]. The contents of the net were placed in a white tray and macroinvertebrates were picked out, placed in a zip-top bag, and frozen immediately. If the number of invertebrates was low, the Surber sampler was deployed a second time to ensure sufficient mass for analysis, thereby allowing for comparison with previous studies in this watershed that used similar methods. Invertebrate samples were stored in a -10 °C freezer before being freeze dried. The dried invertebrates were separated from any detritus with forceps and the dry mass of each sample was recorded. On two occasions (the second visit to P4 and the only visit to LS5), no benthic macroinvertebrates were detected. The smallest mass of benthic macroinvertebrates was

2.2 milligrams at LS1; the largest was 84.6 milligrams at D4. Due to the low mass of some samples, invertebrates from the two sampling dates for D1, D4, LS1, P1, US1, US3, and US4 were combined, for a total of 13 samples. These samples were then digested in a microwave digestor with Fisher Scientific (Waltham, MA, USA) Trace Metal Grade nitric acid.

Both macroinvertebrate and water samples were analyzed for concentrations of 57 cations, trace metals, and REEs using inductively coupled plasma mass spectrometry (ICP-MS). The concentrations of eight commonly measured trace metals (aluminum [Al], cadmium [Cd], copper [Cu], iron [Fe], manganese [Mn], nickel [Ni], lead [Pb], and zinc [Zn]) and 16 rare earth elements (REEs) are reported here. One water sample and one macroinvertebrate sample were run as duplicates to confirm the analytical error of the ICP-MS, which was also confirmed by comparison to standards. The maximum percent error for the water sample was 4.17% (europium), with an average error of 1.60%. The maximum percent error for the macroinvertebrate sample was 11.11% (lutetium), with an average error of 2.01%. The error values were consistently low; thus, metal concentrations are reported without adjustments for error.

In order to address the second objective of this study, the entire suite of bird species in the watershed was observed to gain a broad picture of the community-wide effects of ARD in the terrestrial system. Avian point counts were conducted during three weeks of the avian breeding season in the summer of 2017. Point counts took place between 9 June and 30 June, 2017, based on nesting dates reported in Yang [36]. Water sampling did not take place concurrently with the avian point counts; during the avian breeding season, the water levels were too high to use the Surber sampler. Each site was visited on three separate days for a total of three counts per site. Surveys took place between ten minutes after sunrise and 10 a.m. During each eight-minute count, all individual birds of all species were recorded that were seen or heard. The avian abundance data were converted to presence–absence detection histories for all species that were detected at more than 4 of the 20 sites (>20% naïve occupancy).

The survey-specific covariates recorded for each count included the Julian date, start time, start temperature (from a Garmin tempe wireless temperature sensor), and categorical indices for wind (0 = smoke rises vertically; 1 = smoke drifts; 2 = leaves rustle, wind felt on face; 3 = leaves and small twigs in constant motion; 4 = raises dust, small branches are moved; 5 = small trees in leaf begin to sway), sky (0 = few clouds; 1 = partly cloudy or variable sky; 2 = cloudy), and noise (1 = quiet; 2 = wind rustles leaves; 3 = distant roar of water; 4 = nearby roar of water). In order to reduce the number of modeled parameters, wind was converted to a binary variable with values of 0 and 1 comprising one category and values of 2 or greater comprising the other. Noise was similarly divided into two categories: values of 1 or 2 comprised the first category and values of 3 or 4 comprised the second.

The site-specific covariates were metrics of either habitat or water quality. Habitat variables were derived from remote-sensing data, specifically the USGS GAP/LANDFIRE National Terrestrial Ecosystems dataset (https://gapanalysis.usgs.gov/gaplandcover/data/, accessed on 1 March 2018). Landcover for the state of Colorado was imported into ArcMap along with the study site locations. The first habitat covariate was the landcover type, reclassified as either forest, shrub, or grass. The percent of forest and the percent of wetland shrub were determined around each point at 100-, 200-, and 300-m scales. Finally, elevation (as recorded on a Garmin GPSmap) was included as a habitat variable, because species richness usually decreases with altitude [44]. The Snake River watershed may be at the altitudinal limit of some species. Water-quality covariates were based on the metal and REE concentrations in water or benthic macroinvertebrates.

2.3. Data Analysis

The effects of dissolved metals on biota can be additive [45], and an additive measure of toxicity is more conservative in that it likely overestimates organisms' exposure to harmful conditions. For this study, the "cumulative criterion unit" (CCU) was used and was calculated as the ratio of the measured metal concentration to the U.S. Environmental Protection Agency (EPA) criterion value for that metal, summed for all metals [46]:

$$\text{CCU} = \sum \frac{m_i}{c_i}$$

where *mi* is the total dissolved metal concentration and *ci* is the criterion value for the *i*th metal. CCU values are unitless. Values above 1.0 are likely to cause harm to aquatic organisms [46]. The Colorado Water Quality Control Commission (Regulation No. 33) chronic criterion values were used as the criterion values for Al, Cd, Cu, Mn, Ni, Pb, and Zn, which are hardness dependent and varied with the calcium and magnesium concentrations. The Al standard is for total recoverable metal, but the measurements were made of the dissolved phase and the Al criterion is likely too high. Colorado does not have an Fe standard; thus, the national EPA Criteria Continuous Concentration (CCC), was used—this standard is not hardness dependent and is equal to 1000 μg/L [47]. Chronic values were used instead of acute standards because the chronic values represent a higher threshold, making the CCU a more conservative estimate of toxicity. Only the trace metals were included in the CCU because criterion values for REEs do not exist. Metals that were below the ICP-MS detection limit were not included in the CCU calculation.

In order to evaluate potential spatial autocorrelation, Moran I values were calculated for the dissolved metal and REE concentrations at the stream sites and in the benthic macroinvertebrates, as well as for bird species. The coordinates of the sample sites were used to create a matrix of Euclidean distances for each pair of sites. The inverse of this matrix was used with the Moran I function to determine the Moran I value for each analyte.

A univariate regression was used to determine whether the dissolved metal and REE concentrations were a significant predictor of the invertebrate metal concentration. Separate regressions were conducted for the 9 sites in metal-impacted sub-watersheds and the 5 Deer Creek sites with neutral pH, because these two groups of sites were dominated by invertebrate species in different functional feeding groups. At sites where benthic macroinvertebrate samples from different weeks were combined, the concentrations were averaged across weeks. Data were log-transformed prior to regression. Only regressions with significant p-values (alpha = 0.05) are reported. Regressions, and all other statistical analyses, were conducted in R version 3.4.2 [48].

Occupancy models were used to relate habitat and trace metal and REE exposure variables to the presence of various avian species. In order to avoid the multicollinearity of covariates in occupancy models, aggregate metrics were chosen as site-specific covariates. Principal component analyses (PCAs) were conducted on centered and scaled metal concentrations in water and in invertebrates to reduce the dimensionality of the multivariate dataset while retaining most of the dataset's original variation. The first principal components (PC1) of dissolved metals and REEs in water samples and all metals and REEs in benthic macroinvertebrates were used as site-specific covariates. The CCU (the additive measure of trace-metal concentrations in water) and the concentrations of Pb in invertebrates were also used as predictor variables. Pb was chosen because Pb was the only metal shown to have some cellular-level effects on nesting tree swallows in the Snake River watershed [28]. Occupancy models cannot use sites with missing site-specific covariates, and water and invertebrates were collected for only 14 of the 20 sites; thus, values for the missing sites were interpolated. Table S1 lists all the candidate covariates used in model selection.

The R packages 'unmarked' and 'AICcmodavg' were used to build occupancy models for each bird species using an information theoretic approach [49]. A pre-defined list of candidate models was ranked using the Akaike's Information Criterion (AIC) to account for small sample sizes, such that:

$$AIC_c = -2\log\left(L(\hat{\theta})\right) + 2K\left(\frac{n}{n-K-1}\right)$$

where L refers to the maximum likelihood for the parameter estimates $\hat{\theta}$, given the data and model structure; K is the number of estimated parameters; and n is the sample size [49]. Lower AICc values indicate a more parsimonious model. For a given species, candidate models were ranked according to AICc and the difference was calculated between the lowest AICc value and other models' AICc values, known as the Δ AICc or "AICc difference". Models with a Δ AICc of less than 2 are considered to have substantial support for being the actual best model, while models with a Δ AICc between 4 and 7 have considerably less support [49]. The weights based on the Δ AICc were calculated such that all the individual weights summed to 1 for a given dataset:

$$w_i = \exp(-\frac{1}{2}\Delta_i) / \sum_{r=1}^{R} \exp(-\frac{1}{2}\Delta_i)$$

where R is the number of candidate models. Higher weights indicate a higher "weight of evidence" for a model being the actual best model, given the set of candidate models [49].

A hierarchical method was used in which only the survey-specific covariates were modeled before proceeding to the overall best model selection. For the forest and shrub covariates, the best scale was determined, before proceeding to overall model selection. Site-specific covariates that measured similar characteristics (e.g., PC1 of dissolved metals and CCU, or PC1 of invertebrate metals and Pb in invertebrates) were not included in the same model to avoid multicollinearity. Although these covariates measure similar characteristics, it was necessary to test all covariates since, for example, the CCU does not include information on REEs. The "best" models were defined as those with a Δ AICc of less than 3. These "best" models may have unstable parameters or a low predictive skill; therefore, the model-averaged coefficient estimates (betas) and their 95% confidence intervals for parameters found in the "best" models were determined. A variable was considered to be a stable predictor of occupancy if the 95% confidence interval for the beta estimate did not overlap zero. The predictive skill of the best models was also evaluated by conducting a leave-one-out cross-validation (LOOCV) on both the best model and the null model (the model with no survey- or site-specific parameters). The best model was trained on all data points except one; the occupancy of the left-out point was then predicted based on the observed survey- and site-specific covariates and compared to the observed occupancy. This step was repeated by dropping each point in turn to calculate the root-mean-square error (RMSE) for each model.

3. Results

3.1. Trace-Metal and REE Concentrations in Water and Benthic Invertebrates

All results are presented in downstream order, beginning with the Upper Snake, Deer Creek, the Lower Snake, and finally Peru Creek. The complete results for pH, specific conductance, hardness, and dissolved trace-metal concentrations are presented in Table S1. All dissolved REE concentrations can be found in Table S2. The total mass of the combined benthic invertebrate samples and the sampling effort are presented in Table S3. Trace-metal and REE concentrations in invertebrates can be found in Tables S4 and S5, respectively. All of the Moran I values for dissolved trace metals and REEs and for trace metals and REEs in benthic invertebrates were determined to have an absolute value of less than 0.3, and for most of the trace metals and REEs in invertebrates the p values exceeded 0.05 (Tables S6 and S7). Similar to Pearson coefficients, Moran I values can range from 1 to −1 and the low Moran I values for these analytes indicate a random spatial distribution.

Stoneflies in the genus *Zapata* dominated the Upper Snake and Peru Creek, whereas the most abundant benthic macroinvertebrates in Deer Creek were *Baetis* mayflies, consistent with previous observations in this watershed (Table S8). Invertebrate biomass in the Lower Snake was extremely sparse. As a result of the difference in the functional feeding group of the dominant species in acidic versus neutral pH streams, these streams were analyzed separately.

In the Upper Snake, the pH values were very acidic: between 3.05 and 3.23. The pH values were slightly basic throughout Deer Creek, ranging from 7.66 to 8.05 (Figure 2). In the Lower Snake, below the confluence of Deer Creek, the pH values were less acidic, returning to circum-neutral below the inflow of Saints John Creek (see Figure 1). The pH values along Peru Creek ranged from 4.31 to 5.27, with higher values below tributary inflows.

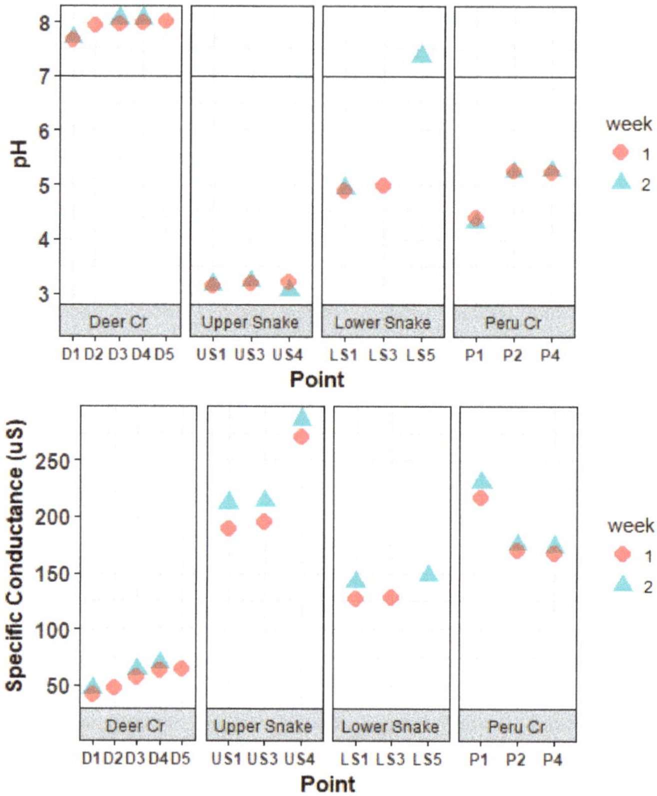

Figure 2. pH values and specific conductance in the Snake River watershed.

The Upper Snake had the highest specific conductance, ranging between 189.5 and 286.5 µS (Figure 2). Specific conductance was much lower in Deer Creek. The Lower Snake had intermediate values between 126.9 and 147.5 µS. Specific conductance in Peru Creek ranged between 167.3 and 230.4 µS.

The CCU values based on the trace-metal concentrations are shown in Figure 3. Dissolved trace-metal and REE concentrations followed a similar pattern throughout the watershed. Figures 4 and 5 present the dissolved concentrations of REEs with high and medium maximum concentrations, respectively. All the REEs in the high category were "light-group" REEs, defined as having an atomic number between 57 and 64.

The concentrations of trace metals in invertebrates are presented in Figure 6, and the concentrations of the more abundant REEs in invertebrates are presented in Figures 7 and 8. In the invertebrates, the concentrations of trace metals and REEs did not follow patterns as clear as were observed in the dissolved concentrations. In general, the REEs that had low dissolved concentrations were also low in invertebrates.

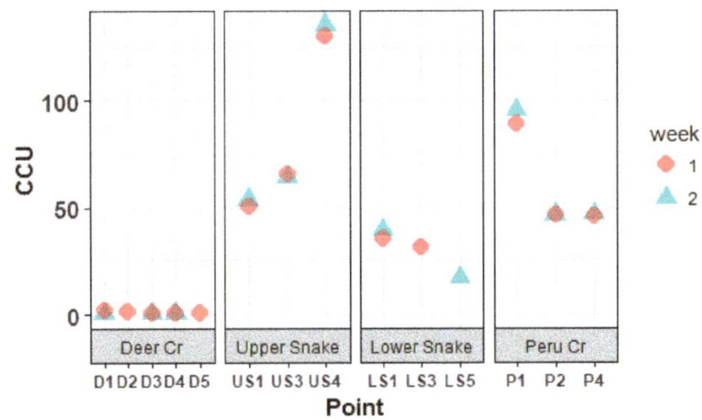

Figure 3. Cumulative Criterion Unit of water samples in the Snake River watershed.

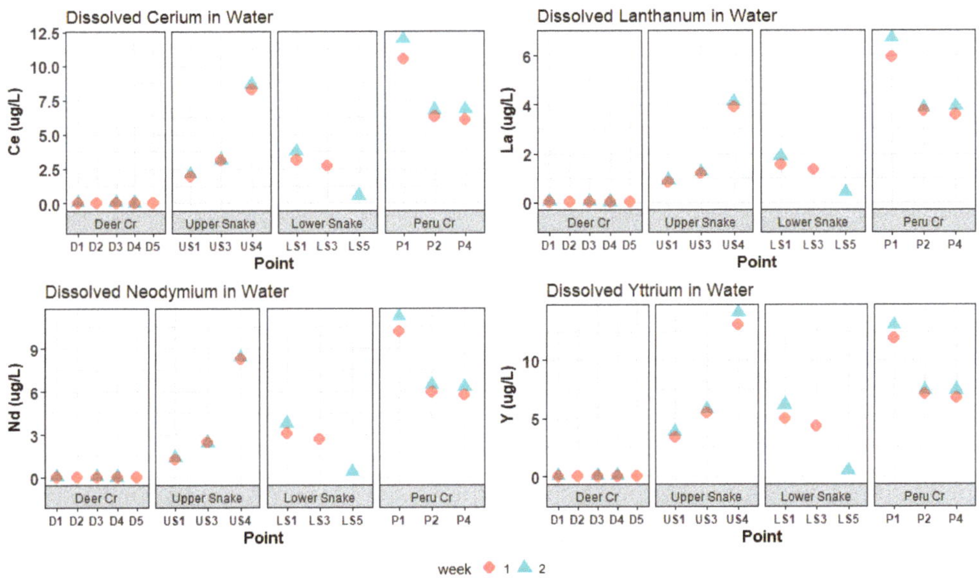

Figure 4. Dissolved Ce, La, Nd, and Y concentrations in water samples. These REEs had high (greater than 6 μg/L) maximum concentrations.

Concentrations of trace metals and REEs in the Upper Snake increased downstream (Figures 4 and 5). A large increase—usually a doubling or tripling in concentration—occurred for Al, Cd, Mn, Ni, Zn, and most REEs between US3 and US4 at the inflow of the "tributary of interest". The concentrations of many elements were higher at US4 than at any other site. Iron concentrations overall were extremely high and decreased from a maximum of 2508 μg/L at US1 to 1696 μg/L at US4. CCU values averaged 58.5 at US1 and US3, and leapt to 132.8 at US4 (Figure 3).

In the Upper Snake, the Mn, Zn, and most REE concentrations in invertebrates increased downstream, but there was no marked increase at US4, below the tributary of interest (Figures 6–8). Meanwhile, Al, Cu, Ni, Pb, Eu, Lu, and Sc peaked at US3 and had lower concentrations at US4. Iron was distinct in that it decreased downstream, with extremely high invertebrate Fe concentrations at US1 of 34,900 μg/g dry mass. Other elements that had maximum concentrations in invertebrates in the Upper Snake were Pb, Ce, Nd, Gd, Sm, and Eu.

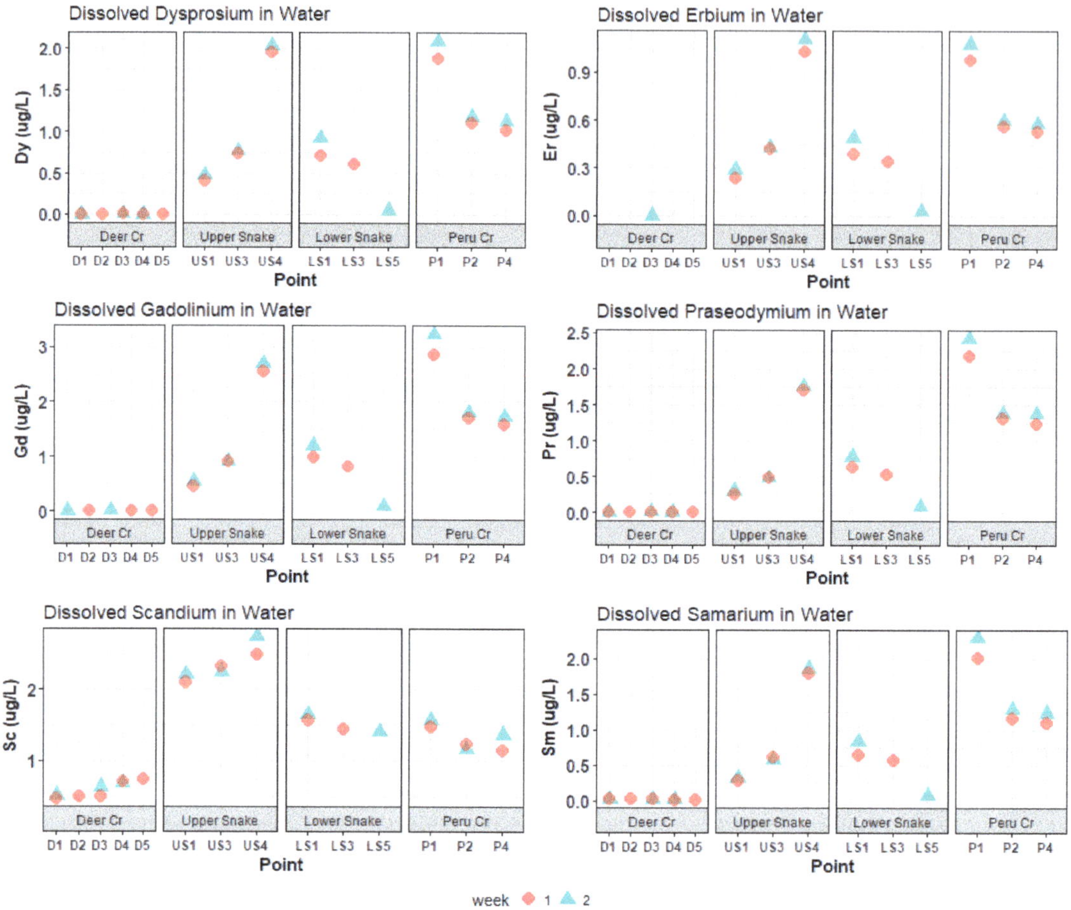

Figure 5. Dissolved Dy, Er, Gd, Pr, Sc, and Sm concentrations in water samples. These REEs had medium (between 1 and 3 μg/L) maximum concentrations. Certain concentrations of gadolinium and erbium in Deer Creek were below ICP-MS detection limits.

In Deer Creek, trace-metal and REE concentrations were relatively low (Figures 3–5). Cadmium, Ni, Pb and all REEs were below 1 μg/L; Al, Cu, Mn, and Zn were all below 11 μg/L; and the highest Fe concentration was 124 μg/L. Metal concentrations were relatively constant going downstream. CCU values at all Deer Creek sites were low (Figure 3).

Despite the relatively low dissolved metal and REE concentrations, the highest concentrations of a few trace metals and REEs in invertebrates in the whole watershed occurred in Deer Creek (Figures 6–8): Cd peaked at 4.12 μg/g dry mass at D5; Pr peaked at 1.93 μg/g at D1; and La peaked at 8.43 μg/g at D1. Manganese and Zn concentrations in invertebrates were also higher in Deer Creek than in the Upper Snake and REE concentrations were within the same range.

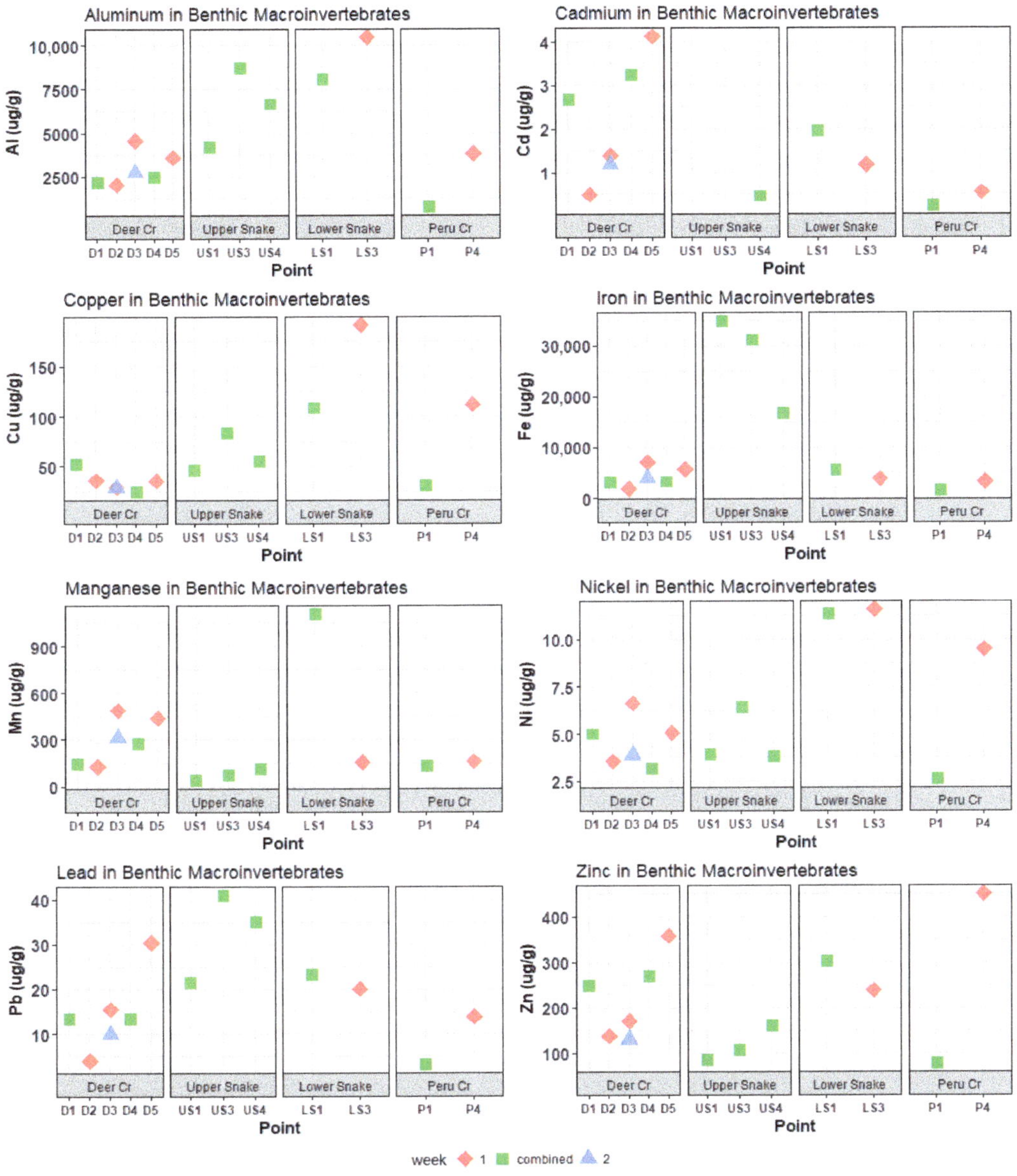

Figure 6. Al, Cd, Cu, Fe, Mn, Ni, Pb, and Zn in benthic macroinvertebrates. Concentrations of cadmium at US1 and US3 were below the ICP-MS detection limits.

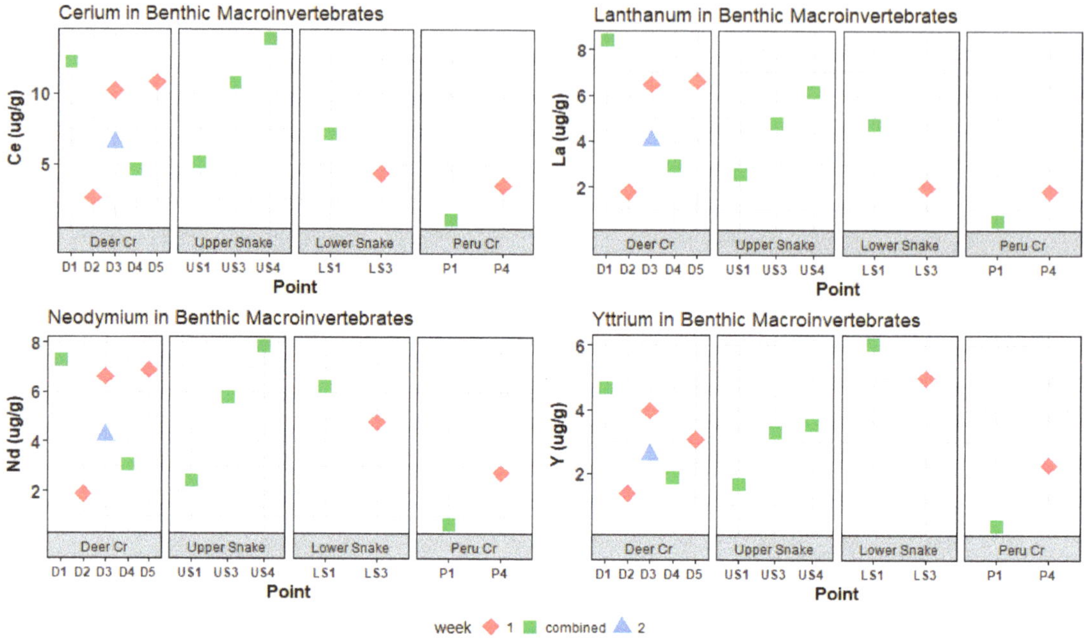

Figure 7. Ce, La, Nd, and Y concentrations in benthic macroinvertebrates. These REEs had high maximum concentrations in water samples. Maximum concentrations in invertebrates were also high (>6 μg/g).

In the Lower Snake, trace-metal and REE concentrations were less than in the Upper Snake (Figures 3–5). White aluminum oxide precipitate was observed on the streambed at LS1. CCU values ranged between 17.2 and 39.7 in the Lower Snake.

In the Lower Snake, invertebrate trace-metal and REE concentrations were intermediate or high compared to those for the headwater streams (Figures 6–8). Concentrations of several metals (Al, Mn, Cu, and Ni) and REEs (Sc, Y, Dy, Er, Yb, Tb, Lu, Ho, and Tm) were higher in the Lower Snake than at any other study site. Most metals and REEs decreased in concentration between LS1 and LS3. No invertebrates were found at LS5, which was the study site below the inflow of Saints John Creek.

In Peru Creek, the dissolved concentrations of many trace metals and REEs were higher at P1 than at any other site in the watershed (Figures 3–5), with Pb concentrations being particularly high. The concentrations of all trace metals and REEs decreased between P1 and P2, where a major tributary joins Peru Creek. P1 had the highest CCU value (95.1). CCU values at P2 and P4 were similar and averaged 46.4.

In Peru Creek, trace-metal and REE concentrations in invertebrates (Figures 6–8) were consistently low at P1 and increased at P4, which is the opposite pattern of the dissolved concentrations. Although dissolved Pb concentrations were the highest in Peru Creek, Pb concentrations in invertebrates were not. Zinc was the only metal that had higher invertebrate concentrations in Peru Creek than in other streams.

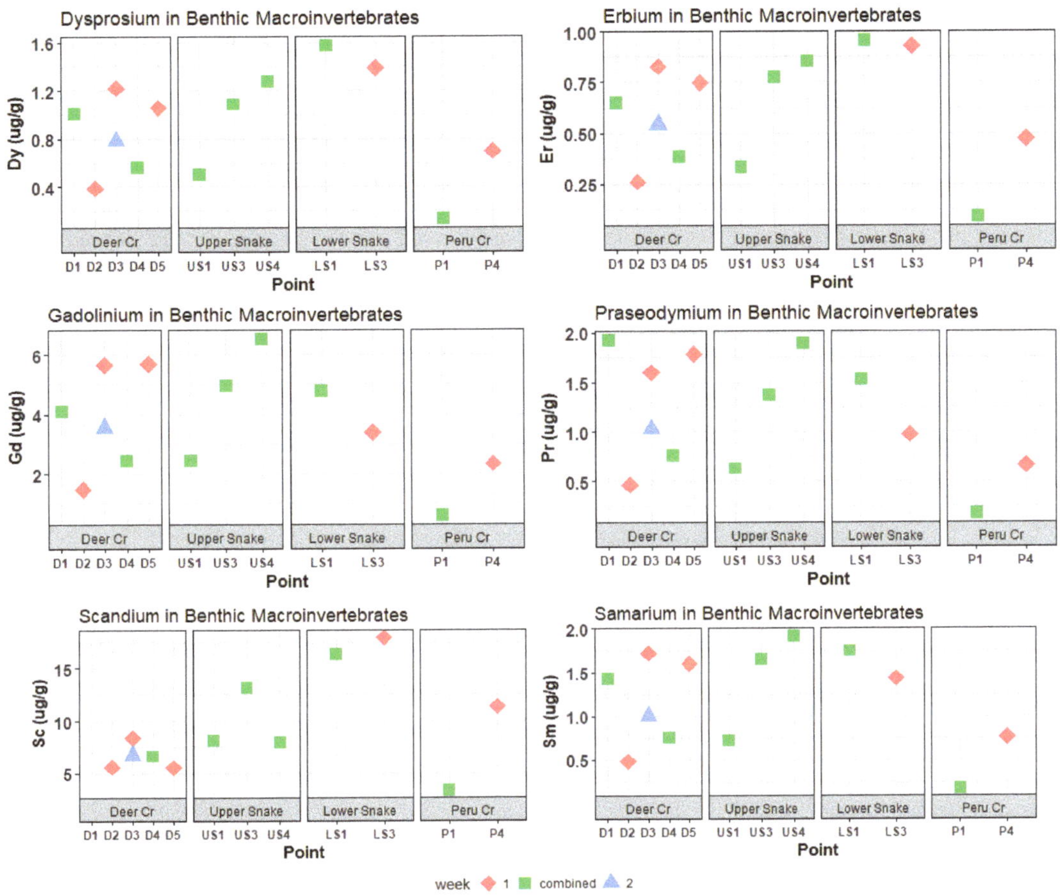

Figure 8. Dy, Er, Gd, Pr, Sc, and Sm concentrations in benthic macroinvertebrates. These REEs had medium maximum concentrations in water samples. Most of these elements also had medium maximum (between 1 and 2 μg/g) concentrations in invertebrates, but Gd and Sc were high (>6 μg/g) relative to other REEs. Concentrations of scandium at D1 were below the ICP-MS detection limits.

3.2. Relationships among Trace-Metal and REE Concentrations in Water and Invertebrates

The PCA of dissolved metal and REE concentrations indicated that each of the four basins had distinct characteristics—sites were neatly clustered by basin (Figure 9A). The first principal component explained 82% of the variation in dissolved metal concentrations and separated out the two sites with the highest metal and REE concentrations (P1 and US4) from the other sites. The loadings on PC1 (Figure 9B) were all positive, indicating that the overall magnitude of metal and REE concentrations drove the variation along PC1. The second PC explained 15% of the variation and separated the Upper Snake from the other basins, especially Peru Creek. The metal loadings on PC2 indicated that differences in Fe and Pb influenced the clusters in Figure 9A. Metal and REE loadings are presented in Table S6.

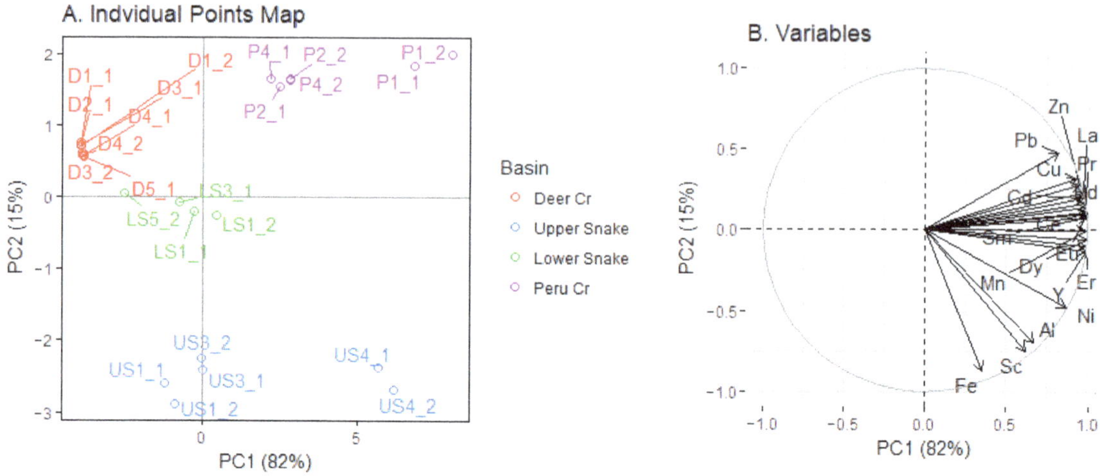

Figure 9. PCA plot of dissolved metals in water. In (**A**), the number after the point indicates which week the point was sampled. In (**B**), loadings have been standardized.

The PCA of trace-metal and REE concentrations in benthic macroinvertebrates did not show a distinct clustering by basin (Figure 10A). The first PC explained 60% of the variation. The loadings on PC1 were all negative, indicating that the overall magnitude of the concentrations drove the variation along PC1. Thus, because the concentrations in Deer Creek and the Upper Snake were similar, the PCA did not separate these two streams. The second PC explained 21% of the variation. The arrows for Fe, Zn, and Mn were the shortest, indicating that they had less of an influence than other elements (Figure 10B). Pb was the trace metal most strongly aligned with the REEs. Metal and REE loadings in invertebrates are presented in Table S7.

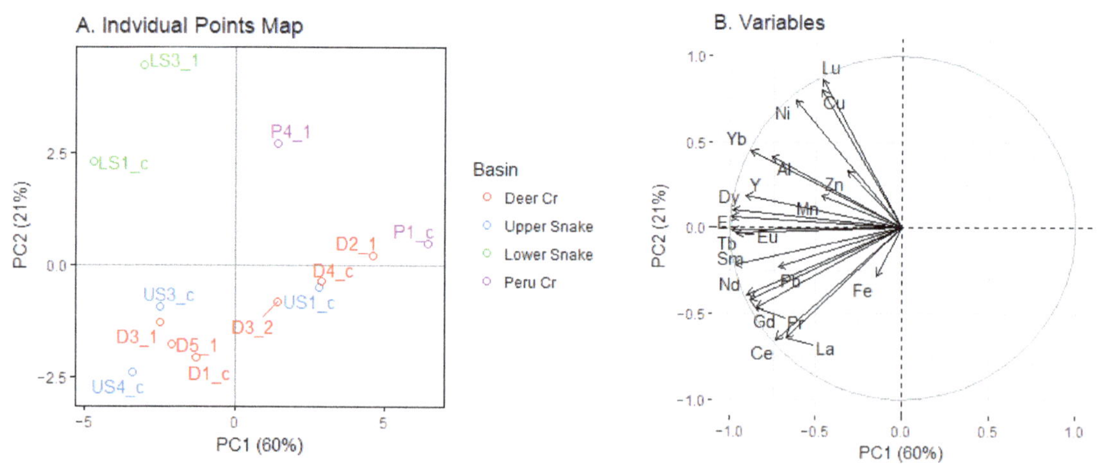

Figure 10. PCA plot of metals in benthic macroinvertebrates. In (**A**), the number after the point indicates which week the point was sampled, or whether it was a combined sample from both weeks. In (**B**), loadings have been standardized.

The clustering of sites in the PCA for tissue concentrations in benthic invertebrates was not similar to the clustering in the PCA for dissolved concentrations. A notable example is the comparison for Cd and Pb, which are known toxicants. Dissolved lead concentrations were lower than those for Cd, but in invertebrates, Pb concentrations were much higher. For REEs, however, high dissolved concentrations (Figures 4 and 5) generally corresponded to high concentrations in invertebrates (Figures 6 and 7). The exceptions were Gd and Sc, which had medium dissolved concentrations (Figure 5) but were high in invertebrates. In fact, concentrations of Ce, Gd, La, Nd, Sc, and Y in invertebrates were as high or higher than concentrations of Cd and Pb.

Regressions for acidic sites in the Upper Snake, Peru Creek, and the Lower Snake showed that dissolved concentrations were generally poor predictors of those metals and REEs in benthic macroinvertebrates (primarily the shredder *Zapata*). One exception was Fe (Table 1), with iron concentrations in invertebrates at acidic sites positively correlated with dissolved iron concentrations ($p = 0.010$). In contrast, the concentrations of Cd and Pb in invertebrates were negatively correlated with the dissolved concentrations of these two metals. This negative relationship may reflect the protective effect of the low pH and high Fe concentrations associated with higher concentrations of Cd and Pb. For all other elements, including REEs, there were no positive or negative significant relationships between the dissolved and invertebrate concentrations (results not shown).

Table 1. r^2 and p-values for significant univariate regressions of log-transformed concentrations of metals in water and invertebrates for acidic sites on the Snake River and Peru Creek ($n = 7$ except for cadmium, where $n = 5$). Water concentrations were averaged across sites for those sites where invertebrate samples were combined; * indicates a significant p-value (alpha = 0.05).

Metal	Direction of Relationship	Adjusted r^2	p-Value
	Relationships for acidic sites		
Cadmium	Negative	0.829	0.020 *
Iron	Positive	0.717	0.010 *
Lead	Negative	0.516	0.042 *

In Deer Creek, where dissolved metal and REE concentrations were low and the dominant invertebrate is a collector–gatherer/scraper (*Baetis*), there were no elements with significant relationships between concentrations in the water and invertebrates.

3.3. Riparian Bird Distribution and Occupancy Models

Twelve bird species were detected during point counts. Table 2 provides scientific and common names for these species, as well as the naïve occupancy (i.e., the number of sites where the species was detected in at least one survey divided by the number of sites). Eight species were detected at more than 20% of the study sites.

Complete abundance data for all twelve species is provided in Tables S11–S13. The Moran I values for the bird species were all less than 0.065, indicating that the data did not exhibit spatial autocorrelation (Table S14). The American robin and dark-eyed junco were detected at all Deer Creek sites, four sites each in the Upper Snake and Peru Creek, and three sites in the Lower Snake. Lincoln's sparrow was found at four of five sites along Deer Creek, three in the Lower Snake, and only two each in the Upper Snake and Peru Creek. The mountain chickadee was found at all sites in both Deer Creek and the Upper Snake, but only three sites each along the Lower Snake and Peru Creek. The ruby-crowned kinglet was detected at all Deer Creek sites, four Upper Snake sites, and only three sites each along the Lower Snake and Peru Creek. The white-crowned sparrow and Wilson's warbler were detected at four of five sites in all basins except Peru Creek, where they were only detected at one site. The yellow-rumped warbler was found at all Upper Snake sites but only three Deer Creek sites, three Peru Creek sites, and two Lower Snake sites.

Table 2. Birds detected in the Snake River watershed. Naïve occupancy refers to the ratio of sites where the species was detected in at least one survey. Occupancy models were not built for the greyed-out species.

Common Name	Scientific Name	Naïve Occupancy
American Crow	*Corvus brachyrhynchos*	0.20
American Robin	*Turdus migratorius*	0.80
Dark-eyed Junco	*Junco hyemalis*	0.80
Lincoln's Sparrow	*Melospiza lincolnii*	0.55
Mountain Chickadee	*Poecile gambeli*	0.80
Northern Flicker	*Colaptes auratus*	0.05
Pine Grosbeak	*Pinicola enucleator*	0.05
Ruby-crowned Kinglet	*Regulus calendula*	0.75
Steller's Jay	*Cyanocitta stelleri*	0.10
White-crowned Sparrow	*Zonotrichia leucophrys*	0.65
Wilson's Warbler	*Cardellina pusilla*	0.65
Yellow-rumped Warbler	*Setophaga coronata*	0.65

Values for site-specific covariates and survey-specific covariates are reported in Tables S15 and S16, respectively. Sites along the Snake River and Deer Creek were a mix of forest and wetland shrub habitat, with one high-elevation site (US1) classified as grass. All Peru Creek sites were dominated by forest habitat. The percent of shrub habitat tended to decrease as the scale increased from 100 m to 300 m, whereas forest habitat tended to increase as the radius increased.

Using these habitat parameters, plus the water and invertebrate parameters, occupancy models were built for the eight species detected at more than 20% of the study sites. One clear relationship emerged from these models. Three wetland-shrub habitat specialists were detected in the watershed: Wilson's warbler, Lincoln's sparrow, and the white-crowned sparrow. None of these three species were detected at US5, LS5, P2, P4, or P5 (among other sites), which were the only five sites with zero percent shrub within 100 m. Otherwise, these species were commonly detected at Deer Creek, Upper Snake, and Lower Snake sites with shrub habitat, indicating that habitat, but not water quality or invertebrate prey quality, is an important factor for these species during the breeding season. There were no obvious patterns in the occurrence of the other five species. The American robin is a habitat generalist while the dark-eyed junco, mountain chickadee, ruby-crowned kinglet, and yellow-rumped warbler all prefer forest habitats. However, all these species were found at sites with a variety of values for percent forest and shrub. The best models for each species (models with a Δ AICc of less than 3) are listed in Table 3, and model-averaged estimates of the coefficients (betas) for parameters in those models are found in Table 4. For certain species, one or two models were clearly better than the others. For others, the AICc weights offered support for a suite of models. The site-specific covariates that appeared in the set of best models varied by species and included both water quality and habitat variables. The 95% confidence intervals for all site-specific covariates, for all species, included zero, indicating that none of these variables were stable predictors of occupancy. However, this outcome may be due to the small sample size. Despite having wide confidence intervals, many of the parameter estimates make sense based on the known habitat preferences for particular species. Finally, the leave-one-out cross-validation (LOOCV) results showed that the best models for four of the eight species did a better job at predicting occupancy than the null models for those species (Table 5).

Table 3. Occupancy models with a Δ AICc of less than 3 for each species. Models are named based on their site-specific (Ψ) and survey-specific covariates (*p*), and Ψ(.), *p*(.) indicates models with no covariates.

Species	Model Names	AICc	Δ AICc	Weight	Cumulative Weight
American Robin	Ψ(Elevation), *p*(Stemp)	77.715	0.000	0.825	0.825
Dark-eyed Junco	Ψ(.), *p*(Stemp)	85.609	0.000	0.318	0.318
	Ψ(.), *p*(.)	86.950	1.341	0.163	0.481
	Ψ(Elevation), *p*(Stemp)	87.948	2.339	0.099	0.579
	Ψ(InvertPb), *p*(Stemp)	88.294	2.685	0.083	0.662
	Ψ(Forest300), *p*(Stemp)	88.590	2.981	0.072	0.734
	Ψ(Shrub100), *p*(Stemp)	88.598	2.989	0.071	0.805
Lincoln's Sparrow	Ψ(Shrub300), *p*(Julian)	54.974	0.000	0.462	0.462
	Ψ(Forest100), *p*(Julian)	56.801	1.827	0.185	0.647
Mountain Chickadee	Ψ(.), *p*(Julian)	67.352	0.000	0.253	0.253
	Ψ(PC1inverts), *p*(Julian)	68.214	0.861	0.165	0.418
	Ψ(Shrub100), *p*(Julian)	68.459	1.106	0.146	0.564
	Ψ(Elevation), *p*(Julian)	68.805	1.453	0.123	0.686
Ruby-crowned Kinglet	Ψ(.), *p*(Julian + Noise)	81.668	0.000	0.376	0.376
	Ψ(PC1inverts), *p*(Julian + Noise)	84.364	2.696	0.098	0.473
White-crowned Sparrow	Ψ(Forest100 + InvertPb), *p*(Julian)	55.342	0.000	0.463	0.463
	Ψ(Shrub100 + InvertPb), *p*(Julian)	55.348	0.006	0.462	0.925
Wilson's Warbler	Ψ(Shrub300), *p*(.)	58.425	0.000	0.417	0.417
	Ψ(Forest100), *p*(.)	58.453	0.028	0.411	0.828
Yellow-rumped Warbler	Ψ(.), *p*(Noise)	80.543	0.000	0.182	0.182
	Ψ(.), *p*(.)	80.747	0.204	0.164	0.346
	Ψ(Forest300 + CCU), *p*(Noise)	81.047	0.504	0.141	0.487
	Ψ(Forest300), *p*(Noise)	81.159	0.617	0.133	0.620
	Ψ(CCU), *p*(Noise)	81.619	1.077	0.106	0.726
	Ψ(PC1water), *p*(Noise)	82.484	1.941	0.069	0.795
	Ψ(PC1inverts), *p*(Noise)	82.659	2.117	0.063	0.858
	Ψ(Elevation), *p*(Noise)	83.096	2.554	0.051	0.909
	Ψ(Shrub300), *p*(Noise)	83.374	2.831	0.044	0.953

Table 4. Modeled-averaged parameter coefficient estimates from occupancy models. Both site-specific (Ψ) and survey-specific (*p*) covariates are included. Only estimates of parameters in occupancy models with a Δ AICc of less than 3 are included. An asterisk (*) indicates a stable estimate, i.e., the confidence interval does not overlap zero.

Species	Parameter	Model-Averaged Beta Estimate	95% Confidence Interval
American Robin	Elevation (Ψ)	34.37	−128.79–197.54
	Stemp (*p*)	0.07	−0.01–0.14

Table 4. *Cont.*

Species	Parameter	Model-Averaged Beta Estimate	95% Confidence Interval
Dark-eyed Junco	Elevation (Ψ)	1.22	−2.19–4.64
	InvertPb (Ψ)	−0.09	−0.42–0.24
	Forest300 (Ψ)	−0.02	−0.17–0.12
	Shrub100 (Ψ)	−0.02	−0.11–0.07
	Stemp (*p*)	−0.07	−0.15–0
Lincoln's Sparrow	Shrub300 (Ψ)	0.15	−0.01–0.3
	Forest100 (Ψ)	−0.03	−0.08–0.03
	Julian (*p*)	0.33 *	0.06–0.59
Mountain Chickadee	PC1inverts (Ψ)	0.78	−0.92–2.48
	Shrub100 (Ψ)	−0.06	−0.18–0.06
	Elevation (Ψ)	1.03	−0.85–2.91
	Julian (*p*)	−0.23 *	−0.35–−0.11
Ruby-crowned Kinglet	PC1inverts (Ψ)	−0.25	−0.81–0.32
	Julian (*p*)	0.1	−0.01–0.21
	Noise1 (*p*)	−2.21 *	−3.85–−0.58
White-crowned Sparrow	InvertPb (Ψ)	3.72	−7.38–14.82
	Shrub100 (Ψ)	3.4	−6.25–13.01
	Forest100 (Ψ)	−1.52	−6.42–3.38
	Julian (*p*)	0.2 *	0.06–0.34
Wilson's Warbler	Shrub300 (Ψ)	2.38	−50.72–55.48
	Forest100 (Ψ)	−0.8	−17.84–16.24
Yellow-rumped Warbler	Forest300 (Ψ)	−0.08	−0.22–0.06
	CCU (Ψ)	0.05	−0.06–0.15
	PC1water (Ψ)	0.33	−0.68–1.34
	PC1inverts (Ψ)	0.50	−0.97–1.97
	Elevation (Ψ)	0.60	−0.82–2.03
	Shrub300 (Ψ)	0.05	−0.12–0.22
	Noise1 (*p*)	−1.35	−2.96–0.25

Table 5. Leave-one-out cross-validation model skill. Lower root-mean-square error (RMSE) indicates that the model did a better job of predicting the occupancy of sites not included in the model's training data. An asterisk (*) indicates that the best model was better than the null model, which has no survey- or site-specific covariates.

Species	RMSE of Null Model	Best Model	RMSE of Best Model
American Robin	0.443	Ψ(Elevation), *p*(Stemp)	0.229 *
Dark-eyed Junco	0.436	Ψ(.), *p*(Stemp)	0.442
Lincoln's Sparrow	0.524	Ψ(Shrub300), *p*(Julian)	0.447 *
Mountain Chickadee	0.423	Ψ(.), *p*(Julian)	0.424

Table 5. *Cont.*

Species	RMSE of Null Model	Best Model	RMSE of Best Model
Ruby-crowned Kinglet	0.465	Ψ(.), *p*(Julian + Noise)	0.469
White-crowned Sparrow	0.510	Ψ(Forest100 + InvertPb), *p*(Julian)	0.358 *
Wilson's Warbler	0.505	Ψ(Shrub300), *p*(.)	0.004 *
Yellow-rumped Warbler	0.507	Ψ(.), *p*(Noise)	0.533

4. Discussion

Overall, the water quality in the study area was similar to many prior studies of this watershed [4,12,28,31,36,37]: acidic sub-watersheds (the Upper Snake and Peru Creek) contained high levels of dissolved trace metals and REEs, while Deer Creek contained low levels of dissolved metals and REEs. As in previous studies, the dominant benthic macroinvertebrate species varied by basin [12,36], which may influence the differences across the sub-watersheds because trace-metal concentrations in freshwater benthic invertebrates have been found to vary with functional feeding guilds [50]. Stoneflies, which dominate the Upper Snake and Peru Creek basins, are shredder–detritivores that consume leaves that have fallen into the stream and have been colonized by bacteria and fungi [51]. As such, the stoneflies may avoid consuming elevated metal concentrations, as demonstrated by the finding that trace-metal and REE concentrations in invertebrates were not well correlated with concentrations in the water. Mayflies, which dominate the relatively pristine Deer Creek basin, are collector–gatherer/scrapers that consume periphyton biomass that may be enriched in trace metals. These mayflies are rarely found in streams with elevated metals, so this taxon is usually considered to be sensitive to poor water quality. However, there were no significant relationships between metal and REE concentrations in the water and invertebrates at Deer Creek sites.

The comparison of invertebrate trace-metal concentrations in this study conducted in 2017 with data for five of the sites studied in 2004 by Yang [36] does show that the concentrations of Al, Cu, Fe, and Ni increased at all five sites. Al concentrations were over an order of magnitude higher in both Deer Creek and the Upper Snake in 2017. Increases in Deer Creek over the thirteen years were smaller, and Cd, Mn, Pb, and Zn concentrations increased at some sites but not others. One of the main patterns that Yang [36] found was a large increase in invertebrate trace-metal concentrations at the Middle Deer site compared to the surrounding sites in Deer Creek. In 2017, Al, Fe, and Mn also peaked at this site (D3) compared to D1 and D4, but other trace metals had decreased. Despite these increases, which may be associated with the increasing concentrations of trace metals over time in the watershed [3,4], it is important to consider that the benthic invertebrate taxa currently present in the watershed may be adapted to these high concentrations of trace metals and REEs that may have occurred during warm periods in the past.

In the context of potential adaptation, it is important to note that the concentrations of REEs in benthic invertebrates found in this ARD-impacted watershed are much higher than other reported values. In comparison to the REE concentrations found by Amyot et al. [18] for invertebrates in pristine temperate lakes in Canada, the REE concentrations for all the stream reaches in the Snake River watershed are 3–4 orders of magnitude higher. Furthermore, the REE concentrations found in this study are 1–2 orders of magnitude higher than the concentrations reported by Pastorino et al. [17] for invertebrates in six stream sites in Italy receiving inputs from nearby agricultural and industrial activities. However, similar to the findings of Pastorino et al. [17], the light REEs generally exhibited higher concentrations than the heavy REEs.

There may be a potential for the bioconcentration of REEs in the aquatic ecosystems of the Snake River watershed. Although there are no native fish populations present in the Snake River, Deer Creek does support native fish and the Lower Snake is occasionally stocked with rainbow trout as a recreational fishery [10]. The concentrations of some REEs in invertebrates were as high or higher than those for Cd or Pb, which are known toxicants; thus, REEs are an emerging concern beyond the widely recognized impacts of ARD.

Of the eight avian species detected at more than 20% of the study sites, the mountain chickadee is the only resident species, although it may move to lower elevations during the winter. The other seven species all migrate south in the fall and arrive back in the Rocky Mountains between mid-March and late-May [52]. All species are reported to eat at least some arthropods, including insects, during the breeding season. The home ranges for these eight species are generally much smaller than the 600-m buffer between the sites. For example, mountain chickadees and white-crowned sparrows have territories of 6–7 ha, which is equivalent to a circle with radius ~150 m [53,54]. Sabo [55] found that territories for the ruby-crowned kinglet, yellow-rumped warbler, and dark-eyed junco ranged from 0.6 to 1.1 ha.

Occupancy models for eight insectivorous avian species provided little evidence that bird presence in the Snake River watershed was influenced by water quality or metal and REE concentrations in invertebrates, which represent potential prey during the breeding season. This is despite large variation in habitat and water-quality parameters across the study area. For three species (Lincoln's sparrow, Wilson's warbler, and the white-crowned sparrow), the best models did predict occupancy better than the null model. All three of these species are habitat specialists—they only breed in willow (*Salix* spp.) wetlands such as those found in parts of the Snake River watershed [56–58]. Therefore, it makes sense that the best models included percent shrub or percent forest landcover. For four species (dark-eyed junco, mountain chickadee, ruby-crowned kinglet, yellow-rumped warbler), the best models did not predict occupancy any better than the null models. This suggests that the factors that are influencing the occupancy of these species were not included in the candidate models, potentially due to the widespread distribution of these species in the study area. The best model for the American robin also predicted occupancy better than the null model, but this may have been an artefact of a small sample size; the model suggested that robin occupancy increases with increasing elevation, but robins are known to thrive at all elevations up to those found in the study area [42].

The only potential indicator of a relationship between avian occupancy and metal concentrations was the best model for the white-crowned sparrow. In addition to percent forest cover within 100 m, the best model for this species included lead concentrations in invertebrates. Given that Custer et al. [28] found that Pb concentrations in the livers of tree swallows had been high enough to inhibit some enzyme activity, the presence of invertebrate Pb concentrations in the best model supports the further study of Pb bioavailability and bioaccumulation.

Besides the white-crowned sparrow, other species did not show any relationship between occupancy and metal and REE concentrations in aquatic invertebrates. This finding, suggesting a limited impact on riparian birds, is consistent with the results of the study by Kraus et al. [29] which included 35 streams and showed that concentrations of Cd and Pb were generally lower in adult aquatic insects compared to the aquatic insect larvae—thereby limiting the trophic transfer to predatory riparian spiders. Furthermore, the results of this study provide a broader context for interpreting the results of a previous study in the Snake River watershed, which examined birds that used nest boxes set up for the study [28]. Ninety percent of the birds using the nest boxes were tree swallows (*Tachycineta bicolor*) or violet–green swallows (*T. thalassina*) [28]. The authors of the previous study found that benthic macroinvertebrates contained elevated concentrations of metals at one site on Deer Creek and that tree swallows did not use nest boxes at that site, indicating a possible link between water quality and avian habitat selection [28]. However, no swallows were observed anywhere in the study area in 2017. Tree swallows are rarely found above 9000 feet in Colorado [42], but they have been known to breed in nest boxes outside their typical range [59]. Thus, the study area may be at too high an elevation for swallows to breed without the incentive of a nest box. This is relevant because swallows primarily obtain food by capturing insects while flying. All of the songbirds considered in the occupancy models of the current study primarily obtain food by gleaning insects from vegetation or the ground [52]. Due to these behavioral differences, it is likely that most of the biomass consumed by swallows is of aquatic origin, i.e., insects that have emerged directly from the stream [60]. The songbirds detected in 2017 have a more diverse diet that includes spiders, caterpillars (*Lepidopetra larvae*), aphids, and some vegetable matter (e.g., [55,61]).

The metal concentrations of non-aquatic avian prey items in the study area are not known but it is certainly possible that they contain lower metal concentrations than aquatic invertebrates. Therefore, by eating a variety of invertebrate prey, as well as seeds and berries, songbirds may be exposed to lower metal concentrations than aerial insectivores such as swallows. On the other hand, studies have shown that predatory invertebrates such as spiders are important vectors for the trophic transfer of metal contaminants. For example, Kraus et al. [62] showed that spiders had higher concentrations of zinc, copper, and cadmium than their aquatic prey. However, because insect emergence decreases as metal contamination in streams increases, especially for mayflies, the total metal exported by insects is the lowest at the most contaminated sites [62]. These studies suggest that understanding avian exposure to contaminants requires a complete picture of the food chain.

There are many other possible reasons beyond a broad diet that water quality, specifically elevated metal and REE concentrations in streams and benthic macroinvertebrates, would have no effect on bird occupancy. First, not all metals are harmful. Of the elements reported in this study, only Pb and Cd are known to be harmful to birds [28], although data on the toxicity of rare earth elements are scarce. Metals such as Cu, Fe, Mn, and Zn are essential or beneficial for birds and are easily metabolized. Additionally, birds may not be exposed to elevated metals for enough time to feel the effects. Except for the mountain chickadee, all of the modeled species are migrants that spend at most half a year in the study area. The winter season could be enough time for birds to metabolize the metals and remove them from their systems. However, philopatry is common in birds [63], so it is probable that most individuals return to the same breeding grounds year after year and are potentially repeatedly exposed to contaminants. Certain species are also arriving to breeding grounds earlier each year due to changing climate cues at low elevations [64], meaning that birds could be exposed to contaminants for a longer time, increasing the chances of a chronic toxicity.

There are also many reasons that birds breeding adjacent to ARD streams could, in fact, be experiencing negative effects that would not be detectable from occupancy modeling. For example, Mulvihill et al. [27] found that Louisiana waterthrushes were still present and breeding along ARD streams, but that their territories had to be larger to account for the unfavorable habitat quality. In addition, while fecundity was the same at ARD sites and control sites, birds at ARD sites produced smaller clutches and nestlings grew more slowly [27]. The point count data obtained in the current study would not have captured all these effects.

5. Conclusions

In the Colorado Mineral Belt and other mountainous regions worldwide, acid rock drainage (ARD) has a pervasive and deleterious influence on water quality. In this study, the uptake of both trace metals and REEs by benthic invertebrates in ARD-impacted streams was considered as a potential factor influencing the presence of breeding birds in the Snake River watershed. One important and novel finding is that the concentrations of some REEs in invertebrates were as high or higher than those for Cd or Pb, which are known toxicants. Furthermore, the PCA indicated that REE concentrations in invertebrates were aligned with Pb concentrations in invertebrates. Overall, given the dearth of information about

REEs in aquatic organisms, further study of the availability and transfer of REEs through ARD-impacted aquatic ecosystems is warranted.

Another important finding was that dissolved metal and REE concentrations were not good predictors of concentrations in benthic macroinvertebrates, even when acidic and pristine sites were considered separately. This lack of a relationship is likely due to differing exposure and uptake by the resident species of benthic macroinvertebrates. Furthermore, while the benthic invertebrates present in the ARD-impacted streams did have high concentrations of metals and REEs in their tissues, this study did not find evidence of an impact on the presence of riparian songbirds in the watershed, although samples sizes were small. The broad diet of the songbirds and potentially lower metal concentrations in the adult invertebrates compared to the aquatic juvenile insects may account for this result. The monitoring of ARD sites during and after remediation often focuses on the stream environment [65]. The finding of a limited impact of ARD on riparian birds may be useful in assessing whether or not monitoring of these sites during remediation of mining impacts may need to expand to include an evaluation of the metal and REE contents of riparian birds and other terrestrial wildlife.

Supplementary Materials: The following supporting information can be downloaded at: https://www.mdpi.com/article/10.3390/10.3390/d15060712/s1, Table S1: Trace-metal concentrations in water, CCU, pH, and specific conductance at all study sites; Table S2: Rare earth element concentrations in water samples for all study sites; Table S3: Mass and sampling effort for benthic macroinvertebrate samples; Table S4: Trace-metal concentrations in benthic macroinvertebrates for all samples; Table S5: Rare earth element concentrations in benthic macroinvertebrates for all samples; Table S6: Moran's I values for dissolved trace metals, rare earth elements, CCU, pH, and specific conductance; Table S7: Moran's I values for dissolved trace metals and rare earth elements in benthic macroinvertebrates; Table S8: Summary of dominant benthic invertebrate taxa in the Snake River watershed and their functional feeding groups; Table S9: Loadings of dissolved metal and REE concentrations in water samples on PC1 and PC2; Table S10: Loadings of metal and REE concentrations in invertebrate samples on PC1 and PC2; Tables S11–S13: Abundance histories for all avian species; Table S14: Moran's I value and the corresponding p value for abundance histories of bird species; Tables S15 and S16: Values of site-specific and survey-specific covariates; Figure S1: Streamflow in the Snake River watershed.

Author Contributions: Study design, field study, statistical analysis, and original draft preparation: K.E.W. with guidance from D.M.M. Manuscript preparation, revision, and editing: K.E.W. and D.M.M. All authors have read and agreed to the published version of the manuscript.

Funding: This research received no external funding.

Institutional Review Board Statement: Not applicable.

Informed Consent Statement: Not applicable.

Data Availability Statement: The data supporting the reported results can be found in the Supplementary Materials.

Acknowledgments: We acknowledge S. Koliavas for help with data collection and J. Carrol for helpful orientation to potential field sites in the watershed. We acknowledge L. Magliozzi and E. Sokol for assistance and advice for evaluating spatial autocorrelation, and we acknowledge helpful comments and discussion provided by P. Newton, A. Todd, and G. Rue.

Conflicts of Interest: The authors declare no conflict of interest.

References

1. Nordstrom, D.K.; Blowes, D.W.; Ptacek, C.J. Hydrogeochemistry and microbiology of mine drainage: An update. *Appl. Geochem.* **2015**, *57*, 3–16. [CrossRef]
2. Government Accountability Office. *Abandoned Mines: Information on the Number of Hardrock Mines, Cost of Cleanup, and Value of Financial Assurances (Statement of Anu K. Mittal, Director, Natural Resources and Environment Team)*; U.S. Government Printing Office: Washington, DC, USA, 2011.

3. Todd, A.S.; Manning, A.H.; Verplanck, P.L.; Crouch, C.; McKnight, D.M.; Dunham, R. Climate-change-driven deterioration of water quality in a mineralized watershed. *Environ. Sci. Technol.* **2012**, *46*, 9324–9332. [CrossRef] [PubMed]
4. Rue, G.P.; McKnight, D.M. Enhanced Rare Earth Element Mobilization in a Mountain Watershed of the Colorado Mineral Belt with Concomitant Detection in Aquatic Biota: Increasing Climate Change-Driven Degradation to Water Quality. *Environ. Sci. Technol.* **2021**, *55*, 14378–14388. [CrossRef]
5. Clements, W.H.; Kiffney, P.M. The influence of elevation on benthic community responses to heavy metals in Rocky Mountain streams. *Can. J. Fish. Aquat. Sci.* **1995**, *52*, 1966–1977. [CrossRef]
6. Hogsden, K.L.; Harding, J.S. Consequences of acid mine drainage for the structure and function of benthic stream communities: A review. *Freshw. Sci.* **2012**, *31*, 108–120. [CrossRef]
7. Lear, G.; Niyogi, D.; Harding, J.; Dong, Y.; Lewis, G. Biofilm bacterial community structure in streams affected by acid mine drainage. *Appl. Environ. Microbiol.* **2009**, *75*, 3455–3460. [CrossRef] [PubMed]
8. Mendez-Garcia, C.; Pelaez, A.I.; Mesa, V.; Sanchez, J.; Golyshina, O.V.; Ferrer, M. Microbial diversity and metabolic networks in acid mine drainage habitats. *Front. Microbiol.* **2015**, *6*, 475.
9. Freda, J. The effects of aluminum and other metals on amphibians. *Environ. Pollut.* **1991**, *71*, 305–328. [CrossRef]
10. Todd, A.S.; McKnight, D.M.; Jaros, C.L.; Marchitto, T.M. Effects of acid rock drainage on stocked Rainbow Trout (*Oncorhynchus mykiss*): An in-situ, caged fish experiment. *Environ. Monit. Assess.* **2006**, *130*, 111–127. [CrossRef]
11. Gray, N.F. Environmental impact and remediation of acid mine drainage: A management problem. *Environ. Geol.* **1997**, *30*, 62–71. [CrossRef]
12. McKnight, D.M.; Feder, G.L. The ecological effect of acid conditions and precipitation of hydrous metal oxides in a Rocky Mountain stream. *Hydrobiology* **1984**, *119*, 129–138. [CrossRef]
13. Niyogi, D.K.; McKnight, D.M.; Lewis, W.M. Influences of water and substrate quality for periphyton in a montane stream affected by acid mine drainage. *Limnol. Oceanogr.* **1999**, *44*, 804–809. [CrossRef]
14. Duarte, S.; Pascoal, C.; Alves, A.; Correia, A.; Cassio, F. Copper and zinc mixtures induce shifts in microbial communities and reduce leaf litter decomposition in streams. *Freshw. Biol.* **2008**, *53*, 91–101. [CrossRef]
15. Niyogi, D.K.; Lewis, W.M.; McKnight, D.M. Litter breakdown in mountain streams affected by mine drainage: Biotic mediation of abiotic controls. *Ecol. Appl.* **2001**, *11*, 506–516. [CrossRef]
16. Cain, D.J.; Luoma, S.N.; Carter, J.L.; Fend, S.V. Aquatic insects as bioindicators of trace element contamination in cobble-bottom rivers and stream. *Can. J. Fish. Aquat. Sci.* **1992**, *49*, 2141–2154. [CrossRef]
17. Pastorino, P.; Brizio, P.; Abete, M.C.; Bertoli, M.; Noser, A.G.O.; Piazza, G.; Prearo, M.; Elia, A.C.; Pizzul, E.; Squadrone, S. Macrobenthic invertebrates as tracers of rare earth elements in freshwater watercourses. *Sci. Total Environ.* **2020**, *698*, 134282. [CrossRef]
18. Amyot, M.; Clayden, M.G.; MacMillan, G.A.; Perron, T.; Arscott-Gauvin, A. Fate and trophic transfer of rare earth elements in temperate lake food webs. *Environ. Sci. Technol.* **2017**, *51*, 6009–6017. [CrossRef]
19. Farag, A.M.; Woodward, D.F.; Goldstein, J.N.; Brumbaugh, W.; Meyer, J.S. Concentrations of metals associated with mining waste in sediments, biofilm, benthic macroinvertebrates, and fish from the Coeur d'Alene Basin, Idaho. *Arch. Environ. Contam. Toxicol.* **1998**, *34*, 119–127. [CrossRef]
20. Pastorino, P.; Pizzul, E.; Bertoli, M.; Perilli, S.; Brizio, P.; Salvi, G.; Esposito, G.; Abete, M.C.; Prearo, M.; Squadrone, S. Macrobenthic invertebrates as bioindicators of trace elements in high-mountain lakes. *Environ. Sci. Pollut. Res.* **2020**, *27*, 5958–5970. [CrossRef]
21. Borga, K.; Campbell, L.; Gabrielsen, G.W.; Norstrom, R.J.; Muir, D.C.G.; Fisk, A.T. Regional and species specific bioaccumulation of major and trace elements in arctic seabirds. *Environ. Toxicol. Chem.* **2006**, *25*, 2927–2936. [CrossRef]
22. Cristol, D.A.; Brasso, R.L.; Condon, A.M.; Fovargue, R.E.; Friedman, S.L.; Hallinger, K.K.; Monroe, A.P.; White, A.E. The movement of aquatic mercury through terrestrial food webs. *Science* **2008**, *320*, 335. [CrossRef]
23. Llacuna, S.; Gorriz, A.; Sanpera, C.; Nadal, J. Metal accumulation in three species of passerine birds (*Emberiza cia, Parus major, and Turdus merula*) subjected to air pollution from a coal-fired power plant. *Arch. Environ. Contamin. Toxicol.* **1995**, *28*, 298–303. [CrossRef]
24. Scheuhammer, A.M. The chronic toxicity of aluminum, cadmium, mercury, and lead in birds: A review. *Environ. Pollut.* **1987**, *46*, 263–295. [CrossRef] [PubMed]
25. Scheuhammer, A.M. Effects of acidification on the availability of toxic metals and calcium to wild birds and mammals. *Environ. Pollut.* **1991**, *71*, 329–375. [CrossRef]
26. Ormerod, S.J.; O'Halloran, J.; Gribbin, S.D.; Tyler, S.J. The ecology of dippers *Cinclus cinclus* in relation to stream acidity in upland wales: Breeding performance, calcium physiology and nestling growth. *J. Appl. Ecol.* **1991**, *28*, 419–433. [CrossRef]
27. Mulvihill, R.S.; Newell, F.L.; Latta, S.C. Effects of acidification on the breeding ecology of a stream-dependent songbird, the Louisiana waterthrush (*Seiurus motacilla*). *Freshw. Biol.* **2008**, *53*, 2158–2169. [CrossRef]
28. Custer, C.M.; Yang, C.; Crock, J.G.; Shearn-Bochster, V.; Smith, K.S.; Hageman, P.L. Exposure of insects and insectivorous birds to metals and other elements from abandoned mine tailings in three Summit County drainages, Colorado. *Environ. Monit. Assess.* **2009**, *153*, 161–177. [CrossRef] [PubMed]
29. Kraus, J.M.; Wanty, R.B.; Schmidt, T.S.; Walters, D.M.; Wolf, R.E. Variation in metal concentrations across a large contamination gradient is reflected in stream but not linked riparian food webs. *Sci. Total Environ.* **2021**, *769*, 144714. [CrossRef] [PubMed]
30. Nordstrom, D.K. Acid rock drainage and climate change. *J. Geochem. Explor.* **2009**, *100*, 97–104. [CrossRef]

31. McKnight, D.M.; Bencala, K.E. The chemistry of Iron, Aluminum, and dissolved organic material in three acidic, metal-enriched, mountain streams, as controlled by watershed and in-stream processes. *Water Resour. Res.* **1990**, *26*, 3087–3100. [CrossRef]
32. Sullivan, A.B.; Drever, J.I.; McKnight, D.M. Diel variation in element concentrations, Peru Creek, Summit County, Colorado. *J. Geochem. Explor.* **1998**, *64*, 141–145. [CrossRef]
33. Crouch, C.M.; McKnight, D.M.; Todd, A.S. Quantifying sources of increasing zinc from acid rock drainage in an alpine catchment under a changing hydrologic regime. *Hydrol. Proc.* **2013**, *27*, 721–733. [CrossRef]
34. Colorado Department of Public Health and Environment Water Quality Control Commission. Regulation #93—Colorado's Section 303(D) List of Impaired Waters and Monitoring and Evaluation List. 5 CCR 1002-93. Available online: https://cdphe.colorado.gov/impaired-waters. (accessed on 19 April 2023).
35. Theobald, P.K.; Lakin, W.; Hawkins, D.B. The precipitation of aluminum, iron and manganese at the junction of Deer Creek with the Snake River in Summit County, Colorado. *Geochim. Cosmochim. Acta* **1963**, *27*, 121–132. [CrossRef]
36. Yang, C. Effects of Acid Mine Drainage on Nesting Tree Swallows. Master's Thesis, University of Colorado, Boulder, CO, USA, 2006.
37. Carrol, J.E. Dynamics of Solute Transport and Rare Earth Element Behavior in Acid Mine Drainage Impacted Alpine Rivers, Snake River CO. Master's Thesis, University of Colorado, Boulder, CO, USA, 2017.
38. Blue River Watershed Group. *Snake River Watershed Plan*; Prepared on behalf of Colorado Department of Public Health & Environment, Water Quality Control Division; Colorado Department of Public Health & Environment, Water Quality Control Division: Boulder, CO, USA, 2013.
39. Boyer, E.W.; McKnight, D.M.; Bencala, K.E.; Brooks, P.D.; Anthony, M.W.; Zellweger, G.W.; Harnish, R.E. *Streamflow and Water Quality Characteristics for the Upper Snake River and Deer Creek Catchments in Summit County, Colorado: Water Years 1980 to 1990*; Institute of Arctic and Alpine Research Occasional Paper No. 53; University of Colorado: Boulder, CO, USA, 1999.
40. Fey, D.L.; Church, S.E.; Unruth, D.M.; Bove, D.J. *Water and Sediment Study of the Snake River Watershed, Colorado, Oct. 9–12, 2001*; U.S. Geological Survey Open-File Report 02-0330; U.S. Geological Survey, U.S. Department of the Interior: Washington, DC, USA, 2001.
41. Todd, A.S.; McKnight, D.M.; Duren, S.M. *Water Quality Characteristics for the Snake River, North Fork of the Snake River, Peru Creek, and Deer Creek in Summit County, Colorado: 2001 to 2002*; Institute of Arctic and Alpine Research, University of Colorado: Boulder, CO, USA, 2005.
42. Andrews, R.; Righter, R. *Colorado Birds: A Reference to Their Distribution and Habitat*; Denver Museum of Natural History: Denver, CO, USA, 1992.
43. Hauer, F.R.; Resh, V.H. Macroinvertebrates. In *Methods in Stream Ecology*, 2nd ed.; Hauer, F.R., Lamberti, G.A., Eds.; Elsevier: New York, NY, USA, 2007; pp. 435–454.
44. Zbinden, N.; Kéry, M.; Keller, V.; Brotons, L.; Herrando, S.; Schmid, H. Species Richness of Breeding Birds along the Altitudinal Gradient—An Analysis of Atlas Databases from Switzerland and Catalonia (NE Spain). In *Data Mining for Global Trends in Mountain Biodiversity*; CRC Press: Boca Raton, FL, USA, 2009; p. 65.
45. Enserink, E.L.; Maas-Diepeveen, J.L.; Van Leeuwen, C.J. Combined effects of metals: And ecotoxicological evaluation. *Water Res.* **1991**, *25*, 679–687. [CrossRef]
46. Clements, W.H.; Carlisle, D.M.; Lazorchak, J.M.; Johnson, P.C. Heavy metals structure benthic communities in Colorado mountain streams. *Ecol. Appl.* **2000**, *10*, 626–638. [CrossRef]
47. Environmental Protection Agency. National Recommended Aquatic Life Criteria Table. Available online: https://www.epa.gov/wqc/national-recommended-water-quality-criteria-aquatic-life-criteria-table. (accessed on 1 March 2018).
48. R Core Team. *R: A Language and Environment for Statistical Computing*; R Foundation for Statistical Computing: Vienna, Austria, 2017; Available online: https://www.R-project.org/ (accessed on 1 October 2017).
49. Burnham, K.P.; Anderson, D.R. *Model Selection and Multimodel Inference: A Practical Information-Theoretic Approach*, 2nd ed.; Springer: New York, NY, USA, 2003.
50. Pastorino, P.; Bertoli, M.; Squadrone, S.; Brizio, P.; Piazza, G.; Noser, A.G.O.; Prearo, M.; Abete, M.C.; Pizzul, E. Detection of trace elements in freshwater microbenthic invertebrates of different functional feeding guilds: A case study in Northeast Italy. *Ecohydrol. Hydrobiol.* **2019**, *19*, 428–440. [CrossRef]
51. Garca, M.A.S.; Cressa, C.; Gessner, M.O.; Feio, M.J.; Callies, K.A.; Barrios, C. Food quality, feeding preferences, survival and growth of shredders from temperate and tropical streams. *Freshw. Biol.* **2001**, *46*, 947–957. [CrossRef]
52. Rodewald, P. (Ed.) . *The Birds of North America*; Cornell Laboratory of Ornithology: Ithaca, NY, USA, 2018.
53. Hill, B.G.; Lein, M.R. Territory overlap and habitat use of sympatric chickadees. *The Auk* **1989**, *106*, 259–268.
54. Baker, M.C.; Mewaldt, R. The use of space by white-crowned sparrows: Juvenile and adult ranging patterns and home range versus body size comparisons in an avian granivore community. *Behav. Ecol. Sociobiol.* **1979**, *6*, 45–52. [CrossRef]
55. Sabo, S.R. Niche and habitat relations in subalpine bird communities of the White Mountains of New Hampshire. *Ecol. Monogr.* **1980**, *50*, 241–259. [CrossRef]
56. Ammon, E.M. Lincoln's Sparrow (*Melospiza lincolnii*), version 2.0. In *The Birds of North America*; Poole, A.F., Gill, F.B., Eds.; Cornell Lab of Ornithology: Ithaca, NY, USA, 1995.
57. Ammon, E.M.; Gilbert, W.M. Wilson's Warbler (*Cardellina pusilla*), version 2.0. In *The Birds of North America*; Poole, A.F., Gill, F.B., Eds.; Cornell Lab of Ornithology: Ithaca, NY, USA, 1999.

58. Chilton, G.; Baker, C.; Barrentine, C.D.; Cunningham, M.A. White-crowned sparrow (*Zonotrichia leucophrys*), version 2.0. In *The Birds of North America*; Poole, A.F., Gill, F.B., Eds.; Cornell Lab of Ornithology: Ithaca, NY, USA, 1995.
59. Winkler, E.W.; Hallinger, K.K.; Ardia, D.R.; Roberson, R.J.; Stutchbury, B.J.; Cohen, R.R. Tree swallow (*Tachycineta bicolor*), version 2.0. In *The Birds of North America*; Poole, A.F., Gill, F.B., Eds.; Cornell Lab of Ornithology: Ithaca, NY, USA, 2011.
60. Mengelkoch, J.M.; Niemi, G.J.; Regal, R.R. Diet of the nestling tree swallow. *Condor* **2004**, *106*, 423–429. [CrossRef]
61. Wheelwright, N.T. The diet of American robins: An analysis of U.S. Biological Survey records. *The Auk* **1986**, *103*, 710–725. [CrossRef]
62. Kraus, J.M.; Schmidt, T.S.; Walters, D.M.; Wanty, R.B.; Zuellig, R.E.; Wolf, R.E. Cross-ecosystem impacts of stream pollution reduce resource and contaminant flux to riparian food webs. *Ecol. Appl.* **2014**, *24*, 235–243. [CrossRef] [PubMed]
63. Greenwood, P.J. Mating systems, philopatry and dispersal in birds and mammals. *Anim. Behav.* **1980**, *28*, 1140–1162. [CrossRef]
64. Inouye, D.W.; Barr, B.; Armitage, K.B.; Inouye, B.D. Climate change is affecting altitudinal migrants and hibernating species. *Proc. Natl. Acad. Sci. USA* **2000**, *97*, 1630–1633. [CrossRef] [PubMed]
65. Johnson, D.B.; Hallberg, K.B. Acid mine drainage remediation options: A review. *Sci. Total Environ.* **2005**, *338*, 3–14. [CrossRef]

diversity MDPI

Article

Climatic and Non-Climatic Drivers of Plant Diversity along an Altitudinal Gradient in the Taihang Mountains of Northern China

Hongzhu Liang [1,2,3,†], Tonggang Fu [1,†], Hui Gao [1], Min Li [3] and Jintong Liu [1,*]

1 Center for Agricultural Resources Research, Key Laboratory of Agricultural Water Resources, Institute of Genetics and Developmental Biology, Chinese Academy of Sciences, No.286 Huaizhong Road, Shijiazhuang 050021, China
2 University of Chinese Academy of Sciences, No.19 (A) Yuquan Road, Beijing 100049, China
3 College of Life Sciences, Hebei Normal University, No.20 Road East. 2nd Ring South, Shijiazhuang 050024, China
* Correspondence: jtliu@sjziam.ac.cn
† These authors contributed equally to this work.

Abstract: Climate is critical for plant altitudinal distribution patterns. Non-climatic factors also have important effects on vegetation altitudinal distribution in mountain regions. The purpose of this study was to explore the current distribution of plant diversity along the altitudinal gradient in the Taihang Mountain range of northern China and to estimate the effects of climatic and non-climatic factors on the elevational pattern. Through a field survey, a total of 480 sampling plots were established in the central Taihang Mountain range. Alpha diversities (the Shannon–Weiner index and Simpson index) and beta diversities (the Jaccard index and Cody index) were measured based on the survey data. Plant community structure change based on the altitudinal gradient was explored by measuring the diversity indices. Canonical correspondence analysis was carried out to determine the factors influencing plant altitudinal distribution. The contributions of climatic and non-climatic factors on plant distribution were determined by partial methods. The results showed that the plant diversity of the elevational gradient complied with a "hump-shaped" pattern, in which communities in the medium altitude area with higher plant diversity had a higher species turnover rate, and non-climatic factors, particularly the anthropogenic factors, had an important influence on the plant altitudinal pattern. In conclusion, climatic and non-climatic factors both had important effects on the plant altitudinal pattern. It is strongly recommended to reduce human interference in mountain vegetation protection and management.

Keywords: α- and β-diversity; vascular plants; altitudinal distribution pattern; canonical correspondence analysis; anthropogenic disturbance; Taihang Mountains

Citation: Liang, H.; Fu, T.; Gao, H.; Li, M.; Liu, J. Climatic and Non-Climatic Drivers of Plant Diversity along an Altitudinal Gradient in the Taihang Mountains of Northern China. *Diversity* **2023**, *15*, 66. https://doi.org/10.3390/d15010066

Academic Editors: Lin Zhang and Jinniu Wang

Received: 28 October 2022
Revised: 18 December 2022
Accepted: 21 December 2022
Published: 5 January 2023

1. Introduction

A mountain is an area with certain elevations, slopes, and relative heights that are a reflection and condensed point of the gradients in natural geographical and ecological features [1]. Due to relatively low human disturbance, mountains provide habitat and shelter for terrestrial biological species [2]. Mountains also represent the most abundant unit of biodiversity on Earth and are key areas for the conservation of biodiversity [3,4]. Mountain areas are extremely sensitive to climate change [5]. The response of mountain ecosystems to climate change is an increasing focus of global change research [6,7].

Biological groups form cluster patterns of vertical and horizontal gradients. Kattan et al., suggested that clustering patterns in dendrograms formed two major patterns of differentiation of the biological groups in Colombia: one horizonal and one elevational [8]. The distribution of vegetation has obvious patterns of horizontal and vertical zonality [9]. This implies that the composition of vegetation varies with altitude [10,11]. Multiple environmental factors drive the altitudinal zonality of vegetation [12]. Natural factors, such as climate, geomorphology, and hydrology, drive obvious changes in altitudinal gradients from the bottom to the top of mountains [13,14]. Generally, changes in a mountain landscape along elevation gradients are 1000 times higher than that in horizontal gradients [15], which can be understood to mean that the landscape change on the vertical scale condenses the change on the horizontal scale. Distribution models of plant species diversity in mountain areas are increasingly used in ecological community research. Such models lead to a better understanding of the mechanism of maintenance of biodiversity and altitudinal change in mountain vegetation [16,17].

Research is still not conclusive about the primary factors that determine the patterns of biodiversity on Earth [18,19]. The gradients of diversity patterns at a large scale have been explained mainly by temperature, productivity, water availability, and geographical area [20]. The study of the altitudinal patterns of species diversity in mountain plant communities is critical for the understanding of zonal variations in vegetation along elevation gradients [21]. Plant diversity varies with altitude gradients [22], showing a variety of altitudinal patterns [23,24]. In the study of altitudinal patterns of plant diversity in mountain terrains, the α-diversity (such as the species richness, Shannon–Weiner index and Simpson index), β-diversity (such as the Jaccard index, Cody index) and γ-diversity indices are widely used [23–25]. The β-diversity index best reflects a community composition and turnover under an environmental gradient [25,26], which is critical for biodiversity conservation [27,28].

The rapid development of quantitative ecological methods and computer technology are innovatively changing data processing and analysis of biodiversity [29]. Studies suggest that surveys in mountain regions provide valuable insights into biological conservation [30,31]. While the relationship between the patterns of biodiversity and elevation gradients depends on various environmental variables [32], altitudinal patterns of mountain plant communities in different geographical regions differ [33–35]. Studies on the process of the development of altitudinal patterns of plant diversity in mountain regions are also attracting much attention [6].

The Taihang Mountain range is a transitional zone from low-elevation plains to high-elevation plateaus, and it is an important ecological barrier to the economic circle of Beijing–Tianjin–Hebei in northern China [36]. Because of the special geographical location of the Taihang Mountain range, it plays an important role as a windbreak, in sand-fixing, and in water conservation. Furthermore, it is a transition zone from economically developed regions to undeveloped regions, where the natural ecosystems are cross-distributed. The Taihang Mountain range is extremely ecologically sensitive. Serious damage to the natural environment and biodiversity has been caused by historical development, which led to serious soil and water loss, frequent droughts, and other significant environmental problems [37]. After decades of overexploitation, along the elevational gradient, the vegetation presents a different distribution pattern. As vegetation provides most of the ecosystem services, the elevational pattern is a reference for evaluating the effects of natural and anthropogenic factors on the vegetation in the Taihang Mountain range. Therefore, the objective of this study was to explore the impact of climate related factors and non-climatic factors on the vertical pattern of plant diversity in the Taihang Mountains.

2. Materials and Methods

2.1. Study Area

The Taihang Mountains (34°36′–40°47′ N, 110°42′–116°34′ E) are a highly heterogeneous geological setting, spreading from the northeast to southwest in northern China (Figure 1). The mountain range acts as a natural boundary between the North China Plain and the Loess Plateau. The altitude across the Taihang Mountains decreases from northwest to southeast, with the highest elevation of 2882 m in the north.

A temperate continental monsoon climate prevails in the study area. From 2009 to 2017, the annual mean temperature was 8.91 °C, and the annual mean precipitation was 529 mm. Both temperature and precipitation increase from the northwest to the southeast.

The Taihang Mountains are also known for their expansive biodiversity in northern China. Warm temperate deciduous broad-leaved forest is the dominant vegetation type in the central Taihang Mountain region.

The eastern slopes (sunny slopes) of the Taihang Mountains are steeper than the western slopes (shady slopes); therefore, the vegetation has a more obvious vertical change on the sunny slopes. The survey plots in this study were mainly on the sunny slopes in the central Taihang Mountain region, where the highest peak (Tuoliang) is 2282 m. The central Taihang Mountain region is divided into three ecological zones: a hilly zone (<500 m), mid-mountain zone (500–1500 m), and sub-alpine zone (>1500 m) [38]. Each zone is characterized by a different set of biodiversity and ecosystem services.

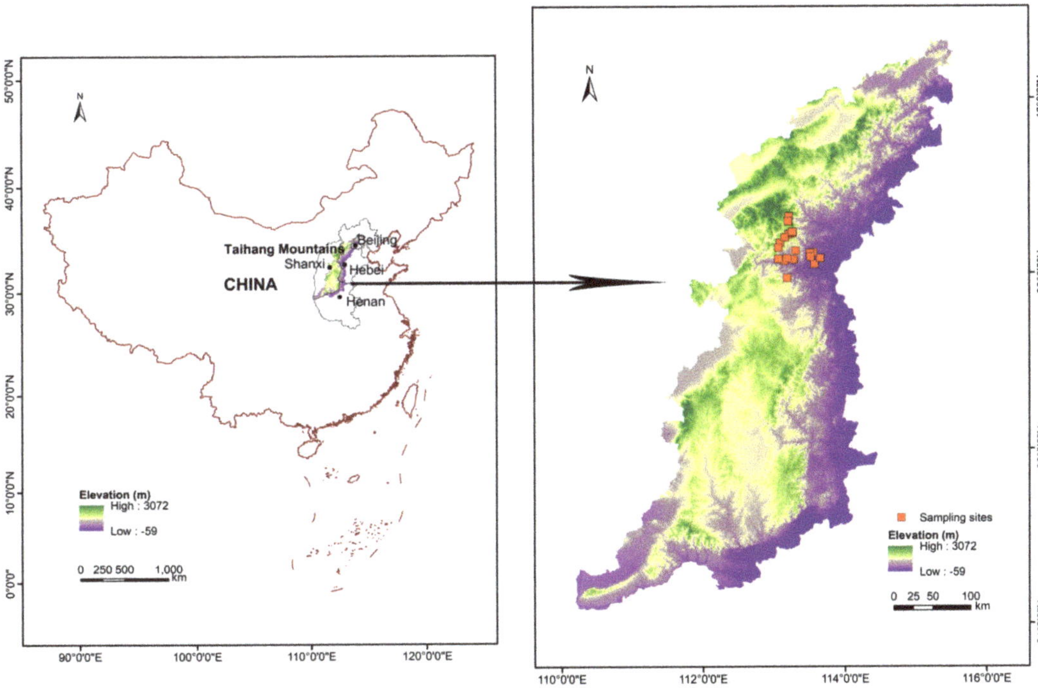

Figure 1. A map depicting the location of Taihang Mountains in northern China (left panel) and an expanded map depicting elevation and sampling sites in the study area.

2.2. Field Survey

This field survey was carried out during the growing season from May 2017 to October 2019 in Tuoliang National Nature Reserve, which is located in the central part of the Taihang Mountain range. A total of 480 survey plots (including tree, shrub, and herb plots) were established at 16 elevations (100 m, 200 m, 300 m, 400 m, 500 m, 600 m, 700 m, 900 m, 1100 m, 1300 m, 1500 m, 1700 m, 1800 m, 1900 m, 2100 m, and 2200 m). At each elevation, five tree plots (10 m × 10 m for each plot), 10 shrub plots (5 m × 5 m), and 15 herb plots (1 m × 1 m) were established, and the species and individual numbers of trees, shrubs and herbs were recorded in each plot. For soil water content (VWC%) and soil pH analyses, three soil samples at a 0–20 cm depth were collected by a cylindrical soil sampler of 5 cm diameter at the same time as the field survey.

2.3. Environmental Data Source

The climatic factors (temperature and precipitation) were collected for the period of 2008–2017 from 101 automatic weather stations installed across the Taihang Mountain range. The temperature and precipitation of each survey plot were derived by kriging interpolation. Non-climatic factors in this study, including elevation, slope, aspect, human footprint index (Hfp), human influence index (Hii), net primary productivity (NPP), human population density, soil pH, and soil water content (VWC%) were derived by synchronizing with the field survey region. The elevation of each plot was measured by GPS, and slope and aspect were detected by a gradiometer. Net primary productivity (NPP) was derived from MOD17A3 data released by the University of Montana, USA (http://ipdaac.usgs.gov, accessed on 28 October 2022). Hfp and Hii were derived from the Socioeconomic Data and Applications Center of NASA (SEDAC, https://sedac.ciesin.columbia.edu/, accessed on 28 October 2022). Human population density is often used as an indicator of vegetation disturbance [39]. The population density was obtained from the Geographical Information Monitoring Cloud Platform maintained by China. The datasets were interpolated for each sampling plot using ordinary kriging. Descriptive statistics of the climatic and non-climatic factors along the altitudinal gradient in the Taihang Mountain study area are given in Table 1.

Table 1. Descriptive statistics of climatic and non-climatic factors along the vertical gradient of Taihang Mountain study area in northern China.

Factors		Min	Max	Mean	Standard Deviation	Skewness	Kurtosis
Climatic factors	Temperature (°C)	7.40	11.56	8.91	1.26	0.43	−0.47
	Precipitation (mm)	491	547	529	15.68	−1.01	0.86
Non-climatic factors	Slope (°)	3.13	43.84	17.65	12.51	0.88	0.46
	Hfp	21	43	28.56	7.31	0.50	−0.80
	Hii	14	28	19.13	4.26	0.57	−0.22
	NPP ($gC{\cdot}m^{-2}{\cdot}a^{-1}$)	139.93	412.47	339.01	62.21	−2.27	7.15
	Population density (p/km^2)	0	258	43.88	93.22	1.88	1.97
	pH	5.31	6.91	6.07	0.41	−0.20	0.52
	VWC (100%)	0.02	0.61	0.20	0.03	1.40	1.61

Notes: Climatic factors: temperature and precipitation; non-climatic factors: factors in addition to temperature and precipitation; temperature: annual mean temperature from 2008 to 2017; precipitation: annual precipitation from 2008 to 2017; Hfp: human footprint index; Hii: human influence index; p in p/km^2 denotes persons; pH: soil pH; VWC%: soil water content to 20 cm depth.

2.4. α-Diversity and β-Diversity Indices

The α-diversity represents species richness within a community. The diversity indices at different altitudes were calculated in terms of the plot survey. The richness index represents the number of species in the sampling plots. The formulas for the Shannon–Weiner index (H) and the Simpson index (D) are as follows:

Shannon-Weiner index:

$$\mathrm{H} = -\sum_{i=1}^{s} P_i ln P_i \quad (1)$$

Simpson index:

$$\mathrm{D} = 1 - \sum_{i=1}^{s} \left(\frac{n_i}{N}\right)^2 \quad (2)$$

where P_i is the proportion of the ith individual to the total number of individuals, n_i represents the number of individuals of the ith species, and N represents the number of individuals of all species in the community.

The β-diversity is often expressed as the ratio of regional (γ-diversity) to α-diversity, and it is often measured as species turnover between different communities. In this study, we used the Jaccard index and Cody index to explore the traits of plant community succession along the altitudinal gradient. While the Jaccard index represents the similarity of different communities and quadrats, the Cody index represents the turnover rate of species along an environmental gradient. The indices are calculated as follows:

Jaccard index:

$$\mathrm{CJ} = \frac{c}{a + b - c} \quad (3)$$

Cody index:

$$\beta_c = \frac{g(H) + l(H)}{2} = \frac{a + b - 2c}{2} \tag{4}$$

where a and b denote the number of species in two communities, c denotes the number of species shared by the two communities, $g(H)$ is the number of species increasing along the gradient (H), and $l(H)$ is the number of species lost along the gradient.

2.5. Data Analysis

Based on the sample data and remote sensing image data for the Taihang Mountains, climatic and non-climatic factors were processed and analyzed by SPSS 23 and mapped in ArcGIS 12.2. Ordination analysis is often used to explain variations in data in relation to species and area [40]. To investigate the distribution patterns of plant diversity and lifeform groups in the survey plots, principal components analysis (PCA) was used to analyze the altitudinal gradient. The influences of climatic and non-climatic factors on species altitudinal distribution were evaluated using canonical correspondence analysis (CCA) in R 3.4.5, and contributions of climatic and non-climatic factors to the plant altitudinal pattern were estimated with partial CCA in R 3.4.5. To verify the significance of environmental factors and the plant species altitudinal distribution, a Monte Carlo permutation test was performed in CANOCO 4.5.

3. Results

3.1. Altitudinal Distribution of A-Diversity in Plants

Based on the field survey, 54 vascular plant species were recorded in the hilly zone, belonging to 32 families and 49 genera; 103 species in the mid-mountain zone, belonging to 47 families and 88 genera; and 58 species in the sub-alpine zone, belonging to 21 families and 48 genera. As the overall elevation was not very high (with the lowest altitude of 0 m and the highest peak of 2282 m in the central Taihang Mountain region), there was not an obvious altitudinal spectrum of vegetation in this region.

Generally, the vascular plants (including trees, shrubs, and herbs) had the same altitudinal pattern, in which the number of plant species increased with increasing elevation. At elevation ranges of 600–900 m and 1500–1900 m, the plant richness (Figure 2d), Shannon-Wiener index (Figure 2a) and Simpson index (Figure 2b) had two peak intervals, and then began to decrease with increasing elevation. The distribution of plant diversity in the central Taihang Mountain region was relatively complicated in terms of the elevation gradient.

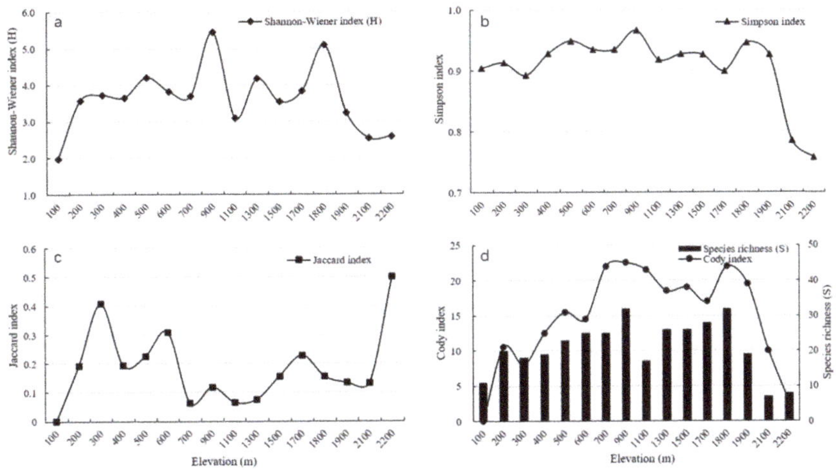

Figure 2. Indices of plant diversity along the altitudinal gradient in the central Taihang Mountain region, northern China.

Quantitative measurement is needed to show differences in communities [41]. PCA analysis (Figure 3) indicated that plant species richness was mainly concentrated in the mid-elevation zone (sites 5–14). Individual plants, species of trees and herbs, and the richness and Shannon–Weiner indices were highest in the area covering sites 5–14, which meant that plant richness was highest in the mid-elevation zone. Shrubs mainly occurred in the low elevation zone, in the area covering sites 1–4.

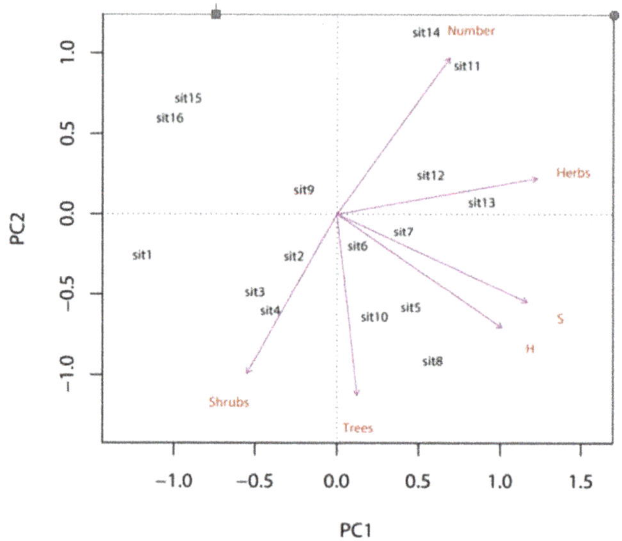

Figure 3. PCA ordination showing the characteristics of altitudinal distribution of plants in the central Taihang Mountain region.

3.2. B-Diversity of Plants along the Altitudinal Gradient

Figure 2 shows that the Shannon–Weiner index, Simpson index and Cody index had coincident altitudinal patterns of plant richness, implying a relatively high plant diversity concentrated in the mid-elevation zone. The patterns of the Jaccard index indicated that the sampling plots in the mid-elevation range had a high similarity in plant community structure.

Low β-diversity can lead to low species turnover rates [42]. The Cody index represents the rate of species turnover between communities, with the highest species turnover always taking place in pioneer and mountain species [43]. In this study, the Jaccard index showed the opposite altitudinal patterns to the richness, Shannon–Weiner index and Simpson index, while the Cody index showed a consistent distribution trend, implying that the altitudinal gradient with lower community similarity had a higher species turnover rate of plant communities.

3.3. Relationships between Plant Diversity and Environmental Factors along the Altitudinal Gradient

In predicting species distribution, canonical correspondence analysis (CCA) is widely used [44]. CCA ordination (Figure 4) showed that the driving factors with positive effects on tree, shrub, and herb richness were soil water content (VWC%), precipitation, and pH. The factors with negative effects on plant richness were temperature, Hii, Hfp, and population density. Slope and pH had the smallest effects on altitudinal distribution of plants in the central Taihang Mountain area.

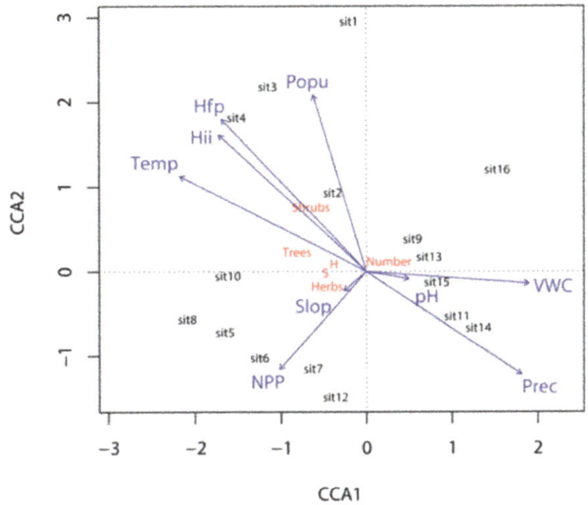

Figure 4. CCA ordination showing the relationship between environmental factors and altitudinal plant patterns in the Taihang Mountain study area, northern China.

4. Discussion

4.1. Plant Diversity Pattern along the Altitudinal Gradient

Based on the field survey and species identification, there were 54 species of vascular plants in the hilly zone, belonging to 32 families and 49 genera; 103 species in the mid-mountain zone, belonging to 47 families and 88 genera; and 58 species in the sub-alpine zone, belonging to 21 families and 48 genera. This conformed with the richness distribution theory on "middle height expansion," consistent with studies of vine plants in other mountain regions [45–48].

In this study, trees were mainly found in the hilly and mid-mountain zones, with few trees in the sub-alpine zone. Shrubs and herbs were widely distributed from the low to the high elevation zones in the central Taihang Mountain region. The richness of herbs was higher than that of shrubs and trees. PCA ordination showed that plant groups were mainly concentrated in the mid-elevation range in the central Taihang Mountain region.

In the sub-alpine zone, temperate herbal plants were the main vegetation type. There was a higher proportion of annual herbaceous plants with more endemic species in this zone, indicating that biodiversity endured to a certain degree even in an area with intense human disturbance. While species richness decreased significantly at elevations above 2000 m, the community similarity index increased sharply. This implied that more common species of vascular plants were concentrated in the sub-alpine areas; therefore, there was relatively low species diversity.

4.2. Characteristics of the B-Diversity Pattern of the Altitudinal Gradient

Patterns of species turnover are vital to the geography of biodiversity [49]. The Cody index for elevations above 2000 m decreased sharply, indicating that the rate of replacement of plant communities decreased to a relatively low degree. Studies show that areas with lower species richness are more easily invaded by exotic species [50], and thus the plant community structure of a low richness area is more easily disturbed. According to this hypothesis, the results in Figure 2 indicated that the community structure of the sub-alpine zone with a lower plant diversity richness was more unstable in the central Taihang Mountain region.

While the processes of biodiversity maintenance and species coexistence are the focus of ecological studies, those of community construction and succession along altitudinal gradients still remain unclear. Ecological niche theories are increasingly becoming a mainstream issue in community construction and species distribution. Species diversity increases with increasing heterogeneity of the environment along vertical or horizontal gradients [6,47]. In β-diversity research, numerous methods of measurement have been proposed. Using combined β-diversity, gradient analysis, and ecological niche modeling, significant and novel insights are made into biological diversity patterns [51]. Among these, similarly and dissimilarity indices are widely used. The application of additive decomposition in β-diversity better reveals the processes of community construction along elevation gradients. In this study, we mainly focused on the Jaccard and Cody indices, which can give insight into the plant community structure change and species turnover rate along the elevational gradient. As mentioned at the beginning, the Taihang Mountain range is a mountainous region that has been overexploited for a long period; thus, human disturbances have created a lasting pressure on the ecosystem of the Taihang Mountains. The altitudinal pattern of α-diversity was found to be relatively complicated, and two peaks of plant richness appeared at the elevation range of 600–900 m and 1500–1900 m, which reflected more suitable natural conditions and less anthropogenic disturbance of the plant diversity. Characteristics of the β-diversity at the elevational gradients with higher richness showed that the plant community had more active succession capacity and a higher species turnover rate. Inner succession activity, natural conditions, and human disturbance formed the current plant altitudinal distribution pattern.

In summary, the altitudinal distribution pattern and species diversity of plant communities on the sunny slopes of the central Taihang Mountain region were a result of the joint actions of community succession and natural and anthropogenic disturbances.

4.3. *Effects of Climatic and Non-Climatic Factors on the Plant Diversity Altitudinal Pattern*

Although environmental factors can influence species distribution, plants can have positive effects on each other [52]. In this study, we focused mainly on the effects of environmental factors on the altitudinal pattern of plant species. Using partial analysis, we evaluated the contribution of climatic (temperature and precipitation) and non-climatic (slope, Hfp, Hii, NPP, population density, pH, and VWC%) factors to the altitudinal patterns of plant diversity. Partial methods are often used to analyze the effects of the main environmental variables and covariates on species distribution. We measured the contribution of climatic and non-climatic factors to plant distribution along the altitudinal gradient by partial CCA. The results showed that the interpretation rate of climatic factors on the altitudinal distribution of plant species (28.89%) was less than that of non-climatic factors (44.51%), which indicated that non-climatic factors were the main driving forces of plant altitudinal distribution in the central Taihang Mountain region. Ohmann and Spies [53] noted that the contribution of climate to species distribution in an Oregon forest was the most significant factor, which contrasted with the results in this study. The joint contribution of climatic and non-climatic factors was 13.72%, and 12.88% of the altitudinal pattern could not be explained by climatic and non-climatic factors (Figure 5). These results implied that the plant distribution pattern was more significantly influenced by altitudinal gradients than by temperature and precipitation gradients.

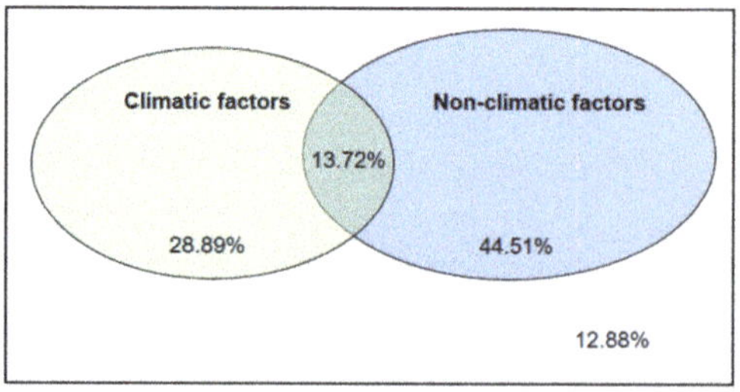

Figure 5. Venn diagram showing the contributions of climatic and non-climatic factors to plant altitudinal distribution in the central Taihang Mountain region, northern China.

By using the Monte Carlo significance test of CCA ordination (Table 2), it was noted that the climatic factors (temperature and precipitation) had the most significant effect on plant altitudinal distribution patterns; moreover, the coefficients of temperature and precipitation reached the most significant and extremely significant levels, respectively. The coefficients of non-climatic factors, including Hfp, Hii, and population density, reached extremely significant levels, implying that these factors also had an important effect on plant altitudinal distribution patterns in the study area. The coefficients of slope, NPP, and soil pH were not significant (Pr > 0.05), implying that these factors were not the dominant drivers of the plant altitudinal distribution in the study area. According to the results of partial CCA, non-climatic factors played a more important role than climatic factors in the plant altitudinal pattern. Among the non-climatic factors, coefficients of Hfp, Hii, and human population density reached an extremely significant level, and the coefficient of water soil content reached a significant level, which implied the anthropogenic factors (Hfp, Hii, and population density) had more important effects than the other non-climatic factors on the plant altitudinal pattern in the central Taihang Mountain region of northern China.

Table 2. Monte Carlo significance test of the environmental factors and plant species altitudinal distribution.

Factors	CCA1	CCA2	r^2	Pr (>r)
Hfp	−0.66	0.75	0.69	0.002 **
Temperature	−0.88	0.48	0.66	0.001 ***
Hii	−0.71	0.70	0.62	0.002 **
Population density	−0.26	0.96	0.55	0.005 **
Precipitation	0.82	-0.58	0.52	0.004 **
VWC (%)	0.99	-0.07	0.40	0.033 *
NPP	−0.64	-0.77	0.28	0.09
pH	0.99	-0.17	0.02	0.85
Slope	-0.74	-0.67	0.01	0.90

Notes: *** represents the most significant level; ** represents an extremely significant level; * represents a significant level.

5. Conclusions

In conclusion, climatic and non-climatic factors both had important effects on the plant altitudinal pattern. Non-climatic factors were the more significant drivers of plant distribution along the altitudinal gradient compared to climatic factors in the central Taihang Mountain region. In addition, among the non-climatic factors, the anthropogenic factors were the main driving forces of the plant altitudinal distribution. To a certain degree, both climatic and non-climatic factors drove the altitudinal distribution of species richness in the study area. The results of the study suggested that in the central Taihang Mountain region, even in the sub-alpine zone, human disturbance was still a critical factor driving the altitudinal distribution of species richness. From the perspective of sustainable development, in the mountain vegetation protection and management, it is strongly recommended to reduce the impact of human interference.

Author Contributions: Conceptualization, J.L. and H.L.; methodology, H.L.; software, T.F.; validation, H.G. and H.L.; formal analysis, H.L.; investigation, H.G.; resources, T.F. and M.L.; data curation, H.L.; writing—original draft preparation, H.L.; writing—review and editing, J.L.; visualization, T.F.; supervision, J.L.; project administration, H.L.; funding acquisition, J.L. All authors have read and agreed to the published version of the manuscript.

Funding: This work was supported by the Key Programme of National Natural Science Foundation of China (No. 41930651).

Institutional Review Board Statement: Not applicable.

Informed Consent Statement: Not applicable.

Data Availability Statement: Data supporting the reported results are included in the manuscript text.

Acknowledgments: We thank Hongjun Li for the statistics analysis, Jiancheng Zhao and Lin Li for their excellent technical support, and Huijun Gao and Fei Qi for the support in the field survey. We also thank LetPub (www.letpub.com, accessed on 28 October 2022) for its linguistic assistance during the preparation of this manuscript.

Conflicts of Interest: The authors declare no conflict of interest. The founding sponsors had no role in the design of the study; in the collection, analyses, or interpretation of data; in the writing of the manuscript, and in the decision to publish the results.

References

1. Dullinger, S.; Gattringer, A.; Thuiller, W.; Moser, D.; Zimmermann, N.E.; Guisan, A.; Willner, W.; Plutzar, C.; Leitner, M.; Mang, T.; et al. Extinction debt of high-mountain plants under twenty-first-century climate change. *Nat. Clim. Change* **2012**, *2*, 619–622. [CrossRef]
2. Gottfried, M.; Pauli, H.; Futschik, A.; Akhalkatsi, M.; Barančok, P.; Benito Alonso, J.L.; Coldea, G.; Dick, J.; Erschbamer, B.; Fernández Calzado, M.R.; et al. Continent-wide response of mountain vegetation to climate change. *Nat. Clim. Change* **2012**, *2*, 111–115. [CrossRef]
3. Beniston, M. Climatic change in mountain regions: A review of possible impacts. *Clim. Change* **2003**, *59*, 5–31. [CrossRef]
4. Engler, R.; Randin, C.F.; Thuiller, W.; Dullinger, S.; Zimmermann, N.E.; Araújo, M.B.; Pearman, P.B.; Le Lay, G.; Piedallu, C.; Albert, C.H.; et al. 21st century climate change threatens mountain flora unequally across Europe. *Glob. Change Biol.* **2011**, *17*, 2330–2341. [CrossRef]
5. McFadden, I.R.; Sandel, B.; Tsirogiannis, C.; Morueta-Holme, N.; Svenning, J.; Enquist, B.J.; Kraft, N.J. Temperature shapes opposing latitudinal gradients of plant taxonomic and phylogenetic β diversity. *Ecol. Lett.* **2019**, *22*, 1126–1135. [CrossRef] [PubMed]
6. Gallardo-Cruz, J.A.; Perez-Garcia, E.A.; Meave, J.A. β-diversity and vegetation structure as influenced by slope aspect and altitude in a seasonally dry tropical landscape. *Landsc. Ecol.* **2009**, *24*, 473–482. [CrossRef]
7. Loarie, S.R.; Duffy, P.B.; Hamilton, H.; Asner, G.P.; Field, C.B.; Ackerly, D.D. The velocity of climate change. *Nature* **2009**, *462*, 1052–1055. [CrossRef]
8. Kattan, G.H.; Franco, P.; Rojas, V.; Morales, G. Biological diversification in a complex region: A spatial analysis of faunistic diversity and biogeography of the Andes of Colombia. *J. Biogeogr.* **2004**, *31*, 1829–1839. [CrossRef]
9. Seipel, T.; Kueffer, C.; Rew, L.J.; Daehler, C.C.; Pauchard, A.; Naylor, B.J.; Alexander, J.M.; Edwards, P.J.; Parks, C.G.; Arevalo, J.R.; et al. Processes at multiple scales affect richness and similarity of non-native plant species in mountains around the world. *Glob. Ecol. Biogeogr.* **2012**, *21*, 236–246. [CrossRef]
10. Vazquez, G.; Givnish, T.J. Altitudinal gradients in tropical forest composition, structure, and diversity in the Sierra de Manantalan. *J. Ecol.* **1998**, *86*, 999–1020.
11. Willig, M.R.; Presley, S.J. The spatial configuration of taxonomic biodiversity along a tropical elevational gradient: α-, β-, and γ-partitions. *Biotropica* **2019**, *51*, 104–116. [CrossRef]
12. Britton, A.J.; Beale, C.M.; Towers, W.; Hewison, R.L. Biodiversity gains and losses: Evidence for homogenization of Scottish alpine vegetation. *Biol. Conserv.* **2009**, *142*, 1728–1739. [CrossRef]
13. Hoorn, C.; Wesselingh, F.P.; Ter Steege, H.; Bermudez, M.A.; Mora, A.; Sevink, J.; Sanmartín, I.; Sanchez-Meseguer, A.; Anderson, C.L.; Antonelli, A. Amazonia through time: Andean uplift, climate change, landscape evolution, and biodiversity. *Science* **2010**, *330*, 927–931. [CrossRef]
14. Zhao, H.; Wang, Q.-R.; Fan, W.; Song, G.-H. The relationship between secondary forest and environmental factors in the southern Taihang Mountains. *Sci. Rep.* **2017**, *7*, 16431. [CrossRef]
15. Walter, H. *Vegetation of the Earth in Relation to Climate and the Eco-Physiological Conditions*; English University Press: London, UK, 1973.
16. Brown, J.H.; Whitham, T.G.; Ernest, S.K.M.; Gehring, C.A. Complex species interactions and the dynamics of ecological systems: Long-term experiments. *Science* **2001**, *293*, 643–650. [CrossRef]
17. Pauli, H.; Gottfried, M.; Dullinger, S.; Abdaladze, O.; Akhalkatsi, M.; Alonso, J.L.B.; Coldea, G.; Dick, J.; Erschbamer, B.; Calzado, R.F.; et al. Recent plant diversity changes on Europe's mountain summits. *Science* **2012**, *336*, 353–356. [CrossRef]
18. Gaston, K.J. Global patterns in biodiversity. *Nature* **2000**, *405*, 220–227. [CrossRef]
19. Fine, P.V.A. Ecological and evolutionary drivers of geographic variation in species diversity. *Annu. Rev. Ecol. Evol. Syst.* **2015**, *46*, 369–392. [CrossRef]
20. Peters, M.; Hemp, A.; Appelhans, T.; Behler, C.; Classen, A.; Detsch, F.; Ensslin, A.; Ferger, S.W.; Frederiksen, S.B.; Gebert, F.; et al. Predictors of elevational biodiversity gradients change from single taxa to the multi-taxa community level. *Nat. Commun.* **2016**, *7*, 13736. [CrossRef]
21. Zimmermann, P.; Tasser, E.; Leitinger, G.; Tappeiner, U. Effects of land-use and land-cover pattern on landscape-scale biodiversity in the European Alps. Agriculture. *Ecosyst. Environ.* **2010**, *139*, 13–22. [CrossRef]
22. Linder, H.P.; De Klerk, H.M.; Born, J.; Burgess, N.D.; Fjeldså, J.; Rahbek, C. The partitioning of Africa: Statistically denied biogeographical regions in sub-Saharan Africa. *J. Biogeogr.* **2012**, *39*, 1189–1205. [CrossRef]
23. Sanders, N.J.; Rahbek, C. The patterns and causes of elevational diversity gradients. *Ecography* **2012**, *35*, 1–3. [CrossRef]

24. Guo, Q.; Kelt, D.A.; Sun, Z.; Liu, H.; Hu, L.-J.; Ren, H.; Wen, J. Global variation in elevational diversity patterns. *Sci. Rep.* **2013**, *3*, srep03007. [CrossRef] [PubMed]
25. Whittaker, R.H. Vegetation of the Siskiyou Mountains, Oregon and California. *Ecol. Monogr.* **1960**, *30*, 279–338. [CrossRef]
26. De Frenne, P.; Rodríguez-Sánchez, F.; Coomes, D.A.; Baeten, L.; Verstraeten, G.; Vellend, M.; Bernhardt-Römermann, M.; Brown, C.D.; Brunet, J.; Cornelis, J.; et al. Microclimate moderates plant responses to macroclimate warming. *Proc. Natl. Acad. Sci. USA* **2013**, *110*, 18561–18565. [CrossRef]
27. Chun, J.H.; Lee, C.B. Partitioning the regional and local drivers of phylogenetic and functional diversity along temperate elevational gradients on an East Asian peninsula. *Sci. Rep.* **2018**, *8*, 2853. [CrossRef]
28. Horák, J.; Materna, J.; Halda, J.P.; Mladenović, S.; Bogusch, P.; Pech, P. Biodiversity in remnants of natural mountain forests under conservation-oriented management. *Sci. Rep.* **2019**, *9*, 89. [CrossRef]
29. Naud, L.; Måsviken, J.; Freire, S.; Angerbjörn, A.; Dalén, L.; Dalerum, F. Altitude effects on spatial components of vascular plant diversity in a subarctic mountain tundra. *Ecol. Evol.* **2019**, *9*, 4783–4795. [CrossRef]
30. Lomolino, M.V. Elevation gradients of species-density: Historical and prospective views. *Glob. Ecol. Biogeogr.* **2001**, *10*, 3–13. [CrossRef]
31. Hansen, A.J.; Spies, T.A.; Swanson, F.J.; Ohmann, J.L. Conserving biodiversity in Managed forests. *BioScience* **1991**, *21*, 382–392. [CrossRef]
32. Zhao, C.; Wang, J.; Yu, F.; Zhang, X.; Yao, Y.; Zhang, B. Altitudinal biodiversity gradient and ecological drives for different lifeforms in the Baotianman Nature Reserve of the eastern Qinling Mountains. *Forests* **2019**, *10*, 332. [CrossRef]
33. Cabrera, O.; Benítez, A.; Cumbicus, N.; Naranjo, C.; Ramón, P.; Tinitana, F.; Escudero, A. Geomorphology and altitude effects on the diversity and structure of the vanishing montane forest of southern Ecuador. *Diversity* **2019**, *11*, 32. [CrossRef]
34. Dhakal, B.; Kattel, R.R. Effects of global changes on ecosystems services of multiple natural resources in mountain agricultural landscapes. *Sci. Total Environ.* **2019**, *676*, 665–682. [CrossRef]
35. Xu, M.; Zhang, S.; Wen, J.; Yang, X. Multiscale spatial patterns of species and biomass together with their correlations along geographical gradients in subalpine meadows. *PLoS ONE* **2019**, *14*, e0211560. [CrossRef]
36. Fan, X. *The Road to the Taihang Mountains*; China Forestry Publishing House: Beijing, China, 2015.
37. Liu, X.; Ge, J.; Feng, X. Study on ecological security of land resources in Taihang Mountain Hebei. *J. Arid Land Resour. Environ.* **2007**, *21*, 68–74.
38. Gao, H.; Fu, T.G.; Liu, J.T.; Liang, H.; Han, L. Ecosystem services management based on differentiation and regionalization along altitudinal gradient in Taihang Mountain, China. *Sustainability* **2018**, *10*, 986. [CrossRef]
39. Ribeiro, E.M.; Arroyo-Rodríguez, V.; Santos, B.A.; Tabarelli, M.; Leal, I.R. Chronic anthropogenic disturbance drives the biological impoverishment of the Brazilian Caatinga vegetation. *J. Appl. Ecol.* **2015**, *52*, 611–620. [CrossRef]
40. Wolf, H.D.J. Diversity patterns and biomass of epiphytic bryophytes and lichens along an altitudinal gradient in the northern Andes. *Ann. Mo. Bot. Gard.* **1993**, *4*, 928–960.
41. Lozupone, C.A.; Hamady, M.; Kelley, S.T.; Knight, R. Quantitative and qualitative β diversity measures lead to different insights into factors that structure microbial communities. *Appl. Environ. Microbiol.* **2007**, *73*, 1576–1585. [CrossRef]
42. Jankowski, J.E.; Ciecka, A.L.; Meyer, N.Y.; Rabenold, K.N. Beta diversity along environmental gradients: Implications of habitat specialization in tropical montane landscapes. *J. Anim. Ecol.* **2009**, *78*, 315–327. [CrossRef]
43. Koellner, T.; Hersperger, A.M.; Wohlgemuth, T. Rarefaction method for assessing plant species diversity on a regional scale. *Ecography* **2004**, *27*, 532–544. [CrossRef]
44. Guisan, A.; Weiss, S.B.; Weiss, A. GLM versus CCA spatial modeling of plant species distribution. *Plant Ecol.* **1999**, *143*, 107–122. [CrossRef]
45. Kessler, M. Patterns of diversity and range size of selected plant groups along an elevational transect in the Bolivian Andes. *Biodivers. Conserv.* **2001**, *10*, 1897–1921. [CrossRef]
46. Wang, G.; Zhou, G.; Yang, L.; Li, Z. Distribution, species diversity and life-form spectra of plant communities along an altitudinal gradient in the northern slopes of Qilianshan Mountains, Gansu, China. *Plant Ecol.* **2002**, *165*, 169–181. [CrossRef]
47. Sanchez-Gonzalez, A.; Lopez-Mata, L. Plant species richness and diversity along an altitudinal gradient in the Sierra Nevada, Mexico. *Divers. Distrib.* **2005**, *11*, 567–575. [CrossRef]
48. Cardelus, C.L.; Colwell, R.K.; Watkins, J.E. Vascular epiphyte distribution patterns: Explaining the mid-elevation richness peak. *J. Ecol.* **2006**, *94*, 144–156. [CrossRef]
49. Buckley, L.B.; Jetz, W. Linking global turnover of species and environments. *Proc. Natl. Acad. Sci. USA* **2008**, *105*, 17836–17841. [CrossRef]
50. Stohlgren, T.J.; Binkley, D.; Chong, G.W.; Kalkhan, M.A.; Schell, L.D.; Bull, K.A.; Otsuki, Y.; Newman, G.; Bashkin, M.; Son, Y. Exotic plant species invade hot spots of native plant diversity. *Ecol. Monogr.* **1999**, *69*, 25–46. [CrossRef]
51. Graham, C.H.; Fine, P.V.A. Phylogenetic beta diversity: Linking ecological and evolutionary processes across space in time. *Ecol. Lett.* **2008**, *11*, 1265–1277. [CrossRef]
52. Callaway, R.M.; Brooker, R.W.; Choler, P.; Kikvidze, Z.; Lortie, C.J.; Michalet, R.; Paolini, L.; Pugnaire, F.I.; Newingham, B.; Aschehoug, E.T.; et al. Positive interactions among alpine plants increase with stress. *Nature* **2002**, *417*, 844–848. [CrossRef]
53. Ohmann, J.L.; Spies, T.A. Regional gradient analysis and spatial pattern of woody plant communities of Oregon forest. *Ecol. Monogr.* **1998**, *68*, 151–182. [CrossRef]

diversity

MDPI

Article

Stochastic Processes Drive Plant Community Assembly in Alpine Grassland during the Restoration Period

Zhaoheng Deng [1,2], Jingxue Zhao [3], Zhong Wang [4], Ruicheng Li [5], Ying Guo [1,2], Tianxiang Luo [1] and Lin Zhang [1,6,*]

1 Institute of Tibetan Plateau Research, Chinese Academy of Sciences, Beijing 100101, China
2 University of Chinese Academy of Sciences, Beijing 100049, China
3 College of Ecology, Lanzhou University, Lanzhou 730000, China
4 College of Life Sciences, Wuhan University, Wuhan 430072, China
5 College of Urban and Environmental Sciences, Peking University, Beijing 100871, China
6 Tibet Autonomous Region Institute of Science and Technology Information, Lhasa 850000, China
* Correspondence: zhanglin@itpcas.ac.cn; Tel.: +86-010-8409-7055

Abstract: Enclosure (prohibition of grazing) is an important process to restore alpine grassland on the Qinghai-Tibetan Plateau. However, few studies have quantified the extent to which the long-term enclosure may contribute to the changes in plant phylogenetic diversity and community assembly in alpine grassland under environmental change. In this study, based on an 11-year fencing experiment along an altitudinal gradient ranging from 4400 m to 5200 m in central Tibet, we conducted an observation of species composition and coverage within and outside the fences in the fifth, eighth and eleventh year, and monitored the related climate and soil factors at 7 sites. Our aim is to quantify the relative effects of environmental change and grassland management on the alpine plant community assemblage. The results were: (1) the overall phylogenetic structure (NRI) of the alpine plant communities, whether inside or outside the enclosure, was divergent at altitudes where the environment was relatively unextreme (4800–5100 m), but aggregative at altitudes with low precipitation (4400–4650 m) or with low temperature (5200 m). (2) The phylogenetic structure of the nearest taxon of species (NTI) was more aggregative along the whole gradient. (3) Precipitation was the dominant factor driving the changes in species richness, phylogenetic diversity and community α-phylogenetic structure indices (NRI and NTI), followed by enclosure duration and soil C:N ratio. (4) The phylogenetic structure of the communities was similar at higher altitudes under grazing or enclosure treatments, and was opposite at lower sites. Stochastic processes have driven the changes in the communities between inside and outside the fences at all altitudes. In addition, homogeneous dispersal occurred in communities at higher sites. In summary, the 11-year enclosure had little effect on community structure of alpine meadows where the grazing pressure is relative lower, whereas it could help restore the community of steppe meadow at lower altitudes where the grazing pressure is extensively higher. This study may provide a vital theoretical support for the formulation of differential management for alpine grassland on the Tibetan Plateau.

Keywords: alpine meadow; community structure; enclosure and grazing; phylogenetic and taxonomic diversity; the Tibetan Plateau

Citation: Deng, Z.; Zhao, J.; Wang, Z.; Li, R.; Guo, Y.; Luo, T.; Zhang, L. Stochastic Processes Drive Plant Community Assembly in Alpine Grassland during the Restoration Period. *Diversity* **2022**, *14*, 832. https://doi.org/10.3390/d14100832

Academic Editor: Julio Peñas de Giles

Received: 2 September 2022
Accepted: 29 September 2022
Published: 3 October 2022

1. Introduction

There is increasing evidence that conserved traits of ecological niches play an important role in ecology and phylogenetic evolution [1,2]. In the context of the long-term debate between ecological niche theory and neutral theory [3–10], researchers have also found that deterministic ecological processes (e.g., environmental filtration and biological interactions) and stochastic ecological processes (e.g., birth/death, species formation/extinction and migration) can both affect the community assembly [11–17] and are considered as complementary in reality ecosystems [18,19]. Therefore, revealing factors that control the relative impact of deterministic and stochastic processes at different spatiotemporal scales has become one of the core issues of community assembly [20]. Based on the neutral hypothesis of biomolecular evolution, Vellend [21] set up a theoretical framework, dividing community dynamic processes into selection, dispersal, speciation and ecological drift. Due to the development and functioning of the above four processes, the impact of the deterministic process gradually decreased, while that of the stochastic process gradually increased [22]. This theoretical framework unifies ecological niche theory and neutral theory and exerts a wide impact on community ecology [19,23–26]. This theoretical innovation also opens a new way to deconstruct the community assembly mechanism in different ecosystems.

Over the past decades, grassland degradation has become one of the vital problems of global changes threatening the stability of the terrestrial ecosystem [27–29] and hinders the sustainable development of regional economies and environments [30,31], and will eventually endanger human society, economic development and earth habitat [29]. On the Qinghai-Tibet Plateau (QTP), alpine grassland is the most widely distributed vegetation type, occupying approximately 60% of the whole area [32]. However, the short growing season length of alpine plants and the increasing number of livestock have led to widespread overgrazing in this region [33–35]. The impact of climate change (e.g., warming and drying) has now degraded up to 4.25 million ha of alpine grassland, accounting for one-third of the grassland area of the QTP [36]. As one of the common methods of vegetation restoration and rehabilitation of degraded grasslands [37,38], fencing refers to the use of fencing facilities to enclose grasslands for a certain period of time to recuperate and regenerate grassland plants, improve productivity and restore biodiversity [37,39]. Different studies reported different conclusions about the impact of enclosure on the phylogenetic community assembly. For example, studies on temperate grasslands in North America found that grassland restoration resulted in increased competition for dominant species, with more closely related species being excluded from the community, leading to an overdispersion pattern of community phylogeny [40]. In the Netherlands, a 7-year enclosure treatment promoted a more stochastic plant community assembly in a floodplain grassland [41]. However, in temperate shrub-grasslands in northern China, Hao found that the long-term enclosure enhanced the negative effect of competition between dominant species and auxiliary species, resulting in a more aggregated community phylogenetic structure [42]. Meanwhile, Sun found that the short-term enclosure did not result in significant differences in the community phylogenetic structure in alpine steppes meadows. It is also found that the effect of grazing on the community phylogenetic structure in alpine grasslands of northern Tibet was more correlated with grazing intensity [43]. High grazing intensity shifted the community phylogenetic structure from dispersion to aggregation, while low grazing intensity had little effect on species richness and community structure [44]. However, the above studies have not yet quantified the extent to which interannual changes of environmental factors during restoration may affect the phylogeny of grassland communities. In addition, whether the community assembly in alpine meadows is dominated by stochastic or deterministic processes during restoration, and whether there exists altitudinal variation is yet poorly known and needs to be answered.

The widely distributed alpine grassland on the QTP, with large altitudinal gradients and fenced versus grazing pastures, provides us an ideal platform to explore whether the long-term enclosure may contribute to the changes in phylogenetic diversity and community assembly in alpine grassland under climate change. We therefore set up an experiment along the altitudinal gradient in central Tibet to reveal the mechanism of alpine grassland community assembly during 11 years of enclosure. Previous studies have shown that species richness and aboveground biomass along the gradient reflected a unimodal pattern, while the limiting climatic factors at the upper and lower limit of the gradient were low temperature and drought, respectively [45]. We hypothesized that environmental filtering dominates the phylogenetic structure of alpine grassland plant communities, and the phylogenetic structure of plant communities at high and low altitudes is relatively clustered, while the phylogenetic structure at medium altitudes is divergent. In addition, we guessed that the change of phylogenetic community structure similarity between inside and outside fences may be related to the grazing intensity and enclosure time period, and it will happen at lower altitudes with high grazing pressure rather than at higher altitudes. Therefore, our aims were to (1) clarify the relative effects of environmental factors and grassland management on plant species richness and phylogenetic community assembly of alpine grassland and (2) reveal the relative impact of stochastic processes and deterministic processes on alpine grassland restoration.

2. Materials and Methods

2.1. Overview of Study Sites

The experiment was set up on the southern slope of Nyenqing-Tanggula Mountain in the central Tibetan Plateau (30°30′–30°32′ N, 91°03′ E; 4400–5200 m), near the grassland station in Dangxumg County, Tibet (Figure S1). This region is dominated by a semi-arid continental climate. According to the data of the Dangxumg meteorological station (4 km from the lowest site and 4288 m a.s.l.) during 1963–2017, the annual mean air temperature and precipitation were 1.94 °C and 479 mm, respectively. There are increasing trends of annual mean air temperature from the 1970s, and annual precipitation from the 1980s to the 2000s (unpublished data). As the only pure animal husbandry county in Lhasa, the capital of Tibet, Dangxumg County is rich in grassland resources (about 70% of the total area is grassland). Due to the high quality and vast area of grassland, the number of yak and sheep in this county reached 2.837×10^5 and 2.215×10^5 in 2017, respectively. In our study area, the resident settlement sits at the lower altitudes (Figure S1) where the grazing intensity is much greater than at higher altitudes.

In 2005, nine HOBO automatic weather stations (Onset Inc., Bourne, MA, USA) were installed at 4400 m, 4500 m, 4650 m, 4800 m, 4950 m, 5100 m, 5200 m, 5300 m and 5500 m (Figure S1). Air temperature (1.5 m above ground) and precipitation dynamics were recorded by the HOBO digger every 30 min. According to the automatic weather station data along the gradient from 2006 to 2016, the multi-year growing season mean air temperature (GST) decreased at a rate of 0.74 °C/100 m. However, growing season precipitation (GSP) showed a unimodal pattern with the maximum GSP (close to 500 mm) occurring at 5100 m and the minimum GSP (less than 370 mm) at and below 4650 m (Figure S2). The soil nutrient content, including soil organic carbon and nitrogen, also showed a unimodal pattern, with the maximum values occurring at 5100 m and minimum values at 4400 m (Figure S2).

Along the slope, the alpine steppe meadows with dominant species of *Stipa pillacea*, *S. purpurea* and *Kobresia humilis* are distributed at 4400–4600 m, while typical alpine meadows with *K. pygmaea* as dominant species are mainly distributed between 4600 and 5200 m. Other companion species include *Androsace tapete*, *K. humilis*, *Potentilla saundersiana* and *Thalictrum alpinum* [45,46]. The grass line, i.e., the upper limit of grassland, locates around 5200 m where the vegetation coverage is about 30–50%. The peak of vegetation above-ground biomass (AGB) occurs in late July to early August.

2.2. Measurement of Species Richness and Coverage of Each Species

In September 2006, we set up a 20 m × 20 m fence at each of the six altitudes below 5200 m (Figure S1). All the fences were set near the automatic weather stations mentioned above. At 5200 m, a large-sized fence was not used due to very low grazing intensity. Four 1 m×1 m fixed quadrats were set up within and without the fence in each altitude. Due to the chronic change of community in alpine grassland, species richness (SR) and coverage of each species were investigated during the peak growing season (from late July to early August) in 2011, 2014 and 2017. The quadrats were divided into 100 small grids of 10 cm × 10 cm, and the species names and coverage of each grid were recorded. Species richness was defined as the number of species per unit area for each quadrat [45].

2.3. Statistical Analysis

Species names were checked by R package plantlist and the assignment of families and genera was based on the Angiosperm Phylogeny Group IV classification [47,48]. The phylogenetic tree was generated by using the ape package [49].

Phylogenetic diversity (PD) has been considered to be a predictor of ecosystem functioning as conserved traits of ecological niches [50] and represents the evolutionary potential of a species setting against environmental changes [51]. Nowadays, PD has been increasingly considered as a strategy to identify geographical regions with maximum diversity and resilience to climatic fluctuations [52,53]. Therefore, to evaluate the total evolutionary history of the species in an assemblage, Faith's PD was used and calculated by ape and picante package [49,54]. We also weighted the coverage of each species to calculate the Net Relatedness Index (NRI) and the Nearest Taxon Index (NTI) for revealing phylogenetic overdispersion (co-occurring species are more distantly related than would be expected by chance) or phylogenetic aggregation (co-occurring species are more closely related than would be expected by chance) by the null model of picante package [54–56].

In detail, a positive value of NRI or NTI indicates phylogenetic aggregation, and if the value is greater than 2, the phylogenetic aggregation pattern is significant; a negative value of NRI or NTI indicates phylogenetic overdispersion, and if the value is less than −2, the phylogenetic overdispersion pattern is significant [57]. The above 3 indices are different aspects of phylogenetic α-diversity in this study. Species richness and phylogenetic α-diversity indices were compared between inside and outside the enclosure at different altitudes using independent sample t-tests. We also used a one-sample t-test to compare the values of NRI or NTI, as well as 0 (the value of total randomness).

To ensure that the environmental factors and experimental treatments had an effect on community taxonomic and phylogenetic structure across the altitudinal gradient, we used the mantel test in the ape package to assess the relationship between standard Euclidean distances of environmental factors and phylogenetic dissimilarity in each quadrant [20].

The effects of management treatments, duration time of enclosure, soil C:N ratio, GST and GSP on SR and phylogenetic α-diversity indices were analyzed using generalized linear mixed-effects models (GLMMs) with generalized response variables being normal or abnormal. Altitude was included as a random effect. The GLMM of the response variables was modeled using a Gaussian error distribution with the lme4 package [58]. Using the following model, X represents a response variable:

$$X \sim \text{management treatment} + \text{duration time} + \text{soil C:N} + \text{GST} + \text{GSP} + (1/\text{altitude})$$

The glmm.hp package was used to estimate the individual fixed effects of the response variables [59]. If the individual fixed effect of factors was negative, we would eliminate the factors and recalculate the equation without the factors.

Finally, to reveal the mechanism of community assembly within and outside the fence at the same altitude, we measured phylogenetic β-diversity metrics (β-nearest taxon index: βNTI, β-net relatedness index: βNRI, and Bray-Curtis-based Raup-Crick index: RC_{bray}) based on the null-model using the iCAMP package and NST package [26,60]. The meaning of each phylogenetic β-diversity index and its threshold are shown below: if the $|\beta NTI|$ or $|\beta NRI| > 2$, the deterministic processes govern community composition and phylogenies are more similar (homogeneous selection, βNTI or $\beta NRI < -2$), or phylogenies are not similar (heterogeneous selection, βNTI or $\beta RTI > +2$). Subsequently, with an $|\beta NTI|$ or $|\beta NRI| < 2$ (stochastic pattern), RC_{bray} was used to further distinguish between homogenized dispersal (taxonomically more similar, RC_{bray} values < -0.95) and dispersal limitation (taxonomically less similar, RC_{bray} values > 0.95). In contrast, with $|\beta NTI|$ or $|\beta NRI| < 2$, and $|RC_{bray}| < 0.95$, the community assembly was treated as an "undominated" pattern, i.e., no single process drove changes in community composition [61]. We also compared the values of βNRI or βNTI with a one-sample t-test, as well as 0 (a completely random value).

SPSS25.0 (SPSS Inc., Chicago, IL, USA) and R 4.1.3 were applied for statistical analysis. Graphpad 7.0 (San Diego, CA, USA) and the online site (https://www.chiplot.online/, accessed on 26 September 2022) was used for drawing figures.

3. Results

3.1. The Overview of the Community Structure

In total, we observed 75 species of vascular plants in our study sites, representing 46 genera from 27 families (Figure 1). Considering the most species-rich families represented, Asteraceae, with the highest number of 16.2%, was the most common family among the seven sites, followed by Poaceae with 9.5%, Cyperaceae with 9.5%, Gentianaceae with 8.1%, and Fabaceae with 5.4%, which together accounted for almost 50% of the total number of plants. In addition, plots at higher sites (>4600 m) were dominated by *Kobresia* and accounted for approximately 60% of the relative coverage, while those at lower altitudes (<4600 m) were mainly dominated by Stipa which accounted for about one-third of the relative coverage (Figure S3).

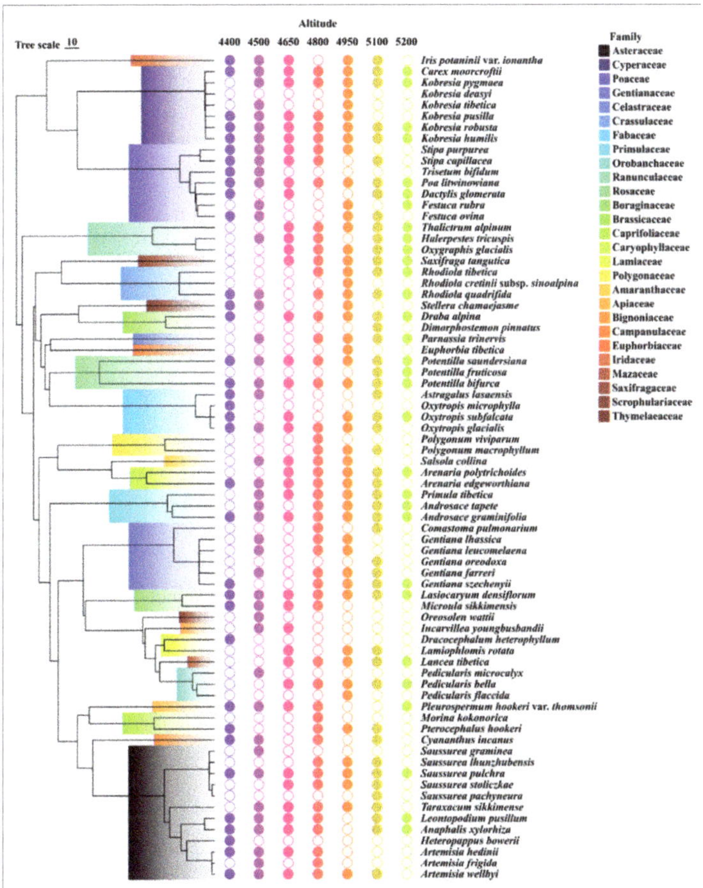

Figure 1. The phylogenetic distribution of all plants in this study. Different colored circles indicate species that occur at different altitudes, blank circles indicate absence, and solid circles indicate presence during the survey. The different colors of the branches correspond to different families in the legend.

3.2. Patterns of SR and α-phylogenetic Indices

There was an approximate unimodal pattern of SR and PD along the altitudinal gradient no matter grazing or not. SR ranged from 7.75 ± 1.70 (at 4400 m in 2017) to 21.75 ± 1.44 (at 4800 m in 2017) within the fences, and from 11.00 ± 1.41 (4400 m in 2014) to 21.50 ± 1.32 (4800 m in 2011) outside the fences (Figure 2a,b). PD ranged from 763.43 ± 119.55 (4400 m in 2017) to 1736.77 ± 55.68 (4800 m in 2017) within the fences, and from 879.79 ± 100.30 (4400 m in 2014) to 1642.67 ± 117.88 (4800 m in 2011) outside the fences, resembled that of SR (Figure 2c,d). However, there was no consistent pattern in NRI and NTI along the gradient. NRI ranged from −0.76 ± 0.20 (4950 m in 2017) to 2.73 ± 0.27 (4650 m in 2017) within the fences, and from −0.91 ± 0.18 (4800 m in 2017) to 2.60 ± 0.54 (4500 m in 2011) outside the fences (Figure 3a,b). NTI ranged from −0.49 ± 0.32 (at 4800 m in 2017) to 2.89 ± 0.27 (at 4500 m in 2011) within the fences and from −0.23 ± 0.28 (4500 m in 2017) to 2.87 ± 0.27 (4500 m in 2011) outside the fences (Figure 3c,d).

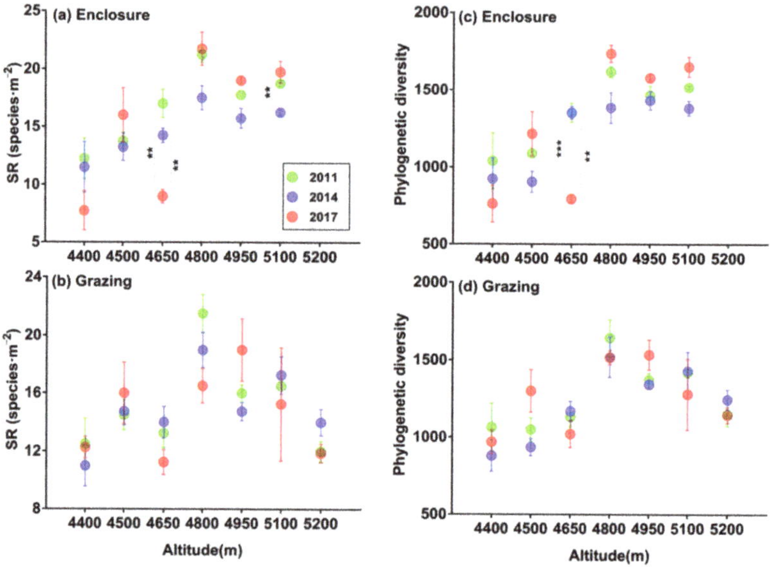

Figure 2. Dynamic variations of effects of enclosure and grazing on (**a**,**b**) species richness (SR) and (**c**,**d**) Faith's phylogenetic diversity along the altitudinal gradient. Asterisk denotes significant difference in SR or phylogenetic diversity among different years. * $P < 0.05$, ** $P < 0.01$, *** $P < 0.001$.

There existed significant interannual variations for these indicators. For example, SR and PD in fenced plots at 4650 m sharply decreased in 2017 compared with 2011 and 2014. NRI in grazing plots at 4500 m decreased significantly in 2017. NTI at 4400 m, 4500 m, 4800 m and 4950 m tended to decrease during the restoration period.

During the investigation period, there was no significant difference in SR between inside and outside the fence at most altitudes, except a few cases (e.g., those at 4950 m and 5100 m in 2011, and 4800 m in 2017, SR inside the fence was higher than that outside) (Figure 4a–c). Similar to the findings of SR, no significant difference was found in PD between inside and outside the fence at most altitudes, except a few cases (e.g., those at 4650 m and 5100 m in 2011, and 4800 m in 2017, PD was higher inside than outside the fence) (Figure 4d–f). There was no significant difference in NRI and NTI between inside and outside the fence at most altitudes (Figure 5).

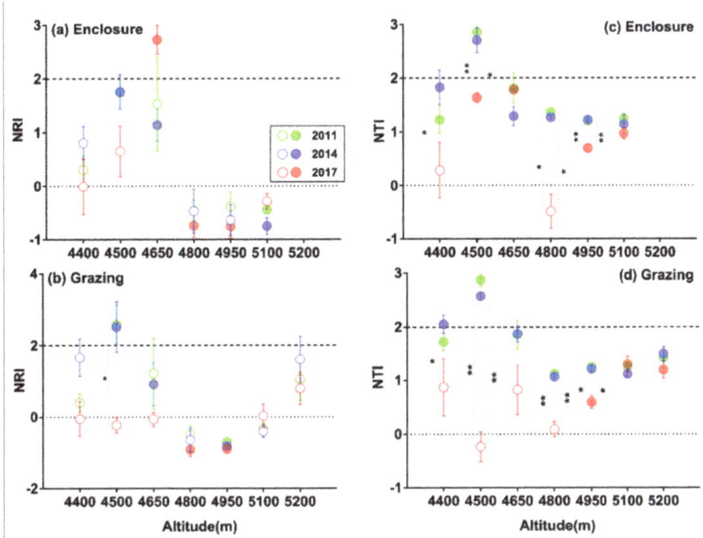

Figure 3. Dynamic variations of effects of enclosure and grazing on (**a**,**b**) net relatedness index (NRI) and (**c**,**d**) nearest taxon index (NTI) along the altitudinal gradient. Blank and solid circles denote insignificant and significant differences from zero in NRI or NTI of each year based on one-sample t test, respectively. Asterisk denotes the significant differences in NRI among different years. * P < 0.05, ** P < 0.01, *** P < 0.001.

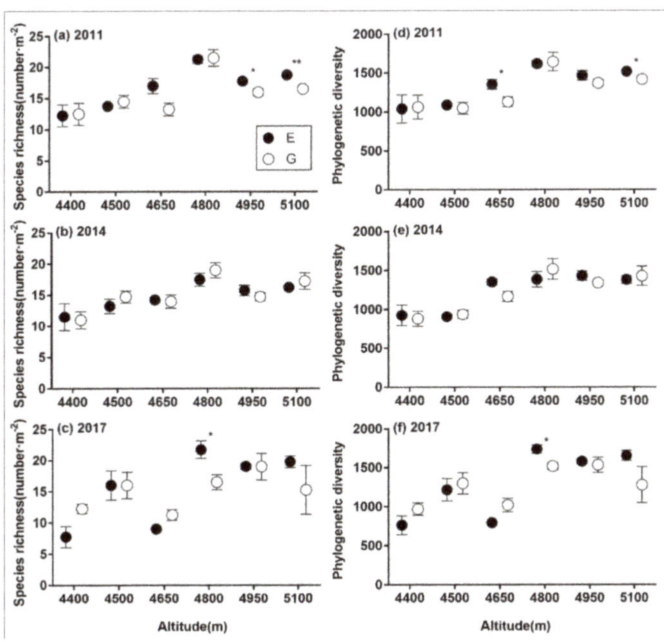

Figure 4. Differences in (**a**–**c**) species richness and (**d**–**f**) Faith's phylogenetic diversity between enclosure and grazing along the altitudinal gradient among different years. Solid circles are for enclosure, and blank circles are for grazing plots. Asterisk denotes the significant difference between enclosure and grazing. * P < 0.05, ** P < 0.01, *** P < 0.001.

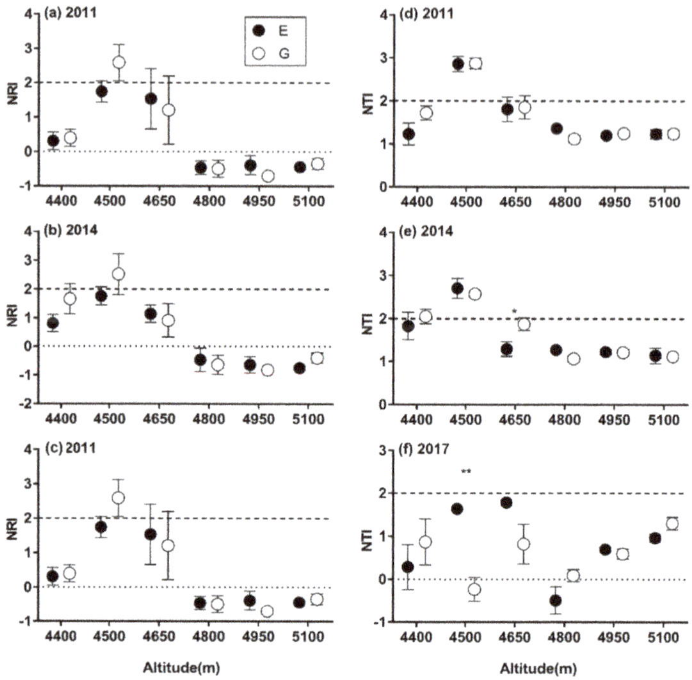

Figure 5. Differences in (**a**–**c**) net relatedness index (NRI) and (**d**–**f**) nearest taxon index (NTI) between enclosure and grazing along the altitudinal gradient among different years. Solid and blank circles are for enclosure and grazing plots, respectively. Asterisk denotes the significant difference between enclosure and grazing. Dashed and dotted line denote an NRI or NTI value of 2 and 0, respectively. * $P < 0.05$, ** $P < 0.01$, *** $P < 0.001$.

In summary, SR and PD were relatively higher at higher altitudes (4800~5100 m). Based on the data of NRI, the pattern of the phylogenetic structure was divergent or random at higher altitudes (4800~5100 m) (NRI $\leq$ 0). At lower altitudes (4400~4650 m) and the highest altitude (5200 m), the pattern of the phylogenetic structure was relatively clustered (NRI > 0). And for NTI, the aggregation structure of closely related species was common (NTI > 0) along the altitudinal gradient.

3.3. The Relative Impacts of Grazing Management, Duration of Enclosure and Environmental Factors on SR and α-phylogenetic Indices

Mantel test found that both taxonomic and phylogenetic signals of alpine grassland community along the gradient reflected a significant correlation with environmental factors (Figure S4), indicating that taxonomic and phylogenetic parameters can indicate the impact of environmental factors on community structure.

According to the established generalized linear model, for the whole altitude gradient, the overall fixed factors impact by the model on SR change was 18.34%. The influence of variation of GSP on SR change was more than 13%, and other factors just explained less than 6.0% (Figure 6a). The overall fixed factors impact by the model on the change of PD was 30.12%, and the influence of variation of GSP on PD change occupied more than half of the total explanations of the model. The second large influential impact was the duration of enclosure and soil C:N ratio which could explain 8.64% and 2.96%, respectively (Figure 6b). The overall fixed factors impact by the model on the change of NRI was 12.86%, and the variance explanations of three influential factors (duration of enclosure, variation of GSP and soil C:N ratio) were 5.33%, 5.10% and 1.66%, respectively (Figure 6c). The fixed factors impact by the model on the change of NTI was 24.40%, and variation of GSP, soil C:N ratio and duration of enclosure had a large impact on the change of NTI, with the variance explanations of 18.96%, 3.81% and 1.63%, respectively (Figure 6d).

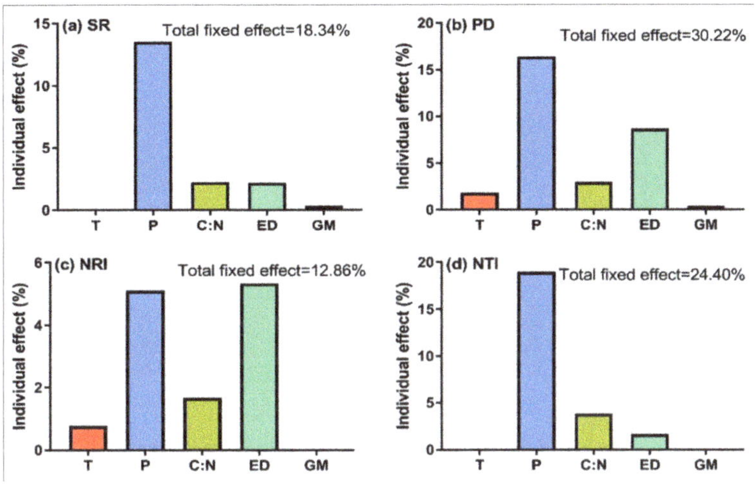

Figure 6. Relative contributions of GST, GSP, soil C:N ratio, enclosure duration f (ED, year) and grassland management (GM, enclosure vs grazing) to the variation of (**a**) species richness (SR), (**b**) Faith's phylogenetic diversity (PD), (**c**) net relatedness index (NRI), and (**d**) nearest taxon index (NTI).

3.4. Patterns of β-phylogenetic Indices between Enclosure and Grazing Plots

In order to estimate the assembly process of plant assemblages in the alpine grassland on the QTP, we calculated the β-phylogenetic indices between the fenced and grazing plots. We found that there were differences in βNRI between 4500 m and other altitudes in 2011 and 2014. The βNRI values of 4500 m in 2011 and 2014 were less than −2, and those of the other plots were between −2~2. Compared with 0 (ecological drift), the values at other altitudes were significantly different from 0 in most cases except that at 5100 m in 2017 (Figure 7a). The βNTI at most sites ranged between −2 and +2 (Figure 7b). Compared with 0 (ecological drift), the frequency of insignificant difference of βNTI was higher at high altitudes. RC_{bray} was negative at all altitudes, and was less than −0.95 from 4650 m to 5100 m in most years (Figure 7c). In addition, the β-phylogenetic indices showed significant inter-annual differences, e.g., βNRI at 4400 m, 4500 m, 4950 m and 5100 m, βNTI from 4650 m to 4950 m, and RC_{bray} at 4800 m (Figure 7).

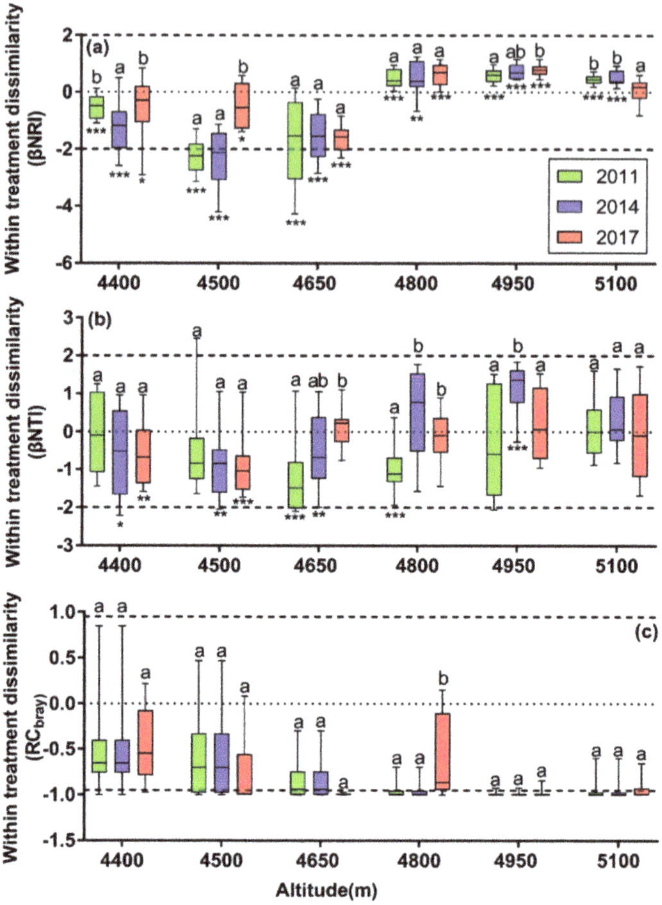

Figure 7. Effects of different types of grazing management on (**a**) net relatedness index based on β-diversity (βNRI), (**b**) nearest taxon index based on β-diversity (βNTI) and (**c**) Raup–Crick-based measure (RC_{Bray}) along the altitudinal gradient among different years. Dashed lines represent the threshold value of each index. Different letters indicate the significant differences ($P < 0.05$) among different years. Asterisk denotes values that are significantly different from zero based on one-sample t test. * $P < 0.05$, ** $P < 0.01$, *** $P < 0.001$.

4. Discussion

Biodiversity plays a key role in maintaining and enhancing ecosystem structure, function and stability [62,63]. Therefore, biodiversity has long been the research hotspot in ecology and evolution. However, most studies concentrated on taxonomic diversity, which would lose some dimensions of biodiversity [64], and taxonomic always introduces polyphyletic groups that lead to frequent failure of taxonomic groupings to map onto ecologically relevant issues [56]. In contrast, phylogenetic diversity combines information from taxonomy and species' evolutionary history, and therefore better represents ecological variation in functional traits and greatly expands the depth of biodiversity research [13,65,66]. In this study, SR was used to characterize the taxonomic diversity of the community, and PD, NRI and NTI were used to characterize the phylogenetic diversity of the community. For the community phylogenetic structure, this study found that NTI was generally more aggregated than NRI. Some researchers believed that the calculation method of the phylogenetic index may explain this difference [61,67], since NRI represents the overall structural pattern of the entire phylogenetic tree, while NTI focuses more on the top of the branches [1]. Therefore, NTI shows a stronger display of a uniform dispersion signal, for the similarity of the close relatives is much smaller than other species from the whole phylogenetic tree in the community. This is exactly supported by the current study. It is suggested that the ecological niche evolution has been relatively conservative and the relatively close species are generally more adapted to similar environments, although rapid uplift processes and repeated ice ages on the QTP have shaped a relatively high species diversity in this region, ultimately leading to aggregation of NTI.

Enclosures are the most common management tool for grassland restoration, and many experiments have shown that enclosures can improve grassland productivity and promote the restoration of species richness [68–70]. In general, large-scale environmental factors (i.e., climate factors of temperature or precipitation) select species from the regional species pool [71], while small-scale environmental factors (i.e., species interaction grazing/enclosure) generate further filtering [72]. In our study, we found that different pasture management had no significant effect on species richness, phylogenetic diversity, and phylogenetic structure across the altitudinal gradient (Figure 6). However, in certain altitudes, there is a controversy of different patterns of SR and NRI/NTI between enclosure and grazing treatment to some degree. Calculation formulas of NRI and NTI need to weigh the relative coverage of each species, while SR and PD do not need to do that, and so highlight rare species [73,74]. After a 11-year restoration, the species interactions might also exert an impact on NRI/NTI differences inside fences. And the high intensity of competition (e.g., light, water and nutrient competition) between dominant species and auxiliary species in the fence may lead to the consequence of NRI/NTI differentiation between outside and inside the fence, which is consistent with the results of enclosure shrub invasion plots in temperate grasslands in northern China [42]. In addition, we found that in 2017, compared to the other two years of the investigation, there was a significant decrease in the degree of aggregation of the phylogenetic structure outside the fence at low altitudes close to the settlement (Figure 3), which may be related to the strengthening of local environmental protection and the reduction of grazing pressure. As for the large difference in SR between inside and outside the high-altitude fences, it may be affected by grazing. And there are more rare species in the fence, which may lead to the differences at higher altitudes in some years. At low altitudes, due to long-term high-intensity grazing, the phenomenon of lack of rare species may be caused by the local seed bank. Even if the coverage of species inside and outside the fence is different, there is little difference in species richness. In brief, compared with SR and PD, NRI and NTI can reflect the impact of grazing on community structure, i.e., the impact of grazing interference at lower altitudes is greater than at higher altitudes.

Strong environmental gradients are powerful ways to study natural selection, especially in alpine ecosystems where environmental filtering is often recognized to be the only main driver of community assembly [75]. In this study, we found that the effect of precipitation on species richness, phylogenetic diversity and community phylogenetic structure were greater than that of temperature, consistent with our previous findings on aboveground biomass (unpublished data). Additionally, studies from alpine meadows in northern Tibet and wetland plants in the QTP concluded that plant species richness and phylogenetic diversity in arid and semi-arid areas of the QTP were mainly driven by precipitation [61,76]. However, Zhang found that temperature was the dominant environmental factor affecting the phylogenetic structure of alpine meadow communities compared to precipitation, and that the low temperature is the dominant force of natural selection [76]. Our study showed that in arid and semi-arid ecosystems of the QTP, although plants are exposed to low-temperature stress, precipitation is more likely a limited resource that may help maintain greater community coverage and species diversity. For the community phylogenetic structure in the middle and low altitudes ($\leq$ 4650 m), relatively higher temperature and lower precipitation as the environmental filter caused a decrease in soil water availability, and the community was mainly composed of species like *Stipa* and Gramineae with relatively deep root systems as the co-dominant species, leading to an aggregated pattern of community phylogenetic structure. However, at middle and high altitudes (4800–5100 m) with more abundant precipitation and higher soil moisture availability, the species richness was also higher, and the interspecific competition effects might be relatively strong, leading to a relatively divergent community phylogenetic structure. In addition, the aggregation pattern of the community phylogenetic structure near the highest altitude of the grass line (5200 m), was more likely to be related to temperature, as cold temperature is often accompanied by more frequent meteorological extremes including frost and hail, especially at the upper edge of the grassland distribution [77]. Furthermore, only a small number of cold-tolerant species that can share similar environmental adaptation strategies were eventually selected by nature. Moreover, it has also been shown that the relative importance of environmental factors on phylogenetic structure increases as the species pool decreases [78], ultimately leading to the aggregation of community phylogenetic structure at lower and higher altitudes. In general, we found that the environmental filtering factors were not influential as proposed to affect species richness, phylogenetic diversity, and phylogenetic indices to a high degree, partly consistent with Qian who argued that shrubs and herbs are less sensitive to climate compared to trees and attributed this to their smaller sizes as they can be sheltered by micro-habitat effects [67]. However, the patterns of NRI and NTI along altitudinal gradient still suggested that the sometimes-questionable dogma-habitat filtering may lead to phylogenetic clustering, and that competitive exclusion will lead to phylogenetic overdispersion [55,79].

Most studies have shown that competitive exclusion dominates community assembly in climax communities in arid and semi-arid regions [80–82], while some researchers believe that the environmental filtering processes drive community assembly in degraded communities [83,84]. Meanwhile, more and more studies have pointed out that stochastic processes affect grassland community composition and structure [85–88]. However, studies on the changes in plant community phylogenetic structure of alpine grassland during the restoration period are still limited. Overall, the deterministic ecological processes affecting community assembly include heterogeneous and homogeneous selection, and the stochastic processes include homogeneous dispersal, dispersal limitation and ambiguous processes [14,20,22,26,60]. In this study, at most altitudes, the pattern of phylogenetic βNTI indicated that the community assembly was mainly determined by stochastic processes. Additionally, according to the values of RC_{bray} and βNRI, the similarity of communities between inside and outside the fence is relatively low at lower altitudes (4400–4650 m), which infers that the stronger grazing pressure leads to differences in the community structure between inside and outside. This is especially true at 4500 m, where grazing pressure was the highest and homogeneous selection occurred in the fifth and eighth year after enclosure. At 4500 m, even after 8 years of enclosures, the environment still showed homogenization whether inside or outside the fences, and the effect of enclosure on community assembly remained insignificant. As homogenization is an important form of biotic impoverishment, it may lead to an increased risk of species extinction and reduce the resilience of the system to large-scale disturbances [89,90]. Until 10 years after enclosure, stochastic processes became dominant in determining the variability of community structure, which reveals that short-term restoration may not restore community structure to a normal state in areas of intense grazing, and the threshold of enclosure duration was similar to the result from an enclosure restoration experiment (13 years from unhealthy stage to health stage) of typical steppe in Northern China [91]. For the middle and high altitudes where grazing pressure was lower, the similarity of the communities between inside and outside the fence was very high, and homogenous dispersal occurred, indicating that the current grazing intensity exerts no significant impact on the community structure in this area. Overall, the results suggested that grassland management should be formulated rationally and depended on different situations. For instance, grazing intensity needs to be reduced at middle and low altitudes, especially at 4500 m which is closest to the settlement (Figure S1), and fencing management can be appropriately decreased at middle and high altitudes.

In most studies, the conserved hypothesis of ecological niche is the basis for inferring ecological processes [14,20,26]. However, we found that phylogenetic signals behaved well in remote relatives rather than in close relatives, albeit with a slight difference under the taxonomic view (Figure S4), which led to some uncertainties in the study. For example, the phylogenetic timing of most species at the genus level (*Saussurea, Kobresia*) and even at the family level may have contributed to this problem. In the future, we need improvements in technology to address the high resolution of the evolutionary history of species.

Although this study pointed out that the stochastic process dominated changes in the community structure of alpine grassland during restoration and identified the specific factors influencing the deterministic processes and the relative effects, the superimposed and antagonistic effects of environmental filtering and biological interactions may also contribute to the emergence of stochastic processes on community assembly [92]. This may depend on the relative importance of ecological niche differences and interspecific differences in competitiveness and/or mutualism [81]. Therefore, future studies need to separate the effects of environmental filtering and interspecific species relationships concisely [72] and refine the effects of deterministic processes on different scales on alpine grassland community assembly.

5. Conclusions

Recently, patterns, processes and mechanisms underlying diversity have raised a colossal interest in ecology. However, there is limited information about human activity regarding plant community assembly in alpine grassland ecosystems. This study attempted to provide insights into different grazing management on the diversity and the community assembly process of alpine grassland on the QTP. We found that the overall phylogenetic structure of the alpine grassland community was divergent at medium and high altitudes (4800–5100 m) where the environment was relatively unextreme. At the lower altitudes (4400–4650 m) with low precipitation, and the highest altitude (5200 m) with low temperature, the phylogenetic structure of the plant community was more aggregative. It seemed that the grassland management exerted little impacts on all the diversity and structure indices. Among the environmental factors, precipitation was the dominant factor affecting these indices. Stochastic processes have driven the changes in the communities between inside and outside the fences. In summary, the 11-year enclosure had little effect on alpine meadow community structure at higher altitudes. However, it could restore the community of steppe meadow at lower altitudes where the grazing pressure is high. For local grazing management, it should be better to relieve the burden of high grazing intensity by reducing the number of livestock below 4700 m, while keeping the current grazing intensity above 4700 m in this area.

Supplementary Materials: The following supporting information can be downloaded at: https://www.mdpi.com/article/10.3390/10.3390/d14100832/s1, Figure S1: Distribution of fenced plots along the altitudinal gradient in Damxung County, Tibet; Figure S2: Variations in (a) growing season mean temperature (GST) and precipitation (GSP) during 2006–2017 and (b) soil total nitrogen (STN) and soil organic carbon (SOC) along the altitude gradient. Different letters indicate significant differences among the seven altitudes (ANOVA and Tukey HSD test, $P < 0.05$). Bars indicate SE of the mean.; Figure S3: The relative coverage of common genera along the altitude gradient.; Figure S4: Relationships between (a) taxonomic and (b) phylogenetic Bray–Curtis similarity of communities and the standardized Euclidean distance of environmental factors (including temperature, precipitation, soil C:N ratio, year of enclosure and grazing management treatments). The Mantel test was used to examine the correlations between pairwise Bray–Curtis similarity and pairwise differences in environmental factors, indicating taxonomic and phylogenetic signal test for ecological niche differences. The solid circles indicate significant signals ($P \leq 0.05$).

Author Contributions: Conceptualization, Z.D.; methodology, Z.D.; software, Z.D.; validation, T.L.; formal analysis, Z.D.; investigation, Z.W., R.L., J.Z. and Z.D.; resources, T.L.; data curation, Z.D.;

writing—original draft preparation, Z.D.; writing—review and editing, Z.D., J.Z., T.L., Z.W. and L.Z.; visualization, Z.D. and Y.G.; supervision, T.L. and L.Z.; project administration, T.L.; funding acquisition, T.L. All authors have read and agreed to the published version of the manuscript.

Funding: The National Natural Science Foundation of China (41830649) and the 2nd Tibetan Plateau Scientific Expedition and Research Program (2019QZKK0106) provided financial support.

Institutional Review Board Statement: Not applicable.

Data Availability Statement: All data used in the manuscript are already publicly accessible, and we provided the download address in the manuscript.

Acknowledgments: We would like to thank X Li and students at Lanzhou University and Tibet University, who engaged in investigation field work, and M Du and X Zhang for their help with the meteorological observations.

Conflicts of Interest: The authors declare no conflict of interest.

References

1. Kraft, N.J.; Cornwell, W.K.; Webb, C.O.; Ackerly, D.D. Trait evolution, community assembly, and the phylogenetic structure of ecological communities. *Am. Nat.* **2007**, *170*, 271–283. [CrossRef] [PubMed]
2. Arnillas, C.A.; Borer, E.T.; Seabloom, E.W.; Alberti, J.; Baez, S.; Bakker, J.D.; Boughton, E.H.; Buckley, Y.M.; Bugalho, M.N.; Donohue, I.; et al. Opposing community assembly patterns for dominant and nondominant plant species in herbaceous ecosystems globally. *Ecol. Evol.* **2021**, *11*, 17744–17761. [CrossRef] [PubMed]
3. Gleason, H.A. The Individualistic Concept of the Plant Association. *Bull. Torrey Bot. Club* **1926**, *53*, 7. [CrossRef]
4. Elton, C.S. *The Ecology of Animals*; Methuen: London, UK, 1933.
5. Sale, P.F. Maintenance of High Diversity in Coral Reef Fish Communities. *Am. Nat.* **1977**, *111*, 337–359. [CrossRef]
6. Chesson, P.L.; Warner, R.R. Environmental variability promotes coexistence in lottery competitive systems. *Am. Nat.* **1981**, *117*, 923–943. [CrossRef]
7. Chesson, P. Mechanisms of maintenance of species diversity. *Annu. Rev. Ecol. Syst.* **2000**, *31*, 343–366. [CrossRef]
8. Hubbell, S.P. *The Unified Neutral Theory of Biodiversity and Biogeography*; Princeton University Press: Princeton, NJ, USA, 2001; Volume 32.
9. Adler, P.B.; Hillerislambers, J.; Levine, J.M. A niche for neutrality. *Ecol. Lett.* **2007**, *10*, 95–104. [CrossRef]
10. Rosindell, J.; Hubbell, S.P.; He, F.; Harmon, L.J.; Etienne, R.S. The case for ecological neutral theory. *Trends Ecol. Evol.* **2012**, *27*, 203–208. [CrossRef]
11. Dini-Andreote, F.; Stegen, J.C.; van Elsas, J.D.; Salles, J.F. Disentangling mechanisms that mediate the balance between stochastic and deterministic processes in microbial succession. *Proc. Natl. Acad. Sci. USA* **2015**, *112*, 1326–1332. [CrossRef]
12. Kembel, S.W. Disentangling niche and neutral influences on community assembly: Assessing the performance of community phylogenetic structure tests. *Ecol. Lett.* **2009**, *12*, 949–960. [CrossRef]
13. Srivastava, D.S.; Cadotte, M.W.; MacDonald, A.A.; Marushia, R.G.; Mirotchnick, N. Phylogenetic diversity and the functioning of ecosystems. *Ecol. Lett.* **2012**, *15*, 637–648. [CrossRef]
14. Stegen, J.C.; Lin, X.; Fredrickson, J.K.; Konopka, A.E. Estimating and mapping ecological processes influencing microbial community assembly. *Front. Microbiol.* **2015**, *6*, 370. [CrossRef]
15. Tripathi, B.M.; Stegen, J.C.; Kim, M.; Dong, K.; Adams, J.M.; Lee, Y.K. Soil pH mediates the balance between stochastic and deterministic assembly of bacteria. *ISME J.* **2018**, *12*, 1072–1083. [CrossRef]
16. Maguire, V.G.; Bordenave, C.D.; Nieva, A.S.; Llames, M.E.; Colavolpe, M.B.; Gárriz, A.; Ruiz, O.A. Soil bacterial and fungal community structure of a rice monoculture and rice-pasture rotation systems. *Appl. Soil Ecol.* **2020**, *151*, 103535. [CrossRef]
17. Mugnai, M.; Trindade, D.P.F.; Thierry, M.; Kaushik, K.; Hrček, J.; Götzenberger, L. Environment and space drive the community assembly of Atlantic European grasslands: Insights from multiple facets. *J. Biogeogr.* **2022**, *49*, 699–711. [CrossRef]
18. Gravel, D.; Canham, C.D.; Beaudet, M.; Messier, C. Reconciling niche and neutrality: The continuum hypothesis. *Ecol. Lett.* **2006**, *9*, 399–409. [CrossRef]
19. Chase, J.M.; Myers, J.A. Disentangling the importance of ecological niches from stochastic processes across scales. *Philos. Trans. R. Soc. Lond. B Biol. Sci.* **2011**, *366*, 2351–2363. [CrossRef]
20. Stegen, J.C.; Lin, X.; Fredrickson, J.K.; Chen, X.; Kennedy, D.W.; Murray, C.J.; Rockhold, M.L.; Konopka, A. Quantifying community assembly processes and identifying features that impose them. *ISME J.* **2013**, *7*, 2069–2079. [CrossRef]
21. Vellend, M. Conceptual synthesis in community ecology. *Q. Rev. Biol.* **2010**, *85*, 183–206. [CrossRef]
22. Zhou, J.; Ning, D. Stochastic Community Assembly: Does It Matter in Microbial Ecology? *Microbiol. Mol. Biol. Rev.* **2017**, *81*, e00002-17. [CrossRef]
23. Anderson, M.J.; Crist, T.O.; Chase, J.M.; Vellend, M.; Inouye, B.D.; Freestone, A.L.; Sanders, N.J.; Cornell, H.V.; Comita, L.S.; Davies, K.F.; et al. Navigating the multiple meanings of beta diversity: A roadmap for the practicing ecologist. *Ecol. Lett.* **2011**, *14*, 19–28. [CrossRef]

24. Hanson, C.A.; Fuhrman, J.A.; Horner-Devine, M.C.; Martiny, J.B. Beyond biogeographic patterns: Processes shaping the microbial landscape. *Nat. Rev. Microbiol.* **2012**, *10*, 497–506. [CrossRef]
25. Fukami, T. Historical contingency in community assembly: Integrating niches, species pools, and priority effects. *Annu. Rev. Ecol. Evol. Syst.* **2015**, *46*, 1–23. [CrossRef]
26. Ning, D.; Yuan, M.; Wu, L.; Zhang, Y.; Guo, X.; Zhou, X.; Yang, Y.; Arkin, A.P.; Firestone, M.K.; Zhou, J. A quantitative framework reveals ecological drivers of grassland microbial community assembly in response to warming. *Nat. Commun.* **2020**, *11*, 4717. [CrossRef]
27. Chen, B.; Zhang, X.; Tao, J.; Wu, J.; Wang, J.; Shi, P.; Zhang, Y.; Yu, C. The impact of climate change and anthropogenic activities on alpine grassland over the Qinghai-Tibet Plateau. *Agric. For. Meteorol.* **2014**, *189*, 11–18. [CrossRef]
28. Li, C.; Peng, F.; Xue, X.; You, Q.; Lai, C.; Zhang, W.; Cheng, Y. Productivity and quality of alpine grassland vary with soil water availability under experimental warming. *Front. Plant. Sci.* **2018**, *9*, 1790. [CrossRef]
29. Bardgett, R.D.; Bullock, J.M.; Lavorel, S.; Manning, P.; Schaffner, U.; Ostle, N.; Chomel, M.; Durigan, G.; Fry, E.L.; Johnson, D.; et al. Combatting global grassland degradation. *Nat. Rev. Earth Environ.* **2021**, *2*, 720–735. [CrossRef]
30. Zhou, L.; Zhu, Y.; Yang, G.; Luo, Y. Quantitative evaluation of the effect of prohibiting grazing policy on grassland desertification reversal in northern China. *Environ. Earth Sci.* **2013**, *68*, 2181–2188. [CrossRef]
31. Liu, S.; Wang, T.; Kang, W.; David, M. Several challenges in monitoring and assessing desertification. *Environ. Earth Sci.* **2015**, *73*, 7561–7570. [CrossRef]
32. Fu, B.; Ouyang, Z.; Shi, P.; Fan, J.; Wang, X.; Zheng, H.; Zhao, W.; Wu, F. Current condition and protection strategies of Qinghai-Tibet Plateau ecological security barrier. *Bull. Chin. Acad. Sci.* **2021**, *36*, 1298–1306.
33. Wei, D.; Ri, X.; Wang, Y.; Wang, Y.; Liu, Y.; Yao, T. Responses of CO_2, CH_4 and N_2O fluxes to livestock exclosure in an alpine steppe on the Tibetan Plateau, China. *Plant Soil* **2012**, *359*, 45–55. [CrossRef]
34. Dong, Q.M.; Zhao, X.Q.; Wu, G.L.; Shi, J.J.; Ren, G.H. A review of formation mechanism and restoration measures of "black-soil-type" degraded grassland in the Qinghai-Tibetan Plateau. *Environ. Earth Sci.* **2013**, *70*, 2359–2370. [CrossRef]
35. Liu, H.; Mi, Z.; Lin, L.; Wang, Y.; Zhang, Z.; Zhang, F.; Wang, H.; Liu, L.; Zhu, B.; Cao, G.; et al. Shifting plant species composition in response to climate change stabilizes grassland primary production. *Proc. Natl. Acad. Sci. USA* **2018**, *115*, 4051–4056. [CrossRef] [PubMed]
36. Shang, Z.; Long, R. Formation causes and recovery of the "Black Soil Type" degraded alpine grassland in Qinghai-Tibetan Plateau. *Front. Agri. China* **2007**, *1*, 197–202. [CrossRef]
37. Wu, G.-L.; Du, G.-Z.; Liu, Z.-H.; Thirgood, S. Effect of fencing and grazing on a Kobresia-dominated meadow in the Qinghai-Tibetan Plateau. *Plant Soil* **2009**, *319*, 115–126. [CrossRef]
38. Lin, X.; Zhao, H.; Zhang, S.; Li, X.; Gao, W.; Ren, Z.; Luo, M. Effects of animal grazing on vegetation biomass and soil moisture on a typical steppe in Inner Mongolia, China. *Ecohydrology* **2022**, *15*, e2350. [CrossRef]
39. Meissner, R.A.; Facelli, J.M. Effects of sheep exclusion on the soil seed bank and annual vegetation in chenopod shrublands of South Australia. *J. Arid Environ.* **1999**, *42*, 117–128. [CrossRef]
40. Barber, N.A.; Farrell, A.K.; Blackburn, R.C.; Bauer, J.T.; Groves, A.M.; Brudvig, L.A.; Jones, H.P.; Gibson, D. Grassland restoration characteristics influence phylogenetic and taxonomic structure of plant communities and suggest assembly mechanisms. *J. Ecol.* **2019**, *107*, 2105–2120. [CrossRef]
41. Alberti, J.; Bakker, E.S.; van Klink, R.; Olff, H.; Smit, C. Herbivore exclusion promotes a more stochastic plant community assembly in a natural grassland. *Ecology* **2017**, *98*, 961–970. [CrossRef]
42. Hao, G.; Yang, N.; Dong, K.; Xu, Y.; Ding, X.; Shi, X.; Chen, L.; Wang, J.; Zhao, N.; Gao, Y. Shrub-encroached grassland as an alternative stable state in semiarid steppe regions: Evidence from community stability and assembly. *Land Degrad. Dev.* **2021**, *32*, 3142–3153. [CrossRef]
43. Sun, W.; Li, S.; Wang, J.; Fu, G. Effects of grazing on plant species and phylogenetic diversity in alpine grasslands, Northern Tibet. *Ecol. Eng.* **2021**, *170*, 106331. [CrossRef]
44. Zhu, J.; Zhang, Y.; Wang, W.; Jiang, L.; Wang, L.; Shen, R.; Chen, N.; Yang, X. Species turnover drives grassland community to phylogenetic clustering over long-term grazing disturbance. *J. Plant. Ecol.* **2020**, *13*, 157–164. [CrossRef]
45. Wang, Z.; Luo, T.; Li, R.; Tang, Y.; Du, M. Causes for the unimodal pattern of biomass and productivity in alpine grasslands along a large altitudinal gradient in semi-arid regions. *J. Veg. Sci.* **2013**, *24*, 189–201. [CrossRef]
46. Zhao, J.X.; Luo, T.X.; Li, R.C.; Wei, H.X.; Li, X.; Du, M.Y.; Tang, Y.H. Precipitation alters temperature effects on ecosystem respiration in Tibetan alpine meadows. *Agric. For. Meteorol.* **2018**, *252*, 121–129. [CrossRef]
47. Zhang, J.L.; Liu, B.; Liu, S.; Feng, Z.H.; Jiang, K.W. Plantlist: Looking Up the Status of Plant Scientific Names based on The Plant List Database. R package version 0.7.2. Available online: https://github.com/helixcn/plantlist/ (accessed on 15 June 2022).
48. The Angiosperm Phylogeny Group. An update of the Angiosperm Phylogeny Group classification for the orders and families of flowering plants: APG IV. *Bot. J. Linn. Soc.* **2016**, *181*, 1–20. [CrossRef]
49. Paradis, E.; Schliep, K. ape 5.0: An environment for modern phylogenetics and evolutionary analyses in R. *Bioinformatics* **2019**, *35*, 526–528. [CrossRef]
50. Gravel, D.; Bell, T.; Barbera, C.; Combe, M.; Pommier, T.; Mouquet, N. Phylogenetic constraints on ecosystem functioning. *Nat. Commun.* **2012**, *3*, 1117. [CrossRef]

51. Forest, F.; Grenyer, R.; Rouget, M.; Davies, T.J.; Cowling, R.M.; Faith, D.P.; Balmford, A.; Manning, J.C.; Procheş, Ş.; van der Bank, M.; et al. Preserving the evolutionary potential of floras in biodiversity hotspots. *Nature* **2007**, *445*, 757–760. [CrossRef]
52. Scherson, R.A.; Faith, D.P. *Phylogenetic Diversity: Applications and Challenges in Biodiversity Science*; Springer International Publishing: Cham, Switzerland, 2018.
53. Miller, J.T.; Jolley-Rogers, G.; Mishler, B.D.; Thornhill, A.H. Phylogenetic diversity is a better measure of biodiversity than taxon counting. *J. Syst. Evol.* **2018**, *56*, 663–667.54. [CrossRef]
54. Kembel, S.W.; Cowan, P.D.; Helmus, M.R.; Cornwell, W.K.; Morlon, H.; Ackerly, D.D.; Blomberg, S.P.; Webb, C.O. Picante: R tools for integrating phylogenies and ecology. *Bioinformatics* **2010**, *26*, 1463–1464. [CrossRef]
55. Webb, C.O.; Ackerly, D.D.; McPeek, M.A.; Donoghue, M.J. Phylogenies and community ecology. *Annu. Rev. Ecol. Syst.* **2002**, *33*, 475–505. [CrossRef]
56. Vamosi, S.M.; Heard, S.B.; Vamosi, J.C.; Webb, C.O. Emerging patterns in the comparative analysis of phylogenetic community structure. *Mol. Ecol.* **2009**, *18*, 572–592. [CrossRef]
57. Webb, C.O. Exploring the Phylogenetic Structure of Ecological Communities: An Example for Rain Forest Trees. *Am. Nat.* **2000**, *156*, 145–155. [CrossRef]
58. Bates, D.; Mächler, M.; Bolker, B.; Walker, S. Fitting Linear Mixed-Effects Models Using lme4. *J. Stat. Softw.* **2015**, *67*, 1–48. [CrossRef]
59. Lai, J.; Zou, Y.; Zhang, J.; Peres-Neto, P.R. Generalizing hierarchical and variation partitioning in multiple regression and canonical analyses using the rdacca.hp R package. *Methods. Ecol. Evol.* **2022**, *13*, 782–788. [CrossRef]
60. Ning, D.; Deng, Y.; Tiedje, J.M.; Zhou, J. A general framework for quantitatively assessing ecological stochasticity. *Proc. Natl. Acad. Sci. USA* **2019**, *116*, 16892–16898. [CrossRef]
61. Cui, S.; Ouyang, J.; Lu, Y.; Liu, W.; Li, W.; Liu, G.; Zhou, W. The diversity and community assembly process of wetland plants from lakeshores on the Qinghai-Tibetan Plateau. *Diversity* **2021**, *13*, 685. [CrossRef]
62. Hopping, K.A.; Knapp, A.K.; Dorji, T.; Klein, J.A. Warming and land use change concurrently erode ecosystem services in Tibet. *Glob. Chang. Biol.* **2018**, *24*, 5534–5548. [CrossRef]
63. Bennett, J.A.; Koch, A.M.; Forsythe, J.; Johnson, N.C.; Tilman, D.; Klironomos, J. Resistance of soil biota and plant growth to disturbance increases with plant diversity. *Ecol. Lett.* **2020**, *23*, 119–128. [CrossRef]
64. Lyashevska, O.; Farnsworth, K.D. How many dimensions of biodiversity do we need? *Ecol. Indic.* **2012**, *18*, 485–492. [CrossRef]
65. Pavoine, S.; Bonsall, M.B. Measuring biodiversity to explain community assembly: A unified approach. *Biol. Rev. Camb. Philos. Soc.* **2011**, *86*, 792–812. [CrossRef] [PubMed]
66. Swenson, N.G. The role of evolutionary processes in producing biodiversity patterns, and the interrelationships between taxonomic, functional and phylogenetic biodiversity. *Am. J. Bot.* **2011**, *98*, 472–480. [CrossRef] [PubMed]
67. Qian, H.; Jin, Y.; Ricklefs, R.E. Phylogenetic diversity anomaly in angiosperms between eastern Asia and eastern North America. *Proc. Natl. Acad. Sci. USA* **2017**, *114*, 11452–11457. [CrossRef] [PubMed]
68. Firincioglu, H.K.; Seefeldt, S.S.; Sahin, B. The effects of long-term grazing exclosures on range plants in the Central Anatolian Region of Turkey. *Environ. Manag.* **2007**, *39*, 326–337. [CrossRef]
69. Wu, J.; Shen, Z.; Zhang, X. Precipitation and species composition primarily determine the diversity–productivity relationship of alpine grasslands on the Northern Tibetan Plateau. *Alpine. Bot.* **2014**, *124*, 13–25. [CrossRef]
70. Wu, J.; Li, M.; Fiedler, S.; Ma, W.; Wang, X.; Zhang, X.; Tietjen, B. Impacts of grazing exclusion on productivity partitioning along regional plant diversity and climatic gradients in Tibetan alpine grasslands. *J. Environ. Manag.* **2019**, *231*, 635–645. [CrossRef]
71. Evan, W.; Paul, K. *Ecological Assembly Rules Perspectives, Advances, Retreats*; Cambridge University Press: New York, NY, USA, 1999.
72. Kraft, N.J.B.; Adler, P.B.; Godoy, O.; James, E.C.; Fuller, S.; Levine, J.M.; Fox, J. Community assembly, coexistence and the environmental filtering metaphor. *Funct. Ecol.* **2015**, *29*, 592–599. [CrossRef]
73. Jost, L. Partitioning diversity into independent alpha and beta components. *Ecology* **2007**, *88*, 2427–2439. [CrossRef]
74. Chao, A.; Gotelli, N.J.; Hsieh, T.C.; Sander, E.L.; Ma, K.H.; Colwell, R.K.; Ellison, A.M. Rarefaction and extrapolation with Hill numbers: A framework for sampling and estimation in species diversity studies. *Ecol. Monogr.* **2014**, *84*, 45–67. [CrossRef]
75. Chalmandrier, L.; Munkemuller, T.; Lavergne, S.; Thuiller, W. Effects of species' similarity and dominance on the functional and phylogenetic structure of a plant meta-community. *Ecology* **2015**, *96*, 143–153. [CrossRef]
76. Zhang, C.; Zhou, H.; Ma, Z. Phylogenetic structure of alpine steppe plant communities along a precipitation and temperature gradient on the Tibetan Plateau. *Glob. Ecol. Conserv.* **2020**, *24*, e01379. [CrossRef]
77. Körner, C. *Alpine Plant Life*; Springer: Berlin, Germany, 2003.
78. González-Caro, S.; Duivenvoorden, J.F.; Balslev, H.; Cavelier, J.; Grández, C.; Macía, M.J.; Romero-Saltos, H.; Sánchez, M.; Valencia, R.; Duque, Á.; et al. Scale-dependent drivers of the phylogenetic structure and similarity of tree communities in northwestern Amazonia. *J. Ecol.* **2020**, *109*, 888–899. [CrossRef]
79. Miller, E.T.; Farine, D.R.; Trisos, C.H. Phylogenetic community structure metrics and null models: A review with new methods and software. *Ecography* **2017**, *40*, 461–477. [CrossRef]
80. Tang, Z.; Fang, J.; Chi, X.; Yang, Y.; Ma, W.; Mohhamot, A.; Guo, Z.; Liu, Y.; Gaston, K.J. Geography, environment, and spatial turnover of species in China's grasslands. *Ecography* **2012**, *35*, 1103–1109. [CrossRef]

81. Odriozola, I.; García-Baquero, G.; Etxeberria, A.; Aldezabal, A. Patterns of species relatedness created by competitive exclusion depend on species niche differences: Evidence from Iberian Atlantic grasslands. *Perspect. Plant Ecol. Evol. Syst.* **2017**, *28*, 36–46. [CrossRef]
82. Zhang, H.; John, R.; Liu, K.; Qi, W.; Long, W. Using functional trait diversity patterns to disentangle the processes influencing the recovery of subalpine grasslands following abandonment of agricultural use. *Front. Ecol. Evol.* **2019**, *7*, 128. [CrossRef]
83. Purschke, O.; Schmid, B.C.; Sykes, M.T.; Poschlod, P.; Michalski, S.G.; Durka, W.; Kühn, I.; Winter, M.; Prentice, H.C.; Fridley, J. Contrasting changes in taxonomic, phylogenetic and functional diversity during a long-term succession: Insights into assembly processes. *J. Ecol.* **2013**, *101*, 857–866. [CrossRef]
84. Dong, L.; Liang, C.; Li, F.Y.; Zhao, L.; Ma, W.; Wang, L.; Wen, L.; Zheng, Y.; Li, Z.; Zhao, C.; et al. Community phylogenetic structure of grasslands and its relationship with environmental factors on the Mongolian Plateau. *J. Arid Land* **2019**, *11*, 595–607. [CrossRef]
85. Price, J.N.; Hiiesalu, I.; Gerhold, P.; Partel, M. Small-scale grassland assembly patterns differ above and below the soil surface. *Ecology* **2012**, *93*, 1290–1296. [CrossRef]
86. Meynard, C.N.; Lavergne, S.; Boulangeat, I.; Garraud, L.; Van Es, J.; Mouquet, N.; Thuiller, W. Disentangling the drivers of metacommunity structure across spatial scales. *J. Biogeogr.* **2013**, *40*, 1560–1571. [CrossRef]
87. Yang, Z.; Guo, H.; Zhang, J.; Du, G. Stochastic and deterministic processes together determine alpine meadow plant community composition on the Tibetan Plateau. *Oecologia* **2013**, *171*, 495–504. [CrossRef]
88. Conradi, T.; Temperton, V.M.; Kollmann, J. Resource availability determines the importance of niche-based versus stochastic community assembly in grasslands. *Oikos* **2017**, *126*, 1134–1141. [CrossRef]
89. Salgado-Luarte, C.; Escobedo, V.M.; Stotz, G.C.; Rios, R.S.; Arancio, G.; Gianoli, E. Goat grazing reduces diversity and leads to functional, taxonomic, and phylogenetic homogenization in an arid shrubland. *Land Degrad. Dev.* **2019**, *30*, 178–189. [CrossRef]
90. Olden, J.D.; Poff, N.L. Toward a mechanistic understanding and prediction of biotic homogenization. *Am. Nat.* **2003**, *162*, 442–460. [CrossRef]
91. Shan, G.L.; Cheng, G.; Liu, Z.I.; Yan, Z.J.; Chu, X.H. VOR and CVOR index for health evaluation of typical steppe in Inner Mongolia. *Acta Agrestia Sin.* **2012**, *20*, 401–406.
92. Soliveres, S.; Torices, R.; Maestre, F.T. Environmental conditions and biotic interactions acting together promote phylogenetic randomness in semi-arid plant communities: New methods help to avoid misleading conclusions. *J. Veg. Sci.* **2012**, *23*, 822–836. [CrossRef]

diversity

MDPI

Article

Siberian Ibex *Capra sibirica* Respond to Climate Change by Shifting to Higher Latitudes in Eastern Pamir

Yingying Zhuo [1,2,3,4,5], Muyang Wang [1,2,3,4,*], Baolin Zhang [6], Kathreen E. Ruckstuhl [7], António Alves da Silva [8], Weikang Yang [1,2,3,4,*] and Joana Alves [8]

1 State Key Laboratory of Desert and Oasis Ecology, Xinjiang Institute of Ecology and Geography, Chinese Academy of Sciences, Urumqi 830011, China
2 Sino-Tajikistan Joint Laboratory for Conservation and Utilization of Biological Resources, Urumqi 830011, China
3 The Specimen Museum of Xinjiang Institute of Ecology and Geography, Chinese Academy of Sciences, Urumqi 830011, China
4 Mori Wildlife Monitoring and Experimentation Station, Xinjiang Institute of Ecology and Geography, Chinese Academy of Sciences, Mori 831900, China
5 University of Chinese Academy of Sciences, Beijing 100049, China
6 Kunming Institute of Zoology, Chinese Academy of Sciences, Kunming 650223, China
7 Department of Biological Sciences, University of Calgary, 2500 University Drive Northwest, Calgary, AB T2N 1N4, Canada
8 Centre for Functional Ecology (CFE), TERRA Associate Laboratory, Department of Life Sciences, University of Coimbra, 3000-456 Coimbra, Portugal
* Correspondence: wangmuyang@ms.xjb.ac.cn (M.W.); yangwk@ms.xjb.ac.cn (W.Y.)

Abstract: Climate change has led to shifts in species distribution and become a crucial factor in the extinction of species. Increasing average temperatures, temperature extremes, and unpredictable weather events have all become a part of a perfect storm that is threatening ecosystems. Higher altitude habitats are disproportionately affected by climate change, and habitats for already threatened specialist species are shrinking. The Siberian ibex, *Capra sibirica,* is distributed across Central Asia and Southern Siberia and is the dominant ungulate in the Pamir plateau. To understand how climate change could affect the habitat of Siberian ibex in the Taxkorgan Nature Reserve (TNR), an ensemble species distribution model was built using 109 occurrence points from a four-year field survey. Fifteen environmental variables were used to simulate suitable habitat distribution under different climate change scenarios. Our results demonstrated that a stable, suitable habitat for Siberian ibex was mostly distributed in the northwest and northeast of the TNR. We found that climate change will further reduce the area of suitable habitat for this species. In the scenarios of RCP2.6 to 2070 and RCP8.5 to 2050, habitat loss would exceed 30%. In addition, suitable habitats for Siberian ibex will shift to higher latitudes under climate change. As a result, timely prediction of the distribution of endangered animals is conducive to the conservation of the biodiversity of mountain ecosystems, particularly in arid areas.

Keywords: ensemble species distribution models; suitable habitat; latitude shift; human distribution; mountain ungulates

Citation: Zhuo, Y.; Wang, M.; Zhang, B.; Ruckstuhl, K.E.; Alves da Silva, A.; Yang, W.; Alves, J. Siberian Ibex *Capra sibirica* Respond to Climate Change by Shifting to Higher Latitudes in Eastern Pamir. *Diversity* **2022**, *14*, 750. https://doi.org/10.3390/d14090750

Academic Editor: Attila D. Sándor

Received: 18 July 2022
Accepted: 8 September 2022
Published: 11 September 2022

1. Introduction

The global ecosystem is being affected by human-induced climate change in unprecedented ways, e.g., phenological change, range shifts, community shifts, and so on [1–3]. For wildlife, climate change will produce a series of terrible outcomes that could further contribute to the sixth mass extinction, such as wildlife population declines and extinction, range distribution shifts, habitat fragmentation, and increased evolutionary pressure for all species [4–6]. According to Thomas, et al. [7], a quarter of terrestrial plants and animals may become extinct by 2050, under the 'middle emission scenario' of climate change. Spooner,

Pearson and Freeman [4] predicted that mammal populations would decline by 1.46% to 1.76% annually under the scenario of 'representative concentration pathways 8.5'. In Asia, about 30% of terrestrial species will be at very high risk of extinction under climate change [8].

Habitat shifts and eventual loss caused by climate change are the most pervasive threats to wildlife causing species to migrate towards higher latitudes and/or higher elevations [9–11] and fragmenting suitable habitats [12,13]. For example, Hickling, et al. [14] predicted that 275 of 329 terrestrial species in Europe would move north in response to climate change. Alpine and plateau species are particularly sensitive to climate change as they are typically cold-adapted or cold-tolerant species [15]. In fact, studies have demonstrated that climate change has caused ibex (*Capra ibex*), chamois (*Rupicapra rupicapra*), and red deer (*Cervus elaphus*) in the Alps to move to significantly higher latitudes (i.e., to the north) [16]. Similarly, due to climate change, four antelope species on the Tibetan Plateau (*Procapra przewalskii, Gazella subgutturosa, P.picticaudata* and *Pantholops hodgsonii*) are predicted to lose up to 53.2% of their habitat and will be forced to move to higher latitudes [17]. In addition, in 90 years' time, the suitable summer range for Alpine ibex will diminish to 26.4% compared with 2011 under the RCP8.5 scenarios, and they will respond to high temperatures by moving to higher altitudes [18], as in the case of chamois who move upslope when it is hotter [19].

The alpine zone of the Pamirs is more sensitive to climate change because the temperature rise is much greater than that in the low-altitude areas [20,21]. Previous studies on ungulates living in the Pamir plateau have shown that climate change will reduce the range of these animals [22–24]. For example, the suitable habitat of Marco Polo sheep (*Ovis ammon polii*) in Eastern Tajikistan may be reduced by more than 60.6% and 63.2% by 2050 and 2070, respectively [22]. Because of climate change, the Chinese population of the same species is losing close to 40% of its suitable habitat, and this is mostly happening at low elevations [24].

The Siberian ibex, *Capra sibirica*, is a typical mountain ungulate, which inhabits the mountains of Southern Siberia, Mongolia, Tianshan, Himalaya, Karakorum, Altai, and Pamir [25,26]. Despite its wide distribution, the species remains poorly studied [27]. Siberian ibex are listed as "near threatened species" by the International Union for Conservation of Nature (IUCN) [25], and as "state second class protection animals" by China [28]. They usually occur at higher elevations from around 1500 m up to 6000 m above sea level. [29].These ibex mostly inhabit mountains, alpine meadows, exposed rock, precipitous terrain interspersed with cliffs, deep valleys, and steep areas with escape terrain [30]. The Siberian ibex mainly feeds on herbs and shrubs in the middle part of the Tianshan Mountains, which is one of seven of the largest mountain systems in the world, located in remote areas of Eurasia [31]. Throughout their distributional range, Siberian ibex show seasonal movement to higher ridgelines in the summer, descending to lower elevations in the winter [32,33]. Such movement can be entirely vertical or reach travel distances of 40 to 100 km [22]. However, Siberian ibex face disturbance from human activities and domestic competition (i.e., hunting and grazing), and climate change, with changes in temperature, wind, and precipitation listed as the main drivers affecting the survival of this species [25,29]. Therefore, it is urgent to study the potential responses of the Siberian ibex to climate change across its distribution range.

In this study, we analyzed the current distribution of Siberian ibex in the Taxkorgan Nature Reserve (TNR), which is located in the southwest of Xinjiang, China, and the factors affecting their distribution by using ensemble species distribution models (eSDMs), which are considered to improve predictive performance compared to individual species models [34,35]. We focused our efforts on determining whether or how the distribution of Siberian ibex will change in response to future global warming. We predict that the suitable habitat for Siberian ibex would decrease in the TNR in the future, and they would be forced to migrate to higher latitudes and/or elevations as a response to climate change. Our study will contribute to understanding the future of ungulates on the Chinese Pamir plateau under climate change. Furthermore, the study provides scientific guidance for the TNR's conservation planning, which needs to take into account potential future changes in the distribution of this species when designing conservation policies and planning the functional zonation of this nature reserve.

2. Methods and Materials

2.1. Study Area

The study area is situated in the Taxkorgan Nature Reserve (TNR) (35.5° to 37.5° N, 74.4 to 77.1° E, Figure 1, an area of ~16,253 km^2) in the eastern part of the Pamir, which is mainly distributed in Taxkorgan County, Xinjiang, China. As part of the Qinghai-Tibetan Plateau, the elevation of the TNR ranges from 2098 m to 8572 m above sea level, with an average elevation above 4000 m. The area has limited precipitation and a dry climate [36]. The average annual precipitation is less than 100 mm, the mean annual temperature is 3 °C, the highest temperature in July reaches 38 °C, and the mean temperature of the coldest quarter is −17 °C. The region is rich in wildlife. Besides Siberian ibex, sympatric ungulates include Marco Polo sheep and blue sheep (*Pseudois nayaur*, AKA bharal). All three species are preyed upon by the snow leopard (*Panthera uncia*). The vegetation is dominated by dwarf shrubs (mainly *Artemisia L.* and *Ceratoides*) and grasses (mainly *Stipa*) [37].

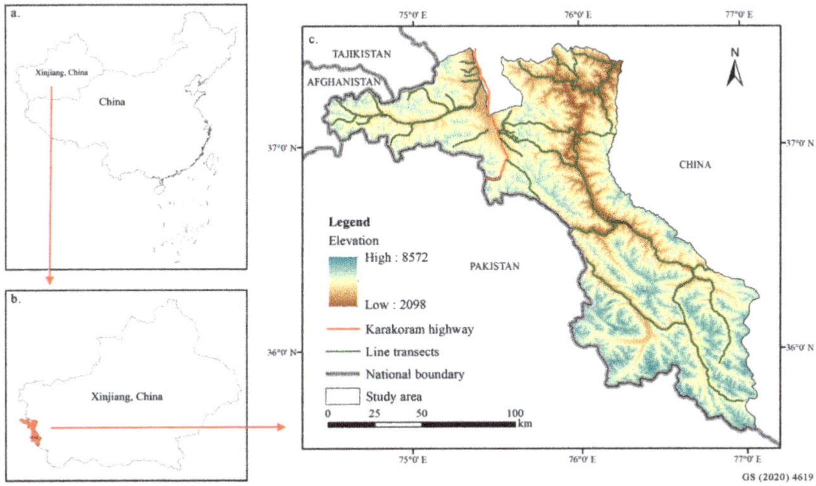

Figure 1. The study area is located in the west of China, which borders Tajikistan, Afghanistan, and Pakistan. (**a**) A map of China. It reflects the location of Xinjiang, a province in China. (**b**) Taxkorgan Nature Reserve (TNR) is located in the southwest of Xinjiang, China. (**c**) The map shows the borders of the Taxkorgan Nature Reserve, with elevation indicated on a green to orange scale. The Karakoram highway is indicated as a red line transect.

Most human settlements are located below 3500 m above sea level, i.e., along both sides of the Karakoram highway, which runs south from the city of Kashi up to the Taxkorgan Valley, through the Marco Polo sheep distribution area, and over the Khunjerab Pass into Pakistan. The southern part of this route lies in the TNR. The reserve is sparsely populated by nomadic Tajik and Kirgiz herding people and their livestock (about 23,000 in total), who live on both sides of the Karakoram highway and in the main valleys inside of the reserve [38].They move yearly between winter (lower elevation areas) and summer pastures (higher elevation areas with productive vegetation) [39,40]. In some areas of the TNR, Siberian ibex share their habitat with domestic animals, which compete with them for food and space. Since 2018, all of the coal and jade mines have been moved out of the reserve, so the herding people and their livestock are the main human disturbance to wild animals in the TNR.

2.2. Occurrence Data

We conducted a survey of Siberian ibex in the TNR from June to November of 2018 to 2021 (for more details on the line transect survey time please see Supplementary Materials Table S1). After interviewing the reserve staff and herding people, we set up 23 line transects with a total length of 1332.74 km, which covers all the habitats occupied by Siberian ibex in the TNR. Each transect was visited once a year. We followed the transects on foot, by car, and horseback according to the terrain, and scanned the surroundings with binoculars (ZEISS 10 × 42, Oberkochen, Germany) at intervals of 2–3 km to spot Siberian ibex. Upon locating Siberian ibex, we used a telescope (ZEISS 20–60×, Oberkochen, Germany) and laser range finder (ZEISS T*RF 10 × 54, Oberkochen, Germany) to observe and record their location (longitude and latitude), their elevation, and their angle and distance from our observation point. A total of 495 locations were set up along the line transects to search for Siberian ibex, and 123 occurrences of ibex were recorded, i.e., 26, 29, 36 and 32 distribution locations were obtained between 2018 and 2021. To reduce any potential spatial sampling bias (only one occurrence point is retained in each 1 km × 1 km grid), we performed spatial thinning of Siberian ibex occurrence data. The spatial thinning was performed using a randomization approach that maximizes the amount of useful information retained [41–43], while simultaneously reducing the sampling bias by removing as few records as possible using the 'thin' function of the "spThin" package (version 0.2.0) [44] in R (version 4.1.0) [45] with a thinning distance of 1 km. The spatial thinning process kept a total of 109 occurrence data points that were used to build the species distribution models.

2.3. Environmental Variables

We considered different environmental factors that can influence the presence of Siberian ibex, including climate, topography, anthropogenic disturbance, and availability of resources (i.e., Normalized Difference Vegetation Index (NVDI)) (Table 1). Climate data was obtained from the WorldClim version2.1 dataset for 1970–2000 (www.worldclim.org (accessed on 20 September 2020)) [46], with a resolution of around 30 arcseconds (roughly 850 m at the study area latitude). To predict the influence of climate change on Siberian ibex, we selected two projection periods for our model: (1) 2050 (average for 2041–2060) and (2) 2070 (average for 2061–2080) and three greenhouse gas emission scenarios: (1) RCP2.6 (low greenhouse gas emissions [47]), (2) RCP4.5 (medium greenhouse gas emissions [48]), and (3) RCP8.5 (high greenhouse gas emissions [49]) based on the Representative Concentration Pathways (RCPs) of Coupled Model Intercomparison Project Phase 5 (CMIP5). We chose the BCC-CSM1-1 as the General Circulation Model (GCM), which more closely conforms to the climatic conditions of our study area [50]. The 30 m resolution elevation data was derived from the Geospatial Data Cloud site of the Chinese Academy of Sciences (www.gscloud.cn (accessed on 2 September 2021)), from which we also calculated the aspect (the range of 0–360), slope (the range of 0–85), and ruggedness of the area, using ArcGIS 10.6 (ESRI). The latest version (1995–2004) of Human Influence Index, covering population density, human infrastructure, and human access was obtained from the Socioeconomic Data and Applications Center (sedac.ciesin.columbia.edu (accessed on 25 August 2021)) [51]. We derived the global land cover data from GlobeLand30 (www.globeland30.org (accessed on 3 July 2022)) and the NDVI from 2018 to 2021 from the DATABANK Remote Sensing Data Engine (databank.casearth.cn (accessed on 3 July 2022)) [52]. NDVI and HII data used in the future scenarios' models remain unchanged from the current dataset. All factor layers were projected to WGS 1984 UTM Zone 43N and resampled to a resolution of 850 m. To avoid collinearity between the environmental variables, we used the "usdm" package (version 1.1-18) [53] in R (version 4.1.0) [45]. We calculated the variance inflation factor (VIF) between predictor variables for each scenario, and then used a stepwise procedure to eliminate the factors with VIF > 10 [54]. All the variables that entered the ensemble species distribution models of the current and future scenarios were selected by the above processes when modeling the ibex habitat were labeled with an asterisk (Table 1).

Table 1. The environmental variables used in the ensemble species distribution models (eSDMs) of Siberian ibex in Taxkorgan Nature Reserve.

Variable Type	Variable Name	Description **	Unit
Climate	Bio1	Annual mean temperature	°C
	Bio2	Mean diurnal range	°C
	Bio3 *	Isothermality	-
	Bio4 *	Temperature seasonality	-
	Bio5	Max temperature of warmest month	°C
	Bio6	Min temperature of coldest month	°C
	Bio7	Temperature annual range	°C
	Bio8 *	Mean temperature of wettest quarter	°C
	Bio9 *	Mean temperature of driest quarter	°C
	Bio10	Mean temperature of warmest quarter	°C
	Bio11	Mean temperature of coldest quarter	°C
	Bio12	Annual precipitation	mm
	Bio13 *	Precipitation of wettest month	mm
	Bio14 *	Precipitation of driest month	mm
	Bio15 *	Precipitation seasonality	-
	Bio16	Precipitation of wettest quarter	mm
	Bio17	Precipitation of driest quarter	mm
	Bio18	Precipitation of warmest quarter	mm
	Bio19	Precipitation of coldest quarter	mm
Topographic	Elevation *	The height above sea level	m
	Slope *	The degree of steepness	°
	Aspect *	Orientation of the topographic slope	-
	Ruggedness *	The difference between the highest and lowest elevations in a given area	m
Human distribution	Human Influence Index *	The extent of human activity	-
Food resources	Land cover *	Land cover type	-
	Distance to water *	Distance from the nearest water source	m
	NDVI *	Normalized Difference Vegetation Index	-

Note: The asterisk (*) means this variable entered the ensemble species distribution models. The future environment variables are consistent with the current environment variables, both of which have been tested for collinearity, and they are all labeled variables in this table *. ** For more detailed information about climate variables see www.worldclim.org (accessed on 20 September 2020).

2.4. Species Distribution Model

The ensemble species distribution models (eSDMs) combine predictions across different modeling methods [55]. There have been studies showing that the performance of eSDMs is better than single models [34,35]. We used eight species distribution algorithms (generalized linear models, generalized boosting models, generalized additive models, artificial neural networks, flexible discriminant analysis, multivariate adaptive regression splines, random forest, and MaxEnt) [55] to build the eSDMs for Siberian ibex in the TNR. Firstly, we used a random strategy [56] to create 5000 pseudo-absence points [57]. We then split our occurrence data into two parts: one part with 80% of the data was used to calibrate the models, and the remaining part (20%) was used for model testing. All algorithms were run ten times and to obtain eighty single models. Then, we took the average TSS value of these eighty models as the threshold of whether to enter the eSDMs (above average TSS were entered in the final model) and built the ensemble model with a weighted average to project the habitat suitability map for Siberian ibex. The continuous probability of the presence of the eSDMs was transformed into binary values using a cut-off value that maximized TSS to estimate species vulnerability [58,59]. The AUC value (the area under the receiver operating characteristic curve; a value greater than 0.9 is considered excellent) and TSS value (a value greater than 0.85 is considered excellent; a value between 0.7 and 0.85 is good) were used to evaluate the performance of the eSDMs [60,61]. All eSDM analyses were performed using the "biomod2" package (version 3.5.1) [62] in R, version 4.1.0 [45].

3. Results

3.1. Current Habitat Distribution

The eSDMs for Siberian ibex in the TNR showed excellent performance with a high AUC (0.94) and TSS (0.77) value. Among the potential factors affecting the current distribution of Siberian ibex, precipitation seasonality, precipitation of the wettest month, Human Influence Index (HII), NDVI, elevation, and temperature seasonality were the main determining factors (Supplementary Materials Table S2). The presence probability of Siberian ibex was positively correlated with NDVI, and elevations of 3000 to 4800 m (Supplementary Materials Figure S1).

The current suitable habitat for Siberian ibex in the TNR is mainly distributed in the northwest and a small part of the northeast (Figure 2). The suitable habitat area is around 2702.15 km^2, accounting for 16.6% of the total area of the natural reserve.

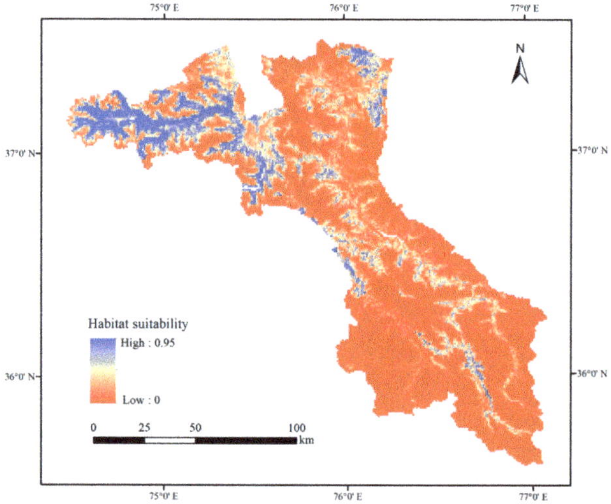

Figure 2. The current habitat suitability of Siberian ibex in Taxkorgan Nature Reserve.

3.2. Future Habitat Distribution

We simulated the potential distribution of Siberian ibex under three greenhouse scenarios for two periods. The performance of the six eSDMs was good (Table 2 and Supplementary Materials Table S3), in which the TSS value range was 0.78–0.81 (average of 0.793), and the AUC value range was 0.94–0.95 (average of 0.947). In these scenarios, temperature seasonality, mean temperature of the wettest quarter, and precipitation of the wettest month were found to be the main climate factors affecting the distribution of Siberian ibex, which will be driving the distribution of the Siberian ibex further in this region. In addition, HII, NDVI, and elevation will also affect the distribution of ibex. Human Influence Index (HII), the indicator of human impact, is also a main factor that will influence the distribution of ibex, since a high HII is not good for the survival of the ibex.

Table 2. Changes in suitable habitat of Siberian ibex in Taxkorgan Nature Reserve under different climate change scenarios.

Period	Future Scenarios	Suitable Habitat (km^2)	The Change of Suitable Habitat (km^2)			The Change of Suitable Habitat (%)	
			Stable	Gain	Loss	Gain	Loss
Current	-	2702.15	-	-	-	-	-
2050	RCP2.6	2290.33	1963.76	326.57	738.40	12.09	27.33
	RCP4.5	2268.65	1911.01	357.64	791.14	13.24	29.28
	RCP8.5	2234.70	1868.39	366.31	833.77	13.56	30.86
2070	RCP2.6	1984.71	1752.79	231.92	948.64	8.58	35.11
	RCP4.5	2577.88	2068.52	509.36	632.91	18.85	23.42
	RCP8.5	2478.90	2015.05	463.85	687.10	17.17	25.43

Our models indicated a tendency for a decrease in the suitable habitat of Siberian ibex in the TNR under the future climate scenarios (RCP2.6, RCP4.5, and RCP8.5 for both 2050 and 2070), while the magnitude of the reduction depends on the emission scenario, although most of the changes do not appear to be spatially substantial (Figure 3). The average decline of suitable habitat predicted in the future is 14.76%, when compared to the present. Under the RCP2.6 for 2070, the suitable habitat decreased the most, reaching 26.55% (717.4 km^2) in some models. Moreover, there was no obvious change in the distribution of Siberian ibex along the altitudinal gradient under climate change (Figure 4). However, we found that the centroid of Siberian ibex distribution tended to move north under different climate scenarios (Figure 5).

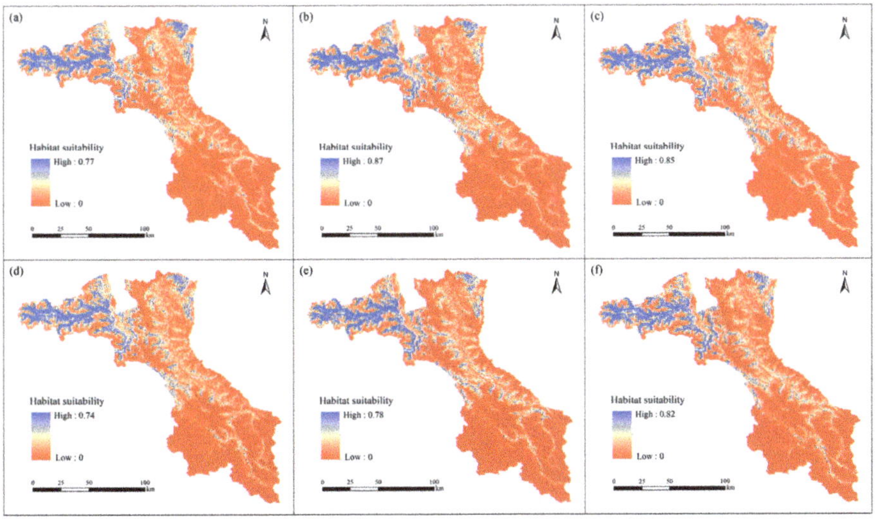

Figure 3. The habitat suitability of Siberian ibex in Taxkorgan Nature Reserve under three greenhouse gas emissions scenarios for two periods; top panel predictions for 2050 and bottom panels for 2070. (**a**) RCP 2.6 for 2050; (**b**) RCP 4.5 for 2050; (**c**) RCP 8.5 for 2050; (**d**) RCP 2.6 for 2070; (**e**) RCP 4.5 for 2070; (**f**) RCP 8.5 for 2070.

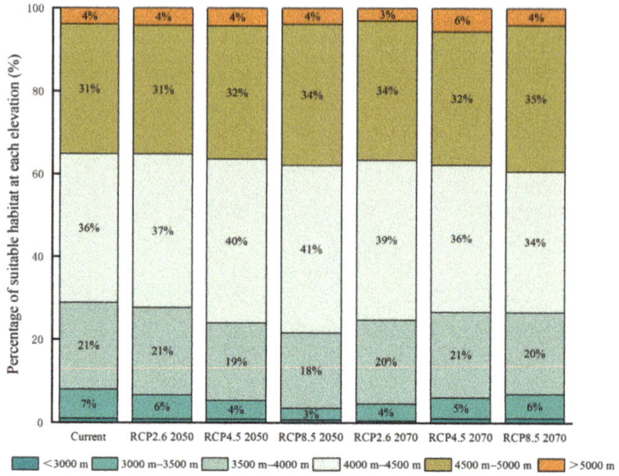

Figure 4. The proportion of suitable habitat area for Siberian ibex at different elevation gradients in Taxkorgan Nature Reserve under current and future climate scenarios.

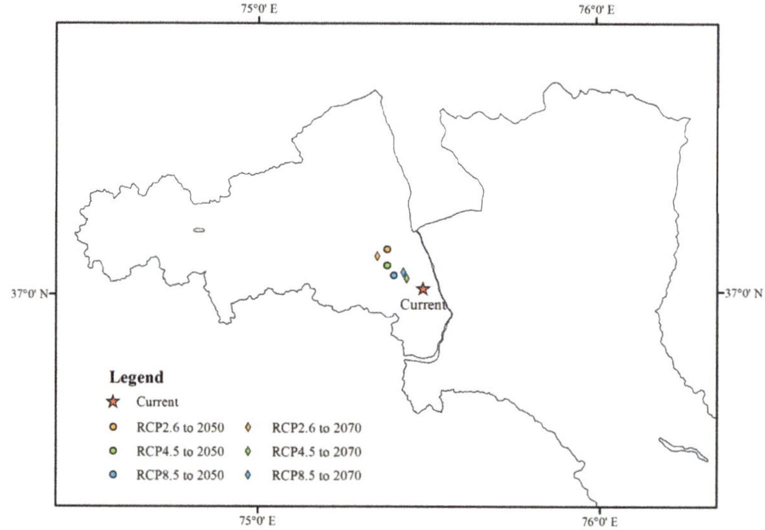

Figure 5. Centroid distribution of suitable habitat for Siberian ibex in Taxkorgan Nature Reserve under current and future climate scenarios. The average shift to the north is 5.81 km when the latitude difference is taken into account, and 10.20 km in the RCP2.6 for 2050.

We found that the northwest of the reserve was the most stable suitable habitat for Siberian ibex under any climate scenario (Figure 6). The area can act as a climate change refuge for Siberian ibex (Supplementary Materials Figure S2), which is the overlapping area of current suitable habitat and future stable suitable habitat. The region with major losses of current suitable habitat was mainly located in the southern part of the reserve, and the area with an increase was mainly located in the central part of the reserve. The loss rate of suitable habitat was highest under the scenarios of RCP2.6 for 2070 and RCP8.5 for 2050, reaching 35.11% and 30.86% loss, respectively. However, we found that the lost suitable habitat and the newly gained suitable habitat were not concentrated in low or high elevations under different climate scenarios. At the same time, we also found that the stable suitable habitats were mainly located in the northwest and a small part of the northeast of the reserve. Under RCP4.5 for 2070, the area of stable suitable habitat was the largest at 2577.88 km^2.

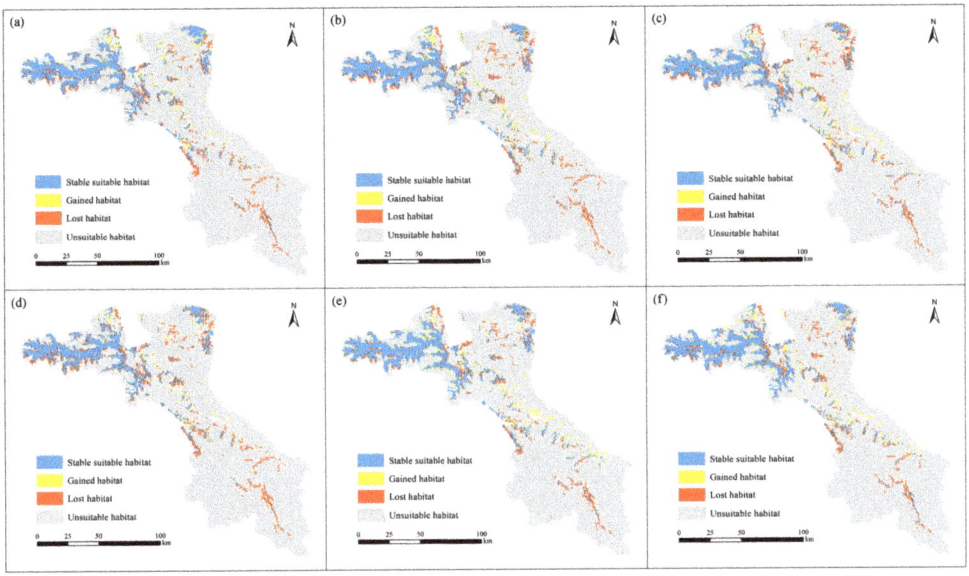

Figure 6. Transfer of suitable habitat for Siberian ibex between current and future climate scenarios in Taxkorgan Nature Reserve. (**a**) RCP 2.6 for 2050; (**b**) RCP 4.5 for 2050; (**c**) RCP 8.5 for 2050; (**d**) RCP 2.6 for 2070; (**e**) RCP 4.5 for 2070; (**f**) RCP 8.5 for 2070.

4. Discussion

Numerous studies have confirmed the profound influence of climate warming on the change in distribution of ungulates [15,63,64]. Such change could be deleterious, leading to a reduction in species abundance or even extinction as habitats deteriorate [65,66]. In the Apennines, Italy, the Apennine chamois may undergo a drastic decline in its historical core range in the next 50 years with a 95% reduction or near-extinction at worst [67]. The Taxkorgan Nature Reserve, located in the Eastern Pamir, is home to several ungulates, including the Siberian ibex. In the present study, we, for the first time, forecast the potential habitat shifts/loss of Siberian ibex in the Eastern Pamir under climate change. We showed that climate change and human disturbance have combined negative impacts on the distribution of Siberian ibex.

Suitable habitat for Siberian ibex currently accounts for only 16.6% of the reserve. Suitable habitat would further contract in the future because any newly gained suitable habitat would be smaller than the habitat being lost. The extent of habitat loss could be up to 29.15% and 27.99% at most by 2050 and 2070, respectively, depending on the climate change scenarios we used in our models. This is partially because the Pamirs are a complex ecosystem that is highly vulnerable to climate change. For example, the glaciers in the Eastern Pamirs that play a crucial role in the water cycle in high elevation areas [68] continue to accelerate their retreat [69] due to a significant upward trend in temperature and precipitation [70]. Meanwhile, reliable water sources remain crucial for the survival of animals [71,72] and the survival and distribution of plant communities. In fact, our models confirmed that the distribution of Siberian ibex in this region is strongly affected by precipitation (precipitation seasonality and the precipitation of the wettest month) and temperature (temperature seasonality). This is consistent with the results for Siberian ibex in Tajikistan [29] and in Pakistan [73]. The rise in temperature will likely cause Siberian ibex and other alpine ungulates to show behavioral and physiological responses, such as heat avoidance and heat stress, leading them to move to higher elevations or seek shade in the short term [16]. Furthermore, the precipitation of the wettest months may affect the distribution of ibex via the change in the phenology and richness of plants [17,74]. In addition, we found that Siberian ibex prefer to inhabit areas with high NDVI, i.e., where there is high vegetation cover allowing species to acquire food without moving too far [29,75]. Ruggedness is also a vital factor that influences the distribution of Siberian ibex in the TNR. The rugged terrain helps ungulates avoid predators [29,76]. In our study, the probability of the presence of ibex decreases accordingly when HII is greater than 15. A higher HII value means more serious human disturbance, which means ibex prefer to inhabit areas with low human disturbance.

Much wildlife has shifted its geographic distribution toward higher elevations or latitudes [5,67,77–79], and Chen et al.'s (2011) meta-analysis showed that the distribution of many species has recently shifted to higher elevations at a median rate of 11.0 m per decade [9]. Interestingly, we have not found such a shift to higher elevations, i.e., the distribution of Siberian ibex at different elevations remained unchanged under future climate scenarios. Similarly, no change in elevation was found for most ungulates in the Tibetan plateau [63]. Non-elevational shifts in those ungulates may be the result of physiological temperature thresholds and precipitation tolerance on the one hand. On the other hand, due to the high elevation in the plateau, there are very few areas available at high elevation for most ungulate species to colonize near their range boundaries and upward migration is thus limited [63,80,81]. As part of the Tibetan Plateau, the Pamir plateau may put the same stressors on wildlife, meaning that Siberian ibex may not have any additional room to climb to higher elevations. Furthermore, another study on ungulates, including Tibetan antelope *Pantholops hodgsoni*, Kiang *Equus kiang*, and wild yak *Bos mutus* living on the Tibetan Plateau, revealed that the distribution of the main forage plants will be reduced by more than 50% in response to climate change [64]. Indeed, research by Walther, et al. [82] showed that there has been an upward shift in alpine plants. Meanwhile, the reduction in mountain surface area with rising elevation [83] may not provide Siberian ibex with contiguous patches of habitat. Therefore, the Siberian ibex may simply run out of space to move to.

In our study, human disturbance may further contribute to the non-elevational shift of Siberian ibex. In the TNR, the lower elevation region experiences strong human disturbances, including settlements and infrastructure (HII in our model) [36], where Tajik herding people graze their livestock [38]. Meanwhile, there is strong food competition between Siberian ibex and domestic sheep, and Siberian ibex show strong avoidance behavior towards domestic sheep, which reduces their range of activity [84]. Thus, a series of human disturbances forces Siberian ibex to abandon potential suitable habitats at lower elevations leading to a reduction in the vertical distribution range of ungulates [85], which is similar to the effects found in primates and Walia ibex (*Capra walie*) [85–87].

In addition to responding to climate change by moving to higher elevations, species may also move to higher latitudes to acquire new habitats [14,88]. Global species distribution shifts under climate change are estimated to move toward higher latitudes at a rate of 16.9 km/decade [5,9,14]. By analyzing the centroid of suitable habitat for Siberian ibex at present and the centroid under different climate scenarios, we found that Siberian ibex in our study area showed a trend of moving northward under future climate scenarios. This is consistent with the study of Marco Polo sheep [22] and ibex [29] on the Pamir plateau in Tajikistan, which have the same shift trend in response to climate change. Similarly, research on Britain's mammals has shown that they have shifted 22 km north in the past 25 years [14]. In addition, previous studies have found that birds in different temperature zones have different distribution mechanisms in response to climate change [89]. Birds from the northern temperate regions migrate to higher latitudes [90–92], while those from the tropics migrate to higher elevations [89,93]. In China, Tibetan antelope and goitered gazelle (*Gazella subgutturosa*), both located in the northern temperate zone, also showed a trend of moving northward under climate change [17]. Although we have no evidence that all animals living in the north temperate zone exhibit poleward migration in response to climate change, our results suggest that this could be one of the strategies of Siberian ibex in response to climate change.

It is a crucial step from theory to practice to put forward conservation management opinions based on study results. According to our study, the northwest part of the reserve may be a refuge which will promote Siberian ibex survival under increasingly adverse climatic conditions. This area should therefore be protected in the future. The key conservation issue in this area is finding an effective way for Siberian ibex to survive under climate change. Meanwhile, reducing human activity and grazing in and around the current suitable habitat will help the Siberian ibex obtain more habitat. In addition, the Siberian ibex is an ungulate widely distributed in Central Asia [25]. Identifying and conserving the ecological corridor linking suitable habitat in the reserve and outside the reserve (including domestic and overseas) will help support gene exchange in the population and promote genetic diversity.

Our study revealed that climate change has a strong effect on the distribution of Siberian ibex in the Eastern Pamir, China. One of the expected effects of recent warming is that it may force animals to move to higher, cooler elevations [94]. Siberian ibex generally show seasonal movement along an elevational gradient, i.e., they migrate to higher elevations in summer and to lower elevations in winter due to phenological changes and variable temperatures at high and low elevations. However, intensive human interference at lower elevations may prevent Siberian ibex from migrating to lower elevations, further limiting their distribution and making the range of their distribution along elevation even narrower [95]. Such distribution patterns may reduce the resistance of Siberian ibex to environmental changes. Although we are incapable of further analyzing whether and how interspecific interactions will alter the distribution of the species due to the lack of data on natural predators (such as snow leopards), domestic animals, and the feeding habits of the species in the region, our research is nevertheless conducive to the conservation of ungulates in the plateau ecosystem. Future studies should focus on the effects of intra- and interspecific interactions on species distribution, and the construction of ecological corridors should be considered to better protect species.

Supplementary Materials: The following supporting information can be downloaded at: https://www.mdpi.com/article/10.3390/d14090750/s1, Figure S1: Response curves of the current ensemble species distribution models for Siberian ibex in Taxkorgan Nature Reserve to each variable, Figure S2: Stable suitable habitat for Siberian ibex in Taxkorgan Nature Reserve under climate change; Table S1: The number of line transects per month for each year (2018–2021) during the survey, Table S2: Ranking of variables' importance affecting the distribution of Siberian ibex in Taxkorgan Nature Reserve, Table S3: The evaluation of ensemble species distribution models (eSDMs, including AUC and TSS) of Siberian ibex in Taxkorgan Nature Reserve under current and different climate change scenarios.

Author Contributions: Conceptualization, Y.Z., M.W. and W.Y.; methodology, Y.Z. and M.W.; software, Y.Z.; validation, M.W. and W.Y.; formal analysis, Y.Z. and M.W.; investigation, M.W. and W.Y.; resources, Y.Z., M.W. and W.Y.; data curation, Y.Z., M.W. and W.Y.; writing—original draft preparation, Y.Z.; writing—review and editing, M.W., B.Z., K.E.R., A.A.d.S., W.Y. and J.A.; visualization, Y.Z.; supervision, M.W. and W.Y.; project administration, M.W. and W.Y.; funding acquisition, M.W., B.Z., A.A.d.S., W.Y. and J.A. All authors have read and agreed to the published version of the manuscript.

Funding: This work was supported by the Western Young Scholar Program-B of the Chinese Academy of Sciences (grant numbers 2021-XBQNXZ-014); the Second Tibetan Plateau Scientific Expedition and Research Program (STEP, grant numbers 2019QZKK0501); Yunnan Applied Basic Research Projects (No. 202101AT070296 to B.Z.) National Natural Science Foundation of China (No. 32101408 to B.Z.), and the Shanghai cooperation organization partnership and international technology cooperation plan of science and technology projects (2021E01020). J.A. and A.A.d.S. were supported by the R&D Unit Centre for Functional Ecology—Science for People and the Planet (CFE), with reference UIDB/04004/2020, financed by FCT/MCTES through national funds (PIDDAC) and by TERRA Associate Laboratory (LA/P/0092/2020).

Institutional Review Board Statement: Not applicable because the study only involved long distance observation from the animals.

Data Availability Statement: Not applicable.

Acknowledgments: We thank the staff of Taxkorgan Nature Reserve for all their help during the field survey. We are grateful to the anonymous reviewers for their thoughtful comments on our manuscript.

Conflicts of Interest: The authors declare no conflict of interest.

References

1. Parmesan, C.; Yohe, G. A globally coherent fingerprint of climate change impacts across natural systems. *Nature* **2003**, *421*, 37–42. [CrossRef]
2. Walther, G.R.; Post, E.; Convey, P.; Menzel, A.; Parmesan, C.; Beebee, T.J.; Fromentin, J.M.; Hoegh-Guldberg, O.; Bairlein, F. Ecological responses to recent climate change. *Nature* **2002**, *416*, 389–395. [CrossRef]
3. Karl, T.R.; Trenberth, K.E. Modern global climate change. *Science* **2003**, *302*, 1719–1723. [CrossRef]
4. Spooner, F.E.B.; Pearson, R.G.; Freeman, R. Rapid warming is associated with population decline among terrestrial birds and mammals globally. *Glob. Chang. Biol.* **2018**, *24*, 4521–4531. [CrossRef]
5. Hetem, R.S.; Fuller, A.; Maloney, S.K.; Mitchell, D. Responses of large mammals to climate change. *Temperature* **2014**, *1*, 115–127. [CrossRef]
6. Bellard, C.; Bertelsmeier, C.; Leadley, P.; Thuiller, W.; Courchamp, F. Impacts of climate change on the future of biodiversity. *Ecol. Lett.* **2012**, *15*, 365–377. [CrossRef]
7. Thomas, C.D.; Cameron, A.; Green, R.E.; Bakkenes, M.; Beaumont, L.J.; Collingham, Y.C.; Erasmus, B.F.; De Siqueira, M.F.; Grainger, A.; Hannah, L.; et al. Extinction risk from climate change. *Nature* **2004**, *427*, 145–148. [CrossRef]
8. IPCC. *Climate Change 2022: Impacts, Adaptation, and Vulnerability. Contribution of Working Group II to the Sixth Assessment Report of the Intergovernmental Panel on Climate Change*; Cambridge University Press: Cambridge, UK, 2022.
9. Chen, I.C.; Hill, J.K.; Ohlemuller, R.; Roy, D.B.; Thomas, C.D. Rapid range shifts of species associated with high levels of climate warming. *Science* **2011**, *333*, 1024–1026. [CrossRef]
10. Thomas, C.D. Climate, climate change and range boundaries. *Divers. Distrib.* **2010**, *16*, 488–495. [CrossRef]
11. Wilson, R.J.; Gutierrez, D.; Gutierrez, J.; Martinez, D.; Agudo, R.; Monserrat, V.J. Changes to the elevational limits and extent of species ranges associated with climate change. *Ecol. Lett.* **2005**, *8*, 1138–1146. [CrossRef]
12. Dirnbock, T.; Essl, F.; Rabitsch, W. Disproportional risk for habitat loss of high-altitude endemic species under climate change. *Glob. Chang. Biol.* **2011**, *17*, 990–996. [CrossRef]
13. Ye, X.; Yu, X.; Yu, C.; Tayibazhaer, A.; Xu, F.; Skidmore, A.K.; Wang, T. Impacts of future climate and land cover changes on threatened mammals in the semi-arid Chinese Altai Mountains. *Sci. Total Environ.* **2018**, *612*, 775–787. [CrossRef] [PubMed]
14. Hickling, R.; Roy, D.B.; Hill, J.K.; Fox, R.; Thomas, C.D. The distributions of a wide range of taxonomic groups are expanding polewards. *Glob. Chang. Biol.* **2006**, *12*, 450–455. [CrossRef]
15. White, K.S.; Gregovich, D.P.; Levi, T. Projecting the future of an alpine ungulate under climate change scenarios. *Glob. Chang. Biol.* **2017**, *24*, 1136–1149. [CrossRef] [PubMed]
16. Büntgen, U.; Greuter, L.; Bollmann, K.; Jenny, H.; Liebhold, A.; Galván, J.D.; Stenseth, N.C.; Andrew, C.; Mysterud, A. Elevational range shifts in four mountain ungulate species from the Swiss Alps. *Ecosphere* **2017**, *8*, e01761. [CrossRef]
17. Zhang, J.; Jiang, F.; Li, G.; Qin, W.; Wu, T.; Xu, F.; Hou, Y.; Song, P.; Cai, Z.; Zhang, T. The four antelope species on the Qinghai-Tibet plateau face habitat loss and redistribution to higher latitudes under climate change. *Ecol. Indic.* **2021**, *123*, 107337. [CrossRef]
18. Brivio, F.; Zurmuhl, M.; Grignolio, S.; von Hardenberg, J.; Apollonio, M.; Ciuti, S. Forecasting the response to global warming in a heat-sensitive species. *Sci. Rep.* **2019**, *9*, 3048. [CrossRef]
19. Mason, T.H.; Stephens, P.A.; Apollonio, M.; Willis, S.G. Predicting potential responses to future climate in an alpine ungulate: Interspecific interactions exceed climate effects. *Glob. Chang. Biol.* **2014**, *20*, 3872–3882. [CrossRef]
20. Yao, T.; Liu, X.; Wang, N.; Shi, Y. Amplitude of climatic changes in Qinghai-Tibetan Plateau. *Chin. Sci. Bull.* **2000**, *45*, 1236–1243. [CrossRef]
21. Ma, R.; Jiang, Z. Impact of global climate change on wildlife. *Acta Ecol. Sin.* **2005**, *25*, 3061–3066.
22. Salas, E.A.L.; Valdez, R.; Michel, S.; Boykin, K.G. Habitat assessment of Marco Polo sheep (*Ovis ammon polii*) in Eastern Tajikistan: Modeling the effects of climate change. *Ecol. Evol.* **2018**, *8*, 5124–5138. [CrossRef] [PubMed]
23. Ali, H.; Din, J.U.; Bosso, L.; Hameed, S.; Kabir, M.; Younas, M.; Nawaz, M.A. Expanding or shrinking? range shifts in wild ungulates under climate change in Pamir-Karakoram mountains, Pakistan. *PLoS ONE* **2021**, *16*, e0260031. [CrossRef] [PubMed]
24. Wang, M.; Zhang, C.; Mi, C.; Han, L.; Li, M.; Xu, W.; Yang, W. Potential impacts of climate change on suitable habitats of Marco Polo sheep in China. *Chin. J. Appl. Ecol.* **2021**, *32*, 3127–3135.
25. Reading, R.; Michel, S.; Suryawanshi, K.; Bhatnagar, Y.V. Capra sibirica in e.T42398A22148720; The IUCN Red List of Threatened Species. 2020. Available online: https://dx.doi.org/10.2305/IUCN.UK.2020-2.RLTS.T42398A22148720.en (accessed on 3 July 2022).
26. Fedosenko, A.; Blank, D. Capra sibirica. *Mamm. Species* **2001**, *2011*, 1–13. [CrossRef]

27. Otgonbayar, B.; Buyandelger, S.; Amgalanbaatar, S.; Reading, R.P. Siberian Ibex (*Capra sibirica*) Neonatal Kid Survival and Morphometric Measurements in Ikh Nart Nature Reserve, Mongolia. *Mong. J. Biol. Sci.* **2017**, *15*, 23–30.
28. National Forestry and Grassland Administration. National Key Protected Wildlife List. 2021. Available online: http://www.forestry.gov.cn/main/5461/20210205/122418860831352.html (accessed on 3 July 2022).
29. Salas, E.A.L.; Valdez, R.; Michel, S.; Boykin, K.G.; Salas, E.A.L.; Valdez, R.; Michel, S.; Boykin, K.G. Response of Asiatic ibex (*Capra sibirica*) under Climate Change Scenarios. *J. Resour. Ecol.* **2020**, *11*, 27–37. [CrossRef]
30. Odonjavkhlan, C.; Alexsander, J.S.; Mishra, C.; Samelius, G.; Sharma, K.; Lkhagvajav, P.; Suryawanshi, K.R. Factors affecting the spatial distribution and co-occurrence of two sympatric mountain ungulates in southern Mongolia. *J. Zool.* **2021**, *314*, 266–274. [CrossRef]
31. Han, L.; Blanks, D.; Wang, M.Y.; Yang, W.K.; Alves Da Silva, A.; Alves, J. Grouping patterns and social organization in Siberian ibex (*Capra sibirica*): Feeding strategy matters. *Folia Zool.* **2019**, *68*, 35–42.
32. Schaller, G.B. *Mountain Monarchs: Wild Sheep and Goats of the Himalaya*; University of Chicago Press: Chicago, IL, USA, 1979.
33. Bhatnagar, Y.V.; Manjrekar, N.; Stuewe, M.; Rawat, G.S.; Johnsingh, A.J.T. Grouping patterns of Asiatic ibex, *Capra ibex sibirica* in Pin Valley National Park, India. In Proceedings of the the 2nd World Conference on Mountain Ungulates, Saint Vincent, Italy, 5–7 May 1997.
34. Marmion, M.; Parviainen, M.; Luoto, M.; Heikkinen, R.K.; Thuiller, W. Evaluation of consensus methods in predictive species distribution modelling. *Divers. Distrib.* **2008**, *15*, 59–69. [CrossRef]
35. Hao, T.; Elith, J.; Guillera-Arroita, G.; Lahoz-Monfort, J.J.; Serra-Diaz, J. A review of evidence about use and performance of species distribution modelling ensembles like BIOMOD. *Divers. Distrib.* **2019**, *25*, 839–852. [CrossRef]
36. Li, M.; Chen, Q.; Wang, M.; Yang, W.; Zhang, C.; Luo, G.; Ding, J.; Lin, Y. Assessment of habitat suitability of *Ovis ammon polii* based on MaxEnt modeling in Taxkorgan Wildlife Nature Reserve. *Chin. J. Ecol.* **2019**, *38*, 594–603.
37. Schaller, G.B.; Kang, A. Status of Marco Polo sheep *Ovis ammon polii* in China and adjacent countries: Conservation of a Vulnerable subspecies. *Oryx* **2008**, *42*, 100–106. [CrossRef]
38. Wang, M.; Blank, D.; Wang, Y.; Xu, W.; Yang, W.; Alves, J. Seasonal changes in the sexual segregation patterns of Marco Polo sheep in Taxkorgan Nature Reserve. *J. Ethol.* **2019**, *37*, 203–211. [CrossRef]
39. Li, M.; Chen, Q.; Han, L.; Wang, P.; Yang, J.; Wang, M.; Yang, W. Habitat suitability assessment of Marco Polo sheep in Taxkorgan Nature Reserve in Xinjiang. *Acta Ecol. Sin.* **2020**, *40*, 3549–3559.
40. Wang, M.; Blank, D.; Liu, W.; Wang, Y.; Yang, W. The group pattern of Marco Polo sheep in the Chinese Pamir plateau. *Eur. J. Wildl. Res.* **2018**, *64*, 75. [CrossRef]
41. Syfert, M.M.; Smith, M.J.; Coomes, D.A. The effects of sampling bias and model complexity on the predictive performance of MaxEnt species distribution models. *PLoS ONE* **2013**, *8*, e55158. [CrossRef]
42. Kiedrzyński, M.; Zielińska, K.M.; Rewicz, A.; Kiedrzyńska, E. Habitat and spatial thinning improve the Maxent models performed with incomplete data. *J. Geophys. Res. Biogeosci.* **2017**, *122*, 1359–1370. [CrossRef]
43. Fourcade, Y.; Besnard, A.G.; Secondi, J. Paintings predict the distribution of species, or the challenge of selecting environmental predictors and evaluation statistics. *Glob. Ecol. Biogeogr.* **2018**, *27*, 245–256. [CrossRef]
44. Aiello-Lammens, M.E.; Boria, R.A.; Radosavljevic, A.; Vilela, B.; Anderson, R.P. spThin: An R package for spatial thinning of species occurrence records for use in ecological niche models. *Ecography* **2015**, *38*, 541–545. [CrossRef]
45. R Core Team. *R: A Language and Environment for Statistical Computing*; R Foundation for Statistical Computing: Vienna, Austria, 2022. Available online: https://www.R-project.org/ (accessed on 19 April 2022).
46. Fick, S.E.; Hijmans, R.J. WorldClim 2: New 1-km spatial resolution climate surfaces for global land areas. *Int. J. Climatol.* **2017**, *37*, 4302–4315. [CrossRef]
47. Van Vuuren, D.P.; Stehfest, E.; den Elzen, M.G.J.; Kram, T.; van Vliet, J.; Deetman, S.; Isaac, M.; Klein Goldewijk, K.; Hof, A.; Mendoza Beltran, A.; et al. RCP2.6: Exploring the possibility to keep global mean temperature increase below 2 °C. *Clim. Chang.* **2011**, *109*, 95–116. [CrossRef]
48. Thomson, A.M.; Calvin, K.V.; Smith, S.J.; Kyle, G.P.; Volke, A.; Patel, P.; Delgado-Arias, S.; Bond-Lamberty, B.; Wise, M.A.; Clarke, L.E.; et al. RCP4.5: A pathway for stabilization of radiative forcing by 2100. *Clim. Chang.* **2011**, *109*, 77–94. [CrossRef]
49. Riahi, K.; Rao, S.; Krey, V.; Cho, C.; Chirkov, V.; Fischer, G.; Kindermann, G.; Nakicenovic, N.; Rafaj, P. RCP 8.5—A scenario of comparatively high greenhouse gas emissions. *Clim. Chang.* **2011**, *109*, 33–57. [CrossRef]
50. Wu, T.W.; Song, L.C.; Li, W.P.; Wang, Z.Z.; Zhang, H.; Xin, X.G.; Zhang, Y.W.; Zhang, L.; Li, J.L.; Wu, F.H.; et al. An Overview of BCC Climate System Model Development and Application for Climate Change Studies. *J. Meteorol. Res.* **2014**, *28*, 34–56. [CrossRef]
51. WCS. *Last of the Wild Project, Version 2, 2005 (LWP-2): Global Human Influence Index (HII) Dataset (Geographic)*; NASA Socioeconomic Data and Applications Center (SEDAC): Palisades, NY, USA, 2005. [CrossRef]
52. Peng, Y.; He, G.J.; Zhang, Z.M.; Yin, R.Y. *Landsat Spectral Indices Products over China*; China Scientific Data: Beijing, China, 2020.
53. Naimi, B.; Hamm, N.A.S.; Groen, T.A.; Skidmore, A.K.; Toxopeus, A.G. Where is positional uncertainty a problem for species distribution modelling? *Ecography* **2014**, *37*, 191–203. [CrossRef]
54. Chatterjee, S.; Hadi, A.S. *Regression Analysis by Example*, 5th ed.; John Wiley & Sons, Inc.: Hoboken, NJ, USA, 2006.
55. Thuiller, W.; Lafourcade, B.; Engler, R.; Araújo, M.B. BIOMOD—A platform for ensemble forecasting of species distributions. *Ecography* **2009**, *32*, 369–373. [CrossRef]

56. Wisz, M.S.; Guisan, A. Do pseudo-absence selection strategies influence species distribution models and their predictions? An information-theoretic approach based on simulated data. *BMC Ecol.* **2009**, *9*, 8. [CrossRef]
57. Lobo, J.M.; Tognelli, M.F. Exploring the effects of quantity and location of pseudo-absences and sampling biases on the performance of distribution models with limited point occurrence data. *J. Nat. Conserv.* **2011**, *19*, 1–7. [CrossRef]
58. Jiménez-Valverde, A.; Lobo, J.M. Threshold criteria for conversion of probability of species presence to either–or presence–absence. *Acta Oecologica* **2007**, *31*, 361–369. [CrossRef]
59. Liu, C.; White, M.; Newell, G.; Pearson, R. Selecting thresholds for the prediction of species occurrence with presence-only data. *J. Biogeogr.* **2013**, *40*, 778–789. [CrossRef]
60. Allouche, O.; Tsoar, A.; Kadmon, R. Assessing the accuracy of species distribution models: Prevalence, kappa and the true skill statistic (TSS). *J. Appl. Ecol.* **2006**, *43*, 1223–1232. [CrossRef]
61. Swets, J.A. Measuring the accuracy of diagnostic systems. *Science* **1988**, *240*, 1285–1293. [CrossRef] [PubMed]
62. Thuiller, W.; Georges, D.; Gueguen, M.; Engler, R.; Breiner, F. biomod2: Ensemble Platform for Species Distribution Modeling. R Package Version 3.5.1. 2021. Available online: https://CRAN.R-project.org/package=biomod2 (accessed on 20 September 2021).
63. Luo, Z.; Jiang, Z.; Tang, S. Impacts of climate change on distributions and diversity of ungulates on the Tibetan Plateau. *Ecol. Appl.* **2015**, *25*, 24–38. [CrossRef]
64. Wu, X.; Dong, S.; Liu, S.; Su, X.; Han, Y.; Shi, J.; Zhang, Y.; Zhao, Z.; Sha, W.; Zhang, X.; et al. Predicting the shift of threatened ungulates' habitats with climate change in Altun Mountain National Nature Reserve of the Northwestern Qinghai-Tibetan Plateau. *Clim. Chang.* **2017**, *142*, 331–344. [CrossRef]
65. Halpin, P.N. Global climate change and natural-area protection: Management responses and research directions. *Ecol. Appl.* **1997**, *7*, 828–843. [CrossRef]
66. Lannoo, M. *Amphibian Declines: The Conservation Status of United States Species*; University of California Press: Berkeley, CA, USA, 2005.
67. Lovari, S.; Franceschi, S.; Chiatante, G.; Fattorini, L.; Fattorini, N.; Ferretti, F. Climatic changes and the fate of mountain herbivores. *Clim. Chang.* **2020**, *162*, 2319–2337. [CrossRef]
68. Radić, V.; Hock, R. Glaciers in the Earth's Hydrological Cycle: Assessments of Glacier Mass and Runoff Changes on Global and Regional Scales. *Surv. Geophys.* **2013**, *35*, 813–837. [CrossRef]
69. Zhang, Z.; Xu, J.-L.; Liu, S.-Y.; Guo, W.-Q.; Wei, J.-F.; Feng, T. Glacier changes since the early 1960s, eastern Pamir, China. *J. Mt. Sci.* **2016**, *13*, 276–291. [CrossRef]
70. Huaming, S.; Yuting, F.; Ruibo, Z.; Shulong, Y.; Tongwen, Z.; Wenshou, W.; Weiping, L. Streamflow variation in the eastern Pamirs and its response to climate change. *Progress. Inquisitiones De Mutat. Clim.* **2021**, *17*, 352–360.
71. Payne, J.C.; Buuveibaatar, B.; Bowler, D.E.; Olson, K.A.; Walzer, C.; Kaczensky, P. Hidden treasure of the Gobi: Understanding how water limits range use of khulan in the Mongolian Gobi. *Sci. Rep.* **2020**, *10*, 2989. [CrossRef]
72. Zhang, Y.; Cao, Q.S.; Rubenstein, D.I.; Zang, S.; Songer, M.; Leimgruber, P.; Chu, H.; Cao, J.; Li, K.; Hu, D. Water Use Patterns of Sympatric Przewalski's Horse and Khulan: Interspecific Comparison Reveals Niche Differences. *PLoS ONE* **2015**, *10*, e0132094. [CrossRef] [PubMed]
73. Khan, B.; Ablimit, A.; Khan, G.; Jasra, A.W.; Ali, H.; Ali, R.; Ahmad, E.; Ismail, M. Abundance, distribution and conservation status of Siberian ibex, Marco Polo and Blue sheep in Karakoram-Pamir mountain area. *J. King Saud Univ.-Sci.* **2016**, *28*, 216–225. [CrossRef]
74. Pettorelli, N.; Pelletier, F.; Von Hardenberg, A.; Festa-Bianchet, M.; Cote, S.D. Early onset of vegetation growth vs. rapid green-up: Impacts on juvenile mountain ungulates. *Ecology* **2007**, *88*, 381–390. [CrossRef]
75. Scillitani, L.; Sturaro, E.; Monaco, A.; Rossi, L.; Ramanzin, M. Factors affecting home range size of male Alpine ibex (*Capra ibex ibex*) in the Marmolada massif. *Hystrix* **2012**, *23*, 19–27.
76. Han, L.; Wang, Z.; Blank, D.; Wang, M.; Yang, W. Different environmental requirements of female and male Siberian ibex, *Capra sibirica*. *Sci. Rep.* **2021**, *11*, 6064. [CrossRef]
77. Forero-Medina, G.; Joppa, L.; Pimm, S.L. Constraints to species' elevational range shifts as climate changes. *Conserv. Biol.* **2011**, *25*, 163–171. [CrossRef]
78. Martinez-Lopez, O.; Koch, J.B.; Martinez-Morales, M.A.; Navarrete-Gutierrez, D.; Enriquez, E.; Vandame, R. Reduction in the potential distribution of bumble bees (Apidae: *Bombus*) in Mesoamerica under different climate change scenarios: Conservation implications. *Glob. Chang. Biol.* **2021**, *27*, 1772–1787. [CrossRef]
79. Moritz, C.; Patton, J.L.; Conroy, C.J.; Parra, J.L.; White, G.C.; Beissinger, S.R. Impact of a century of climate change on small-mammal communities in Yosemite National Park, USA. *Science* **2008**, *322*, 261–264. [CrossRef]
80. Hickling, R.; Roy, D.B.; Hill, J.K.; Thomas, C.D. A northward shift of range margins in British Odonata. *Glob. Chang. Biol.* **2005**, *11*, 502–506. [CrossRef]
81. MacLean, S.A.; Beissinger, S.R. Species' traits as predictors of range shifts under contemporary climate change: A review and meta-analysis. *Glob. Chang. Biol.* **2017**, *23*, 4094–4105. [CrossRef]
82. Walther, G.-R.; Beißner, S.; Burga, C.A. Trends in the upward shift of alpine plants. *J. Veg. Sci.* **2005**, *16*, 541–548. [CrossRef]
83. Elsen, P.R.; Tingley, M.W. Global mountain topography and the fate of montane species under climate change. *Nat. Clim. Chang.* **2015**, *5*, 772–776. [CrossRef]

84. Han, L. *Sexual Segregation of Siberian ibex (Capra sibirica) in Middle Tianshan Mountains, Xinjiang, China*; University of Chinese Academy of Sciences: Beijing, China, 2021.
85. Gebremedhin, B.; Chala, D.; Flagstad, Ø.; Bekele, A.; Bakkestuen, V.; van Moorter, B.; Ficetola, G.F.; Zimmermann, N.E.; Brochmann, C.; Stenseth, N.C. Quest for New Space for Restricted Range Mammals: The Case of the Endangered Walia Ibex. *Front. Ecol. Evol.* **2021**, *9*, 611632. [CrossRef]
86. Huang, P.Z.; Bian, K.; Huang, Z.P.; Li, Q.; Dunn, D.W.; Fang, G.; Liu, J.H.; Wang, M.Y.; Yang, X.F.; Pan, R.L.; et al. Human activities and elevational constraints restrict ranging patterns of snub-nosed monkeys in a mountainous refuge. *Integr. Zool.* **2020**, *16*, 202–213. [CrossRef]
87. Yang, L.; Shi, K.C.; Ma, C.; Ren, G.P.; Fan, P.F. Mechanisms underlying altitudinal and horizontal range contraction: The western black crested gibbon. *J. Biogeogr.* **2020**, *48*, 321–331. [CrossRef]
88. Root, T.L.; Price, J.T.; Hall, K.R.; Schneider, S.H.; Rosenzweig, C.; Pounds, J.A. Fingerprints of global warming on wild animals and plants. *Nature* **2003**, *421*, 57–60. [CrossRef]
89. La Sorte, F.A.; Jetz, W. Avian distributions under climate change: Towards improved projections. *J. Exp. Biol.* **2010**, *213*, 862–869. [CrossRef]
90. Zuckerberg, B.; Woods, A.M.; Porter, W.F. Poleward shifts in breeding bird distributions in New York State. *Glob. Chang. Biol.* **2009**, *15*, 1866–1883. [CrossRef]
91. Thomas, C.D.; Lennon, J.J. Birds extend their ranges northwards. *Nature* **1999**, *399*, 213. [CrossRef]
92. Hitch, A.T.; Leberg, P.L. Breeding distributions of north American bird species moving north as a result of climate change. *Conserv. Biol.* **2007**, *21*, 534–539. [CrossRef]
93. Peh, K.S.H. Potential Effects of Climate Change on Elevational Distributions of Tropical Birds in Southeast Asia. *Condor* **2007**, *109*, 437–441. [CrossRef]
94. Colwell, R.K.; Brehm, G.; Cardelus, C.L.; Gilman, A.C.; Longino, J.T. Global warming, elevational range shifts, and lowland biotic attrition in the wet tropics. *Science* **2008**, *322*, 258–261. [CrossRef]
95. Manes, S.; Costello, M.J.; Beckett, H.; Debnath, A.; Devenish-Nelson, E.; Grey, K.-A.; Jenkins, R.; Khan, T.M.; Kiessling, W.; Krause, C.; et al. Endemism increases species' climate change risk in areas of global biodiversity importance. *Biol. Conserv.* **2021**, *257*, 109070. [CrossRef]

diversity

MDPI

Article

Biogeographic Patterns and Richness of the *Meconopsis* Species and Their Influence Factors across the Pan-Himalaya and Adjacent Regions

Ning Shi [1,2,3,†], Chunya Wang [1,3,4,†], Jinniu Wang [1,3,*], Ning Wu [1], Niyati Naudiyal [5], Lin Zhang [6], Lihua Wang [7], Jian Sun [6], Wentao Du [8], Yanqiang Wei [8], Wenkai Chen [9] and Yan Wu [1]

1 Chengdu Institute of Biology, Chinese Academy of Sciences, Chengdu 610041, China
2 College of Life Sciences, University of Chinese Academy of Sciences, Beijing 100049, China
3 Mangkang Biodiversity and Ecological Station, Tibet Ecological Safety Monitor Network, Changdu 854500, China
4 Earth Sciences College, Chengdu University of Technology, Chengdu 610059, China
5 Independent Researcher, 99 Old Nehru Colony, Dehradun 248001, India
6 Institute of Tibetan Plateau Research, Chinese Academy of Sciences, Beijing 100101, China
7 College of Resources and Environment, Aba Teachers University, Wenchuan 623002, China
8 Key Laboratory of Remote Sensing of Gansu Province, Northwest Institute of Eco-Environment and Resources, Chinese Academy of Sciences, Lanzhou 730000, China
9 Chengdu Botanical Garden, Chengdu 610503, China
* Correspondence: wangjn@cib.ac.cn
† These authors contributed equally to this work.

Abstract: Understanding the potential habitat of *Meconopsis*, their species richness distribution patterns, and their influencing factors are critical for the conservation and rational exploitation of this valuable resource. In this study, we applied the MaxEnt model to predict their potential distribution, mapped the distribution pattern of species richness, and analyzed the variation of species richness along environmental gradients. Finally, we calculated the landscape fragmentation indices between the five subregions. Our results found that: (1) the medium- and high-suitable habitats of *Meconopsis* were mainly distributed in the central and eastern Himalaya, the Hengduan Mountains, and the southeast edge of the plateau platform, with suitable habitats ranged from 3200 m to 4300 m, whose most important factor is precipitation of the warmest quarter; (2) species richness showed a hump pattern along the environmental gradients except for longitude that showed an increasing trend, mainly concentrated in the south and southeast; and (3) the subregions are in the descending order of species richness: plateau platform, Hengduan Mountains, central, eastern, and western Himalaya; the highest and lowest degree of landscape fragmentation were in the western Himalaya and eastern Himalaya, respectively. Our study provides a theoretical background for the conservation and sustainable exploitation of *Meconopsis* in the wild.

Keywords: *Meconopsis*; potential distribution; species richness; environmental factors; landscape fragmentation; Himalaya; Qinghai-Tibet Plateau; Hengduan Mountains

Citation: Shi, N.; Wang, C.; Wang, J.; Wu, N.; Naudiyal, N.; Zhang, L.; Wang, L.; Sun, J.; Du, W.; Wei, Y.; et al. Biogeographic Patterns and Richness of the *Meconopsis* Species and Their Influence Factors across the Pan-Himalaya and Adjacent Regions. *Diversity* **2022**, *14*, 661. https://doi.org/10.3390/d14080661

Academic Editor: Michel Baguette

Received: 22 June 2022
Accepted: 9 August 2022
Published: 16 August 2022

Publisher's Note: MDPI stays neutral with regard to jurisdictional claims in published maps and institutional affiliations.

1. Introduction

Climate change is an indisputable reality today, evidenced by glacial melting from mountain areas, greening of alpine tundra or arctic, upward shifting of alpine tree lines, and shrub encroachment into alpine grassland [1–3]. Many scientists believe that mountains act as early warning systems and can provide direct evidence to understand potential changes in the lowland environment [4,5]. Mountains around the world vary in terms of shape, extension, elevation, climate change impact, and biodiversity because of the differences in their geographic locations coupled with the complex region-specific hydrothermal condition [5,6]. Most studies have shown that the Qinghai-Tibet Plateau sensu lato (QTPsl) is not a natural tectonic association, but rather comprises a "plateau platform" (i.e., Qinghai-Tibet Plateau sensu stricto; QTPss) [7,8], the Himalaya, and the Hengduan Mountains [9,10]. Together these three regions are also referred to as the pan-Himalayan region [10,11] which boasts of complex topography, diverse geomorphological types, a wide altitudinal range, and different soil textures and geological conditions. This region, where one can observe diverse habitats within a short distance, supports a variety of endemic and endangered species [11,12]. The pan-Himalayan region is an important biodiversity hotspot with the world's highest diversity of alpine plants, attracting extensive scientific attentions [10,13,14].

Biodiversity, especially plant diversity, plays an important role in regulating climate, stabilizing the structure and functions of ecosystems, and providing ecosystem services for humans [15,16]. However, this diversity of plant resources has been constantly threatened by climate change and excessive anthropogenic disturbance since the last century [17,18]. As one of the most severe threats to global biodiversity, climate change has interfered with the composition, structure, and functioning of many mountainous ecosystems, resulting in an ecosystem imbalance that eventually affects human well-being [19–21]. Climate variables can determine the geographic distribution patterns of plant species [22] in addition to several non-climatic factors, such as topography, soil types, and land use, which equally impact species distribution [23]. The geographical distribution of alpine plants, in particular, is sensitive to the interaction between climate and topography [11,24]. Furthermore, human disturbances in mountainous landscapes not only accelerate climate change but also cause severe habitat loss and fragmentation and support the invasion of alien species, which are often directly or indirectly related to the loss of biodiversity [18,25–27]. Many native plant habitats have suffered habitat loss and fragmentation [23,25,27] that leads to the creation of isolated habitat patches resulting in an eventual decline in species viability across the landscape and ultimately reduction of species diversity across the ecosystem [18,28]. Numerous studies have shown the negative impact of fragmentation at the landscape scale on species diversity; however, the ecological effects of fragmentation at larger macroscopic scales are not clearly demonstrated [29,30]. Habitat loss and fragmentation are the primary drivers for the extinction of plant species [31]. Maintaining plant diversity while ensuring their sustainable utilization for human well-being is a common global concern [32]. Species richness is a fundamental scale for measuring regional diversity and the basis for constructing evolutionary and ecological models and conservation strategies [33]. Species diversity at the local scale is influenced by habitat heterogeneity such as geographic differences, climate, topography (elevation, slope, slope direction, etc.), and latitude and longitude, among other factors [34,35]. Thus, understanding the geographic distribution patterns of species richness in a particular area is of great scientific significance for the conservation, development, and sustainable utilization of plant resources.

The genus *Meconopsis* (Himalayan or blue poppies) belongs to the Papaveraceae family and contains over 70 species [36]. These plants are mainly localized in the Qinghai-Tibet Plateau, Hengduan Mountains, and Himalaya between the elevational range of 2000 m to 5800 m, where habitat changes from temperate forests to alpine meadows to screes and nival zones [37,38]. The East Himalaya-Hengduan Mountains region is the center of diversity for *Meconopsis* genus. *Meconopsis* plays a unique role in the alpine ecosystem. It is an important component of the regional biodiversity and participates in primary production, which is critical for ecosystem functioning [39]. *Meconopsis* is well-known for its colorful, attractive, and gracefully postured flower with high ornamental value and is widely used in horticultural gardening [40]. Species of this genus are symbols of tenacious vitalities since they bloom with beautiful flowers despite the extremely cold and harsh environment, inspiring those who live in the same extreme conditions [41]. These flowers are often shown in frescoes and thangkas, being closely related to Tibetan Buddhism, being the prototype of the ubala flower held by the *Green Tara* for relieving suffering [42]. *Meconopsis* is valued for its medicinal properties with a long clinical history in China and other Asian countries. The medicinal properties of the genus were first recorded in the Tibetan medicine book *Yue Wang Yao Zhen*. Famous masterpieces of traditional Tibetan herbal medicine such as

The Four Medical Tantras and *Jing Zhu Materia Medica* have described the use of *Meconopsis* for its anti-inflammatory or analgesic properties [41]. For instance, the flowers or whole plant of *M. integrifolia* can be used for curing hepatitis, pneumonia, liver heat, lung heat, and edema [41]. Recently, a variety of isoquinoline alkaloids with bioactivities have been isolated from *Meconopsis*, which partly explains its unique therapeutic effects [41]. However, only a few species of this genus have been cultivated for floriculture successfully in a controlled environment [41,43]. With the development of the pharmaceutical economy in these regions, overexploitation and anthropogenic habitat destruction are increasingly threatening the survival of many wild *Meconopsis* species, and some *Meconopsis* species have been placed under protection by law [41,42]. Hence, it is of great necessity to map the habitat distribution and ensure the sustainable development of *Meconopsis* from both natural and social perspectives for their habitat conservation and sustainable development.

Predicting the potential species distribution is a significant step toward habitat protection, and the species distribution model (SDM) has become one of the most widely used tools for simulating the potential distribution of organisms [44,45]. The basic principle of SDM is to correlate current species distribution with corresponding environmental variables to estimate the ecological requirements of a species based on ecological niches, thus predicting the suitable habitat [46,47]. The MaxEnt model is the most widely used among the many SDMs because it not only provides stable and reliable prediction results even with small sample data sizes but can work with presence-only data unlike some other models [42,48,49]. Due to the high prediction accuracy, it has been widely used in studies on the spatial distribution of species in response to climate change [9], suitable planting areas for important economic crops, and priority conservation areas for endangered and rare species [50–52]. The screening of suitable habitats and identifying priority conservation areas is critical for habitat management. However, to develop effective landscape management plans, the role of landscape fragmentation in determining the distribution and survival of species cannot be ignored.

Landscape fragmentation is the process by which the surface of a landscape is transformed from a regular homogeneous entity into smaller, complex, and poorly connected patches, mainly as a result of human activities and environmental disturbances [53]. The landscape index can reflect the landscape information well and is a common method to study landscape fragmentation quantitatively. Landscape indices are generally calculated using FragStats (Fragment Statistic), a software program for calculating indices of different types of landscape patterns in classified map patterns or patch mosaics, and analysis of spatial patterns for quantifying landscape structure (i.e., composition and configuration) [54]. It offers a comprehensive choice for the calculation of landscape pattern indices, and three levels (individual patch, patch class, and landscape) of the landscape index can be obtained after the calculation [54]. It is of practical significance to clarify the relationship between landscape pattern and species richness for the conservation and sustainable development of diversity.

In this study, we used MaxEnt to predict the potential distribution pattern of *Meconopsis* and used regression analysis to explore the geospatial pattern of species richness for *Meconopsis* in the pan-Himalaya and its adjacent regions. In addition, we divided the study area into five subregions and used landscape index as a variable to measure landscape fragmentation, and compared the species richness of *Meconopsis* in each subregion (Figure 1). This paper aimed to: (1) predict the potential distribution of *Meconopsis* and determine the key factors affecting the distribution of these species in the pan-Himalaya and adjacent regions; (2) clarify the species richness pattern of *Meconopsis* and analyze the distribution characteristics of species richness under different altitude, latitude and longitude, and other topographical factors (aspect and slope); and (3) assess the degree of landscape fragmentation and species richness *Meconopsis* in five subregions. The results will contribute to identifying the appropriate geographical space available for *Meconopsis* species and help in ensuring sustainable utilization and management of the genus.

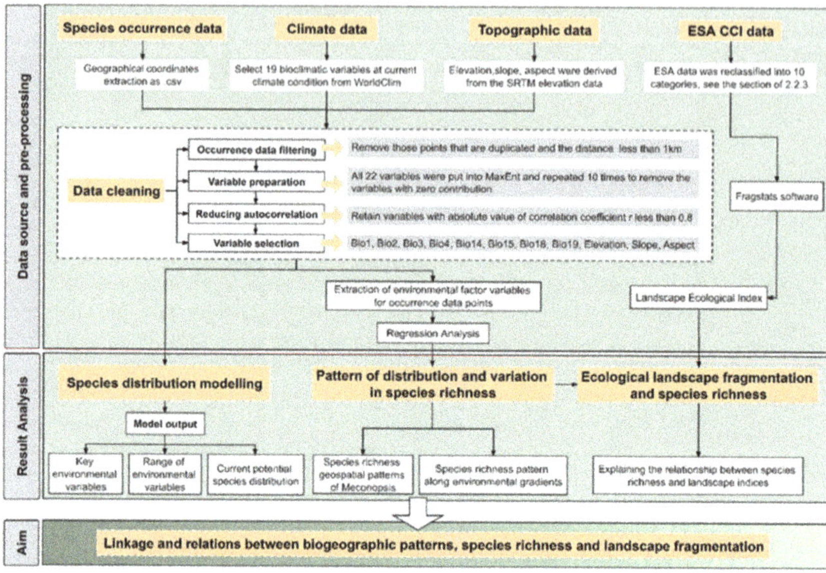

Figure 1. The conceptual framework of the study.

2. Materials and Methods

2.1. Study Area

The study area lies between 68°04′00″~104°40′00″ E and 27°25′30″~40°08′00″ N (Figure 2), which extends from the southern margin of the Himalaya to the northern margin of the Kunlun and Qilian Mountains, and the Pamirs Plateau and Karakorum Mountains in the west to the Hengduan Mountains in the east [55]. This area has the highest elevation (more than 4000 m above sea level) and diverse topography, which support a variety of vegetation types at different elevations. From low to high elevations, these vegetation types include dry and hot river valley scrubs, broadleaved forests, mixed coniferous broadleaved forests, coniferous forests, shrubs, meadows, cold subnival belts, and nival belts (glaciers). This region harbors the world's most species-rich temperate alpine flora driven by orogenetic movements and climate change (monsoon intensification) [56]. It is currently home to approximately 12,000 species of vascular plants belonging to 1500 genera, of which more than 20% are endemic [57]. The mountain region is a diversity center for many species-rich genera, such as rhododendron, gentiana, saussurea, and pedicularis, all of which contain numerous endemic species [14]. We divided the entire study area into five subregions based on the boundaries of the Qinghai-Tibet Plateau, Hengduan Mountains, and Hindu-Kush Himalaya: plateau platform, Hengduan Mountains, eastern Himalaya, central Himalaya, and western Himalaya (Figure 2) [10,11,55]. Of these, the alpine plant biota in Hengduan Mountains is the richest [56].

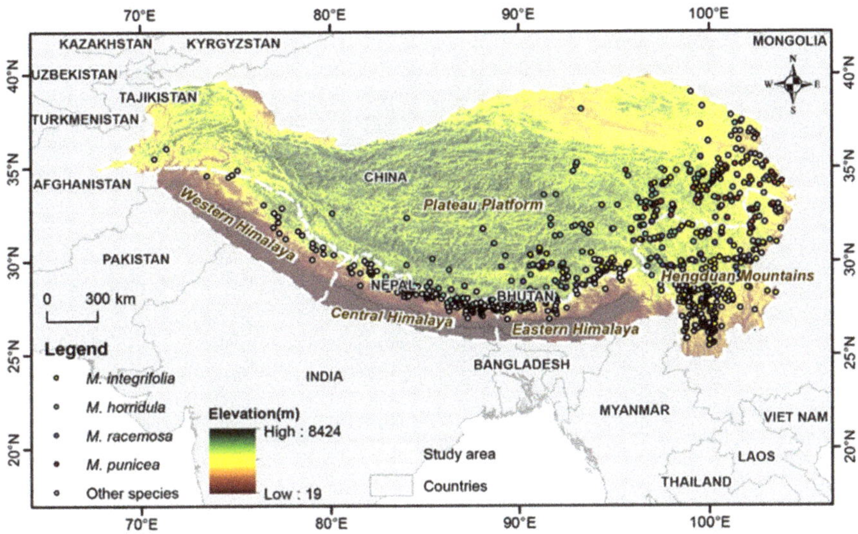

Figure 2. The presence points of *Meconopsis* species in the study area.

2.2. Data Sources and Preprocessing

2.2.1. Species Occurrence Data

We collected the occurrence data for *Meconopsis*, *M. integrifolia*, *M. horridula*, *M. racemosa*, and *M. punicea* from the following three sources: (1) Chinese Virtual Herbarium (CVH, http://www.cvh.ac.cn/; accessed on 22 December 2021), (2) Global Biodiversity Information Facility (GBIF.org, https://www.gbif.org/; accessed on 22 December 2021), and (3) published literature (China National Knowledge Infrastructure, CNKI, https://www.cnki.net/; Web of Science, https://www.webofscience.com/; accessed on 22 December 2021). All data points were carefully evaluated to exclude duplicate data points and then validated via Google Earth to eliminate possible errors; finally, ArcGIS 10.8 was used to remove points that were not in the study region. We also ensured that the selected presence data points were evenly distributed throughout the study area to avoid sampling bias. Following this selection and elimination criteria, the selected presence locations of each species involved in our study are as follows: *Meconopsis* (1445), *M. integrifolia* (296), *M. horridula* (304), *M. racemosa* (111), and *M. punicea* (90) (Figure 2, Table S1).

2.2.2. Bioclimatic and Topographic Data

Bioclimatic data used for the study included 19 bioclimatic variables with the spatial resolution of 30 arc-seconds (~1 km at the equator) representing current climatic conditions (average for 1950–2000) [45], obtained from the WorldClim dataset (https://www.worldclim.org; accessed on 5 January 2022). Besides, three topographic variables (elevation, slope, and aspect) derived from digital elevation model (DEM) data downloaded from WorldClim were also used in the study. To improve the prediction accuracy by eliminating multicollinearity, we performed a Pearson correlation analysis on all environmental variables and retained the environmental factors with low correlation coefficients ($|r| < 0.8$) [58]. Among the environmental factors with high correlation coefficients ($|r| > 0.8$), we retained only one factor of the two variables [58]. Finally, a total of 11 environmental variables, including 8 bioclimatic variables and 3 topographic variables, were obtained to simulate the current potential habitat of *Meconopsis* (Table 1).

Table 1. Environmental variables for modelling the habitat suitability for *Meconopsis* species and four typical species.

Type	Variable Name	Code	Data Source	Unit	Resolution
Bio-Climatic	Annual Mean Temperature	Bio1	WorldClim	°C	30″
	Mean Diurnal Range	Bio2	WorldClim	°C	30″
	Isothermality (Bio2/Bio7) (×100)	Bio3	WorldClim	°C	30″
	Temperature Seasonality (Standard Deviation × 100)	Bio4	WorldClim	°C	30″
	Precipitation of Driest Month	Bio14	WorldClim	mm	30″
	Precipitation Seasonality (Coefficient of Variation)	Bio15	WorldClim	1	30″
	Precipitation of Warmest Quarter	Bio18	WorldClim	mm	30″
	Precipitation of Coldest Quarter	Bio19	WorldClim	mm	30″
Topographic	Elevation	Elevation	WorldClim	m	30″
	Slope	Slope	DEM	°	30″
	Aspect	Aspect	DEM	°	30″

2.2.3. Land Type Data

ESA CCI-LC land use/cover data (300 m spatial resolution, 1992–2015) were downloaded from the European space agency (http://maps.elie.ucl.ac.be/CCI/viewer/; accessed on 11 January 2022). The ESA data (including 22 categories) were reclassified into 10 categories: agriculture, forest, grassland, wetland, settlement, shrubland, sparse vegetation, bare area, water, and permanent snow and ice (Figure 3).

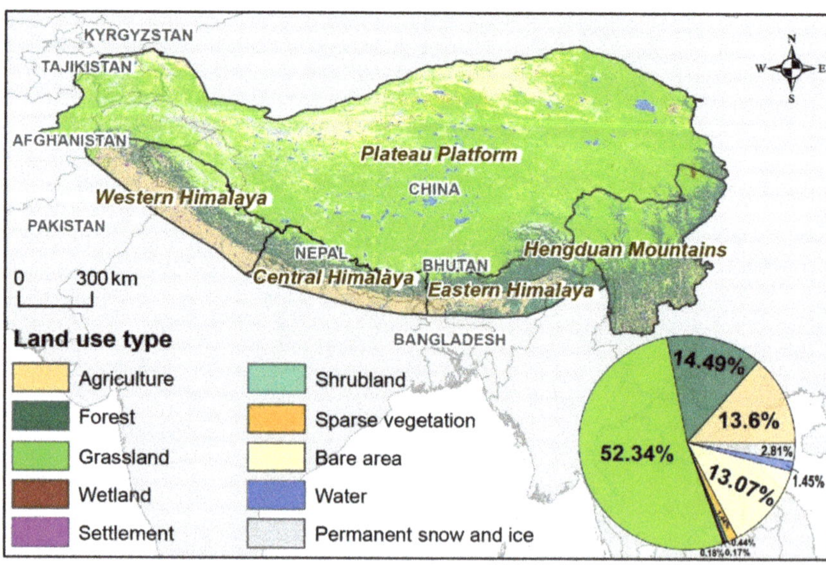

Figure 3. The land-use type maps in the study area.

2.3. Methods

2.3.1. MaxEnt Modeling

MaxEnt is a popular open-access species distribution model. It employs the maximum entropy algorithm and species occurrence points to predict the probability of species occurrence in potential distribution areas. In the study, we used MaxEnt 3.4.4 to model the current habitat distribution of typical *Meconopsis* species and the whole genus. Species occurrence data were divided into two data sets, 75% of total occurrence records were used for training the model, while the remaining 25% were used for validating the model. The number of iterations to run the algorithm is set to 1000 and the model was repeated for 10 runs, which allows adequate time for the model to converge [59], and the rest of the settings were the default settings.

Receiver operating characteristic (ROC) analysis was used to calibrate and validate the robustness of the evaluation of the MaxEnt model [58]. The area under the curve (AUC) of the ROC curve was used to estimate the model performance, which varies from 0 to 1 and is a diagnostic evaluation index with high recognition at present. Generally, the model performance can be categorized according to the value of AUC as fair (0.6–0.7), good (0.7–0.8), very good (0.8–0.9), and excellent (>0.9) [58]. A jackknife test was applied to identify variables that made a significant contribution to model output. The results of this test showed the training gains of each variable when the model was run in isolation and was compared with the training gains of all variables. Additionally, species response curves were created to explore the relationships between the probability of occurrence and environmental variables.

A threshold value of 25% was set based on the logistic output from the Maxent model, below which the probability of species occurring can be considered almost negligible. The final suitable habitat predictions were regrouped into three classes: low suitable (25–50%), medium suitable (50–75%), high suitable (>75%), and the values below 25% as nonsuitable.

2.3.2. Spatial Pattern Distribution of Species Richness

A total of 62 *Meconopsis* species were included in this study, which represents the species richness of *Meconopsis* genus in the study area. The species distribution pattern was affected by different spatial scales or areas, and to eliminate their influence, when calculating the species richness, the fishnet tool was used to divide the study area according to the equal area grid of 50 km $\times$ 50 km, and a total of 1314 grids were obtained. Then, the species distribution points of *Meconopsis* were associated with the grids, and the number of species in each equal-area grid was counted based on the actual recorded points of *Meconopsis*, and this value was taken as the species richness value of each grid. Finally, the species richness was divided into 15 levels using the classification function of ArcGIS 10.8, and the spatial pattern of species richness was visualized.

2.3.3. Distribution Pattern of Species Richness along Environmental Gradients

We extracted the topographic factor data of 1445 occurrence data points for 62 species of the genus *Meconopsis* in the study area, including elevation, slope, and aspect, and explored the relationship among topographic factors, longitude and latitude, and species richness. In this study, elevation was divided into 32 bands, each with an interval of 100 m, and the species richness was counted according to the altitudinal band. In terms of slope, aspect, longitude, and latitude, we divided them into 30, 36, 33, and 14 bands at 1° intervals, and calculated the species richness of each band in turn. The influence of environmental factors on *Meconopsis* plant species richness geospatial patterns were determined by running polynomial regression or nonlinear regression (GaussMod) models.

2.3.4. Landscape Fragmentation

Four landscape indices were used to measure landscape fragmentation in this paper (Table 2). Based on the reclassified land cover data grid-by-grid, the landscape indices in the grid based on the landscape scale were calculated, the spatial data were processed in ArcGIS 10.8, and the landscape indices were calculated by software Fragstats 4.2.1.

Table 2. Selected landscape indices and their ecological significance.

Landscape Indices	Range of Value	Ecological Significance
Patch density index (PD)	PD > 0	It can reflect the degree of fragmentation of the landscape. The higher the value, the higher the degree of fragmentation.
Patch richness index (PR)	PR >= 1	It indicates the total number of all patch types in the landscape and is one of the key indicators of landscape components, as well as spatial heterogeneity, and has implications for many ecological processes.
Contagion index (CONTAG)	0 < CONTAG <= 100	It describes the degree of clustering or extension trend of different patch types in the landscape. Since this indicator contains spatial information, it is one of the most important indices to describe the landscape pattern. Generally, high spreading values indicate that some dominant patch types in the landscape form a good connectivity; conversely, it indicates that the landscape is a dense pattern with multiple elements and a high degree of fragmentation in the landscape.
Shannon's diversity index (SHDI)	SHDI >= 0	An increase in SHDI indicates an increase in patch types or an equalizing trend in the distribution of each patch type in the landscape.

3. Results

3.1. Model Performance and Key Variables to Predict Typical Meconopsis Species

Models for the *Meconopsis*, *M. integrifolia*, *M. horridula*, *M. racemosa*, and *M. punicea* performed better than random, with the given set of training and test data. The average AUC values for *Meconopsis*, *M. integrifolia*, *M. horridula*, *M. racemosa*, and *M. punicea* were 0.847, 0.888, 0.852, 0.914, and 0.951, respectively, indicating these five models performed well and generated very good (excellent) evaluations.

The results showed that precipitation of warmest quarter (Bio 18, 43.8%), temperature seasonality (Bio 4, 20%), elevation (13.5%) and annual mean temperature (Bio 1, 7.8%) made the greatest contributions to the distribution model for *Meconopsis* relative to other variables (Table 3), and the cumulative contributions of these factors reached values as high as 85.1%. Among the 11 environmental variables, precipitation of the warmest quarter (Bio 18, 59.2%), annual mean temperature (Bio 1, 18.9%), precipitation seasonality (Bio 15, 8.3%), and precipitation of the coldest quarter (Bio 19, 5.5%) made a greater contribution to the species distribution model for *M. integrifolia* than other environmental variables (Table 2), accounting for 91.9% of variation in total. For *M. horridula*, the most important factors were precipitation of the warmest quarter (Bio 18, 36%), elevation (11.8%), temperature seasonality (Bio 4, 11.5%), and isothermality (Bio 3, 10.2%) (Table 2); the cumulative contribution value accounted for 69.5% of the total contribution value of all environmental factors to the model. As for *M. racemosa*, precipitation of the warmest quarter (Bio 18, 37.7%), temperature seasonality (Bio 4, 24.4%), annual mean temperature (Bio 1, 16.7%), and precipitation of the coldest quarter (Bio 19, 4.4%) totally contributed 83.2% in the model (Table 2), which means these four variables contain the most significant and useful information to predict species distribution. Similarly, precipitation of the warmest quarter (Bio 18, 43.6%) had the highest contribution in the *M. punicea* model and, followed by precipitation seasonality (Bio 15, 15.8%), temperature seasonality (Bio 4, 8.6%), and elevation (8.5%) (Table 3), they totally accounted for 76.5% of the contribution value and were identified as main factors influencing the species' spatial distribution.

The thresholds (presence probability > 0.25) of the main environmental parameters were obtained from the response curve (Figure 4). The presence probability of all species is the greatest with 300–400 mm precipitation of the warmest quarter (Figure 4). In *Meconopsis*, temperature seasonality (>500) and annual mean temperature (>8 °C) beyond thresholds affect the habitat suitability, and the most suitable elevation was about 3700 m (Figure 4a). *M. integrifolia* had the highest probability of existence when the annual mean temperature was about 10 °C and precipitation seasonality was 60–95. With the maximum probability of species occurrence when precipitation value reached 20 mm in the coldest quarter for *M. integrifolia* (Figure 4b). *M. horridula* was most likely to occur at elevations between 3200 and 4800 m, with the most suitable temperature seasonality ranging from 500 to 800, and isothermality had a sigmoid trend with the maximum species presence probability when its value up to 54 (Figure 4c). For *M. racemosa*, the optimum ranges of temperature seasonality, annual mean temperature, and precipitation of the coldest quarter were 540–690, 0–13 °C, and 10–30 mm, respectively (Figure 4d). The response curves of *M. punicea* indicated that precipitation seasonality, temperature seasonality and elevation followed a Gaussian shape, and they were within a certain range when the probability of species occurrence was higher (Figure 4e).

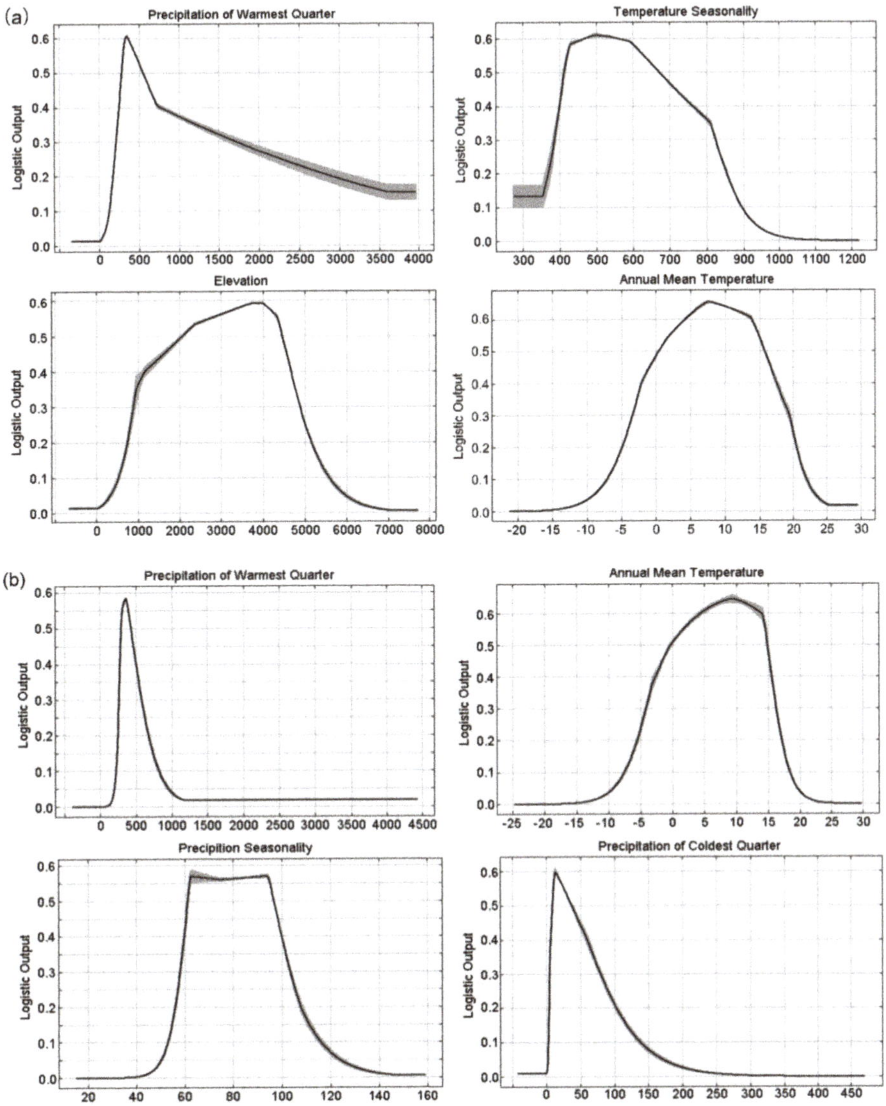

Figure 4. *Cont.*

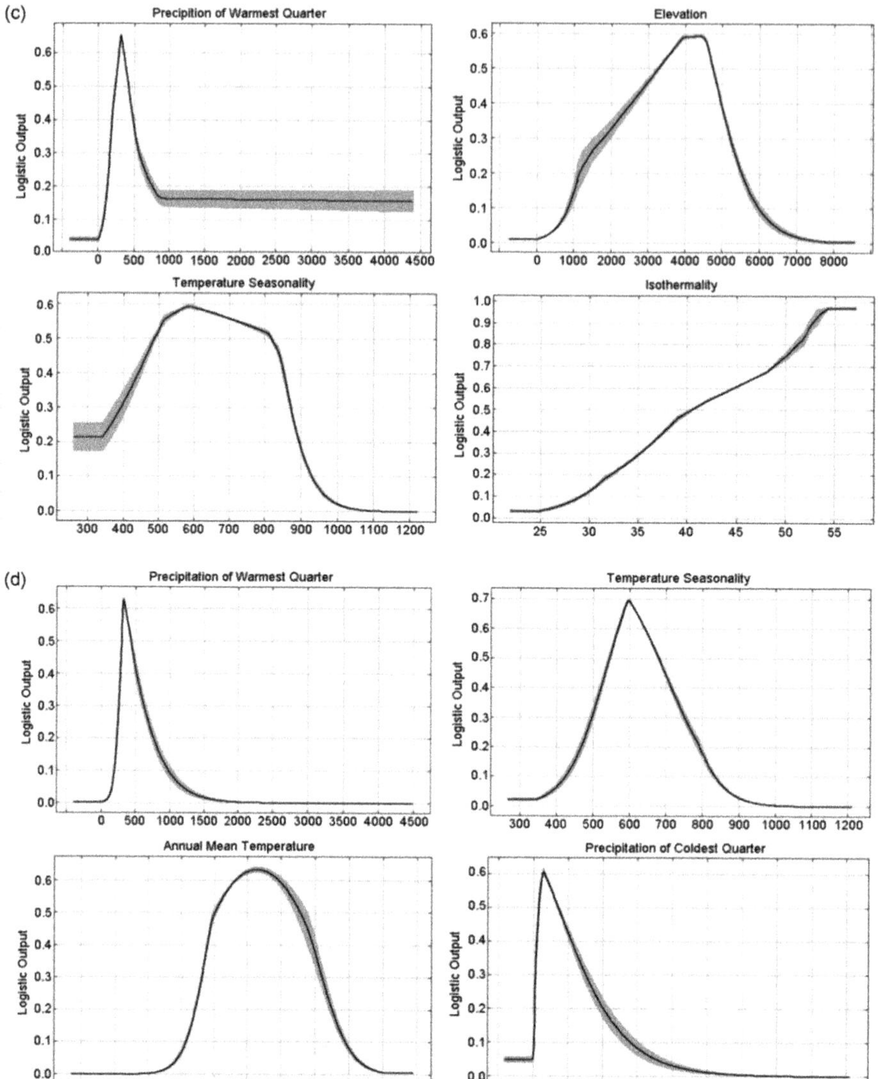

Figure 4. *Cont.*

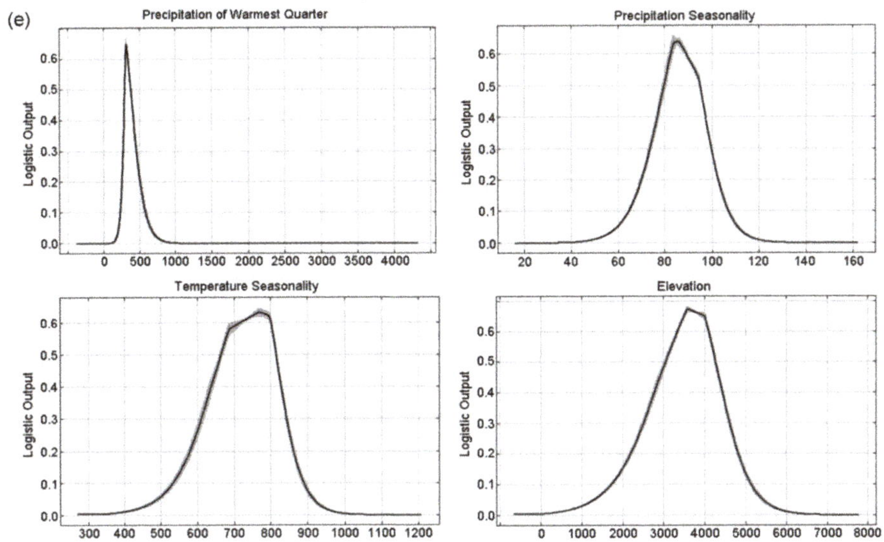

Figure 4. Relationships between key predictor variables and probability of presence of (**a**) *Meconopsis* species, (**b**) *M. integrifolia*, (**c**) *M. horridula*, (**d**) *M. racemose*, and (**e**) *M. punicea*.

Table 3. Relative contribution (%) of environmental variables to the MaxEnt model output for (a) *Meconopsis* species, (b) *M. integrifolia*, (c) *M. horridula*, (d) *M. racemosa*, and (e) *M. punicea*.

Variable	Code	a	b	c	d	e
Precipitation of Warmest Quarter	Bio18	43.8	59.2	36	37.7	43.6
Temperature Seasonality	Bio4	20	2.7	11.5	24.4	8.6
Elevation	Elevation	13.5	1.8	11.8	4.4	8.5
Annual Mean Temperature	Bio1	7.8	18.9	10	16.7	7.5
Precipitation Seasonality	Bio15	5.3	8.3	4.9	1.4	15.8
Isothermality	Bio3	3.9	0.2	10.2	4.3	1.4
Slope	Slope	2.6	2.3	5.4	2.4	2.3
Mean Diurnal Range	Bio2	1.5	0.6	3.7	1.6	3.2
Precipitation of Driest Month	Bio14	1.1	0.4	4	0.3	4.9
Precipitation of Coldest Quarter	Bio19	0.3	5.5	1.6	4.4	2.6
Aspect	Aspect	0.1	0.3	1	2.4	1.6

3.2. Current Potential Geographical Distribution Patterns of Species

The grade of the suitability zone of species has been presented in Figure 4 under the current climatic conditions across the whole study area. The potentially suitable habitats of *Meconopsis* are mainly distributed in the Himalaya and Hengduan Mountains, with an area of 1417.92 × 10^3 km^2, 31.74% of the total area. The high-suitability area accounted for only 0.23% of the study area, which is mainly distributed in the central Himalaya and the Nujiang and Jinsha River Basins in Yunnan Province (Figure 5a, Table 4). The potentially suitable distribution area of *M. integrifolia* extended from the western Himalaya to the east of Hengduan Mountains, with 1105.33 × 10^3 km^2 in total, accounting for 24.74% of the study area. Among these areas, the low and medium suitable areas were dominant, while the high suitable regions were concentrated in the Jinsha River Basins in Yunnan Province, with an area of only 9.14 × 10^3 km^2, accounting for 0.2% of the total study area (Figure 5b, Table 4). The most suitable region for *M. horridula* was located from the central Himalaya to the Hengduan Mountains, including the Minshan and Qilian Mountains, with a total area of 1428.40 × 10^3 km^2, accounting for 31.98% of the study area, while the area of the middle- and low-suitability zone (31.52%) was much larger than that of high suitable area

(0.46%) (Figure 5c, Table 4). Unlike *M. integrifolia* and *M. horridula*, the potentially suitable habitat area for *M. racemosa* was much smaller, accounting for only 15.78% of the study area, but with a slightly larger high suitable area than *M. integrifolia* and *M. horridula* (0.55%, 24.51×10^3 km^2), mainly scattered in the eastern Himalaya and the Hengduan Mountains (Figure 5d, Table 4). The suitable habitat for *M. punicea* covered an area of 407.19×10^3 km^2, accounting for less than 10% of the study area, and it was concentrated in the eastern part of the Hengduan Mountains, while regions with middle and high suitability for this species were mainly concentrated in the East Kunlun Mountains (Figure 5e, Table 4).

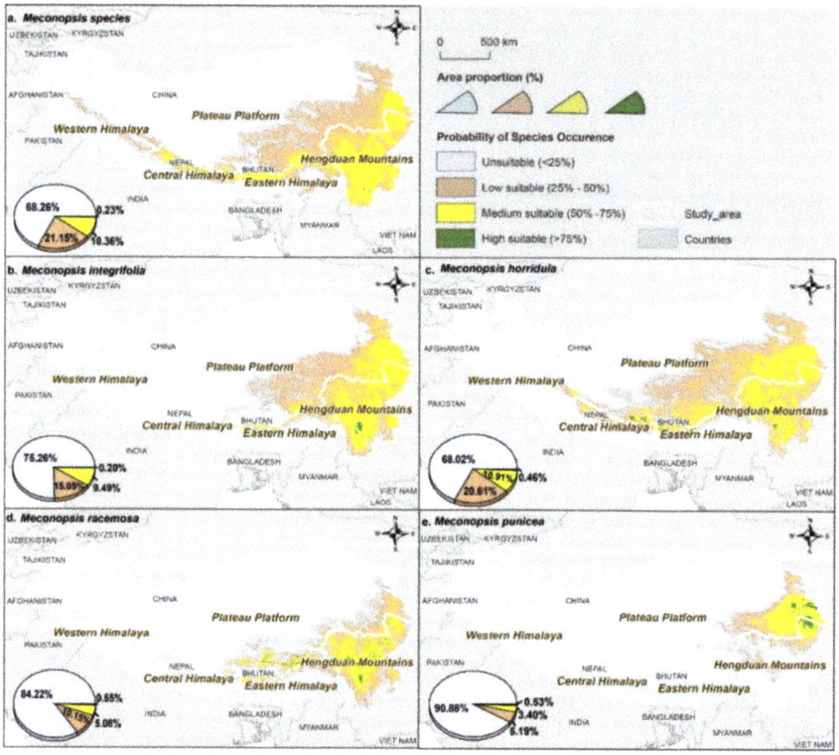

Figure 5. Potential geographic distributions of (**a**) *Meconopsis* species, (**b**) *M. integrifolia*, (**c**) *M. horridula*, (**d**) *M. racemosa*, and (**e**) *M. punicea* under current climate conditions in the study area. In this case, the probability of occurrence of *Meconopsis* species is categorized as low-, medium-, and high-suitable areas based on results of niche modeling.

Table 4. The areas of different suitable habitat of *Meconopsis* species based on model predictions under current climate scenario. (Unit: 10^3 km^2).

Probability of Occurrence	*Meconopsis*	*M. integrifolia*	*M. horridula*	*M. racemosa*	*M. punicea*
Unsuitable (<25%)	3049.07	3361.65	3038.57	3761.98	4059.79
Low suitable (25–50%)	944.89	672.12	920.74	453.50	231.63
Medium suitable (50–75%)	462.92	424.07	486.94	226.99	152.11
High suitable (>75%)	10.11	9.14	20.72	24.51	23.45

We found that the mean elevation distribution of suitable habitats for all species ranged from 3200 m to 4300 m (Table 5). For the genus *Meconopsis*, the distribution elevation range of the middle- and high-suitable zones was narrower than that of the low-suitable zones, and the elevation range of the high suitable zone was only 381 m, distributed in the area over 4000 m above sea level. The elevation widths of the low-suitable habitats of *M. integrifolia*, *M. horridula*, *M. racemosa*, and *M. punicea* were larger than those of the medium- and high-suitable habitats. In the high-suitable habitats, the elevational width of *M. integrifolia* was the narrowest at 843 m. In the medium-suitable habitats, the elevation width of *M. punicea* was 2320 m, which was the narrowest among them. *M. punicea* still had the narrowest elevation width of 2777 m in the low suitability habitat, and *M. horridula* had the widest in all suitability habitats, which means that it was better adapted than the other three species.

Table 5. Elevational pattern in species potential distribution in climate change scenarios under current climate conditions, in regions exhibiting high, medium, and low probability of species occurrence. All values in meters above mean sea level. (a: *Meconopsis*, b: *M. integrifolia*, c: *M. horridula*, d: *M. racemosa*, e: *M. punicea*).

Species	Low Suitable (25–50%)			Medium Suitable (50–75%)			High Suitable (>75%)		
	Mean ± SD	Min	Max	Mean ± SD	Min	Max	Mean ± SD	Min	Max
a	4111 ± 791	2014	5983	3586± 737	2049	4988	4194 ± 142	4043	4424
b	4139 ± 760	2033	5983	3628 ± 713	2049	4925	3620 ± 367	3279	4122
c	4265 ± 807	2049	5953	4027 ± 643	2126	5406	4118 ± 540	2599	4617
d	3877 ± 766	2049	5142	3718 ± 647	2058	4707	3697 ± 391	2928	4259
e	4084 ± 565	2180	4958	3817 ± 454	2326	4646	3234 ± 484	2237	3677

3.3. Species Richness Geospatial Patterns of Meconopsis

We found an uneven spatial distribution of *Meconopsis*, with species richness ranging from 0 to 14 species per grid; most of the species were concentrated in the southeastern and southern regions of the study area (Figure 6a), i.e., the Hengduan Mountains, the central and eastern Himalaya, and the southeastern part of the plateau platform adjacent to the Himalaya and Hengduan Mountains, with relatively higher the species richness than other parts. According to the number of species per grid, we divided species richness into three categories: low (0–3), medium (3–8), and high (8–14). From Figure 6b, we could find that the areas with high species richness were mainly forests, while the areas with moderate species richness were forests and grasslands. In the Hengduan Mountains, central and eastern Himalaya, the areas with high and medium species richness were mainly forests, and in the plateau were mainly grasslands. The area with medium–high species richness in all subregions accounted for less than 10%, except for 18.3% in the Hengduan Mountains. Particularly, there was no area with medium–high species richness.

3.4. Species Richness Pattern along Environmental Gradients

The species richness gradient in the study area exhibited a hump-shaped pattern with increasing richness with elevation, peaking at around 3900–4100 m and declining toward the ends of the elevational gradient (Figure 7a). The pattern of *Meconopsis* plant richness along the latitudinal gradient increased remarkably to the maximum and gradually decreased with Gaussian shape fitted (R^2 = 0.96, Figure 7b). The richness of *Meconopsis* species showed a more or less monotonous increasing trend along the longitudinal gradient and was found absent beyond 104° (Figure 7c). In our study area, their richness varied strongly with the slope gradient with peaks occurring between 6° and 11°, and decreasing in both directions (Figure 7d). The *Meconopsis* plant richness presented a clear hump-shaped pattern along the aspect gradient, reaching its peak at the range of 150° to 200° (south slope or sunny slope), and it was similar at both the highest (north and northwest slopes) and lowest (north and northeast slopes) aspect belts (Figure 7e). In a word, the polynomial curve fitted precisely for richness patterns of *Meconopsis* plant along the elevation, longitude, slope, and aspect, with reasonable R^2 values ($R^2 > 0.5$).

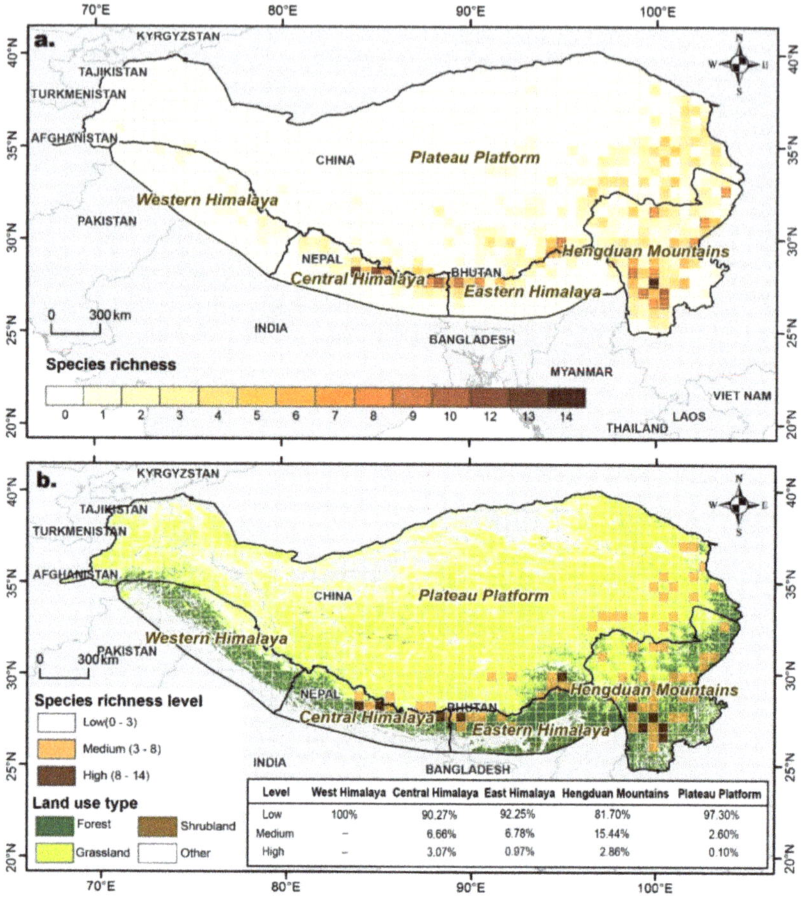

Figure 6. Spatial distribution of species richness of *Meconopsis* per 50 km × 50 km grid cell. (**a**) species richness distribution of each grid in the study area; (**b**) the distribution of species richness of *Meconopsis* in different vegetation.

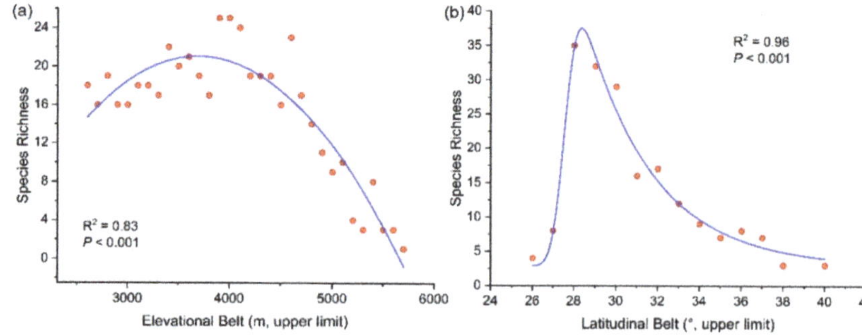

Figure 7. *Cont.*

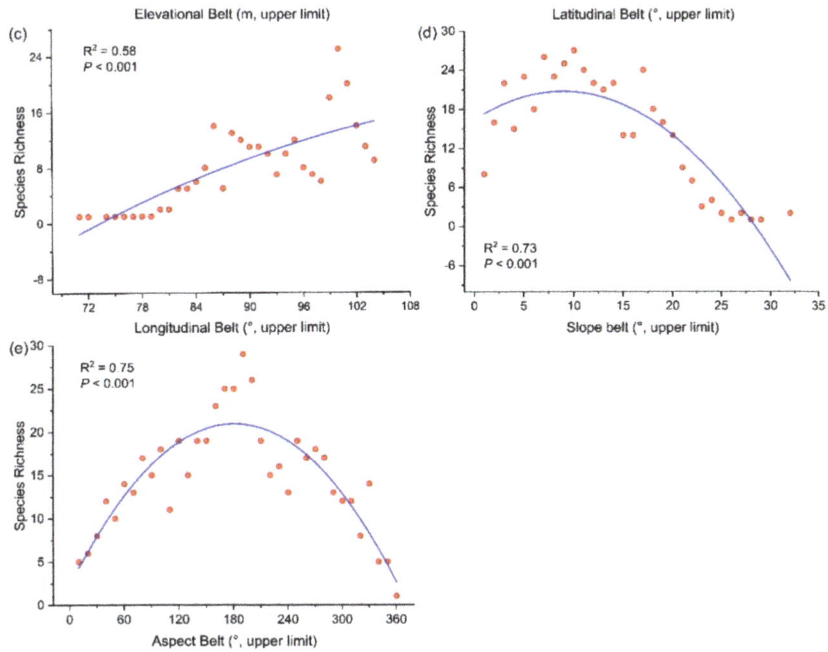

Figure 7. Species richness pattern of *Meconopsis* along the studied slope (**a**), latitudinal (**b**), longitudinal (**c**), elevational (**d**), and aspect (**e**) gradient.

3.5. Ecological Landscape Fragmentation and Species Richness

High PD in all subregions indicated a relatively large degree of landscape fragmentation across the pan-Himalaya that was quite similar in all subregions (Figure 8). Specifically, the plateau platform occupied the largest proportion of the study area with the highest PD (152.4069), while the central Himalayan region had the lowest PD (103.3529). In terms of the CONTAG, the value was the highest in eastern Himalaya (66.2219), indicating better connectivity between dominant patches in the landscape, while the CONTAG of the western Himalaya was the lowest (60.1523). The results of PR illustrated that each subregion covered all land-use types. For SHDI, all subregions were greater than 0.8, indicating high landscape heterogeneity in the study area, and the highest SHDI value was in the western Himalayan region (1.3280). According to the four landscape indices, the degree of landscape fragmentation in each subregion from high to low was the western Himalaya, plateau platform, the Hengduan Mountains, the Central Himalaya, and eastern Himalaya. Areas with the highest species richness were the Hengduan Mountains (35) and plateau platform (34); the central Himalaya (20) and eastern Himalaya (19) had similar richness, while the lowest richness was in the western Himalaya.

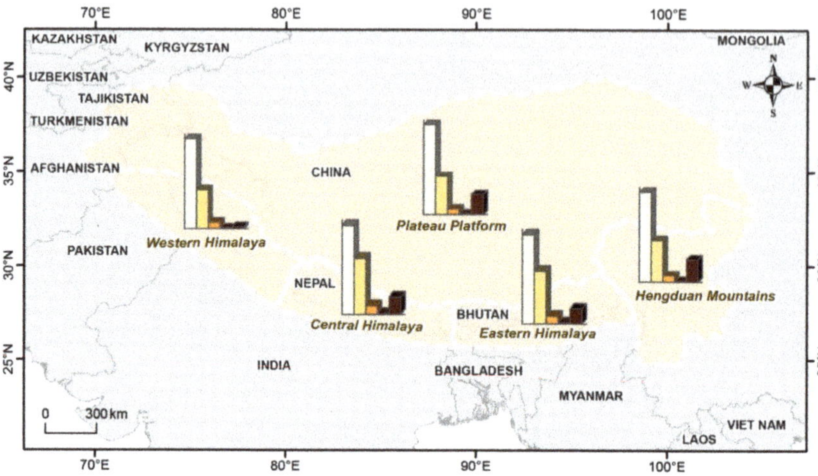

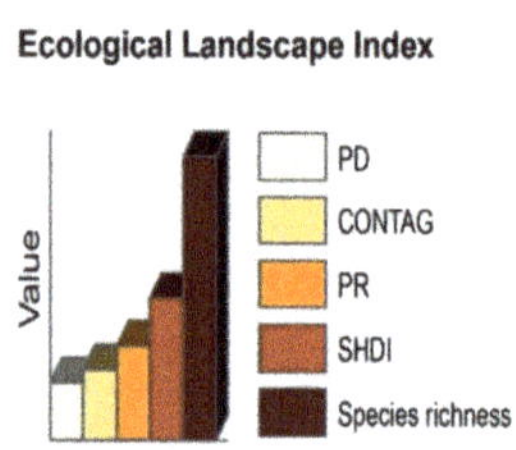

Subregion	Ecological landscape index				Species richness
	PD (*10^7)	CONTAG	PR	SHDI	
Western_Himalaya	139.7002	60.1523	10.0000	1.3280	3
Central_Himalaya	103.3529	65.2103	10.0000	1.1771	20
Eastern_Himalaya	111.8434	66.2219	10.0000	1.1548	19
Hengduan_Mountains	141.1810	65.1070	10.0000	1.1012	35
Plateau_platform	152.4069	65.0408	10.0000	1.1080	34

Figure 8. Ecological landscape index and species richness in each sub-region.

4. Discussion

4.1. Divergent Environmental Factors Affect the Spatial Distribution of Meconopsis Species

The long-term joint influence of several factors resulted in the geographical distribution of plants. Climate is the most important factor determining the geographical distribution of plants on a regional scale [60,61]. The findings for MaxEnt indicated that the precipitation of the warmest quarter (Bio 18) had the greatest effect on the distribution of all *Meconopsis* species amongst the 11 environmental variables involved in modeling, and this is consistent with our previous study in Minjiang headwater region [42]. Since precipitation can affect plant growth, morphology, phenology, and biomass accumulation [62–66], in particular for seedling emergence and establishment [67], an appropriate degree of precipitation can supply sufficient water and promote plant growth; however, excessive rainfall can deteriorate the soil permeability [68], creating an anaerobic environment that inhibits the regular respiration of roots [69], and limits plant growth and development by influencing their metabolism [70]. Moreover, waterlogging can also cause stomata to close and photosynthesis to decrease, and increases the energy consumption for respiration, affecting the accumulation of organic matter [71]. In addition, high humidity caused by excess water favors the rapid multiplication of pathogens and the formation of serious diseases that not only threaten the survival of plants but also their distribution patterns on a geographical scale [72]. The species response curves demonstrated that the ecological amplitude of precipitation is small, indicating the narrow tolerance range of *Meconopsis* to precipitation and limiting the geographical distribution range.

Temperature variation affects plant distribution by influencing their germination, water absorption, photosynthesis, transpiration, respiration, reproduction, and growth [73]. Our results showed that temperature seasonality (Bio4) and mean annual temperature (Bio1) were the two dominant temperature-related factors affecting the habitat suitability of *Meconopsis* species. Elevation was another key variable that had a strong indirect impact on the distribution of *Meconopsis* species. It integrates the effect of temperature, humidity, light, and other indicators to make secondary allocation of resources for influencing species composition and distribution [74,75]. Under the current climate, a potentially suitable area for *Meconopsis* was mainly distributed in the southeast Qinghai-Tibet Plateau and Hengduan Mountains where have similar climate conditions, particularly in terms of hydrothermal coupling. Monsoons, with sufficient rainfall with suitable temperature conditions, bolster the growth and development of alpine vegetation [76], accumulating assimilation products for overwintering and reproduction. Furthermore, high mountains and valleys are distributed in tandem with tremendous elevation variation, which creates a unique climate with cold temperatures at the mountain top and dry-hot conditions at the foot of the mountain. The suitable habitats of *Meconopsis* are mostly distributed in the alpine regions above 3500 m, consistent with their preference for cold and cool ecological habits [77,78]. This comprehensive analysis found that elevation and variables related to precipitation and temperature were the key factors restricting the distribution of *Meconopsis*.

4.2. The Richness of Meconopsis *Species Varies along Geospatial Gradients*

The decline of plant species richness from the equator to the poles is one of the most striking ecological patterns on Earth [79]. However, along the longitudinal gradient, from west to east, plant species richness increases with increasing wetness in different regions of the study area [80–83]. The highest species richness of *Meconopsis* was founded in Hengduan Mountain, which stands at the intersection of latitudinal and longitudinal curves (Figure 5) providing appropriate hydrothermal conditions, high habitat heterogeneity, and complex terrain [84,85] for the growth of *Meconopsis* species. Interestingly, many areas with medium and high species richness were forests. This contradicted our common knowledge that their favorable habitats are shrublands or grasslands, which might be due to scale variations or spatial resolution since shrub and grassland patches in forests are often classified as forests at a 300 m spatial resolution. Moreover, it is generally accepted that forests, shrubs, deserts, and other plant communities with woody plant coverage of less than 40% should be classified as grassland [86]. However, the grassland area calculated based on the area of herbaceous vegetation here is underestimated.

Topography (elevation, slope, and aspect) has a significant impact on the spatial distribution of plants by influencing habitat temperature and humidity [87,88]. *Meconopsis* species richness revealed a unimodal distribution pattern along the elevational gradients in mountainous regions (Figure 5), similar to prior studies [89–91]. This may be attributed to the appropriate mild habitats for plant growth in mid-elevation [92], and this mid-domain effect was highly pronounced in the Himalaya [91], which brings in more rainfall than higher elevations in monsoon and decreases from the southernmost to the northernmost slopes [93]. The optimum temperature decreases along with the elevation increase, and only a few species can adapt to harsh environments [91,94]. Lower *Meconopsis* species richness beyond mid-elevations can be attributed to the decrease of soil cover with more rocks and slow biogeochemical cycle in contrast with favorable mid-elevation providing higher atmospheric moisture and cooler summer temperature [95]. Some species have specific requirements for slope and aspect [96], which directly or indirectly influence the solar radiation, temperature, and soil conditions [97,98] In concordance with other studies, we found that species richness of *Meconopsis* was higher on the sunlit southern slope as compared to the northern slope [99]. Meanwhile, the steeper slope triggers soil erosion [97] which stunts seed settling and plant growth [100]. Therefore, gentle slopes have higher species richness and plant growth as compared to steeper slopes. [97]. All the aforementioned topographical variables play essential and unique roles in the distribution of *Meconopsis*.

4.3. Linkage and Relations between Landscape Heterogeneity and Species Richness

The complex spatial heterogeneity of mountain areas provides advantageous conditions for biodiversity with a variety of patch types [101]. However, our study did not find a relatively clear relationship between landscape fragmentation and species richness, which might be relevant to inconsistent scale and metastability [29,30,102]. At the landscape scale, the degree of landscape fragmentation seems to have little impact on the species richness of *Meconopsis*. Given that spatial heterogeneity such as topography can significantly affect the richness [103], the effects of landscape fragmentation are probably more pronounced at comparatively smaller scales.

A larger habitat can harbor more species and support richer plant communities with more available resources [104]. Therefore, it is easy to understand why the species richness is higher on the plateau platform than in other parts, whose higher species richness is mainly concentrated in the southeastern part. Blocked by the Himalaya and Gangdise mountains, warm moist air from the Indian Ocean flows into China along the Hengduan Mountains, thus bringing abundant rain to the southeastern of the Qinghai-Tibetan Plateau. While the western Himalaya has the lowest species richness with a quite fragmented landscape, the region is inherently less suitable for *Meconopsis* because of poor hydrothermal conditions [105], and this is consistent with the current potential distribution modeling results. The lower latitudes of the central and eastern Himalaya on the southern slope of the Himalaya with abundant summer rainfall have a higher species richness. [105,106]. The Hengduan Mountains are climatically influenced by the westerly circulation and the Indian and Pacific monsoon circulation, with dry winters and rainy summers, which are ideal for the more abundant *Meconopsis*; despite their smaller area than the plateau platform, the species richness of *Meconopsis* is more abundant.

5. Conclusions

Our study discovered that the potential distribution regions with medium- and high-suitable habitats for *Meconopsis*, under the current climate scenario, were mainly located in the central and eastern Himalaya, Hengduan Mountains, and the southeast edge of the plateau platform. Precipitation of the warmest quarter had the greatest impact on the distribution of *Meconopsis* species. We also found species richness within the genus *Meconopsis* was distributed in a unimodal pattern along geospatial gradients except for the longitudinal gradient in pan-Himalaya. However, no obvious consistent relationships exist between landscape fragmentation and species richness for *Meconopsis*. Our findings not only promote an understanding of the distribution and diversity of *Meconopsis* species but also provide an indispensable foundation for future studies on *Meconopsis* plant functions and the sustainability of alpine ecosystems. This study also provides data and theoretical support for species diversity protection policies in pan-Himalaya and adjacent regions.

Supplementary Materials: The following supporting information can be downloaded at https://www.mdpi.com/article/10.3390/10.3390/d14080661/s1: Table S1: The number of presence locations of each species.

Author Contributions: Conceptualization, N.S., C.W. and J.W.; methodology, N.S. and C.W.; software, N.S. and C.W.; validation, N.S., C.W. and J.W.; formal analysis, N.S. and C.W.; investigation, N.S. and C.W.; resources, N.S. and C.W.; data curation, N.S. and C.W.; writing—original draft preparation, N.S.; writing—review and editing, N.S., C.W., J.W., N.W., N.N., L.Z., L.W., J.S., W.D., Y.W. (Yanqiang Wei), W.C. and Y.W. (Yan Wu); visualization, N.S. and C.W.; supervision, J.W.; project administration, J.W.; funding acquisition, J.W. All authors have read and agreed to the published version of the manuscript.

Funding: This research was funded by the National Natural Science Foundation of China (31971436), CAS "Light of West China" Program (2021XBZG_XBQNXZ_A_007), State Key Laboratory of Cryospheric Science, Northwest Institute of Eco-Environment and Resources, Chinese Academy Sciences (SKLCS-OP-2021-06), Talented Young Scientist Program (Indian-18-008) by China Science and Technology Exchange Centre, Ministry of Science and Technology, and China Biodiversity Observation Networks (Sino BON).

Institutional Review Board Statement: Not applicable.

Informed Consent Statement: Not applicable.

Data Availability Statement: The raw data supporting the conclusions of this article will be made available by the authors, without undue reservation.

Acknowledgments: Special thanks to Li Yike from the Chengdu Institute of Biology, Chinese Academy of Sciences for his valuable discussion about Fragstats and other issues.

Conflicts of Interest: The authors declare no conflict of interest.

References

1. Huss, M.; Bookhagen, B.; Huggel, C.; Jacobsen, D.; Bradley, R.S.; Clague, J.J.; Vuille, M.; Buytaert, W.; Cayan, D.R.; Greenwood, G.; et al. Toward mountains without permanent snow and ice. *Earth's Future* **2017**, *5*, 418–435. [CrossRef]
2. Nie, Y.; Pritchard, H.D.; Liu, Q.; Hennig, T.; Wang, W.; Wang, X.; Liu, S.; Nepal, S.; Samyn, D.; Hewitt, K.; et al. Glacial change and hydrological implications in the Himalaya and Karakoram. *Nat. Rev. Earth Environ.* **2021**, *2*, 91–106. [CrossRef]
3. Shi, H.; Zhou, Q.; He, R.; Zhang, Q.; Dang, H. Climate warming will widen the lagging gap of global treeline shift relative to densification. *Agric. For. Meteorol.* **2022**, *318*, 108917. [CrossRef]
4. Pepin, N.; Bradley, R.S.; Diaz, H.F.; Baraer, M.; Caceres, E.B.; Forsythe, N.; Fowler, H.; Greenwood, G.; Hashmi, M.Z.; Liu, X.D.; et al. Elevation-dependent warming in mountain regions of the world. *Nat. Clim. Chang.* **2015**, *5*, 424–430. [CrossRef]
5. Rogora, M.; Frate, L.; Carranza, M.; Freppaz, M.; Stanisci, A.; Bertani, I.; Bottarin, R.; Brambilla, A.; Canullo, R.; Carbognani, M.; et al. Assessment of climate change effects on mountain ecosystems through a cross-site analysis in the Alps and Apennines. *Sci. Total Environ.* **2018**, *624*, 1429–1442. [CrossRef]
6. Beniston, M. Climatic Change in Mountain Regions: A Review of Possible Impacts. *Clim. Chang.* **2003**, *59*, 5–31. [CrossRef]
7. Li, S.; Zhang, Y.; Wang, Z.; Li, L. Mapping human influence intensity in the Tibetan Plateau for conservation of ecological service functions. *Ecosyst. Serv.* **2018**, *30*, 276–286. [CrossRef]
8. Zhang, Y.; Qian, L.; Spalink, D.; Sun, L.; Chen, J.; Sun, H. Spatial phylogenetics of two topographic extremes of the Hengduan Mountains in southwestern China and its implications for biodiversity conservation. *Plant Divers.* **2020**, *43*, 181–191. [CrossRef]
9. Liang, Q.; Xu, X.; Mao, K.; Wang, M.; Wang, K.; Xi, Z.; Liu, J. Shifts in plant distributions in response to climate warming in a biodiversity hotspot, the Hengduan Mountains. *J. Biogeogr.* **2018**, *45*, 1334–1344. [CrossRef]
10. Mao, K.; Wang, Y.; Liu, J. Evolutionary origin of species diversity on the Qinghai–Tibet Plateau. *J. Syst. Evol.* **2021**, *59*, 1142–1158. [CrossRef]
11. Liu, Y.; Ye, J.; Hu, H.; Peng, D.; Zhao, L.; Lu, L.; Zaman, W.; Chen, Z.; Ebach, M. Influence of elevation on bioregionalisation: A case study of the Sino-Himalayan flora. *J. Biogeogr.* **2021**, *48*, 2578–2587. [CrossRef]
12. Körner, C.; Jetz, W.; Paulsen, J.; Payne, D.; Rudmann-Maurer, K.; Spehn, E.M. A global inventory of mountains for bio-geographical applications. *Alp. Bot.* **2016**, *127*, 1–15. [CrossRef]
13. Liu, J.; Li, J.; Lai, Y. Plant diversity and ecology on the Qinghai–Tibet Plateau. *J. Syst. Evol.* **2021**, *59*, 1139–1141. [CrossRef]
14. Sun, H.; Zhang, J.; Deng, T.; Boufford, D.E. Origins and evolution of plant diversity in the Hengduan Mountains, China. *Plant Divers.* **2017**, *39*, 161–166. [CrossRef]
15. Chen, S.; Wang, W.; Xu, W.; Wang, Y.; Wan, H.; Chen, D.; Tang, Z.; Tang, X.; Zhou, G.; Xie, Z.; et al. Plant diversity enhances productivity and soil carbon storage. *Proc. Natl. Acad. Sci. USA* **2018**, *115*, 4027–4032. [CrossRef]
16. Quijas, S.; Schmid, B.; Balvanera, P. Plant diversity enhances provision of ecosystem services: A new synthesis. *Basic Appl. Ecol.* **2010**, *11*, 582–593. [CrossRef]
17. Guo, F.; Lenoir, J.; Bonebrake, T.C. Land-use change interacts with climate to determine elevational species redistribution. *Nat. Commun.* **2018**, *9*, 1–7. [CrossRef]
18. Wu, J.; Liu, Z.M. Effect of habitat fragmentation on biodiversity: A review. *Chinese J. Ecol.* **2014**, *33*, 1946–1952.
19. Bálint, M.; Domisch, S.; Engelhardt, C.H.M.; Haase, P.; Lehrian, S.; Sauer, J.; Theissinger, K.; Pauls, S.U.; Nowak, C. Cryptic biodiversity loss linked to global climate change. *Nat. Clim. Chang.* **2011**, *1*, 313–318. [CrossRef]
20. Boddy, L.; Büntgen, U.; Egli, S.; Gange, A.C.; Heegaard, E.; Kirk, P.M.; Mohammad, A.; Kauserud, H. Climate variation effects on fungal fruiting. *Fungal Ecol.* **2014**, *10*, 20–33. [CrossRef]
21. Hooper, D.U.; Adair, E.C.; Cardinale, B.J.; Byrnes, J.; Hungate, B.A.; Matulich, K.L.; Gonzalez, A.; Duffy, J.E.; Gamfeldt, L.; O'Connor, M.I. A global synthesis reveals biodiversity loss as a major driver of ecosystem change. *Nature* **2012**, *486*, 105–108. [CrossRef]
22. Zhang, K.; Yao, L.; Meng, J.; Tao, J. Maxent modeling for predicting the potential geographical distribution of two peony species under climate change. *Sci. Total Environ.* **2018**, *634*, 1326–1334. [CrossRef]
23. Lenoir, J.; Gégout, J.-C.; Guisan, A.; Vittoz, P.; Wohlgemuth, T.; Zimmermann, N.E.; Dullinger, S.; Pauli, H.; Willner, W.; Svenning, J.-C. Going against the flow: Potential mechanisms for unexpected downslope range shifts in a warming climate. *Ecography* **2010**, *33*, 295–303. [CrossRef]
24. Alatalo, J.M.; Jägerbrand, A.K.; Molau, U. Impacts of different climate change regimes and extreme climatic events on an alpine meadow community. *Sci. Rep.* **2016**, *6*, 21720. [CrossRef]

25. Ceballos, G.; Ehrlich, P.R.; Barnosky, A.D.; García, A.; Pringle, R.M.; Palmer, T.M. Accelerated modern human–induced species losses: Entering the sixth mass extinction. *Sci. Adv.* **2015**, *1*, e1400253. [CrossRef]
26. Di Giulio, M.; Holderegger, R.; Tobias, S. Effects of habitat and landscape fragmentation on humans and biodiversity in densely populated landscapes. *J. Environ. Manag.* **2009**, *90*, 2959–2968. [CrossRef]
27. Fahrig, L. Effects of Habitat Fragmentation on Biodiversity. *Annu. Rev. Ecol. Evol. Syst.* **2003**, *34*, 487–515. [CrossRef]
28. Fischer, J.; Lindenmayer, D.B. Landscape modification and habitat fragmentation: A synthesis. *Glob. Ecol. Biogeogr.* **2007**, *16*, 265–280. [CrossRef]
29. Reino, L.; Beja, P.; Araújo, M.B.; Dray, S.; Segurado, P. Does local habitat fragmentation affect large-scale distributions? The case of a specialist grassland bird. *Divers. Distrib.* **2013**, *19*, 423–432. [CrossRef]
30. Xu, C.; Huang, Z.Y.X.; Chi, T.; Chen, B.J.W.; Zhang, M.; Liu, M. Can local landscape attributes explain species richness patterns at macroecological scales? *Glob. Ecol. Biogeogr.* **2013**, *23*, 436–445. [CrossRef]
31. Huang, H. Plant diversity and conservation in China: Planning a strategic bioresource for a sustainable future. *Bot. J. Linn. Soc.* **2011**, *166*, 282–300. [CrossRef] [PubMed]
32. Brooks, T.M.; Mittermeier, R.A.; Da Fonseca, G.A.B.; Gerlach, J.; Hoffmann, M.; Lamoreux, J.F.; Mittermeier, C.G.; Pilgrim, J.D.; Rodrigues, A.S.L. Global Biodiversity Conservation Priorities. *Science* **2006**, *313*, 58–61. [CrossRef] [PubMed]
33. Xin, L.; Zhi-Heng, W.; Zhi-Yao, T.; Shu-Qing, Z.; Jing-Yun, F. Geographic patterns and environmental correlates of terrestrial mammal species richness in China. *Biodivers. Sci.* **2009**, *17*, 652–663. [CrossRef]
34. Stiles, A.; Scheiner, S.M. A multi-scale analysis of fragmentation effects on remnant plant species richness in Phoenix, Arizona. *J. Biogeogr.* **2010**, *37*, 1721–1729. [CrossRef]
35. Giladi, I.; Ziv, Y.; May, F.; Jeltsch, F. Scale-dependent determinants of plant species richness in a semi-arid fragmented agro-ecosystem. *J. Veg. Sci.* **2011**, *22*, 983–996. [CrossRef]
36. Xiao, W.; Simpson, B.B. A New Infrageneric Classification of *Meconopsis* (Papaveraceae) Based on a Well-supported Molecular Phylogeny. *Syst. Bot.* **2017**, *42*, 226–233. [CrossRef]
37. Xie, H.; Ash, J.E.; Linde, C.; Cunningham, S.; Nicotra, A. Himalayan-Tibetan Plateau Uplift Drives Divergence of Polyploid Poppies: *Meconopsis* Viguier (Papaveraceae). *PLoS ONE* **2014**, *9*, e99177. [CrossRef]
38. Egan, P.A. *Meconopsis autumnalis* and *M. manasluensis* (Papaveraceae), two new species of Himalayan poppy endemic to central Nepal with sympatric congeners. *Phytotaxa* **2011**, *20*, 47–56. [CrossRef]
39. Duffy, J.E.; Godwin, C.M.; Cardinale, B.J. Biodiversity effects in the wild are common and as strong as key drivers of productivity. *Nature* **2017**, *549*, 261–264. [CrossRef]
40. Li, X.X.; Tan, W.; Sun, J.Q.; Du, J.H.; Zheng, C.G.; Tian, X.X.; Zheng, M.; Xiang, B.B.; Wang, Y. Comparison of four complete chloroplast genomes of medicinal and ornamental *Meconopsis* species: Genome organization and species discrimination. *Sci. Rep.* **2019**, *9*, 112. [CrossRef]
41. Guo, Q.; Bai, R.; Zhao, B.; Feng, X.; Zhao, Y.; Tu, P.; Chai, X. An Ethnopharmacological, Phytochemical and Pharmacological Review of the Genus *Meconopsis*. *Am. J. Chin. Med.* **2016**, *44*, 439–462. [CrossRef]
42. Shi, N.; Naudiyal, N.; Wang, J.; Gaire, N.P.; Wu, Y.; Wei, Y.; He, J.; Wang, C. Assessing the Impact of Climate Change on Potential Distribution of *Meconopsis punicea* and Its Influence on Ecosystem Services Supply in the Southeastern Margin of Qinghai-Tibet Plateau. *Front. Plant Sci.* **2022**, *12*, 3338. [CrossRef]
43. Xiao, W.; Simpson, B.B. The Role of Allotriploidy in the Evolution of *Meconopsis* (Papaveraceae): A Preliminary Study of Ancient Polyploid and Hybrid Speciation. *Lundellia* **2014**, *17*, 5–17. [CrossRef]
44. Forester, B.R.; DeChaine, E.G.; Bunn, A.G. Integrating ensemble species distribution modelling and statistical phylogeography to inform projections of climate change impacts on species distributions. *Divers. Distrib.* **2013**, *19*, 1480–1495. [CrossRef]
45. Booth, T.H.; Nix, H.A.; Busby, J.R.; Hutchinson, M.F.; Franklin, J. Bioclim: The first species distribution modelling package, its early applications and relevance to most current MaxEnt studies. *Divers. Distrib.* **2014**, *20*, 1–9. [CrossRef]
46. Tang, Y.; Winkler, J.A.; Vina, A.; Liu, J.; Zhang, Y.; Zhang, X.; Li, X.; Wang, F.; Zhang, J.; Zhao, Z. Uncertainty of future projections of species distributions in mountainous regions. *PLoS ONE* **2018**, *13*, e0189496. [CrossRef]
47. Liao, Z.; Nobis, M.P.; Xiong, Q.; Tian, X.; Wu, X.; Pan, K.; Zhang, A.; Wang, Y.; Zhang, L. Potential distributions of seven sympatric sclerophyllous oak species in Southwest China depend on climatic, non-climatic, and independent spatial drivers. *Ann. For. Sci.* **2021**, *78*, 5. [CrossRef]
48. Liu, X.-T.; Yuan, Q.; Ni, J. Research advances in modelling plant species distribution in China. *Chin. J. Plant Ecol.* **2019**, *43*, 273–283. [CrossRef]
49. Naudiyal, N.; Wang, J.; Ning, W.; Gaire, N.P.; Peili, S.; Wei, Y.; He, J.; Ning, S. Potential distribution of *Abies*, *Picea*, and *Juniperus* species in the sub-alpine forest of Minjiang headwater region under current and future climate scenarios and its implications on ecosystem services supply. *Ecol. Indic.* **2021**, *121*, 107131. [CrossRef]
50. Du, Z.; He, Y.; Wang, H.; Wang, C.; Duan, Y. Potential geographical distribution and habitat shift of the genus *Ammopiptanthus* in China under current and future climate change based on the MaxEnt model. *J. Arid Environ.* **2020**, *184*, 104328. [CrossRef]
51. Wu, X.; Dong, S.; Liu, S.; Liu, Q.; Han, Y.; Zhang, X.; Su, X.; Zhao, H.; Feng, J. Identifying priority areas for grassland endangered plant species in the Sanjiangyuan Nature Reserve based on the MaxEnt model. *Biodivers. Sci.* **2018**, *26*, 138–148. [CrossRef]
52. Li, J.; Fan, G.; He, Y. Predicting the current and future distribution of three Coptis herbs in China under climate change conditions, using the MaxEnt model and chemical analysis. *Sci. Total Environ.* **2019**, *698*, 134141. [CrossRef]

53. Saunders, D.A.; Hobbs, R.J.; Margules, C.R. Biological Consequences of Ecosystem Fragmentation: A Review. *Conserv. Biol.* **1991**, *5*, 18–32. [CrossRef]
54. McGarigal, K.; Marks, B.J. *FragStats: Spatial Pattern Analysis Program for Quantifying Landscape Structure*; PNW-GTR-351; U.S. Department of Agriculture, Forest Service, Pacific Northwest Research Station: Portland, OR, USA, 1995; p. 122.
55. Zhang, Y.; Li, B.; Liu, L.; Zheng, D. Redetermine the region and boundaries of Tibetan Plateau. *Geogr. Res.* **2021**, *40*, 1543–1553.
56. Ding, W.-N.; Ree, R.H.; Spicer, R.A.; Xing, Y.-W. Ancient orogenic and monsoon-driven assembly of the world's richest temperate alpine flora. *Science* **2020**, *369*, 578–581. [CrossRef] [PubMed]
57. Wen, J.; Zhang, J.-Q.; Nie, Z.-L.; Zhong, Y.; Sun, H. Evolutionary diversifications of plants on the Qinghai-Tibetan Plateau. *Front. Genet.* **2014**, *5*, 4. [CrossRef]
58. Merow, C.; Smith, M.J.; Silander, J.A., Jr. A practical guide to MaxEnt for modeling species' distributions: What it does, and why inputs and settings matter. *Ecography* **2013**, *36*, 1058–1069. [CrossRef]
59. Pearson, R.G.; Raxworthy, C.J.; Nakamura, M.; Peterson, A.T. Predicting species distributions from small numbers of occurrence records: A test case using cryptic geckos in Madagascar. *J. Biogeogr.* **2007**, *34*, 102–117. [CrossRef]
60. Lambers, H.; Chapin, F.S.; Pons, T.L. *Plant Physiology Ecology*, 2nd ed.; Springer-Verlag: New York, NY, USA, 2008.
61. Peng, J.; Gou, X.; Chen, F.; Li, J.; Liu, P.; Zhang, Y.; Fang, K. Difference in Tree Growth Responses to Climate at the Upper Treeline: Qilian Juniper in the Anyemaqen Mountains. *J. Integr. Plant Biol.* **2008**, *50*, 982–990. [CrossRef]
62. Chmielewski, F.-M.; Rötzer, T. Response of tree phenology to climate change across Europe. *Agric. For. Meteorol.* **2001**, *108*, 101–112. [CrossRef]
63. Molina-Montenegro, M.A.; Atala, C.; Gianoli, E. Phenotypic plasticity and performance of *Taraxacum officinale* (dandelion) in habitats of contrasting environmental heterogeneity. *Biol. Invasions* **2010**, *12*, 2277–2284. [CrossRef]
64. Muhanguzi, H.D.R.; Obua, J.; Oryem-Origa, H.; Vetaas, O.R. Tree fruiting phenology in Kalinzu Forest, Uganda. *Afr. J. Ecol.* **2003**, *41*, 171–178. [CrossRef]
65. Pol, R.G.; Pirk, G.I.; Marone, L. Grass seed production in the central Monte desert during successive wet and dry years. *Plant Ecol.* **2010**, *208*, 65–75. [CrossRef]
66. Zhang, F.; Quan, Q.; Song, B.; Sun, J.; Chen, Y.; Zhou, Q.; Niu, S. Net primary productivity and its partitioning in response to precipitation gradient in an alpine meadow. *Sci. Rep.* **2017**, *7*, 15193. [CrossRef]
67. Perring, M.P.; Hovenden, M.J. Seedling survivorship of temperate grassland perennials is remarkably resistant to projected changes in rainfall. *Aust. J. Bot.* **2012**, *60*, 328–339. [CrossRef]
68. Koster, R.D.; Dirmeyer, P.A.; Guo, Z.; Bonan, G.; Chan, E.; Cox, P.; Gordon, C.T.; Kanae, S.; Kowalczyk, E.; Lawrence, D.; et al. Regions of Strong Coupling Between Soil Moisture and Precipitation. *Science* **2004**, *305*, 1138–1140. [CrossRef]
69. Hopkins, F.; Gonzalez-Meler, M.A.; Flower, C.E.; Lynch, D.J.; Czimczik, C.; Tang, J.; Subke, J.-A. Ecosystem-level controls on root-rhizosphere respiration. *New Phytol.* **2013**, *199*, 339–351. [CrossRef]
70. Brant, A.N.; Chen, H.Y. Patterns and Mechanisms of Nutrient Resorption in Plants. *Crit. Rev. Plant Sci.* **2015**, *34*, 471–486. [CrossRef]
71. Yan, X.; Wang, S.; Duan, Y.; Han, J.; Huang, D.; Zhou, J. Current and future distribution of the deciduous shrub *Hydrangea macrophylla* in China estimated by MaxEnt. *Ecol. Evol.* **2021**, *11*, 16099–16112. [CrossRef]
72. Xin, X.-F.; Nomura, K.; Aung, K.; Velásquez, A.C.; Yao, J.; Boutrot, F.; Chang, J.; Zipfel, C.; He, S.Y. Bacteria establish an aqueous living space in plants crucial for virulence. *Nature* **2016**, *539*, 524–529. [CrossRef]
73. Zhang, K.; Sun, L.; Tao, J. Impact of Climate Change on the Distribution of *Euscaphis japonica* (Staphyleaceae) Trees. *Forests* **2020**, *11*, 525. [CrossRef]
74. Li, Z.; Han, H.; You, H.; Cheng, X.; Wang, T. Effects of local characteristics and landscape patterns on plant richness: A multi-scale investigation of multiple dispersal traits. *Ecol. Indic.* **2020**, *117*, 106584. [CrossRef]
75. Swenson, N.G.; Anglada-Cordero, P.; Barone, J.A. Deterministic tropical tree community turnover: Evidence from patterns of functional beta diversity along an elevational gradient. *Proc. R. Soc. B Biol. Sci.* **2011**, *278*, 877–884. [CrossRef] [PubMed]
76. Li, X.; Zhang, L.; Luo, T. Rainy season onset mainly drives the spatiotemporal variability of spring vegetation green-up across alpine dry ecosystems on the Tibetan Plateau. *Sci. Rep.* **2020**, *10*, 8797. [CrossRef]
77. Ren, Z.S. The effects of climate on the growth of *Meconopsis* seedlings in Kunming. *Acta Bot. Yunnanica* **1993**, *15*, 110–112.
78. Still, S.; Kitto, S.; Swasey, J.; Harbage, J. Influence of Temperature on Growth and Flowering of Four *Meconopsis* Genotypes. *Acta Hortic.* **2003**, *620*, 289–298. [CrossRef]
79. Willig, M.R.; Kaufman, D.M.; Stevens, R.D. Latitudinal Gradients of Biodiversity: Pattern, Process, Scale, and Synthesis. *Annu. Rev. Ecol. Evol. Syst.* **2003**, *34*, 273–309. [CrossRef]
80. Behera, M.D.; Roy, P.S.; Panda, R.M. Plant species richness pattern across India's longest longitudinal extent. *Curr. Sci. India* **2016**, *111*, 1220–1225. [CrossRef]
81. O'Brien, E.M. Climatic Gradients in Woody Plant Species Richness: Towards an Explanation Based on an Analysis of Southern Africa's Woody Flora. *J. Biogeogr.* **1993**, *20*, 181–198. [CrossRef]
82. Chou, C. Land–sea heating contrast in an idealized Asian summer monsoon. *Clim. Dyn.* **2003**, *21*, 11–25. [CrossRef]
83. Carpenter, C. The environmental control of plant species density on a Himalayan elevation gradient. *J. Biogeogr.* **2005**, *32*, 999–1018. [CrossRef]

84. Zhang, Y.-Z.; Qian, L.-S.; Chen, X.-F.; Sun, L.; Sun, H.; Chen, J.-G. Diversity patterns of cushion plants on the Qinghai-Tibet Plateau: A basic study for future conservation efforts on alpine ecosystems. *Plant Divers.* **2021**, *44*, 231–242. [CrossRef]
85. Shrestha, N.; Su, X.; Xu, X.; Wang, Z. The drivers of high *Rhododendron* diversity in south-west China: Does seasonality matter? *J. Biogeogr.* **2018**, *45*, 438–447. [CrossRef]
86. Dong, S.K. Views on distinguishing the concepts of rangeland and grassland and proposing proper use of their terminology. *Chinese J. Ecol.* **2022**, *41*, 992–1000. [CrossRef]
87. Nagamatsu, D.; Hirabuki, Y.; Mochida, Y. Influence of micro-landforms on forest structure, tree death and recruitment in a Japanese temperate mixed forest. *Ecol. Res.* **2003**, *18*, 533–547. [CrossRef]
88. Palmer, M.W.; Dixon, P.M. Small-scale environmental heterogeneity and the analysis of species distributions along gradients. *J. Veg. Sci.* **1990**, *1*, 57–65. [CrossRef]
89. Guo, Q.; Kelt, D.A.; Sun, Z.; Liu, H.; Hu, L.-J.; Ren, H.; Wen, J. Global variation in elevational diversity patterns. *Sci. Rep.* **2013**, *3*, 03007. [CrossRef]
90. Manish, K.; Pandit, M.K.; Telwala, Y.; Nautiyal, D.C.; Koh, L.P.; Tiwari, S. Elevational plant species richness patterns and their drivers across non-endemics, endemics and growth forms in the Eastern Himalaya. *J. Plant Res.* **2017**, *130*, 829–844. [CrossRef]
91. Pandey, B.; Nepal, N.; Tripathi, S.; Pan, K.; Dakhil, M.A.; Timilsina, A.; Justine, M.F.; Koirala, S.; Nepali, K.B. Distribution Pattern of Gymnosperms' Richness in Nepal: Effect of Environmental Constrains along Elevational Gradients. *Plants* **2020**, *9*, 625. [CrossRef]
92. Zhang, W.; Huang, D.; Wang, R.; Liu, J.; Du, N. Altitudinal Patterns of Species Diversity and Phylogenetic Diversity across Temperate Mountain Forests of Northern China. *PLoS ONE* **2016**, *11*, e0159995. [CrossRef]
93. Miehe, G. Vegetation patterns on Mount Everest as influenced by monsoon and föhn. *Vegetatio* **1988**, *79*, 21–32. [CrossRef]
94. Gao, J.; Liu, Y. Climate stability is more important than water–energy variables in shaping the elevational variation in species richness. *Ecol. Evol.* **2018**, *8*, 6872–6879. [CrossRef]
95. Rai, H.; Upreti, D.K.; Gupta, R.K.; Rai, H. Diversity and distribution of terricolous lichens as indicator of habitat heterogeneity and grazing induced trampling in a temperate-alpine shrub and meadow. *Biodivers. Conserv.* **2012**, *21*, 97–113. [CrossRef]
96. González-Tagle, M.A.; Schwendenmann, L.; Pérez, J.J.; Schulz, R. Forest structure and woody plant species composition along a fire chronosequence in mixed pine–oak forest in the Sierra Madre Oriental, Northeast Mexico. *For. Ecol. Manag.* **2008**, *256*, 161–167. [CrossRef]
97. Zhang, Q.-P.; Wang, J.; Wang, Q. Effects of abiotic factors on plant diversity and species distribution of alpine meadow plants. *Ecol. Inform.* **2021**, *61*, 101210. [CrossRef]
98. Burt, T.P.; Butcher, D.P. Topographic controls of soil moisture distributions. *J. Soil Sci.* **1985**, *36*, 469–486. [CrossRef]
99. Dauber, J.; Hirsch, M.; Simmering, D.; Waldhardt, R.; Otte, A.; Wolters, V. Landscape structure as an indicator of biodiversity: Matrix effects on species richness. *Agric. Ecosyst. Environ.* **2003**, *98*, 321–329. [CrossRef]
100. Marden, M.; Rowan, D.; Phillips, C. Stabilising Characteristics of New Zealand Indigenous Riparian Colonising Plants. *Plant Soil* **2005**, *278*, 95–105. [CrossRef]
101. Gaston, K.J. Global patterns in biodiversity. *Nature* **2000**, *405*, 220–227. [CrossRef]
102. Tsianou, M.A.; Koutsias, N.; Mazaris, A.D.; Kallimanis, A.S. Climate and landscape explain richness patterns depending on the type of species' distribution data. *Acta Oecologica* **2016**, *74*, 19–27. [CrossRef]
103. Katayama, N.; Amano, T.; Naoe, S.; Yamakita, T.; Komatsu, I.; Takagawa, S.-I.; Sato, N.; Ueta, M.; Miyashita, T. Landscape Heterogeneity–Biodiversity Relationship: Effect of Range Size. *PLoS ONE* **2014**, *9*, e93359. [CrossRef] [PubMed]
104. Sun, L.; Luo, J.; Qian, L.; Deng, T.; Sun, H. The relationship between elevation and seed-plant species richness in the Mt. Namjagbarwa region (Eastern Himalayas) and its underlying determinants. *Glob. Ecol. Conserv.* **2020**, *23*, e01053. [CrossRef]
105. Li, J.J. The Seed Plants Differentiation and It's Response to the Uplifts in Himalayans. Ph.D. Thesis, Chengdu University of Technology, Chengdu, China, 2014.
106. Boral, D.; Moktan, S. Predictive distribution modeling of Swertia bimaculata in Darjeeling-Sikkim Eastern Himalaya using MaxEnt: Current and future scenarios. *Ecol. Process.* **2021**, *10*, 26. [CrossRef]

diversity

MDPI

Article

Biodiversity and Variations of Arbuscular Mycorrhizal Fungi Associated with Roots along Elevations in Mt. Taibai of China

Mengge Zhang [1,2,3,†], Mei Yang [1,2,3,†], Zhaoyong Shi [1,2,3,*], Jiakai Gao [1,2,3] and Xugang Wang [1,2,3]

1 College of Agriculture, Henan University of Science and Technology, Luoyang 471023, China
2 Luoyang Key Laboratory of Symbiotic Microorganism and Green Development, Luoyang 471023, China
3 Henan Engineering Research Center of Human Settlements, Luoyang 471023, China
* Correspondence: 9903105@haust.edu.cn
† These authors contributed equally to this work.

Abstract: (1) Background: environmental gradient strongly affects microbial biodiversity, but which factors drive the diversity of arbuscular mycorrhizal fungi (AMF) associated with roots at relatively large spatial scales requires further research; (2) Methods: an experiment on large spatial scales of Mt. Taibai was conducted to explore the biodiversity and drivers of AMF-associated with roots using high-throughput sequencing; (3) Results: a total of 287 operational taxonomic units (OTUs) belong to 62 species representing 4 identified and 1 unclassified order were identified along different altitudinal gradients. With increasing altitude, AMF colonization could be simulated by a quadratic function trend, and altitude has a significant impact on colonization. AMF alpha diversity, including the Sobs and Shannon indexes, tended to be quadratic function trends with increasing altitude. The highest diversity indices occurred at mid-altitudes, and altitude had a significant effect on them. AMF communities have different affinities with soil and root nutrient, and *Glomus* is most affected by soil and root nutrient factors through the analysis of the heatmap. *Glomus* are the most dominant, with an occurrence frequency of 91.67% and a relative abundance of 61.29% and 53.58% at the level of species and OTU, respectively. Furthermore, AMF diversity were mostly associated with soil and root nutrients; (4) Conclusions: in general, AMF molecular diversity is abundant in Mt. Taibai, and altitude and nutrient properties of soil and root are the main influencing factors on AMF diversity and distribution.

Keywords: arbuscular mycorrhizal fungi (AMF); diversity; community; altitude; mountain

Citation: Zhang, M.; Yang, M.; Shi, Z.; Gao, J.; Wang, X. Biodiversity and Variations of Arbuscular Mycorrhizal Fungi Associated with Roots along Elevations in Mt. Taibai of China. *Diversity* **2022**, *14*, 626. https://doi.org/10.3390/d14080626

Academic Editors: Michael Wink, Lin Zhang and Jinniu Wang

Received: 1 July 2022
Accepted: 4 August 2022
Published: 6 August 2022

1. Introduction

Mountain ecosystems are rich in species diversity, and the climatic gradients are obvious within a relatively short distance, so it provides more possibilities for the research of biodiversity [1]. The diversity and community distribution of plants and animals are most frequently investigated in mountain ecosystems because of the environmental gradients and slant characteristics on a small spatial scale [2]. Peters et al. (2016) reported that the species richness of nearly half of the plant and animal taxa showed a decreasing trend with increasing altitude while the other half showed hump-shaped or bimodal distribution patterns in Mt. Kilimanjaro [3]. However, most of the current studies have focused on plants and animals, ignoring the interaction between soil microorganisms and plants [4]. In addition, soils are believed to be exceptionally biodiverse parts of ecosystems [5]. As widespread mutualists, fungi are symbiotic with plant roots and affect the growth and distribution of plants [6,7], and the effect may be different in different environment gradients [8], which play an important ecological role in ecosystem functions.

As the most widespread mutualists, arbuscular mycorrhizal (AM) fungi can form symbionts with 80% of plant species [9–11], which play important ecological functions in maintaining ecosystem balance in all kinds of ecosystems [11,12]. Research showed that AM fungi promote root growth and have positive effects on aboveground plant productivity through direct and indirect interactions [2]. Furthermore, AMF diversity was a key factor in maintaining plant biodiversity and ecosystem function [13–15]. Moreover, studying AMF biodiversity and distribution is the basis for predicting the evolution and succession of mountain ecosystems [16]. At present, more and more attention has been paid to the study of AMF diversity in mountain ecosystems. Yang et al. reported that elevations had a significant effect on AMF diversity and community distribution in the Qinghai-Tibet Plateau [17]. Gough et al. suggested that AMF are a ubiquitous group of soil microorganisms [18]. However, the measurement results of AMF diversity might be different using different methods in the same region. For example, Shi et al. (2014) researched AMF diversity and identified 63 AMF belonging to 12 genera by the traditional morphological identification method in Mt. Taibai [15], while Zhang et al. (2021) found 103 AMF species from soil samples, which belong to 19 genera using molecular identification method in Mt. Taibai [19]. It can be suggested that more AMF taxa are identified by the molecular method. Therefore, in this study, high-throughput sequencing molecular methods were used to explore AMF diversity and distribution in plant roots and investigate its influencing factors.

As the intersection of the flora of North China, Central China, and West China, Qinling Mountain is the natural dividing line between North and South China, with abundant species and resources. As the main peak of Qinling Mountain, Mt. Taibai is dominated by forest landscapes, rich in biological species, and is known as a green pearl in Western China [20]. The plant species is very rich in Mt. Taibai, which is one of the most abundant plant species in the temperate zone in China. Due to the different climatic gradients and the particularity of the vertical distribution of vegetation along altitudes, Mt. Taibai has become a natural place to study biodiversity [21].

Therefore, this study used molecular identification methods to explore the diversity and distribution mechanism of AMF associated with roots at different altitudes in Mt. Taibai, aiming to determine the biodiversity and variations of AMF with altitudes in the mountain ecosystem. It is expected to enrich the ecological theory of AMF by providing supporting data on different altitudes of mountain ecosystems.

2. Materials and Methods

2.1. Description of Study Region

This study was conducted in Mt. Taibai (23°49′31″–34°08′11″ N and 107°41′23″–107°51′40″ E) of the Qinling Mountain, which lies in the ecological transition zone between the subtropical zone and the warm temperate zone and is an important east-west mountain range across the central part in China. As the main peak of the Qinling Mountains, Mt. Taibai is the first peak in the east of the Qinghai-Tibet Plateau in China, with the highest altitude of 3771.2 m. The climate zone of Mt. Taibai is obvious, which is divided into a temperate monsoon climate zone (800–1500 m), a cold temperate monsoon climate zone (1500–3000 m), a subarctic climate zone (3000–3350 m), and frigid climate zone (>3350 m). Besides, Mt. Taibai is rich in plant species and has a special geographic location, complex and diverse climate, and large altitude gradient, as well as one of the most abundant temperate plant species in China. In addition, the distribution of vegetation in the vertical zone of Mt. Taibai is also very special, which is of great significance to the study of the distribution of vegetation in the north and south of China and provides an ideal environment to conduct scientific research [22]. The plant distribution from the bottom to the top of Mt. Taibai can be divided into deciduous broad-leaved forest belt, coniferous forest belt, alpine shrub belt, and meadow belt. The deciduous broad-leaved forest belt is mainly distributed by *Quercus variorum* forest, *Quercus aliena* var. *Acuteserrata* forest and *Betula albo-sinensis* forest. The coniferous forest belt is mainly distributed by *Abies fabri* forest and *Larix gmelinii* forest. The alpine shrub belt and meadow belt are above 3350 m, mainly distributed dwarf creeping shrub, dwarf meadow, mossy community, and lichen community (Table S1). Meanwhile, it has always been a hotspot for biodiversity research.

2.2. Collection of Samples

Twelve different altitudes were selected within the range of 663–3511 m in Mt. Taibai. At every target altitude, three 20 m × 20 m sample squares were set up, and the distance between each sample square was at least 50 m. In each sample square, a five-point sampling method was used to collect 0–30 cm soil, including all the plant roots and soil, and they were mixed as one sample. Three squares were considered three replicates. Finally, we separated all the mixed roots from the soil and put the mixed root samples and soil into different sealed bags, respectively.

Soil samples were used to determine the physical properties and nutrient elements after air-drying. All plant root samples were divided into the following three parts. The first part of each plant root sample was immediately stored in a −80 °C freezer for DNA extraction. The second part of each plant root sample was transported to the laboratory to carry out the determination of AMF colonization. The third part was stored in a 4 °C refrigerator to determine the nutrient elements, such as C, N, P, and C/N.

2.3. Bioinformatics Analysis of Sequence Data

Genomic DNA was extracted from plant root samples using the Fast DNA SPIN Kit for Soil (MP Biomedicals LLC, Santa Ana, CA, USA) according to the manufacturer's protocols. The extracted DNA was subjected to nested PCR by a thermocycler PCR system (GeneAmp 9700, ABI, Foster City, CA, USA). PCR amplification was performed with primers AML1F (5′-ATCAACTTTCGATGGTAGG ATAGA-3′) and AML2R (5′-GAACCCAAACACTTTGGTTTCC-3′) by an ABI GeneAmp® 9700 PCR thermocycler (ABI, CA, USA).

Purified barcoded amplicons were pooled in equimolar concentrations and paired-end sequenced on an Illumina MiSeq PE300 platform/NovaSeq PE250 platform (Illumina, San Diego, CA, USA) according to the standard protocols by Majorbio Bio-Pharm Technology Co., Ltd. (Shanghai, China). Microbial community sequencing was conducted by Shanghai Majorbio Bio-pharm Technology using the Illumina-MiSeq sequencing platform. The data were analyzed on a free online platform (Majorbio I-Sanger Cloud Platform, available online: http://www.i-sanger.com, accessed on 3 August 2021). Used Uparse (version 7.1) software platform to perform taxonomic analysis of OTU representative sequences at a 97% similar level.

2.4. Measurement of AM Colonization and Parameters of Soil and Plant Roots

The colonization of AMF was determined according to the method of Phillips et al. [23]. First, selected fresh fine roots were cleaned and wiped dry, then immersed in a test tube with a mass fraction of 10% KOH solution, and heated in a water bath at 90 °C for 30 min until the roots became relatively transparent. When the roots were relatively transparent, the lye on the roots was cleaned and the roots were soaked in a 5% CH3COOH for 5 min. They were dyed with 5% volume fraction acid acetic ink and heated in a water bath at 90 °C for about 30 min. As the roots were fully dyed by the acetic acid ink, they were cleaned and then put in lactic acid for color separation for approximately 30 min. Finally, the roots were cut at about 1 cm and placed on a glass slide (each glass slide has 15 roots). The glass slide was observed under a Motic BA310 microscope at 100–400 times magnification to survey the colonization status.

The concentration of total carbon and nitrogen were measured by an elemental analyzer (GC IsolinkFlash 2000; Thermo Scientific, Waltham, MA, USA) analyzer. The phosphorus content in plant roots is determined by the molybdenum-antimony colorimetric method [24].

2.5. Calculation of AM Colonization, Relative Abundance, and Occurrence Frequency

The percentage of root length colonized by AM fungal structures was estimated and calculated according to the grading criteria of Trouvelot et al. [25].

The relative abundance (species/OTU) of AM fungal genus was calculated as the percentage of the number of species/OTUs in each genus divided by the total number of species/OTUs in all genera, then take the average of the three samples.

The occurrence frequency of AM fungal genus was defined as the percentage of the number of samples where this genus was observed to the number of all samples in this genus.

Shannon–Weiner index: the P_i of AMF species or OTUs was defined as the percentage of the sequences for each species or OTUs detected to total species or OTUs sequences in a sample;

$$H = -\sum [P_i \log_2 (P_i)]$$

The Sobs index of AMF species or OTUs was defined as the numbers of species or OTUs in a sample.

2.6. Statistics and Analysis of Data

The colonization ratio, colonization density, AMF diversity indies including Sobs and Shannon index, stoichiometric characteristics of plant roots, and edaphic factors were all statistically analyzed and curved estimated by SPSS 25.0. Then made the scatter charts through Origin 21.0. AMF relative abundance and occurrence frequency were analyzed based on genus level by Excel 2019. The heat map was presented to explore the relationships between the AMF community and environmental variables based on correlation analyses. And it is conducted on MajorBio Cloud's bioinformatics analysis cloud platform. Based on multiple linear regression analysis of the relationship between altitude, environmental factors, and AMF diversity.

3. Results

3.1. Arbuscular Mycorrhizal Colonization in Plant Roots at Different AltitudesSubsection

The colonization of AM varied from 0 to 100%, with an average of 55.03% from 663 m to 3511 m (Figure 1a). AM colonization showed a quadratic function trend with R^2 was 53.46% and P was less than 0.01. It suggested that altitude has a significant effect on AM colonization. The highest colonization of AM occurred at 1170 m and 1450 m. Meanwhile, the highest colonization density also occurred at 1170 m with 41.09% (Figure 1b). And the colonization density changed from 0 to 41.09%, with an average of 10.39% in Mt. Taibai. Besides, the colonization density formed the quadratic function with R^2 was 34.18%. Elevation had a significant effect on AM colonization density.

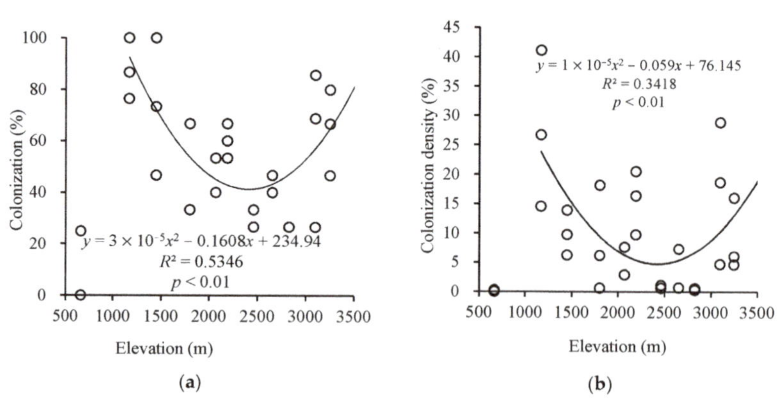

Figure 1. Change of arbuscular mycorrhizal colonization (**a**) and colonization density (**b**) in plant roots among different altitudes. Note: There are overlapping data points in (**a**,**b**).

3.2. AMF Community Composition and Distribution at Different Altitudes

A total of 287 OTUs belong to 62 species belonging to 8 identified and 1 unclassified genus representing four identified and one unidentified order (Table S2). Among them, *Glomus* was the dominant genus with the largest number of species and OTUs with 39 and 104, respectively. However, only 1 OTU was identified in the genus of *Pacispora*. In addition, the maximum number of species and OTU occurred at the altitude of 2190 m, with 38 and 166, respectively. While the minimum number of species occurred at 3097 m, and the minimum number of OTU was 25 and occurred at 2828 m.

The highest relative abundance of *Glomus* was 99.14%, which occurred at 1800 m (Figure 2). In addition, the relative abundance of *Glomus* exceeded 90% at the altitudes of 663 m, 1170 m, 1450 m, 1800 m, and 2070 m. At the higher altitudes of 2828 and 3250 m, the relative abundance of *Acaulospora* was higher with 49.14% and 56.52%. Besides, the genus of *Pacispora* only appeared above 2000 m and the relative abundance was the highest with 17.85% at the highest altitude of 3511 m.

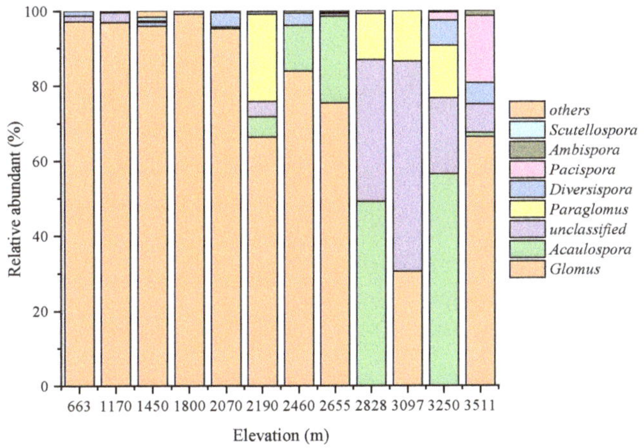

Figure 2. Relative abundance of genera identified based on different altitudes at the genus level. Note: The genus, whose abundance was less than 1%, were collectively classified as others.

3.3. Diversity of AMF in Mt. Taibai

AMF alpha diversity is expressed by the Sobs and Shannon indices (Figure 3). Both Sobs and Shannon indices showed the trends of increased first and then decreased with the increase of altitude based on the species and OTU level. Whether at species or OTU level, the highest Sobs and Shannon index both occurred at 2460 m. The highest Sobs index were 26 and 49 on the level of species and OTU, respectively (Figure 3a). At the same time, the highest Shannon indices were 2.68 and 2.08 at the level of species and OTU (Figure 3b). And the lowest alpha diversity indices both appeared at higher altitudes. With the increase in altitude, the changing trend of Sobs and Shannon indices could be simulated with quadratic function at the species and OTU level. What's more, elevation had a significant effect on alpha diversity indices.

AMF beta diversity was revealed by nonmetric multidimensional scaling analysis (NMDS) (Figure 4). The NMDS ordination resulted in a final stress value of 0.14 and 0.18 on species and OTU level, respectively. The results indicated that beta diversity also differed among the altitudes based on species and OTU level in Mt. Taibai.

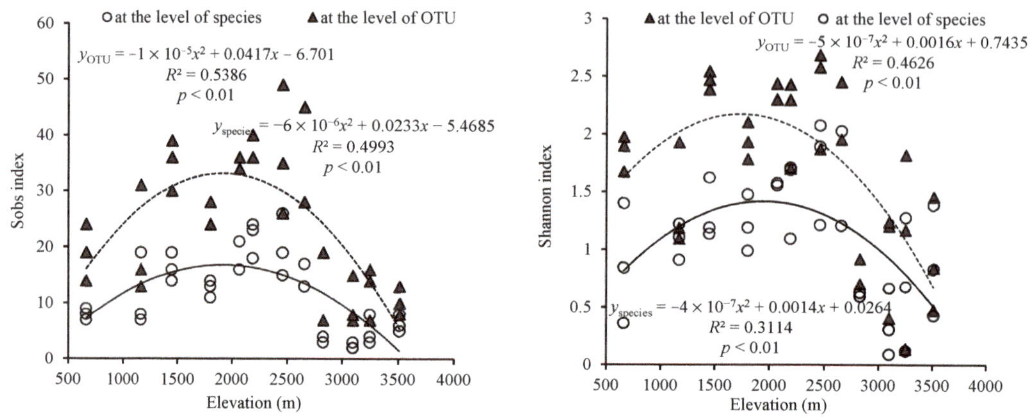

(a) (b)

Figure 3. The variation of AMF is based on the species and OTU level among different elevations with the Sobs index of species (**a**) and the Shannon index of species (**b**).

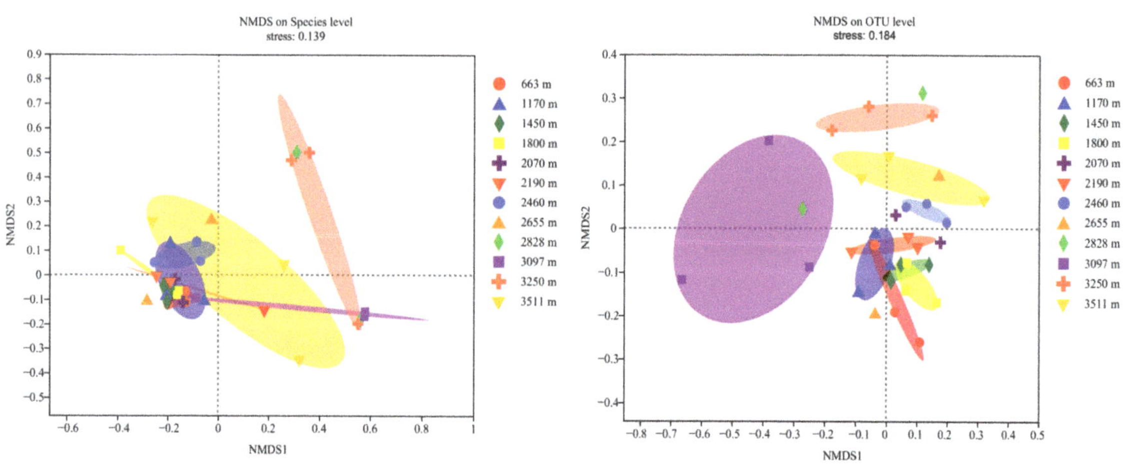

(a) (b)

Figure 4. Nonmetric multidimensional scaling (NMDS) ordination of symbiosis AMF community composition of plant roots at different altitudes in Mt. Taibai ((**a**,**b**) stands for the species and OTU levels, respectively).

3.4. The Relative Abundance and Occurrence Frequency of AMF Genus in Mt. Taibai

It was found that the relative abundance ranged from 0.76% to 61.29% and from 0.24% to 53.58% based on species and OTU levels, respectively (Table 1). The fungi in the genus of *Glomus* were the most dominant, with the highest relative abundance of 61.29% and 53.58% based on species and OTU level, which was significantly higher than other genera. Meanwhile, the occurrence frequency of *Glomus* also was higher at 91.67%. *Unclassified in Glomeromycetes* was found in all altitudes with the highest occurrence frequency of 100%. The second abundant genus is *Acaulospora*, with a relative abundance of 9.46% and 5.32% at species and OTU levels, and the occurrence frequency was 66.67%. At the same time, the relative abundance of *Pacispora* was the same as *Ambispora* based on species level with 2.18%. In addition, the occurrence frequency of *Pacispora* was the same as *Scutellospora* with 33.33%, and the occurrence frequency of *Diversispora*, *Acaulospora*, and *Paraglomus* were the same with 66.67%.

Table 1. Relative abundance and occurrence frequency of AMF genus in Mt. Taibai.

Genus	Relative Abundance (Species)/%	Relative Abundance (OTU)/%	Occurrence Frequency/%
Acaulospora	9.46b	5.32	66.67
Ambispora	2.18	1.09	50.00
Archaeospora	5.72	1.82	41.67
Diversispora	4.36	3.58	66.67
Glomus	61.29	53.58	91.67
Pacispora	2.18	0.72	33.33
Paraglomus	4.96	5.45	66.67
Scutellospora	2.57	0.65	33.33
no rank	0.81	0.34	8.33
unclassified in Diversisporaceae	0.76	0.24	8.33
unclassified in Archaeosporales	2.18	0.85	16.67
unclassified in Diversisporales	1.36	0.37	16.67
unclassified in Glomeromycetes	2.18	25.97	100.00

3.5. The Drive Factors of AMF Community and Diversity

Among different genera, *Glomus* was most affected by soil and root nutrient factors (Figure 5). Both soil factors (pH and C/N) and root nutrients (C, N, and C/N) had a significant effect on *Glomus* through correlation analysis by heatmap at the genus level (Figure 5). *Unclassified in Diversisporaceae* and *unclassified in Archaeosporales* were both affected by root nutrient of N ($p < 0.01$) and ecological stoichiometry of C/N ($p < 0.01$). As for *unclassified in Glomeromycetes*, it was influenced by C/N ($p < 0.01$) and available phosphorus ($p < 0.05$) in soil. And soil and root nutrients had no significant effect on other AMF genera.

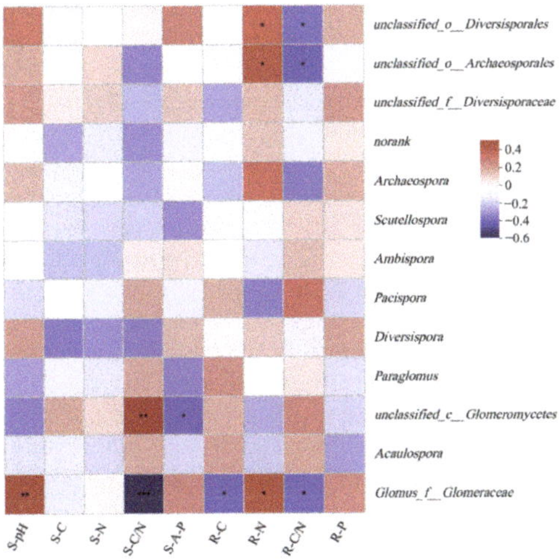

Figure 5. Influence of environmental factors on AMF genus. **Note:** *, **, *** indicate significant correlation at $p < 0.05$, $p < 0.01$, $p < 0.001$ confidence level, respectively.

Elevation had a positive and prominent effect on AMF Shannon diversity ($r = 0.493$) based on OTU level (Figure 6). Meanwhile, elevation had a positive effect on soil factors of pH ($r = 0.651$) and ecological stoichiometry of C/N ($r = 0.605$), plant factors, such as C ($r = 0.364$), N ($r = 0.379$) and C/N ($r = 0.345$). Besides, the AMF diversity index of Shannon was greatly affected by soil and root nutrients, while the Sobs index was affected by soil and root ecological stoichiometry C/N and root N content.

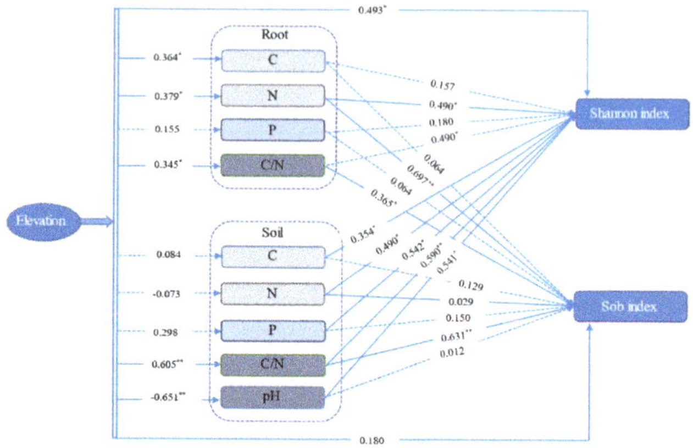

Figure 6. The relationships among elevation, soil factors (C, N, P, C/N, and pH), plant factors (C, N, P, and C/N), and AMF diversity indices are based on OTU level. Note: The effect of altitude on soil and root is expressed as a correlation coefficient, while the effects of soil and root factors on diversity indices are represented by standard regression coefficients. Blue solid represents significant positive or negative effects. Blue dashed represent nonsignificant paths. ** means $p < 0.01$; * means $p < 0.05$, respectively.

4. Discussion

The changes and laws of biodiversity along different environmental gradients are the important topic of biodiversity research [26,27]. Many environmental factors vary with altitude in mountain systems, so altitude is often used as an integrated factor to study plant and animal distribution patterns in mountain systems. In recent years, people have become more and more interested in knowing how microbes respond to the changes in environmental conditions because of their critical role in ecosystem functions [28–30]. Luo et al. also suggested that understanding the diversity of fungi in ecosystems may have predictive implications for biodiversity and ecosystem evolution processes [31]. Therefore, AMF diversity and community distribution along different altitudes were studied to explore the role of AMF in the mountain ecosystem and the responses to clime change.

In this study, the colonization rate and colonization density of AMF showed a trend of first increasing, then decreasing trends with the elevation. However, Gai et al. and Kotilinek et al. believed that AMF colonization showed a downward trend with increasing altitudes [32,33]. There are even studies that there were no significant differences in AMF colonization between high-altitude and low-altitude areas in the southeast of the Qinghai-Tibet Plateau [34]. The different results may be due to the differences in research sites or environmental factors, such as plant species, soil types, and so on. Liu showed that the arbuscular abundance of AMF was significantly influenced by altitude gradients [35], which was consistent with the results of our research. These different results also suggest that AMF could form a good symbiotic relationship with plant roots in the mountain ecosystem.

In the present study, the 287 OTUs and 62 species of AMF were identified and represented 8 identified genera, 5 unidentified genera in Mt. Taibai, which supported that AMF had a wide ecological range and was an important part of the ecosystem. Our research showed that 39 species belonged to the genus of *Glomus*, followed by *Acaulospora* with 5 species. Whether at the species or OTU levels, the relative abundance and the occurrence frequency of *Glomus* were the highest, which was consistent with the conclusions of most previous studies on the molecular diversity of AMF that *Glomus* was the dominant genus in the AMF community [19,36,37]. This may be due to its wide ecological range and the certain resistance in complex environments [38]. Besides, *Glomus* can usually produce large numbers of spores and hypha fragments, which can extensively spread and colonize the roots of plants [37,39]. In terms of different altitudes, *Glomus* dominated at lower altitudes, whereas *Acaulospora* were more abundant at the higher altitudes of 3250–3511 m. This result was consistent with Oehl et al., who suggested that the genus of *Acaulospora* was more abundant in the highlands than in the lowlands in Switzerland [40]. Moreover, Haug et al. and Yang et al. also came to similar conclusions [8,41]. The different distribution at low and high altitudes in AMF explains the correlation of AMF species with altitude and suggests that there may be potential niche differentiation along the altitudinal gradient.

In addition to different distribution, the diversity of AMF was also different with the altitude change. It was discovered that AMF diversity indices of Sobs and Shannon showed the trends of quadratic function increasing first and then decreasing, whether on the level of species or OTU. However, Guo et al. and Egan et al. studied the AMF diversity and suggested that the alpha diversity decreased monotonically with the increase in altitude [21,42]. The reason for this phenomenon may be that this study was conducted in a large-scale altitude range of 663–3511 m, while Guo et al. and Egan et al. explored the AMF diversity in a relatively small altitude range. Therefore, in this study, the AMF diversity varies in different climatic environments. Moreover, some studies have proved that AMF diversity is closely related to plant richness [43], and plant richness is also different on different altitude gradients in Mt. Taibai (Table S1). This may also be the reason why AMF diversity shows different trends with increasing altitude in Mt. Taibai.

In this study, it was also found that the higher Shannon and Sobs indices appeared at mid-altitudes, whether based on species or OTU level. The highest diversity indices occurred at 2460 m, and altitude has a significant effect on them. Bonfim et al. showed that AMF diversity at higher altitudes was higher than at lower altitudes in the Atlantic forest system [44]. The reason for this phenomenon may be that the mid-altitudes have less human disturbance than the low altitude region and a less extreme climate environment than the high altitude region [45]. Therefore, the mid-elevation area is favorable for AMF sporulation and growth. Gai et al. and Shi et al. supported that altitude has no significant influence on AMF diversity [32,39]. Because altitude is a comprehensive factor, the effects of altitude on AMF diversity may be caused by differences in geographic location and environmental factors [15,46,47]. Besides, previous studies have shown that environmental factors, especially the geographical environment and soil factors, have an important impact on AMF diversity. Different ecological factors would affect the growth, development, colonization, and reproduction of AMF, which would cause differences in AMF diversity in different ecosystems [48–51]. Therefore, it is necessary to study the environmental factors of different altitudes further.

Determination of soil and plant nutrients found a significant impact on AMF diversity indices of Sob and Shannon. Our results were consistent with previous research conclusions that altitudes and soil variables had a significant impact on AMF diversity and richness [36]. Besides, Montiel-Rozas et al. [52] showed that AMF diversity and richness were only affected by soil properties, and soil factors were the main driving force for AM fungal communities. Our research found that whether it was in soil or plant roots, P concentration significantly affects the AMF Shannon index. This result was consistent with Maitra et al., who confirmed that the AMF Shannon diversity index showed a positive response to P [53]. Ceulemans et al. showed that AMF diversity decreased with the increase in soil P utilization [54]. Therefore, the change of P concentration is an important predictor of the response of AMF diversity to soil nutrients [55]. Previous studies have also shown that the addition of N increased AMF diversity in N-deficient soil [56–58], which was consistent with our study that N content of plant roots has a significant effect on AMF diversity, but N content in the soil had no effect on it. This result showed that plant roots had a greater impact on AMF diversity than soil. This also suggests that AMF tends to be symbiotic with plants to absorb nutrients, thereby increasing the diversity of mycorrhiza to increase the plant's own competitive advantage [59]. In addition, studies have suggested that the identity of the host plant has been considered to be one of the most important factors in shaping AMF community composition [60–62]. It was speculated that the vital effects of the host plant on the AMF community might be related to the C/N in plant roots.

Moreover, it was found that the AMF community have different affinities with soil and root nutrient. And glomus is most affected by soil and root nutrient factors through the analysis of Heatmap. The genus of *unclassified in Diversisporales* and *unclassified in Archaeosporales* were significantly affected by root N concentration and C/N. However, the genus of unclassified *Glomeromycetes* were correlated with soil P concentration and C/N. The results revealed that there were different relationships between soil characteristics and the AMF genus, which is consistent with Kim et al. [63]. These also suggested that AMF taxa have different environmental preferences in tropical montane rainforests. Therefore, this also explained the different distribution of AMF communities at different altitudes. Besides, soil nutrient concentration also is an important factor in the AMF community. Zhao et al. [37] confirmed that soil nutrients have an impact on AMF communities, as a lack of nutrients inhibits spore germination and dissociation. Therefore, soil factors play an important role in AMF diversity and community in the mountain ecosystem.

5. Conclusions

In this study, the biodiversity and variation of AMF with plant roots at different altitudes were explored by molecular identification methods in Mt. Taibai. And it was found that there is abundant AMF diversity in the mountain ecosystem. Altitude has a significant impact on AMF diversity and community distribution. Whether it is species level or out level, Sob and Shannon indices show a quadratic equation changing trends with the increase of altitudes. In addition, whether in soil or plant roots, the ecological stoichiometry of C/N has a major impact on AMF diversity. Further, *Glomus*, as a dominant genus, was most affected by root and soil nutrients. These findings suggest that soil and root nutrient are important factors affecting AMF diversity and variation among different altitudes in the mountain forest.

Supplementary Materials: The following supporting information can be downloaded at: https://www.mdpi.com/article/10.3390/d14080626/s1, Table S1: Plant species at different altitudes in Mt. Taibai. Table S2: The distribution of AMF order, family, genus, and the number of AMF species aoutOTU among different altitudes.

Author Contributions: Conceptualization, M.Z., M.Y. and Z.S.; methodology, M.Z., M.Y. and Z.S.; software, M.Z.; validation, M.Z., M.Y., Z.S. and J.G.; formal analysis, M.Z.; investigation, M.Z., Z.S., J.G. and X.W.; resources, M.Z. and Z.S.; data curation, M.Z. and Z.S.; writing—original draft preparation, M.Z. and M.Y.; writing—review and editing, M.Z. and Z.S.; visualization, M.Z., Z.S. and X.W.; supervision, Z.S.; project administration, Z.S.; funding acquisition, Z.S. All authors have read and agreed to the published version of the manuscript.

Funding: This research was funded by NSFC (32171620, 31670499), Scientific and technological research projects in Henan province (192102110128), Program for Science & Technology Innovation Talents in Universities of Henan Province (18HASTIT013), Key Laboratory of Mountain Surface Processes and Ecological Regulation, CAS (20160618).

Institutional Review Board Statement: Not applicable.

Informed Consent Statement: Not applicable.

Data Availability Statement: The datasets presented in this study can be found in online repositories. The names of the repository/repositories and accession number(s) can be found below: SRA database and PRJNA843859.

Conflicts of Interest: The authors declare no conflict of interest.

References

1. Deng, W.; Wang, J.L.; Scott, M.B.; Fang, Y.H.; Liu, S.R.; Yang, X.Y.; Xiao, W. Sampling methods affect nematode-trapping fungi biodiversity patterns across an elevational gradient. *BMC Microbiol.* **2020**, *20*, 15. [CrossRef]
2. Wagg, C.; Husband, B.C.; Green, D.S.; Massicotte, H.B.; Peterson, R.L. Soil microbial communities from an elevational cline differ in their effect on conifer seedling growth. *Plant Soil* **2011**, *340*, 491–504. [CrossRef]
3. Peters, M.K.; Hemp, A.; Appelhans, T.; Behler, C.; Classen, A.; Detsch, F.; Ensslin, A.; Ferger, S.W.; Frederiksen, S.B.; Gebert, F.; et al. Predictors of elevational biodiversity gradients change from single taxa to the multi-taxa community level. *Nat. Commun.* **2016**, *7*, 13736. [CrossRef]
4. Bryant, J.A.; Lamanna, C.; Morlon, H.; Kerkhoff, A.J.; Enquist, B.J.; Green, J.L. Microbes on mountain sides: Contrasting elevational patterns of bacterial and plant diversity. *Proc. Natl. Acad. Sci. USA* **2008**, *105*, 11505–11511. [CrossRef] [PubMed]
5. Fitter, A.H. Darkness visible: Reflections on underground ecology. *J. Ecol.* **2005**, *93*, 231–243. [CrossRef]
6. Jing, X.; Sanders, N.J.; Shi, Y.; Chu, H.Y.; Classen, A.T.; Zhao, K.; Chen, L.T.; Shi, Y.; Jiang, Y.X.; He, J.S. The links between ecosystem multifunctionality and above-and belowground biodiversity are mediated by climate. *Nat. Commun.* **2015**, *6*, 8159. [CrossRef] [PubMed]
7. Bauer, J.T.; Koziol, L.; Bever, J.D. Local adaptation of mycorrhizae communities changes plant community composition and increases aboveground productivity. *Oecologia* **2020**, *192*, 735–744. [CrossRef]
8. Yang, W.; Zheng, Y.; Gao, C.; Duan, J.C.; Wang, S.P.; Guo, L.D. Arbuscular mycorrhizal fungal community composition affected by original elevation rather than translocation along an altitudinal gradient on the Qinghai-Tibet Plateau. *Sci. Rep.* **2016**, *6*, 36606. [CrossRef]
9. Smith, S.E.; Read, D. Mycorrhizal Symbiosis. *Q. Rev. Biol.* **2008**, *3*, 273–281. [CrossRef]
10. Davison, J.; Moora, M.; Opik, M.; Adholeya, A.; Ainsaar, L.; Ba, A.; Burla, S.; Diedhiou, A.G.; Hiiesalu, I.; Jairus, T.; et al. Global assessment of arbuscular mycorrhizal fungus diversity reveals very low endemism. *Science* **2015**, *349*, 6251. [CrossRef] [PubMed]
11. Shi, Z.Y.; Zhang, J.C.; Lu, S.C.; Li, Y.; Wang, F.Y. Arbuscular mycorrhizal fungi improve the performance of sweet sorghum grown in a mo-contaminated soil. *J. Fungi* **2020**, *6*, 44. [CrossRef] [PubMed]
12. Qiang, W.; He, X.; Wang, J.; Zhao, L. Temporal and spatial variation of arbuscular mycorrhizal fungi under the canopy of Hedysarum scoparium in the northern desert, China. *Appl. Soil Ecol.* **2019**, *136*, 139–147. [CrossRef]
13. van der Heijden, M.G.A.; Klironomos, J.N.; Ursic, M.; Moutoglis, P.; Streitwolf-engel, R.; Boller, T.; Wiemken, A.; Sanders, L.R. Mycorrhizal fungal diversity determines plant biodiversity, ecosystem variability and productivity. *Nature* **1998**, *396*, 69–72. [CrossRef]
14. van der Heijden, M.G.A.; Bardgett, R.D.; van Straalen, N.M. The unseen majority: Soil microbes as drivers of plant diversity and productivity in terrestrial ecosystems. *Ecol. Lett.* **2008**, *11*, 296–310. [CrossRef]
15. Shi, Z.Y.; Wang, F.Y.; Zhang, K.; Chen, Y.L. Diversity and distribution of arbuscular mycorrhizal fungi along altitudinal gradients in Mount Taibai of the Qinling Mountains. *Can. J. Microbiol.* **2014**, *60*, 811–818. [CrossRef]
16. Li, X.L.; Xu, M.; Christie, P.; Li, X.L.; Zhang, J.L. Large elevation and small host plant differences in the arbuscular mycorrhizal communities of montane and alpine grasslands on the Tibetan Plateau. *Mycorrhiza* **2018**, *28*, 605–619. [CrossRef] [PubMed]
17. Yang, M.; Shi, Z.Y.; Mickan, B.S.; Zhang, M.G.; Cao, L.B. Alterations to arbuscular mycorrhizal fungal community composition is driven by warming at specific elevations. *Peer J.* **2021**, *9*, e11792. [CrossRef]
18. Gough, E.C.; Owen, K.J.; Zwart, R.S.; Thompson, J.P. A systematic review of the effects of arbuscular mycorrhizal fungi on root-lesion nematodes, *Pratylenchus* spp. *Front. Plant Sci.* **2020**, *11*, 923. [CrossRef]
19. Zhang, M.G.; Shi, Z.Y.; Yang, M.; Lu, S.C.; Wang, X.G. Molecular diversity and distribution of arbuscular mycorrhizal fungi at different elevations in Mt. Taibai of Qinling Mountain. *Front. Microbiol.* **2021**, *12*, 609386. [CrossRef] [PubMed]
20. Shuai, L.Y.; Ren, C.L.; Yan, W.B.; Song, Y.L.; Zeng, Z.G. Different elevational patterns of rodent species richness between the southern and northern slopes of a mountain. *Sci. Rep.* **2017**, *7*, 8743. [CrossRef]
21. Guo, Y.X.; Ren, C.J.; Yi, J.J.; Doughty, R.; Zhao, F.Z. Contrasting responses of rhizosphere bacteria, fungi and arbuscular mycorrhizal fungi along an elevational gradient in a temperate montane forest of China. *Front. Microbiol.* **2020**, *11*, 2042. [CrossRef] [PubMed]
22. Zhao, F.Z.; Feng, X.X.; Guo, Y.X.; Ren, C.J.; Wang, J.; Doughty, R. Elevation gradients affect the differences of arbuscular mycorrhizal fungi diversity between root and rhizosphere soil. *Agric. For. Meteorol.* **2020**, *284*, 107894. [CrossRef]
23. Phillips, J.M.; Hayman, D.S. Improved procedures for clearing roots and staining parasitic and vesicular-arbuscular mycorrhizal fungi for rapid assessment of infection. *Trans. Br. Mycol. Soc.* **1970**, *55*, 158–161. [CrossRef]
24. Bao, S.D. *Soil Agrochemical Analysis*; China Agricultural Press: Beijing, China, 2000; pp. 176–185.
25. Koske, R.E.; Tessier, B. A convenient, permanent slide mounting medium. *Mycol. Soc. Am. Newsl.* **1983**, *34*, 59.
26. Bhople, P.; Samad, A.; Sisic, A.; Antonielli, L.; Sessitsch, A.; Keiblinger, K.; Djukic, I.; Zehetner, F.; Zechmeister-Boltenstern, S.; Joergensen, R.G.; et al. Variations in fungal community structure along elevation gradients in contrasting Austrian Alpine ecosystems. *Appl. Soil Ecol.* **2022**, *177*, 104508. [CrossRef]
27. Xu, D.P.; Kong, H.J.; Yang, E.J.; Wang, Y.; Li, X.R.; Sun, P.; Jiao, N.Z.; Lee, Y.J.; Jung, J.Y.; Cho, K.H. Spatial dynamics of active microeukaryotes along a latitudinal gradient: Diversity, assembly process, and co-occurrence relationships. *Environ. Res.* **2022**, *212*, 113234. [CrossRef] [PubMed]

28. Cotton, T.E.A. Arbuscular mycorrhizal fungal communities and global change: An uncertain future. *FEMS Microbiol. Ecol.* **2018**, *94*, 14. [CrossRef] [PubMed]
29. Lu, Y.W.; Liu, X.; Zhou, S.R. Nitrogen addition altered the plant-arbuscular mycorrhizal fungi network through reducing redundant interactions in an alpine meadow. *Soil Biol. Biochem.* **2022**, *171*, 108727. [CrossRef]
30. Thangavel, P.; Anjum, N.A.; Muthukumar, T.; Sridevi, G.; Vasudhevan, P.; Maruthupandian, A. Arbuscular mycorrhizae: Natural modulators of plant-nutrient relation and growth in stressful environments. *Arch. Microbiol.* **2022**, *204*, 264. [CrossRef] [PubMed]
31. Luo, L.; Guo, M.; Wang, E.; Yin, C.Y.; Wang, Y.J.; He, H.L.; Zhao, C.Z. Effects of mycorrhiza and hyphae on the response of soil microbial community to warming in eastern Tibetan Plateau. *Sci. Total Environ.* **2022**, *837*, 155498. [CrossRef] [PubMed]
32. Gai, J.P.; Tian, H.; Yang, F.Y.; Christie, P.; Li, X.L.; Klironomos, J.N. Arbuscular mycorrhizal fungal diversity along a Tibetan elevation gradient. *Pedobiologia* **2012**, *55*, 145–151. [CrossRef]
33. Kotilinek, M.; Hiiesalu, I.; Kosnar, J.; Smilauerov, M.; Smilauer, P.; Altman, J.; Dvorsky, M.; Kopecky, M.; Dolezal, J. Fungal root symbionts of high-altitude vascular plants in the Himalayas. *Sci. Rep.* **2017**, *7*, 6562. [CrossRef] [PubMed]
34. Li, X.L.; Xu, M.; Li, X.L.; Christie, P.; Wagg, C.; Zhang, J.L. Linkages between changes in plant and mycorrhizal fungal community composition at high versus low elevation in alpine ecosystems. *Environ. Microbiol. Rep.* **2020**, *12*, 229–240. [CrossRef] [PubMed]
35. Liu, M. Research on the Arbuscular Mycorrhizal Fungal Diversity in the Rhizosphere of *Clematis fruticosa* and the Mechanism of Mycorrhizal Seedlings Response to Drought Stress. Ph.D. Thesis, Inner Mongolia Agricultural University, Hohhot, China, 2017.
36. Li, X.L.; Gai, J.P.; Cai, X.B.; Li, X.L.; Christie, P.; Zhang, F.S.; Zhang, J.L. Molecular diversity of arbuscular mycorrhizal fungi associated with two co-occurring perennial plant species on a Tibetan altitudinal gradient. *Mycorrhiza* **2014**, *24*, 95–107. [CrossRef] [PubMed]
37. Zhao, H.; Li, X.Z.; Zhang, Z.M.; Zhao, Y.; Yang, J.T.; Zhu, Y.W. Species diversity and drivers of arbuscular mycorrhizal fungal communities in a semi-arid mountain in China. *Peer J.* **2017**, *5*, e4155. [CrossRef]
38. Camenzind, T.; Hempel, S.; Homeier, J.; Horn, S.; Velescu, A.; Wilcke, W.; Rillig, M.C. Nitrogen and phosphorus additions impact arbuscular mycorrhizal abundance and molecular diversity in a tropical montane forest. *Glob. Change Biol.* **2014**, *20*, 3646–3659. [CrossRef]
39. Shi, Z.Y.; Yin, K.J.; Wang, F.Y.; Mickan, B.S.; Wang, X.G.; Zhou, W.L.; Li, Y.J. Alterations of arbuscular mycorrhizal fungal diversity in soil with elevation in tropical forests of China. *Diversity* **2019**, *11*, 181. [CrossRef]
40. Oehl, F.; Sykorova, Z.; Redecker, D.; Wiemken, A.; Sieverding, E. Acaulospora alpina, a new arbuscular mycorrhizal fungal species characteristic for high mountainous and alpine regions of the Swiss Alps. *Mycologia* **2006**, *98*, 286–294. [CrossRef]
41. Haug, I.; Setaro, S.; Suarez, J.P. Species composition of arbuscular mycorrhizal communities changes with elevation in the Andes of south Ecuador. *PLoS ONE* **2019**, *14*, e0221091. [CrossRef]
42. Egan, C.P.; Callaway, R.M.; Hart, M.M.; Pither, J.; Klironomos, J. Phylogenetic structure of arbuscular mycorrhizal fungal communities along an elevation gradient. *Mycorrhiza* **2017**, *27*, 273–282. [CrossRef]
43. Hiiesalu, I.; PärtelM, D.J.; Gerhold, P.; Metsis, M.; Moora, M.; Öpik, M.; Vasar, M.; Zobel, M.; Wilson, S.D. Species richness of arbuscular mycorrhizal fungi: Associations with grassland plant richness and biomass. *New Phytol.* **2014**, *203*, 233–244. [CrossRef] [PubMed]
44. Bonfim, J.A.; Vascon, R.L.F.; Gumiere, T.; Mescolotti, D.D.L.C.; Oehl, F.; Cardoso, E.J.B.N. Diversity of arbuscular mycorrhizal fungi in a Brazilian atlantic forest toposequence. *Microb. Ecol.* **2016**, *71*, 164–177. [CrossRef]
45. Moora, M.; Davison, J.; Opik, M.; Metsis, M.; Saks, U.; Jairus, T.; Vasar, M.; Zobel, M. Anthropogenic land use shapes the composition and phylogenetic structure of soil arbuscular mycorrhizal fungal communities. *FEMS Microbiol. Ecol.* **2014**, *90*, 609–621. [CrossRef] [PubMed]
46. Geml, J. Altitudinal gradients in mycorrhizal symbioses. In *Biogeography of Mycorrhizal Symbiosis. Ecological Studies (Analysis and Synthesis)*; Tedersoo, L., Ed.; Springer International Publishing: Berlin/Heidelberg, Germany, 2017; Volume 230, pp. 107–123. [CrossRef]
47. Tunnisa, R.; Ezawa, T. Nestedness in arbuscular mycorrhizal fungal communities in a volcanic ecosystem: Selection of disturbance-tolerant fungi along an elevation gradient. *Microbes. Environ.* **2019**, *34*, 327–333. [CrossRef] [PubMed]
48. Zhang, X.M.; Chen, B.D.; Yin, R.B.; Xing, S.P.; Fu, W.; Wu, H.; Hao, Z.P.; Ma, Y.B.; Zhang, X. Long-term nickel contamination increased soil fungal diversity and altered fungal community structure and co-occurrence patterns in agricultural soils. *J. Hazard. Mater.* **2022**, *436*, 129113. [CrossRef]
49. Ma, X.C.; Geng, Q.H.; Zhang, H.G.; Bian, C.Y.; Chen, H.Y.H.; Jiang, D.L.; Xu, X. Global negative effects of nutrient enrichment on arbuscular mycorrhizal fungi, plant diversity and ecosystem multifunctionality. *New Phytol.* **2021**, *229*, 2957–2969. [CrossRef]
50. Jia, Y.Y.; Zhang, T.; Walder, F.; Sun, Y.; Shi, Z.Y.; Wagg, C.; Tian, C.Y.; Feng, G. Can mycorrhizal fungi alleviate plant community instability caused by increased precipitation in arid ecosystems? *Plant Soil* **2022**. [CrossRef]
51. Adnan, M.; Islam, W.; Gang, L.; Chen, H.Y.H. Advanced research tools for fungal diversity and its impact on forest ecosystem. *Environ. Sci. Pollut. Res.* **2022**, *29*, 45044–45062. [CrossRef]
52. Montiel-Rozas, M.D.M.; Lopez-Garcia, A.; Madejon, P.; Madejon, E. Native soil organic matter as a decisive factor to determine the arbuscular mycorrhizal fungal community structure in contaminated soils. *Biol. Fertil. Soils* **2017**, *53*, 327–338. [CrossRef]
53. Maitra, P.; Zheng, Y.; Wang, Y.L.; Mandal, D.; Lu, P.P.; Gao, C.; Babalola, B.J.; Ji, N.N.; Li, X.C.; Guo, L.D. Phosphorus fertilization rather than nitrogen fertilization, growing season and plant successional stage structures arbuscular mycorrhizal fungal community in a subtropical forest. *Biol. Fertil. Soils* **2021**, *57*, 685–697. [CrossRef]

54. Ceulemans, T.; Van, G.M.; Jacquemyn, H.; Boeraeve, M.; Plue, J.; Saar, L.; Kasari, L.; Peeters, G.; van Acker, K.; Crauwels, S.; et al. Arbuscular mycorrhizal fungi in European grasslands under nutrient pollution. *Glob. Ecol. Biogeogr.* **2019**, *28*, 1796–1805. [CrossRef]
55. Ji, B.; Bever, J.D. Plant preferential allocation and fungal reward decline with soil phosphorus enrichment: Implications for mycorrhizal mutualism. *Ecosphere* **2016**, *7*, e01256. [CrossRef]
56. Wang, Q.Q.; Gao, W.; Bol, R.; Xiao, Q.; Wu, L.; Zhang, W.J. Microbial regulation of net N mineralisation is driven by C, N, P content and stoichiometry. *Eur. J. Soil Sci.* **2022**, *73*, e13257. [CrossRef]
57. He, D.; Xiang, X.J.; He, J.S.; Wang, C.; Cao, G.M.; Adams, J.; Chu, H.Y. Composition of the soil fungal community is more sensitive to phosphorus than nitrogen addition in the alpine meadow on the Qinghai-Tibetan Plateau. *Biol. Fertil. Soils* **2016**, *52*, 1059–1072. [CrossRef]
58. Garo, G.; Van Geel, M.; Eshetu, F.; Swennen, R.; Honnay, O.; Vancampenhout, K. Arbuscular mycorrhizal fungi community composition, richness and diversity on enset (*Ensete ventricosum* (Welw.) Cheesman) in Ethiopia is influenced by manure application intensity in low-input farming systems. *Plant Soil* **2022**. [CrossRef]
59. Wang, G.Z.; Koziol, L.; Foster, B.L.; Bever, J.D. Microbial mediators of plant community response to long-term N and P fertilization: Evidence of a role of plant responsiveness to mycorrhizal fungi. *Glob. Change Biol.* **2022**, *28*, 2721–2735. [CrossRef] [PubMed]
60. Turrini, A.; Bedini, A.; Loor, M.B.; Santini, G.; Sbrana, C.; Giovannetti, M.; Avio, L. Local diversity of native arbuscular mycorrhizal symbionts differentially affects growth and nutrition of three crop plant species. *Biol. Fertil. Soils* **2018**, *54*, 203–217. [CrossRef]
61. Chen, K.; Huang, G.; Li, Y.; Zhang, X.R.; Lei, Y.H.; Li, Y.; Xiong, J.; Sun, Y.F. Illumina MiSeq sequencing reveals correlations among fruit ingredients, environmental factors, and amf communities in three lycium barbarum producing regions of China. *Microbiol. Spectr.* **2022**, *10*, e0229321. [CrossRef]
62. Wang, J.P.; Wang, G.G.; Zhang, B.; Yuan, Z.M.; Fu, Z.Y.; Yuan, Y.D.; Zhu, L.J.; Ma, S.L.; Zhang, J.C. Arbuscular mycorrhizal fungi associated with tree species in a planted forest of Eastern China. *Forests* **2019**, *105*, 424. [CrossRef]
63. Kim, K.; Neuberger, P.; Daly, E.J.; Gorzelak, M.; Hernandez-Ramirez, G. Arbuscular mycorrhizal fungi community linkages to soil nutrient availability across contrasting agroecosystems. *Appl. Soil Ecol.* **2022**, *176*, 104464. [CrossRef]

diversity

MDPI

Article

Spatial-Temporal Change for Ecological Intactness of Giant Panda National Park and Its Adjacent Areas in Sichuan Province, China

Chuan Luo [1,2], Hao Yang [1,*], Peng Luo [1,*], Shiliang Liu [3,*], Jun Wang [4], Xu Wang [5], Honglin Li [1,2], Chengxiang Mou [6], Li Mo [7], Honghong Jia [1,2], Sujuan Wu [1,2], Yue Cheng [1,2], Yu Huang [1,2] and Wenwen Xie [1,2]

1 CAS Key Laboratory of Mountain Ecological Restoration and Bioresource Utilization & Ecological Restoration and Biodiversity Conservation Key Laboratory of Sichuan Province, Chengdu Institute of Biology, Chinese Academy of Sciences, Chengdu 610041, China; luochuan@cib.ac.cn (C.L.); lihl@cib.ac.cn (H.L.); jiahh@cib.ac.cn (H.J.); wusujuan18@mails.ucas.ac.cn (S.W.); chengyue20@mails.ucas.ac.cn (Y.C.); huangyu21@mails.ucas.ac.cn (Y.H.); xiewenwen21@mails.ucas.ac.cn (W.X.)
2 University of Chinese Academy of Sciences, Beijing 100049, China
3 State Key Laboratory of Water Environment Simulation, School of Environment, Beijing Normal University, Beijing 100875, China
4 College of Environmental Science and Engineering, China West Normal University, Nanchong 637000, China; drjunw@163.com
5 Ya'an Branch of Giant Panda National Park Administration, Ya'an 625000, China; wangxuexu888.student@sina.com
6 College of Chemistry and Life Sciences, Sichuan Provincial Key Laboratory for Development and Utilization of Characteristic Horticultural Biological Resources, Chengdu Normal University, Chengdu 611130, China; mouchengxiang@126.com
7 Sichuan Key Laboratory of Conservation Biology for Endangered Wildlife, Chengdu Research Base of Giant Panda Breeding, Chengdu 610081, China; jasmineae@126.com
* Correspondence: yanghao@cib.ac.cn (H.Y.); luopeng@cib.ac.cn (P.L.); shiliangliu@bnu.edu.cn (S.L.)

Citation: Luo, C.; Yang, H.; Luo, P.; Liu, S.; Wang, J.; Wang, X.; Li, H.; Mou, C.; Mo, L.; Jia, H.; et al. Spatial-Temporal Change for Ecological Intactness of Giant Panda National Park and Its Adjacent Areas in Sichuan Province, China. *Diversity* **2022**, *14*, 485. https://doi.org/10.3390/d14060485

Academic Editor: Michel Baguette

Received: 18 April 2022
Accepted: 8 June 2022
Published: 15 June 2022

Abstract: Human activities change the natural ecosystem and cause the decline of the intact ecosystem. Establishing an applicable and efficient human activity monitoring indicator system benefits China's ambitious national park system construction. In this study, we established a refined technique for ecological intactness scores (EIS) and applied it in the area of Giant Panda National Park (GPNP) from 1980 to 2020 by quantifying four types of human interferences including land use and cover change (LUCC), road construction, water reservoir and hydropower construction, and mining. The results show the following: (1) Under the ecological intactness score range of 0–10, the GPNP with about 92.6% area of the EIS was above 6.0, and the mean baseline level of intactness was 7.1 when it was established in 2018. (2) The EIS in the east of Qionglaishan and south of Minshan were relatively lower than the rest of the study area. (3) During the past 40 years, 80% of the GPNP's ecological intactness has remained stable. (4) In total, 14% of the GPNP was degraded mainly in the areas below 1200 m with severe human activities. (5) LUCC and road construction were the main driving factors for the decrease of ecological intactness in the GPNP. (6) The habitat of the giant panda is mainly distributed in the areas with an EIS above 6.0, and this is a key link between ecological intactness and habitat suitability. Our research proved that the ecological intactness score (EIS) is an effective indicator for monitoring and assessing the impact of human activities on the regional natural ecosystem and could be helpful for ecological restoration and human activities management GPNP in the future.

Keywords: national park; ecological intactness; human activity; land use and cover change; habitat suitability

1. Introduction

China has launched an ambitious plan to build a national park system to protect its most important natural ecosystems and their associated biodiversity in response to the current climate change risks and biodiversity crises at the regional and global scales. The newly established Giant Panda National Park (GPNP) and its adjacent areas are one of the hotspots for global biodiversity conservation [1] and the priority areas for biodiversity conservation in China [2]. Many studies have revealed that human disturbance is one of the main reasons for the fragmentation and degradation of giant panda habitats [3–5]. Conservation of the giant panda's habitat and maintenance of its intact ecosystem have been set as the core goals for the new GPNP. However, a report about the regional ecosystem status or comprehensive effects of human disturbance on giant panda habitat is rare [5]. Selecting appropriate and efficient evaluation protocols and indicators is the key issue for achieving conservation management goals [6].

With the widespread of nature conservation practices, monitoring and evaluation systems for conservation have been continuously enriched. Representative evaluation tools include the following: intactness evaluation [7,8], naturalness evaluation [9,10], and integrity evaluation [11,12]. In the broad field of conservation biology "intactness"," naturalness", and "integrity" can be treated as synonymous [10,13]. From our point of view, these three indicator systems have their own focus: (1) naturalness, commonly defined as the similarity of a current ecosystem state to its natural state, focuses on "compositional", "structural", and "functional" indicators of the ecosystem [10]; (2) intactness is the state of an ecosystem that develops in response to natural processes without human interference, with an emphasis on measuring the degree of human disturbance to natural ecosystems [7]; (3) ecosystem integrity is a huge notion with meanings ranging from intactness, wholeness, health, and functioning to quality and resilience, which pays attention to the absence of ecosystem elements (wholeness) [6,14,15]. Naturalness, intactness, and ecosystem integrity are continuity concepts for supporting ecosystem and biodiversity conservation aims of sustainable development [10,16]. Based on our knowledge, we chose the term "intactness" to construct an indicator that can measure the effectiveness of national parks in managing human activities at the landscape scale.

In this study, we define the degree to which an ecosystem retains its natural state unmodified by remote sensing identifiable human activity as ecological intactness. Traditionally, China and other countries have adopted forest management practices, such as improvement of forest cover, tree stock volume, and carbon storage as the goals of natural ecosystem preservation [17,18]. These forest management goals may ignore the benefits of ecosystem services and natural values other than wood fiber production [19], such as biodiversity conservation, water regulation, water retention, and climate change mitigation. Many studies have shown that an intact ecosystem supports significant environmental values including biodiversity [20,21], carbon sinks [22], water regulation and provision [23,24], and the maintenance of human health [25]. Hence, ecological intactness could be one valuable indicator for the management of GPNP to reflect human activity and the intact degree of an ecosystem.

The assessment of ecological intactness requires the measurement of human activities and their caused ecosystem changes. Methods to quantify human activity in an ecosystem include human footprint [26,27] and human modification [28,29]. Those methods collect human activity indicators including building, croplands, pasture lands, nighttime lights, railways, etc. to assess human modification or footprint in continental-scale and larger regional areas. Those studies revealed that 60% of all land changes are associated with direct human activities, and the protection of the intact ecosystem is needed [26,28,30–34]. Human activities are not mutually exclusive and usually co-occur in the same location [35]. The regional assessment of anthropogenic impact is generally conducted by remote sensing [36]. The critical step for assessing the ecological intactness is to identify the main disturbance factors in Giant Panda National Park, and the identifiability of the image is an important criterion for screening the types of human activities involved in the evaluation of ecological intactness. Land Use and Cover Change (LUCC) is an important material for regional ecological intactness assessment, which contains basic anthropogenic activity information such as villages, cultivated land, etc. According to the evaluation of the 3rd and 4th National Panda Surveys [37–39] and the World Natural Heritage Outlook Report of Giant Panda Habitat [40], road interference, mining interference, water reservoirs, and hydropower construction are identified as the major human disturbances affecting the regional ecological environment.

Previous work aimed to protect the quality of the giant panda habitat. The indicators for evaluating and monitoring giant pandas' habitat quality came from the habitat suitability index (HSI), which is based on environmental factors including vegetation types, elevation, slope, and bamboo distribution [37,41,42]. While the new GPNP takes ecological intactness as a conservation goal in management, it is interesting and important to test if the "intactness" can cover the spatial area of "suitable habitat". With these regards, this study focuses on the following aims: (1) studying the current status of the ecological intactness so as to provide a baseline for GPNP, (2) detecting the changes of ecological intactness in GPNP during the past 40 years and identifying specific areas in need of management, and (3) testing the relationship between ecological intactness and the quality of the panda habitat.

2. Materials and Methods

2.1. Study Area

The Giant Panda National Park lies in the mountainous area of southwest China, spanning the Sichuan, Gansu, and Shanxi Provinces. We considered all counties of Sichuan Province that are partially or fully designated as national parks as our study area (Figure 1). Sichuan has 1387 giant panda individuals accounting for 74.4% of the total population, and the GPNP Sichuan part composed 74.6% of the total GPNP. The study area can be divided into two parts, one is called GPNP, and the other is the study area outside the GPNP called the exterior area (EA). The research area covered 52,473.3 km^2 and the GPNP occupied 20,177 km^2 (Figure 1). This study area occupies the QionglaiShan, Minshan, Qinling, and Daxiangling mountain regions. The elevation of the study area ranges from 421 m and 6132 m, and the northwest is higher than the southeast.

The Sichuan GPNP involves 119 townships and 20 counties that belong to 7 prefectures and has a population of 89,900 inhabitants. The local economic structure is relatively simple, with mining, hydropower, and other resource development enterprises generating industrial revenue. The main source of income for community residents is agriculture, and some residents are also engaged in mining and processing labor. According to the Fourth National Giant Panda Census, the common disturbances affecting wild giant pandas and their habitat are livestock, roads, and farming (cultivation of row crops), all of which are significantly increasing, while disturbance from logging has been significantly decreasing after the implementation of the Natural Forest Conservation Program (NFCP) and the Grain-to-Green Program (GTGP) [37].

2.2. Materials and Method

2.2.1. Data Preprocessing

Taking the regional characteristics of GPNP and data availability into consideration, we finally tailored four categories of human activity (LUCC, roads, mining, and water reservoirs and hydropower construction) and employed the "increasing" fuzzy sum function to map the ecological intactness in a spatially explicit way [30]. The data timing, resolution, and source information of the four categories of human activity used in this study are shown in Table 1. Based on the local studies concerning the effects of roads on the distribution of giant pandas [43,44] and expert judgment, we set the maximum impact range as 3 km and established four different widths of road influence radius, which were 100 m, 500 m, 1000 m, and 3000 m for assigning different human interference scores(Table 2.).

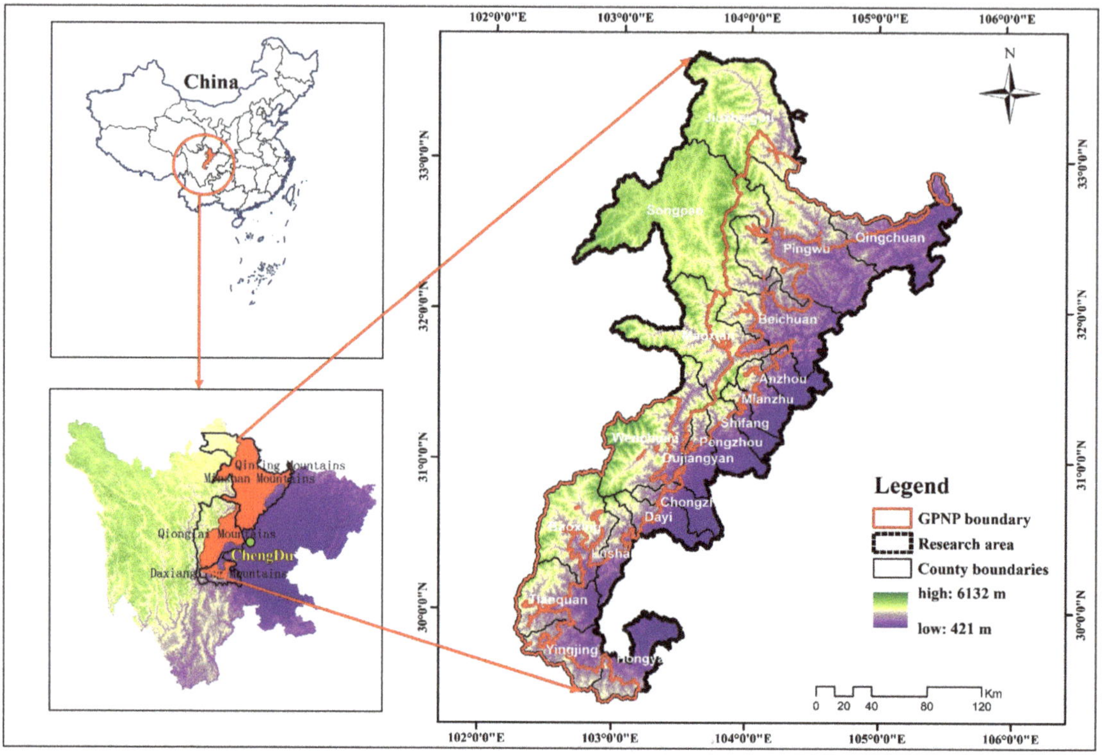

Figure 1. Research area and the Giant Panda National Park boundary.

Table 1. Overview of human activity, spatial resolution, and the years of data used in the calculation of ecological intactness.

Human Activity	Year	Resolution	Data Source
Road interference	1980, 1990, 2000, 2010, 2018, and 2020	-	The Data Center of Resources and Environmental Sciences, Chinese Academy of Sciences (https://www.resdc.cn/data.aspx?DATAID = 237) (accessed on 6 September 2020); the Traffic Yearbook of Sichuan Province (http://cnki.nbsti.net/CSYDMirror/Trade/yearbook/single/N2021080079?z=Z014) (accessed on 6 September 2020)
Mining interference	1980, 1990, 2000, 2010, 2018, and 2020	30 m × 30 m	local authorities; field investigation
Water reservoirs and hydropower construction	1980, 1990, 2000, 2010, 2018, and 2020	30 m × 30 m	local authorities; field investigation
Land Use and Cover Change	1980, 1990, 2000, 2010, 2018, and 2020	30 m × 30 m	The Data Center of Resources and Environmental Sciences, Chinese Academy of Sciences (https://www.resdc.cn) (accessed on 31 December 2020)

Mine development data are not available online. We collected the data regarding mine development from local authorities. Those data included the location, area, and times of mining activity. We sorted out six mine development sub-datasets for 1980, 1990, 2000, 2010, 2018, and 2020 according to the date of initial activity. Then, the 2020 sub-dataset was validated with field investigation.

Table 2. Road grade * influences degree and scope parameter based on the previous study [30,31] and modified according to related research [43–45] and expert judgment.

Road Grade	Road Facility (0–100 m)	100–500 m	500–1000 m	1000–3000 m
Highway	10	3	2	1
First Grade	10	3	2	1
Second grade	8	2	1	0
Third grade	6	2	0	0
Fourth grade	4	1	0	0
Railway	10	3	1	0

* Note: The road grade standards refer to "Road Engineering Technical Standards" (JTGB01-2014).

Human influence scores for mining interference were determined from relevant studies [30,46] and based on onsite investigations. We adjusted the scope of mining interference from 0.5 km to 1 km, and the impact score was adjusted according to our field evaluations on the degree of interference for different mining methods and mining sizes. Mining areas exceeding 1 km^2 are considered large, and mining areas less than 1 km^2 are considered small. Because mining areas tend to destroy local vegetation and cause slope erosions, mining areas yield the largest impact score of 10, and the impacts gradually fade to zero with increasing distance to 800 m or 1 km. The impact range of mining interference was set at 200 m, 500 m, 800 m, and 1000 m beyond the boundary of the mining area to establish buffers relative to what was being impacted (Table 3).

Table 3. Parameter table of influence degree and scope of mining type determined from relevant studies [30,46], and expert judgment based on onsite investigations.

Mining Type	Mining Area	0–200 m	200–500 m	500–800 m	800–1000 m
Large open pit mine	10	8	4	2	0
Small open pit mine	10	4	2	1	0
Large underground mine	10	4	2	1	0
Small underground mine	10	2	1	0	0

The hydrological project data source and production process were the same as we used for assessing mining interference. We adopted the ecological impact assessment procedure for hydroelectric project construction developed for the Nuozhadu Nature Reserve [47] and set the maximum distance of influence at 5 km. For the construction of water reservoirs projects and hydropower development, the impact degree and scope of the project were evaluated according to the size of the project or the installed capacity. The criteria for classifying the size of hydropower stations were as follows: small-scale station under 5000 KW/h, medium-sized station of 5000–100,000 KW/h, large-scale station of 100,000–1 million KW/h, and mega-sized station over 1 million KW/h. Most hydropower stations were small and medium sized, with very few large hydropower stations and no mega hydropower stations. To obtain accurate vector data regarding the size of the areas impacted by water reservoirs and hydropower dams, this study assigned a width of 500 m from the center point of each project. From the circle described by this radius, 4 buffers of different widths were established as concentric circles at 1000 m, 1500 m, 2500 m, and 5000 m. The human influence scores assigned to each type of water reservoir and hydropower construction were based on our field investigation, relevant studies [47–49], and expert judgment, which accounted for the characteristics of each study area. Because of destruction caused by inundation, reservoirs behind dams were given the largest impact score of 10. This impact diminished gradually to zero approximately 2.5 km or 5 km upstream (Table 4).

China's multi-period Land Use and Cover Change data classified land use into 6 categories and 25 subcategories [50]. We assigned human interference scoring values for each subcategory of the Land Use and Cover Change data based on relevant studies [30,31] and expert judgment (Table 5). Shrubland and alpine grassland occur above the timberline, which begins generally at 3800 m above sea level [51,52]. The types of shrubs and grasses that appear below 3800 m are mostly secondary types colonizing after the abandonment of land use [52] and were assigned human influence scores of 3 and 4. Grazing is common at altitudes above 3800 m, and grasslands were more disturbed by grazing than shrubland. Human influence scores of 2 and 1 were assigned to them, respectively. No new canals were excavated in the research area during the research period. Previously excavated canals were all modified by sand removal, quarrying, river dredging, etc. The canal mentioned in this study was either excavated or may have been a natural stream that was modified by human impact, and we assigned it a score of 2. Cropland, including dry land and paddy fields, was assigned scores of 7 in the practice of Venter et al. (2016) and Li et al. (2018) [31,53]. In mountain areas, paddy fields were consistently located in low elevations and dry land at higher elevations nearer the forest. Paddy fields were more disturbed by humans than dry land, and human influence scores of 7 and 5 were assigned to them, respectively.

Table 4. Parameter table of influence degree and scope of hydropower and water reservoirs facilities.

Type	Dam Region	500–1000 m	1000–1500 m	1500–2500 m	2500–5000 m
large-scale hydraulic project	10	8	4	2	1
Small and medium hydraulic project	10	4	2	1	0
Large hydropower	10	8	4	2	1
Small and medium hydropower	10	2	1	0	0

Table 5. The human influence scores assigned to each land use type based on relevant studies [30,31] and expert judgment.

Land-Use Type	Sub-Type	Score
Cultivated Land	paddy field	7
	dry land	5
Woodland	forested land (Altitude $\leq$ 3800)	1
	forested land (Altitude > 3800)	0
	shrubland (Altitude $\leq$ 3800)	3
	shrubland (Altitude > 3800)	1
	sparse forested land (Altitude $\leq$ 3800)	2
	sparse forested land (Altitude > 3800)	0
	other woodlands	6
Grassland	grassland (Altitude $\leq$ 3800)	4
	grassland (Altitude > 3800)	2
Water Area	canal	2
	reservoirs/ponds	3
	Lake, permanent glacier snow area, tidal flat, floodplain	0
Industrial and Residential	urban land	10
	rural residential area	10
	other construction lands	10
Unused Land	desert, the Gobi Desert, saline-alkali land, swamp, bare land, bare rock and gravel fields, others	0

2.2.2. Calculation of Ecological Intactness Scores

All data sets were computed with the fuzzy sum function (Function (1)) to obtain the comprehensive spatial distribution of Human Modification (HM) [30]. The ecological intactness score (EIS) was calculated on a 10-point scale, using an ArcGIS 10.5 grid calculator using EIS = 10-HM.

$$\mathrm{HM} = 1.0 - \prod\nolimits_{i=1}^{k}(1 - Fi/10) \quad (1)$$

where HM is the score of all human interference, and Fi is the score of the i-th type of interference.

$$\mathrm{EIS} = 10(1.00 - \mathrm{HM}) \quad (2)$$

An ecosystem that is entirely natural and not at all impacted would have an EIS of 10.0, and one that was completely destroyed would have an EIS of 0.0.

2.2.3. Analysis of the Spatiotemporal Change Trend and Driving Force of EIS

The method of unary linear regression analysis combined with Theil–Sen median trend analysis was used to calculate the trend of EIS changes in the study area. The results of the trend analyses are coupled with the Mann–Kendall test of significance for the trends evident during 1980, 1990, 2000, 2010, 2018, and 2020 [54].

$$\text{Slope} = \frac{n \times \sum_{i=1}^{n} i \times SI_i - \sum_{i=1}^{n} i \sum_{i=1}^{n} SI_i}{n \times \sum_{i=1}^{n} i^2 - \left(\sum_{i=1}^{n} i\right)^2} \tag{3}$$

In the formula, Slope is the EIS trend slope of a certain grid in 1980, 1990, 2000, 2010, 2018, and 2020; n is the total number of years; and i is the annual ordinal. Slope > 0 indicates that the EIS trend increases, and Slope < 0 indicates that the EIS tends to decrease.

The Mann–Kendall test was used to determine whether the time series data had an upward or downward trend. The calculation formula is as follows:

$$Z = \begin{cases} \frac{S-1}{\sqrt{var(S)}}, S > 0 \\ 0, S = 0 \\ \frac{S+1}{\sqrt{var(S)}}, S < 0 \end{cases} \tag{4}$$

$$S = \sum_{i=1}^{n-1} \sum_{j=i+1}^{n} Sign(SI_j - SI_i) \tag{5}$$

$$\text{Var}(S) = \frac{n(n-1)(2n+5)}{18} \tag{6}$$

$$\text{Sign}(SI_j - SI_i) = \begin{cases} 1, SI_j - SI_i > 0 \\ 0, SI_j - SI_i = 0 \\ -1, SI_j - SI_i < 0 \end{cases} \tag{7}$$

We assumed that changes for the EIS were significant at the α = 0.05 confidence level. We divided the test result Z_C into significant change ($|Z_C| > 1.96$) and insignificant change $|Z_C| < 1.96$. Combining the analysis of slope and Z value, we classified the ecological intactness change trends into five categories: significant degradation, slight degradation, stable, slight improvement, and significant improvement (see Table 6).

Table 6. The classification standard of ecological intactness changes categories.

Slope	Z Value	Category
≥0.0005	>1.96 or <−1.96	significant improvement
≥0.0005	−1.96–1.96	slight improvement
−0.0005–0.0005	−1.96–1.96	stable
<−0.0005	>1.96 or <−1.96	significant degradation
<−0.0005	−1.96–1.96	slight degradation

We employed the Geodetector to analyze the driving force of EIS. The Geodetector is a statistical method to detect spatial stratified heterogeneity and reveal the driving factors behind it; this statistical method with no linear hypothesis has an elegant form and definite physical meaning [55,56].

2.2.4. Analysis of the Relationship between Ecological Intactness and Giant Panda Habitat Suitability

To understand the environmental characteristic of giant panda signs and the coupling relationship between ecological intactness and habitat suitability, we analyzed the habitat suitability of giant pandas within the study area. Currently, species distribution models (SDMs) are commonly used to predict the geographic range of a species [57,58]. As one of the SDMs, the MaxEnt algorithm, was a proven powerful model when modeling rare species with narrow ranges and scarce presence-only occurrence data, and has been widely used in giant panda habitat assessment work [59–61].

This study employed MaxEnt models to evaluate the habitat suitability index (HSI) for the study area, with the habitat suitability index (HSI) ranging from 0 (unsuitable) to 1 (most suitable) [59]. The input variables included altitude, slope, vegetation type, annual average temperature, maximum yearly temperature, minimum yearly temperature, and annual rainfall [61]. We analyzed habitat suitability without considering human disturbance to avoid autocorrelation with the results of EIS. The parameters for model running were set as follows: we used 1066 giant panda signs, which came from the Fourth National Giant Panda Census, and selected 75% of the total 1066 presence data (n = 800) as training data, and the remaining 25% of presence data (n = 266) were used for testing the model. The habitat suitability index (HSI) map was obtained by the average training AUC (0.963) and test AUC (0.960), indicating that the performance of the model was reliable.

Next, we used linear regression analysis to analyze the correlations between EIS and HIS using the method of random point sampling [2]. Two types of points data were analyzed, which were giant panda signs that came from the Fourth National Giant Panda Census and 1000 random sample points from the study area. To further understand the ecological intactness characteristic of the giant panda habitat, we analyzed the EIS components by randomly sampling 1000 points from each type of giant panda realistic habitat, giant panda suitable habitat, and marginally suitable habitat, which were obtained from the Fourth National Giant Panda Census [62]. The technical framework of the study is shown in Figure 2.

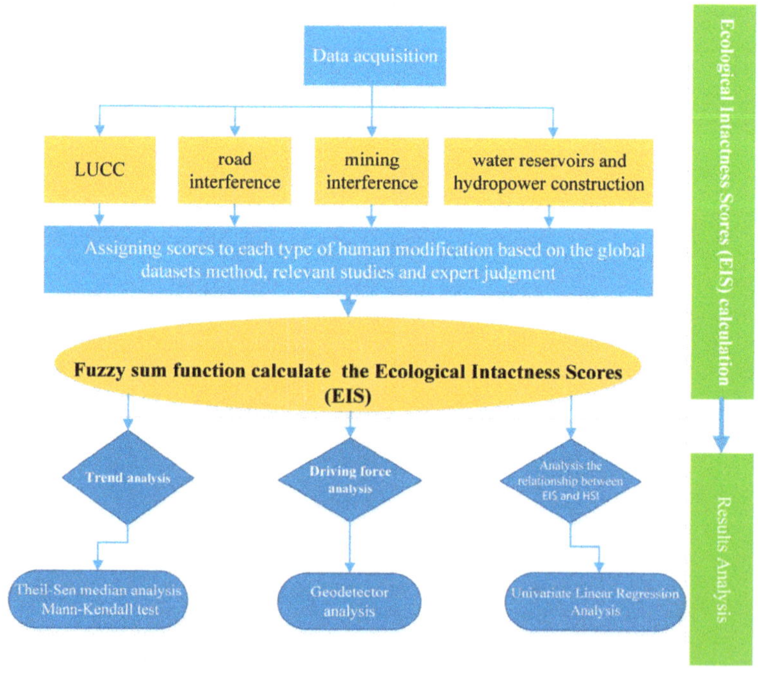

Figure 2. Flow chart of EIS calculation and technical framework of this study.

3. Results

3.1. Spatial Characteristics of EIS

In general, the study area has a high level of intactness (6.4 ± 2.2) (Figure 3), and the GPNP (7.1 ± 1.6) is more intact than exterior areas (6.0 ± 2.5). Most areas (92.6%) of the GPNP are above 6.0 (Figure 4A,B). One-fifth of the GPNP (about 4171.8 km^2) has an EIS of more than 9.0, which belongs to the intact ecosystem in the strict sense (Figure 5B).

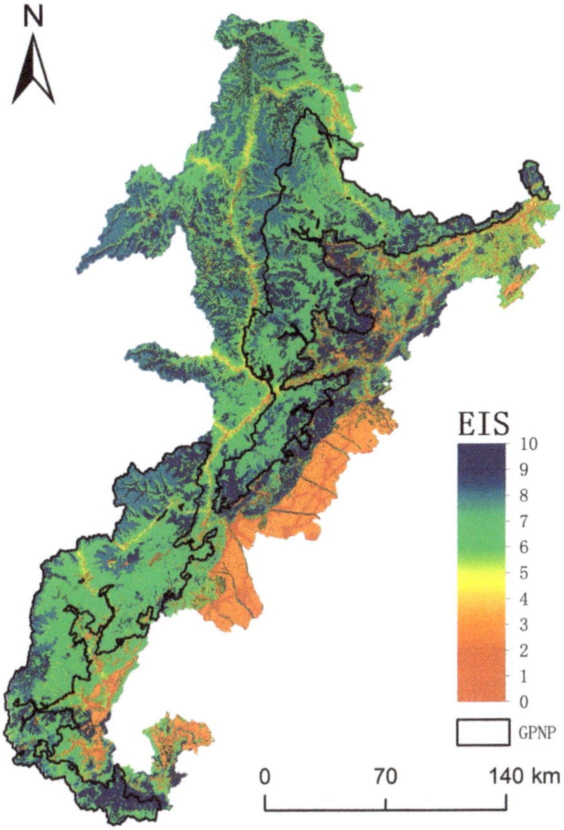

Figure 3. Spatial distribution pattern of different intactness levels in the study area in 2018.

In terms of spatial characteristics, from the comparison of mountain systems in the study area, the order of intactness levels is the mountains of Minshan (6.5 ± 2.2) > Daxiangling (6.3 ± 2.6) > Qionglaishan (6.1 ± 2.2) > Qinling (6.0 ± 2.3). The EIS in GPNP in the mountains of Qionglaishan, Minshan, Qinling, and Daxiangling was 6.9 ± 1.5, 7.2 ± 1.5, 7.1 ± 2.2, and 7.9 ± 1.5, respectively.

Within GPNP, the EIS in the east of Qionglaishan and south of Minshan was relatively lower than in other regions. Additionally, the ecological intactness scores in the north of Qionglaishan, south of Daxiangling, and middle of Minshan were relatively high (Figure 3).

Comparing the EIS levels of different altitude ranges, the mean EIS at elevations less than 1200 m was 6.0. The mean EIS values from 1200 m to 4000 m were between 6.5 and 7.5, and the EIS above 4000 m mean scores exceeded 7.5. Mean intactness scores below 1200 m in the GPNP were significantly higher than scores at the same altitudinal range in exterior areas (Figure 5A).

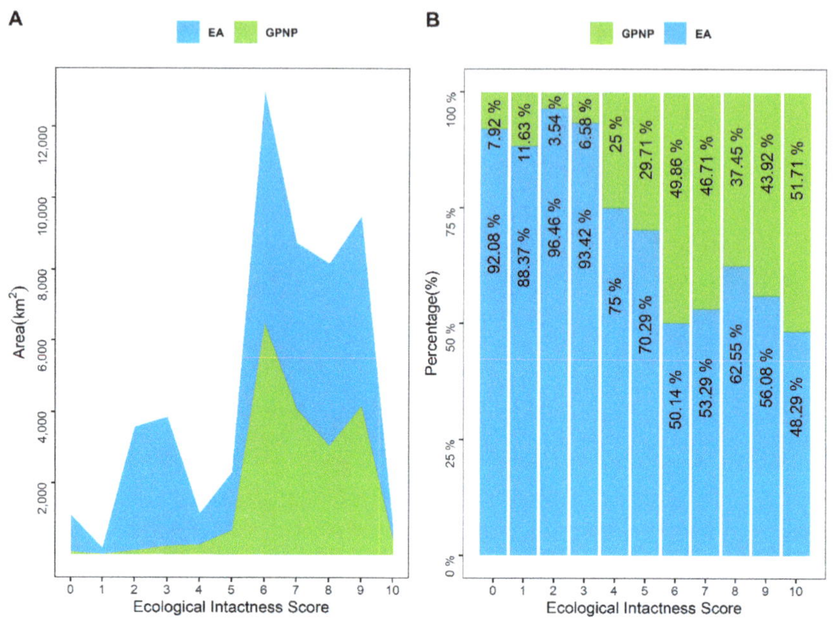

Figure 4. The area (**A**) and area proportion (**B**) of different ecological intactness scores within GPNP and exterior areas (EA) in the study area.

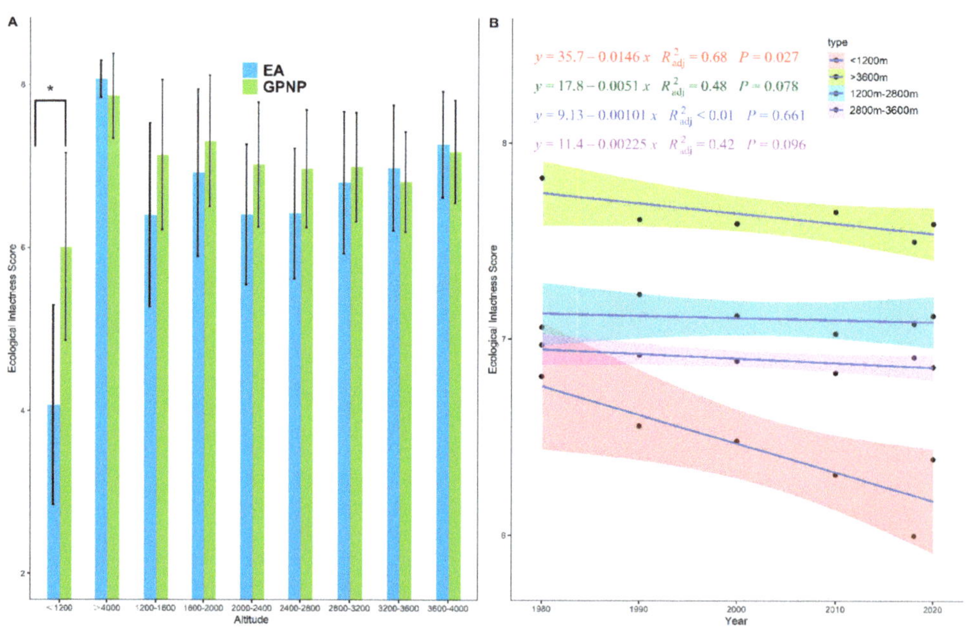

Figure 5. Ecological intactness score of different elevations between the GPNP and exterior areas (EA) for 2018 (**A**) and 40 years time-series dynamics for different elevations (**B**). * indicates a significant difference, and no mark indicates an insignificant difference.

3.2. Fourty-Year Intactness Trends and Driving Force in GPNP

The mean intactness scores in the GPNP from 1980 to 2020 are shown in Figure 6. The mean EIS for 1980, 1990, 2000, 2010, 2018, and 2020 were 7.2 ± 1.4, 7.2 ± 1.4, 7.1 ± 1.4, 7.1 ± 1.5, 7.1 ± 1.5, and 7.1 ± 1.4, respectively. Scores remained stable during the past 40 years in 79.88% of the GPNP (Figure 7). Ecological intactness underwent degradation in approximately 14.77% of the GPNP, of which about 0.99% (96 km^2) declined significantly. The areas showing a downward trend are mainly distributed along valleys and low-altitude areas (Figure 7). In 5.35% of the GPNP, intactness scores improved, and significantly so in 0.42% (82.79 km^2) of the park.

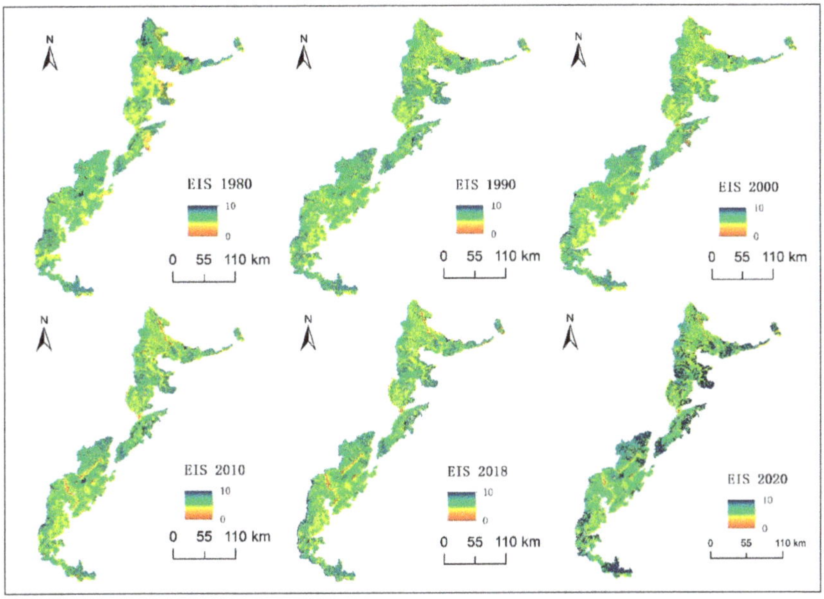

Figure 6. Characteristics of the ecological intactness change of GPNP from 1980 to 2020.

The mean value of ecological intactness below 1200 m was <7.0, and from 1980 to 2018 ecological intactness diminished (Figure 5B). EIS was stable at all elevations above 1200 m.

The driving force of ecological intactness change was analyzed by the geographic detector. The result shows that the explanatory rate of land use, road interference, mining interference, and water reservoirs and hydropower construction was 84.25%, 22.45%, 0.44%, and 0.017%, respectively. Land use and road penetration (not only road construction, but also for road usage) are the main factors affecting ecological intactness.

3.3. Relationship between Ecological Intactness and Giant Panda Habitat Suitability Index (HSI)

The correlation was not significant between ecological intactness scores (EIS) and the habitat suitability index (HIS), which means that a site with a high level of ecological intactness score does not mean a high level of habitat quality (Figure 8A,B). It can be seen from Figure 8A that the giant panda signs have a high level of suitability, and the linear correlation between the HSI and EIS at giant panda signs is not significant. Visual inspection of the 3D scatter plot is sufficient to detect not only the spatial distribution characteristics between the EIS and HIS, but also distribution differences concerning altitude (Figure 8C,D). Giant panda signs are mainly distributed in the areas with an EIS between 6.0 and 9.0. The realistic habitat of the giant panda is mainly distributed in the regions with an EIS above 6.0 (Figure 9).

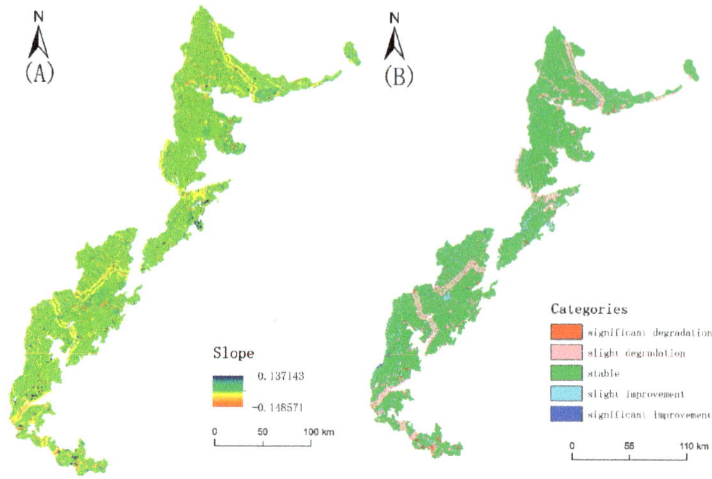

Figure 7. Change slope (**A**) and change category (**B**) of ecological intactness in the GPNP.

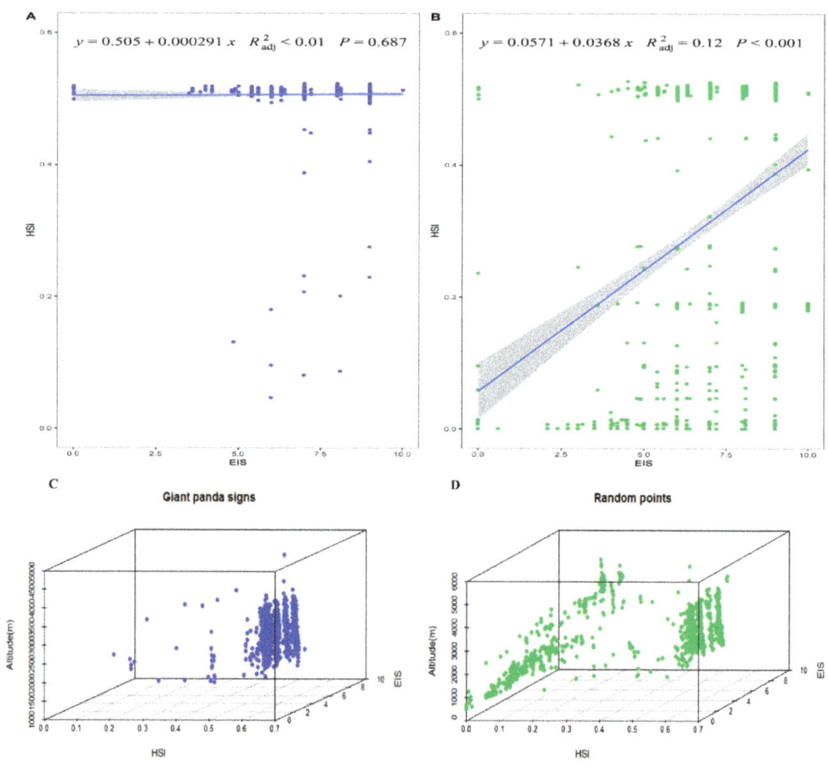

Figure 8. Linear regression analysis of the relationship between the ecological intactness score(EIS) and giant panda habitat suitability index (HSI) for giant panda signs (**A**) and random points (**B**); 3D spatial distribution relationship of altitude, ecological intactness score (EIS) and giant panda habitat suitability index (HSI) for giant panda signs (**C**) and random points (**D**).

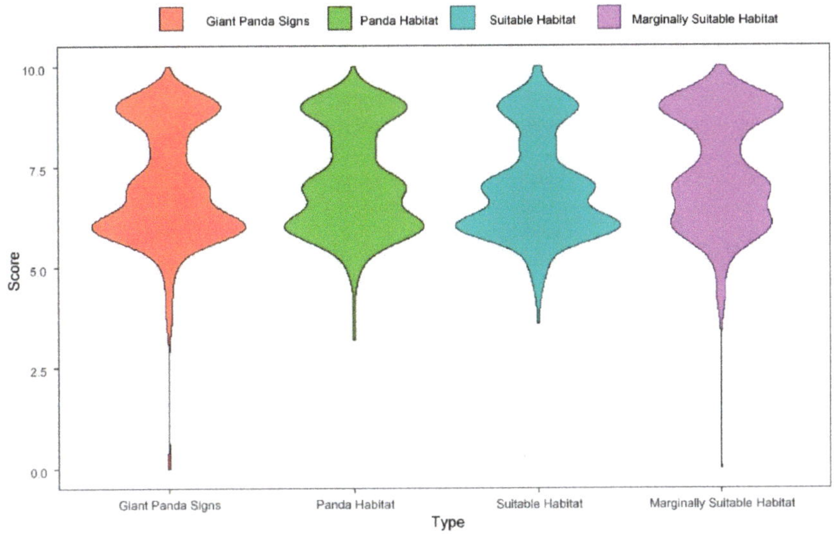

Figure 9. The ecological intactness characteristic of giant panda signs, giant panda realistic habitat, giant panda suitable habitat, and marginally suitable habitat.

4. Discussion

4.1. Ecological Intactness in Giant Panda National Park

This study reveals the ecological intactness status of giant panda habitat ecosystems and the spatio-temporal changes of human disturbance at the landscape scale. Sichuan GPNP consists of two parts, one part is the 26 former natural reserves comprising 10,985.36 km^2, and the other is the adjacent area to connect these former natural reserves, comprising about 9191.64 km^2. By comparing the EIS in the GPNP and exterior areas (EA), we found that the currently designated national parks have a higher level of ecological intactness than their exterior areas (EA). This result is not surprising because the earlier giant panda reserves were all included in national park management [63]. The GPNP has about 4171.8 km^2 (one-fifth of the GPNP) intact ecosystem, where the EIS is higher than 9.0. About 80% of the GPNP the ecological intactness has remained stable, but14% of the area has shown signs of slow degradation over the past 40 years. The degraded areas of the GPNP are mainly located in the adjacent area, which was not strictly managed in the early stage and has strong human activity. Systematic planning of ecological restoration and wildlife corridor construction should be carried out in this area.

EIS varies between mountain ranges in national parks; the EIS in the east of Qionglaishan and south of Minshan were relatively lower than in other regions. This part of GPNP is close to the populated area of Chengdu Plain and existing intensive human activities. On the other hand, the ecological intactness in the north of Qionglaishan, south of Daxiangling, and middle of Minshan were relatively high. These areas are the original giant panda nature reserves.

We further analyzed the status quo characteristics of regional ecological intactness in different altitude dimensions; EIS was especially low at elevation >1200 m, where human interference is intense. Human activities, such as hamlet construction and road construction, are mostly located in surrounding foothills and valleys, causing fragmentation of the panda habitat [4]. Li and others reported that free-ranging livestock threatened the long-term survival of giant pandas, which means that human activities in low-altitude areas, especially derivative disturbances, such as grazing and wild collection, drove giant pandas up to more elevated habitats [64–66].

This study has a close assessment result with global human modification (GHM) for the ecological intactness of the study area. The EIS result of GHM is twice that of the global human footprint (GHF), and the EIS of our study is in between. Differences in scale for resolution for buildings, arable land, night lights, and pastures (1000 m × 1000 m) were recorded by the global human footprint method [27,53], compared with the data for land use and coverage used in this study (30 m × 30 m), probably accounted for part of these discrepancies. GHF mapping of the area land cover types, which does not differentiate the intensity of the impact of those cover types (crop and pasture), may be the main reason for the difference [28].

4.2. *Ecological Intactness and Habitat Suitability for Giant Panda*

The GPNP takes ecological intactness as a conservation goal, whether or not the new goal can strengthen the protection of giant pandas and giant panda habitats. This study can give some positive information and strategies for this goal. Analysis shows that giant panda habitats and panda signs are not located in completely intact or natural areas; this does not mean that giant pandas do not like the natural habitat environment, but rather, the result reveals the broad impact of human activities on giant panda habitats. Because the suitable habitat for giant pandas is distributed on gentle slopes between 1100 m and 2800 m above sea level [44,67], these areas are also vulnerable to human activities such as roads, farming, and planting. Our result indicates that the EIS scores for the giant pandas' current habitat and their suitable habitats are at a relatively high level (>6); this level could be set as a threshold level for management of the human activity. It can also be used as a standard to adjust the boundaries of national parks; if so, most of the optimum and suitable habitats for pandas would be included in the protection of the GPNP.

From the spatial relationship between the ecological intactness score (EIS) and giant panda habitat suitability index (HSI), if a suitable EIS threshold is set, we can see that the aims to protect the ecological intactness may not only fulfill the habitat suitability goals, but also cover other giant panda habitats, and this is a potentially comprehensive conservation strategy.

4.3. *Application of EIS in National Parks Management*

As an example, this study evaluated the GPNP EIS baseline and measured the 40-year change. Through our case study in the GPNP, we suggest that ecological intactness will be an effective indicator for monitoring and assessing the impact of human activities on regional natural ecosystems. Protected areas with different human pressure baselines should establish different management measures to improve conservation effectiveness [68]. National parks need the establishment a comprehensive monitoring and evaluation framework including ecological intactness for managing human activities. Feng, Cao et al. developed a "baseline + change" framework to assess the effectiveness of protected areas [68]. This valuation framework of "baseline + change" could be taken in the National Parks monitoring system.

With the development of remote sensing techniques, fine-scale image was easy to obtain. The data resolution can shift to different levels to fit the management's needs. It is recommended to establish a land use management database in national parks combine with the periodic national land survey and to use high-resolution remote sensing or drone technology to achieve refined management of national parks. Managers can evaluate different scales by selecting images with suitable resolutions. Compared with traditional management indicators, such as forest coverage and forest volume, ecological intactness may be more comprehensive, and the monitoring costs will be lower.

Setting reasonable scientific goals and selecting appropriate evaluation protocols are key issues for the management of natural ecosystems preserves everywhere [6]. Otherwise, landscape patterns may change, and natural ecosystems may undergo undesirable modifications. National parks are the most strictly managed protected areas in China, and human activities are strictly restricted in this kind of area. Managers of the new GPNP are faced with resolving many issues, among them the selection of ecological intactness thresholds for different managing zones. The GPNP is divided into two main zones including the core protection zone and the general control zone [69]. The EIS thresholds of the general control zone could be lower than the core protection zone, allowing for certain types of human activities such as ecological restoration, habitat improvement, and new corridor construction [63]. This idea of differentiated control target setting may help the effective carry out of post-2020 biodiversity conservation work.

5. Conclusions

This study provided an evaluation method combining remote sensing and field investigation to reveal ecological intactness. We synthesized a map of the ecological intactness, which spatially and intuitively showed the intensity of human activities in Giant Panda National Park and its adjacent areas. We assessed the current status and identified the historical changes from 1980 to 2020 of the ecological intactness caused by human impacts on the landscape scale. This study found that Giant Panda National Park has a high ecological intactness baseline but still can be improved. Low-elevation areas (<1200 m) where a significant decrease occurred should be regarded as the focus of conservation management. From the mountain system point of view, the ecological intactness in the north of Qionglaishan, south of Daxiangling, and middle of Minshan were relatively higher than east of Qionglaishan and south of Minshan. During the past 40 years, the ecological intactness in most areas of the GPNP remained stable, and about 14% of the area was degraded. LUCC and road construction were the main driving factors for the decrease of ecological intactness in the GPNP. The habitat of the giant panda is mainly distributed in the regions with an EIS above 6; this is the key link between ecological intactness and habitat suitability. Our research recommends that the ecological intactness score (EIS) can be used as one of the evaluation and monitoring indicators for the protection effectiveness of the national park.

Author Contributions: Conceptualization, C.L. and P.L.; methodology, C.L. and H.Y.; software, C.L.; validation, C.L. and H.Y.; formal analysis, C.L.; investigation, C.L., X.W., J.W., H.L., C.M., L.M., H.J., S.W., Y.C., Y.H. and W.X.; writing—original draft preparation, C.L.; writing—review and editing, P.L. and S.L.; visualization, C.L.; supervision, P.L.; project administration, P.L.; funding acquisition, P.L. All authors have read and agreed to the published version of the manuscript.

Funding: This research was funded by the National Key R&D Program of China (2016YFC0503305), Management Framework and Capability Building for Development of Ya'an Giant Panda National Park (NOR/15/301/16/002), and the Sichuan Science and Technology Program of Key Technology and Demonstration for Biodiversity Conservation in Giant Panda National Park, Grant/Award Number: 2018SZDZX0036.

Institutional Review Board Statement: Not applicable.

Informed Consent Statement: Not applicable.

Data Availability Statement: The data presented in this study are available in the article.

Acknowledgments: Thanks to Andre F. Clewell for his contribution to improving the language of the manuscript. We also thank Shaoyao Zhang and Guyue Hu for their help in data acquisition and analysis for this work.

Conflicts of Interest: The authors declare no conflict of interest.

References

1. Mittermeier, R.A.; Myers, N.; Mittermeier, C.G.; Robles Gil, P. *Hotspots: Earth's Biologically Richest and Most Endangered Terrestrial Ecoregions*; CEMEX: Mexico City; Graphic Arts Center Publishing Company: Portland, OR, USA, 2002; pp. 0022–2372.
2. Zhang, J.; Xu, W.; Kong, L.; Hull, V.; Xiao, Y.; Xiao, Y.; Ouyang, Z. Strengthening protected areas for giant panda habitat and ecosystem services. *Biol. Conserv.* **2018**, *227*, 1–8. [CrossRef]
3. Wei, F.; Swaisgood, R.; Hu, Y.; Nie, Y.; Yan, L.; Zhang, Z.; Qi, D.; Zhu, L. Progress in the ecology and conservation of giant pandas. *Conserv. Biol.* **2015**, *29*, 1497–1507. [CrossRef] [PubMed]

4. Xu, W.; Viña, A.; Kong, L.; Pimm, S.L.; Zhang, J.; Yang, W.; Xiaodong, C.; Zhang, L.; Chen, X.; Liu, J.; et al. Reassessing the conservation status of the giant panda using remote sensing. *Nat. Ecol. Evol.* **2017**, *1*, 1635–1638. [CrossRef] [PubMed]
5. Kang, D.W. A review of the impacts of four identified major human disturbances on the habitat and habitat use of wild giant pandas from 2015 to 2020. *Sci. Total Environ.* **2021**, *763*, 142975. [CrossRef]
6. Tierney, G.L.; Faber-Langendoen, D.; Mitchell, B.R.; Shriver, W.G.; Gibbs, J.P. Monitoring and evaluating the ecological integrity of forest ecosystems. *Front. Ecol. Environ.* **2009**, *7*, 308–316. [CrossRef]
7. Haurez, B.; Daïnou, K.; Vermeulen, C.; Kleinschroth, F.; Mortier, F.; Gourlet-Fleury, S.; Doucet, J.-L. A look at Intact Forest Landscapes (IFLs) and their relevance in Central African forest policy. *For. Policy Econ.* **2017**, *80*, 192–199. [CrossRef]
8. Beyer, H.L.; Venter, O.; Grantham, H.S.; Watson, J.E. Substantial losses in ecoregion intactness highlight urgency of globally coordinated action. *Conserv. Lett.* **2019**, *13*. [CrossRef]
9. Machado, A. An index of naturalness. *J. Nat. Conserv.* **2003**, *12*, 95–110. [CrossRef]
10. Winter, S. Forest naturalness assessment as a component of biodiversity monitoring and conservation management. *Forestry* **2012**, *85*, 293–304. [CrossRef]
11. Hansen, A.J.; Noble, B.P.; Veneros, J.; East, A.; Goetz, S.J.; Supples, C.; Watson, J.E.M.; Jantz, P.A.; Pillay, R.; Jetz, W.; et al. Toward monitoring forest ecosystem integrity within the post-2020 Global Biodiversity Framework. *Conserv. Lett.* **2021**, *14*, e12822. [CrossRef]
12. Carter, S.K.; Fleishman, E.; Leinwand, I.I.F.; Flather, C.H.; Carr, N.B.; Fogarty, F.A.; Leu, M.; Noon, B.R.; Wohlfeil, M.E.; Wood, D.J.A. Quantifying Ecological Integrity of Terrestrial Systems to Inform Management of Multiple-Use Public Lands in the United States. *Environ. Manag.* **2019**, *64*, 1–19. [CrossRef] [PubMed]
13. CBD. *Ecosystem Integrity and International Policy*; Wildlife Conservation Society (WCS): New York, NY, USA, 2021.
14. Reza, M.I.H.; Abdullah, S.A. Regional Index of Ecological Integrity: A need for sustainable management of natural resources. *Ecol. Indic.* **2011**, *11*, 220–229. [CrossRef]
15. Roche, P.K.; Campagne, C.S. From ecosystem integrity to ecosystem condition: A continuity of concepts supporting different aspects of ecosystem sustainability. *Curr. Opin. Environ. Sustain.* **2017**, *29*, 63–68. [CrossRef]
16. Karr, J.R.; Larson, E.R.; Chu, E.W. Ecological integrity is both real and valuable. *Conserv. Sci. Pr.* **2021**, *4*, e583. [CrossRef]
17. Chiarucci, A.; Piovesan, G. Need for a global map of forest naturalness for a sustainable future. *Conserv. Biol.* **2019**, *34*, 368–372. [CrossRef]
18. *The Forest Law of the People's Republic of China Revised in 2019*; Congress The National People's: Beijing, China, 2019.
19. Chazdon, R.L.; Brancalion, P.H.S.; Laestadius, L.; Bennett-Curry, A.; Buckingham, K.; Kumar, C.; Moll-Rocek, J.; Vieira, I.C.G.; Wilson, S.J. When is a forest a forest? Forest concepts and definitions in the era of forest and landscape restoration. *Ambio* **2016**, *45*, 538–550. [CrossRef]
20. Jones, G.M.; Keane, J.J.; Gutiérrez, R.J.; Peery, M.Z. Declining old-forest species as a legacy of large trees lost. *Divers. Distrib.* **2017**, *24*, 341–351. [CrossRef]
21. Morales-Hidalgo, D.; Oswalt, S.; Somanathan, E. Status and trends in global primary forest, protected areas, and areas designated for conservation of biodiversity from the Global Forest Resources Assessment 2015. *For. Ecol. Manag.* **2015**, *352*, 68–77. [CrossRef]
22. Luyssaert, S.; Schulze, E.-D.; Börner, A.; Knohl, A.; Hessenmöller, D.; Law, B.; Ciais, P.; Grace, J. Old-growth forests as global carbon sinks. *Nature* **2008**, *455*, 213–215. [CrossRef]
23. Vertessy, R.A.; Watson, F.G.; O'sullivan, S.K. Factors determining relations between stand age and catchment water balance in mountain ash forests. *For. Ecol. Manag.* **2001**, *143*, 13–26. [CrossRef]
24. Brookhuis, B.; Hein, L. The value of the flood control service of tropical forests: A case study for Trinidad. *For. Policy Econ.* **2016**, *62*, 118–124. [CrossRef]
25. Watson, J.E.M.; Evans, T.; Venter, O.; Williams, B.; Tulloch, A.; Stewart, C.; Thompson, I.; Ray, J.C.; Murray, K.; Salazar, A.; et al. The exceptional value of intact forest ecosystems. *Nat. Ecol. Evol.* **2018**, *2*, 599–610. [CrossRef] [PubMed]
26. McGowan, P.J.K. Mapping the terrestrial human footprint. *Nature* **2016**, *537*, 172–173. [CrossRef] [PubMed]
27. Venter, O.; Sanderson, E.W.; Magrach, A.; Allan, J.R.; Beher, J.; Jones, K.R.; Possingham, H.P.; Laurance, W.F.; Wood, P.; Fekete, B.M.; et al. Sixteen years of change in the global terrestrial human footprint and implications for biodiversity conservation. *Nat. Commun.* **2016**, *7*, 12558. [CrossRef] [PubMed]
28. Theobald, D.M.; Kennedy, C.; Chen, B.; Oakleaf, J.; Baruch-Mordo, S.; Kiesecker, J. Earth transformed: Detailed mapping of global human modification from 1990 to 2017. *Earth Syst. Sci. Data* **2020**, *12*, 1953–1972. [CrossRef]
29. Theobald, D.M. Estimating natural landscape changes from 1992 to 2030 in the conterminous US. *Landsc. Ecol.* **2010**, *25*, 999–1011. [CrossRef]
30. Theobald, D.M. A general model to quantify ecological integrity for landscape assessments and US application. *Landsc. Ecol.* **2013**, *28*, 1859–1874. [CrossRef]
31. Li, S.; Wu, J.; Gong, J.; Li, S. Human footprint in Tibet: Assessing the spatial layout and effectiveness of nature reserves. *Sci. Total Environ.* **2018**, *621*, 18–29. [CrossRef]
32. Song, X.-P.; Hansen, M.C.; Stehman, S.V.; Potapov, P.V.; Tyukavina, A.; Vermote, E.F.; Townshend, J.R. Global land change from 1982 to 2016. *Nature* **2018**, *560*, 639–643. [CrossRef]
33. Lovejoy, T.E. Eden no more. *Sci. Adv.* **2019**, *5*, eaax7492. [CrossRef]

34. Potapov, P.; Hansen, M.C.; Laestadius, L.; Turubanova, S.; Yaroshenko, A.; Thies, C.; Smith, W.; Zhuravleva, I.; Komarova, A.; Minnemeyer, S.; et al. The last frontiers of wilderness: Tracking loss of intact forest landscapes from 2000 to 2013. *Sci. Adv.* **2017**, *3*, e1600821. [CrossRef] [PubMed]
35. Liu, J.; Dietz, T.; Carpenter, S.R.; Alberti, M.; Folke, C.; Moran, E.; Pell, A.N.; Deadman, P.; Kratz, T.; Lubchenco, J.; et al. Complexity of Coupled Human and Natural Systems. *Science* **2007**, *317*, 1513–1516. [CrossRef] [PubMed]
36. Lück-Vogel, M.; O'Farrell, P.J.; Roberts, W. Remote sensing based ecosystem state assessment in the Sandveld Region, South Africa. *Ecol. Indic.* **2013**, *33*, 60–70. [CrossRef]
37. Wei, W.; Swaisgood, R.; Dai, Q.; Yang, Z.; Yuan, S.; Owen, M.A.; Pilfold, N.W.; Yang, X.; Gu, X.; Zhou, H.; et al. Giant panda distributional and habitat-use shifts in a changing landscape. *Conserv. Lett.* **2018**, *11*, e12575. [CrossRef]
38. State Forestry Administration. *The Giant Pandas of China: Status Quo, Major Findings of the Fourth National Survey on Giant Panda*; Science Press: Beijing, China, 2015.
39. State Forestry Administration. *The 3th National Survey Report on Giant Panda in China*; Science Press: Beijing, China, 2006.
40. IUCN. *Sichuan Giant Panda Sanctuaries—Wolong, Mount Siguniang & Jiajin Mountains—2020 Conservation Outlook Assessment*; IUCN: Gland, Switzerland, 2020.
41. Zhang, Y.; Mathewson, P.D.; Zhang, Q.; Porter, W.P.; Ran, J. An ecophysiological perspective on likely giant panda habitat responses to climate change. *Glob. Chang. Biol.* **2017**, *24*, 1804–1816. [CrossRef] [PubMed]
42. Kong, L.; Xu, W.; Zhang, L.; Gong, M.; Xiao, Y.; Ouyang, Z. Habitat conservation redlines for the giant pandas in China. *Biol. Conserv.* **2016**, *210*, 83–88. [CrossRef]
43. He, K.; Dai, Q.; Gu, X.; Zhang, Z.; Zhou, J.; Qi, D.; Gu, X.; Yang, X.; Zhang, W.; Yang, B.; et al. Effects of roads on giant panda distribution: A mountain range scale evaluation. *Sci. Rep.* **2019**, *9*, 1–8. [CrossRef]
44. Xu, W.; Ouyang, Z.; Viña, A.; Zheng, H.; Liu, J.; Xiao, Y. Designing a conservation plan for protecting the habitat for giant pandas in the Qionglai mountain range, China. *Divers. Distrib.* **2006**, *12*, 610–619. [CrossRef]
45. Kang, D.; Wang, X.; Yang, H.; Duan, L.; Li, J. Habitat use by giant pandas (Ailuropoda melanoleuca) in relation to roads in the Wanglang Nature Reserve, People's Republic of China. *Can. J. Zool.* **2014**, *92*, 715–719. [CrossRef]
46. Li, J.; Zhang, T.T.; Yang, W.; Zhang, Y. The Environmental Impact of Mining and Its Countermeasures 2016. *MATEC Web Conf.* **2016**, *63*, 7. [CrossRef]
47. Liu, S.L.; Zhao, Q.H.; Wen, M.X.; Deng, L.; Dong, S.; Wang, C. Assessing the impact of hydroelectric project construction on the ecological integrity of the Nuozhadu Nature Reserve, southwest China. *Stoch. Environ. Res. Risk Assess.* **2013**, *27*, 1709–1718. [CrossRef]
48. Chong, X.Y.; Vericat, D.; Batalla, R.J.; Teo, F.Y.; Lee, K.S.P.; Gibbins, C.N. A review of the impacts of dams on the hydromorphology of tropical rivers. *Sci. Total Environ.* **2021**, *794*, 148686. [CrossRef] [PubMed]
49. Lu, W.W.; Lei, H.M.; Yang, D.W.; Tang, L.; Miao, Q. Quantifying the impacts of small dam construction on hydrological alterations in the Jiulong River basin of Southeast China. *J. Hydrol.* **2018**, *567*, 382–392. [CrossRef]
50. Liu, J.; Liu, M.; Tian, H.; Zhuang, D.; Zhang, Z.; Zhang, W.; Tang, X.; Deng, X. Spatial and temporal patterns of China's cropland during 1990–2000: An analysis based on Landsat TM data. *Remote Sens. Environ.* **2005**, *98*, 442–456. [CrossRef]
51. Hu, J.C. Giant pandas at Qionglai Mountain(in Chinese). *J. Nanchong Norm. Univ. Nat. Sci. Ed.* **1986**, *1*, 21–28.
52. Sichuan Vegetation Cooperative Group, Sichuan. *Sichuan Vegetation*; Sichuan People's Publishing House: Chengdu, China, 1980.
53. Venter, O.; Sanderson, E.W.; Magrach, A.; Allan, J.; Beher, J.; Jones, K.R.; Possingham, H.; Laurance, W.F.; Wood, P.; Fekete, B.M.; et al. Global terrestrial Human Footprint maps for 1993 and 2009. *Sci. Data* **2016**, *3*, 160067. [CrossRef]
54. He, Y.; Fan, G.F.; Zhang, X.W.; Liu, M.; Gao, D. Variation of vegetation NDVI and its response to climate change in Zhejiang Province. *Acta Ecol. Sin.* **2012**, *32*, 4352–4362. [CrossRef]
55. Wang, J.-F.; Li, X.-H.; Christakos, G.; Liao, Y.-L.; Zhang, T.; Gu, X.; Zheng, X.-Y. Geographical Detectors-Based Health Risk Assessment and its Application in the Neural Tube Defects Study of the Heshun Region, China. *Int. J. Geogr. Inf. Sci.* **2010**, *24*, 107–127. [CrossRef]
56. Wang, J.-F.; Zhang, T.-L.; Fu, B.-J. A measure of spatial stratified heterogeneity. *Ecol. Indic.* **2016**, *67*, 250–256. [CrossRef]
57. Dakhil, M.A.; Halmy, M.W.A.; Liao, Z.; Pandey, B.; Zhang, L.; Pan, K.; Sun, X.; Wu, X.; Eid, E.M.; El-Barougy, R.F. Potential risks to endemic conifer montane forests under climate change: Integrative approach for conservation prioritization in southwestern China. *Landsc. Ecol.* **2021**, *36*, 3137–3151. [CrossRef]
58. Hirzel, A.H.; Lay, G.L. Habitat suitability modelling and niche theory. *J. Appl. Ecol.* **2008**, *45*, 1372–1381. [CrossRef]
59. Li, C.; Connor, T.; Bai, W.; Yang, H.; Zhang, J.; Qi, D.; Zhou, C. Dynamics of the giant panda habitat suitability in response to changing anthropogenic disturbance in the Liangshan Mountains. *Biol. Conserv.* **2019**, *237*, 445–455. [CrossRef]
60. Zhen, J.; Wang, X.; Meng, Q.; Song, J.; Liao, Y.; Xiang, B.; Guo, H.; Liu, C.; Yang, R.; Luo, L. Fine-Scale Evaluation of Giant Panda Habitats and Countermeasures against the Future Impacts of Climate Change and Human Disturbance (2015–2050): A Case Study in Ya'an, China. *Sustainability* **2018**, *10*, 1081. [CrossRef]
61. Kong, L.Q.; Xu, W.H.; Xiao, Y.; Pimm, S.L.; Shi, H.; Ouyang, Z.Y. Spatial models of giant pandas under current and future conditions reveal extinction risks. *Nat. Ecol. Evol.* **2021**, *5*, 1309–1316. [CrossRef] [PubMed]
62. Sichuan Forestry Department. *The 4th Survey Report on Giant Panda in Sichuan Province*; Sichuan Science and Technology Press: Chengdu, China, 2015.

63. Li, C.; Yu, J.; Wu, W.; Hou, R.; Yang, Z.; Owens, J.R.; Gu, X.; Xiang, Z.; Qi, D. Evaluating the efficacy of zoning designations for national park management. *Glob. Ecol. Conserv.* **2021**, *27*, e01562. [CrossRef]
64. Li, B.V.; Pimm, S.L.; Li, S.; Zhao, L.; Luo, C. Free-ranging livestock threaten the long-term survival of giant pandas. *Biol. Conserv.* **2017**, *216*, 18–25. [CrossRef]
65. Liu, J.G.; Ouyang, Z.; Taylor, W.W.; Groop, R.; Tan, Y.; Zhang, H. A Framework for Evaluating the Effects of Human Factors on Wildlife Habitat: The Case of Giant Pandas. *Conserv. Biol.* **1999**, *13*, 1360–1370. [CrossRef]
66. Liu, J.; Viña, A. Pandas, Plants, and People. *Ann. Mo. Bot. Gard.* **2014**, *100*, 108–125. [CrossRef]
67. Xu, W.; Wang, X.; Ouyang, Z.; Zhang, J.; Li, Z.; Xiao, Y.; Zheng, H. Conservation of giant panda habitat in South Minshan, China, after the May 2008 earthquake. *Front. Ecol. Environ.* **2009**, *7*, 353–358. [CrossRef]
68. Feng, C.; Cao, M.; Liu, F.; Zhou, Y.; Du, J.; Zhang, L.; Huang, W.; Luo, J.; Jun-Sheng, L.; Wang, W. Improving protected area effectiveness through consideration of different human pressure baselines. *Conserv. Biol.* **2022**. [CrossRef]
69. Huang, Q.; Fei, Y.; Yang, H.; Gu, X.; Songer, M. Giant Panda National Park, a step towards streamlining protected areas and cohesive conservation management in China. *Glob. Ecol. Conserv.* **2020**, *22*, e00947. [CrossRef]

diversity

MDPI

Article

Detection and Quantification of Forest-Agriculture Ecotones Caused by Returning Farmland to Forest Program Using Unmanned Aircraft Imagery

Bin Wang [1,2], Hu Sun [1], Arthur P. Cracknell [3], Yun Deng [2], Qiang Li [1], Luxiang Lin [2], Qian Xu [1], Yuxin Ma [1], Wenli Wang [1,*] and Zhiming Zhang [1,*]

[1] Institute of Ecology and Geobotany, School of Ecology and Environmental Science, Yunnan University, Kunming 650091, China; wb931022@hotmail.com (B.W.); sunlei1071872920@163.com (H.S.); cvpphhct8j@126.com (Q.L.); ynsdbXU@163.com (Q.X.); myx9837@mail.ynu.edu.cn (Y.M.)

[2] Key Laboratory of Tropical Forest Ecology, Xishuangbanna Tropical Botanical Garden, Chinese Academy of Sciences, Menglun 666303, China; dy@xtbg.org.cn (Y.D.); linluxa@xtbg.ac.cn (L.L.)

[3] Division of Electronic Engineering and Physics, University of Dundee, Dundee DD1 4HN, Scotland, UK; apcracknell774787@yahoo.co.uk

* Correspondence: wwl@ynu.edu.cn (W.W.); zzming76@ynu.edu.cn (Z.Z.); Tel.: +86-138-8846-1491 (W.W.); +86-136-6974-9143 (Z.Z.)

Abstract: The 'Returning Farmland to Forest Program' (RFFP) in China has become an essential factor in land cover changes and forest transition, especially in terms of the ecological processes between two adjacent ecosystems. However, accurately delineating ecotones is still a big challenge for vegetation and landscape ecologists. Acquiring high spatial resolution imagery from a small, unmanned aircraft system (UAS) provides new opportunities for studying ecotones at a small scale. This study aims to extract forest-agriculture ecotones by RGB ultrahigh-resolution images from a small UAS and quantify the small biotopes in 3D space. To achieve these objectives, a canopy height model (CHM) is constructed based on a UAS-photogrammetric-derived point cloud, which is derived from the digital surface model (DSM) minus the digital terrain model (DTM). Afterward, according to the difference of plant community height between abandoned farmland ecosystem and forest ecosystem, the ecotones are delineated. A landscape pattern identified with ecotones and other small biotopes at the fine scale. Furthermore, we assess the accuracy of the ecotones' delineation based on the transects method with the previous situ work we carried out and quantify the landscape structure using common landscape metrics to describe its spatial and geometric characteristics. Through transect-based analysis at three transects, the overall accuracy of the width of UAS-derived delineation is greater than 70%, and the detection accuracy for the occurrence location is 100%. Finally, we conclude that ecotones extraction from UAS images would also provide the possibility to gain a comprehensive understanding of the entire ecological process of agricultural abandoned land restoration through continuous investigation and monitoring.

Keywords: returning farmland to forest program; ecotone; unmanned aircraft system; CHM; mountainous landscape; transect-based analysis

Citation: Wang, B.; Sun, H.; Cracknell, A.P.; Deng, Y.; Li, Q.; Lin, L.; Xu, Q.; Ma, Y.; Wang, W.; Zhang, Z. Detection and Quantification of Forest-Agriculture Ecotones Caused by Returning Farmland to Forest Program Using Unmanned Aircraft Imagery. *Diversity* **2022**, *14*, 406. https://doi.org/10.3390/d14050406

Academic Editors: Lin Zhang, Jinniu Wang and Michael Wink

Received: 30 April 2022
Accepted: 18 May 2022
Published: 20 May 2022

1. Introduction

Ecotone was first introduced by Clements [1] as transition zones where principal species from adjacent communities meet their limits. Ecotones are transitional zones between different habitats, and they exist at all spatial and temporal scales [2–4]. Generally, these zones have a set of characteristics (e.g., physiognomy, species composition, etc. [5]) uniquely defined by space and time scales and by the strength of the interactions between adjacent ecological systems [6]. Hence, they exist beside the boundaries between biomes or ecosystems [7]. Due to highly sensitive spatiotemporal dynamics, ecotones play various vital roles in community ecology, landscape ecology, and biodiversity conservation, including the implementation of landscape functions, patterns, and ecological processes, as well as the provision of evidence that "indicates" or "forewarns of" the impacts of global climate change [8–12]. The intrinsic factors driving the spatiotemporal dynamic changes in ecotones are determined by the ecological processes of species migration, settlement, reproduction, and growth at the community scale [13,14].

The Yangtse River flood of 1998 resulted in the 'Returning Farmland to Forest Program' (RFFP) in China, also known as the Sloping Land Conversion Program (SLCP). It aims to restore some farmlands to forests gradually. This policy was initiated in 1999 and expanded in 2002 to cover most of China's provinces and has received immense attention [15,16]. The policy has become an essential factor in land cover changes and forest transition [17], especially in terms of the ecological processes that are responsible for material flows and species movement [18–20]. Studies of land cover change at regional and provincial scales find gains in forest cover in jurisdictions in which the RFFP was implemented [21,22]. Yet land cover change patterns vary sharply at smaller scales [21,23]. A previous study confirmed that ecotones often occur on the local scale [24]. These ecotones are dynamic entities whose changes are always influenced by the interactions of ecological processes between the abandoned agricultural lands and the adjacent ecosystems [25]. In vegetation (or plant community) science, knowledge of ecotones is crucial in understanding basic ecological patterns and processes [26]. Ecotones (or boundary) delineation is a critical issue in vegetation ecology, which can help identify the organization rules of communities and help understand the ecological processes between two adjacent ecosystems [27].

Formerly, several edge-detecting methods were used for detecting ecotones and transition zones using one-dimensional transect data [26]. The moving split-window (MSW) technique is the most used ecotone delineation method. It is hard to detect the spatial features of ecotones using the one-dimensional ecotone delineation technique, such as location, shapes, and spatial distribution patterns [8]. The delineation of ecotones at the community level is greatly affected by topography, data sampling directions, and the size and numbers of samples using one-dimensional MSW. However, these edge detection methods often need an excessive amount of effort and meticulous sampling to obtain ecological data in the same manner (grid, regular lattice) and amount (exhaustive study area) as pixel data based on transects. Delineating ecotones accurately continues to be a significant difficulty for vegetation and landscape ecologists.

Currently, for mapping the continuous spatial distribution and quantifying the degree of ecotone occurrence, algorithms also exist for detecting edges from the aerial view, and the satellite remotely sensed image data [28,29]. Remote sensing data acquired from satellites and piloted aircraft are potent tools for measuring the state and trends of environmental changes associated with natural processes and human disturbances. However, the conventional airborne and satellite remote sensing platforms upon which most sensors are mounted have not always met the needs of researchers and environmental professionals [30,31]. Not only do data gathered using conventional remote sensing systems lack operational flexibility and variety, but they also have a poor spatial and temporal resolution. With the fast growth of photogrammetry and 3D visualization technology in recent years, approaches for creating 3D models from regular digital images have evolved [32,33]. Researchers have used conventional RGB digital cameras mounted on a UAS to determine tree heights [34,35] and crop heights [36], and for biomass estimations [37,38]. Such applications have raised the possibility of using UAS-mounted ordinary digital cameras to obtain information on the heights and 3D structures of plant communities, thereby providing new research ideas and means to determine the spatial structures of ecotones. Hou and Walz [39] developed a method for detecting ecotones based on change in plant height and landcover by integrating several remote sensing data. However, data with varying resolutions may result in estimating errors.

Previous studies have suggested that 3D or even multidimensional in space and time information are key components of ecotones [11] and incorporating landscape 3D structures into modeling and monitoring will produce outcomes that better approximate reality [40].

An inherent property of forest-agricultural ecotones is higher structural heterogeneity (changes in vegetation height) than adjacent plant communities [5,41], permitting UAS-based 3D photographic techniques to be used for ecotone delineation. The objective of this study is comparing the efficacy of UAS-derived canopy height versus traditional transect methods to extract the width of forest-abandoned-land ecotones at known RFFP sites. The UAS acquired photogrammetry point cloud and high-resolution orthoimage will be derived using UAS photogrammetry techniques in addition to field investigations using the transect method as ground truth. Optimizing limited field time is conducive to monitoring and researching landscape pattern changes and monitoring the associated ecological processes, thereby leading to better assessments of the mountainous ecological restoration process after farmland abandonment.

2. Materials and Methods

2.1. Study Area

The study area is located in Yunnan Province, China. Weixi, in the County of the Diqing Tibetan Autonomous Prefecture (98°54′–99°34′ E, 26°53′–28°02′ N) (Figure 1). The study area belongs to the temperate mountain monsoon climate, with the annual temperature difference of 11.3 °C, annual sunshine duration of 2071.3 h, annual precipitation of 954 mm, frost-free period of 195 days, and the vertical change of mountain climate is obvious. The forest soil in Weixi county showed an obvious vertical distribution, from high altitude to low altitude, alpine meadow soil, subalpine shrub meadow soil, bleaching earth, dark brown soil, yellow brown soil, and red soil. The soil in this study area is dark brown soil. It is a steep mountain area, and the prominent disturbances are grazing and firewood collection. There is a typical landscape with ecotones across the original Yunnan Pinus (*Pinus yunnanensis*) community and an abandoned land under the influence of the RFFP. Since 2003, the impact of this policy has been here for more than a decade. Before that, the study area was mainly planted with corn. In addition to Yunnan Pine, other dominant species in this study area include *Coriaria nepalensis* and *Desmodium yunnanense*.

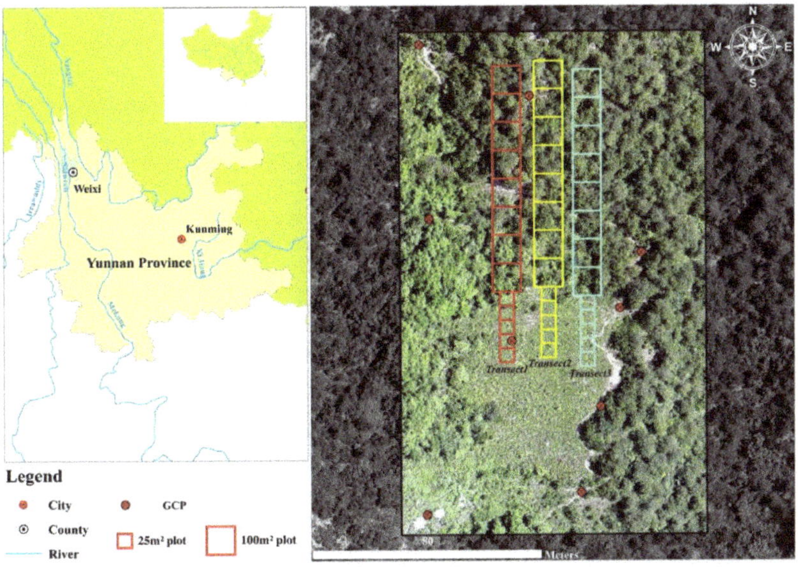

Figure 1. Location of the sample plot. The right pic is the Baijixun sample plot, Weixi, with 19,203 m^2. Furthermore, here are three sample transects that were be set up along the slope.

2.2. In-Situ Data

Three parallel sample transects were set along the direction perpendicular to the forest edge, with an interval of 20 m. Taking the forest margin as the starting point, quadrats were set in two opposite directions, respectively. Starting from the edge of the forest, five quadrats with an area of 5 m × 5 m were set in the direction of the abandoned land. The original Yunnan pine forest was set with eight quadrats with an area of 10 m × 10 m continuously (Figure 1 and Figure S1). The area of each quadrat on abandoned land or forest has been confirmed using the nested sampling approach, which in earlier work considered the species–area relationship [42]. Based on the sampling strategies (S1), the work of plant species investigation in each of the quadrats was completed in May 2013. At the same time, in each quadrat, multiple points were randomly and uniformly selected to collect soil, and the soil was mixed to ensure its representativeness (see S1). The pH value and the organic matter (OM) content of soil were be measured using the potentiometric method and oil-bath heating potassium dichromate volumetric method, respectively (see Table S1) [42].

To detect the ecotones based on photogrammetric technology, a DJI M600Pro UAS platform (DJI, Shenzhen, China) was used to collect aerial imagery with a spatial resolution of 0.1 m. The platform uses an intelligent flight control system, a professional tripod head, and a Sony a7r (Sony, Tokyo, Japen) digital camera taking red (R), green (G), and blue (B) images for aerial photography. To assure comprehensive and practical remote sensing data, this research covers the following important steps: (1) The camera's lens length, shutter speed, and orientation were manually tuned depending on ground surface reflectance and ambient light conditions. (2) Nine GCPs (Ground Control Point) were put in open regions of the Baijixun sample plot, and their precise locations were gathered using RTK (Real-time kinematic) (CHCNAV, Shanghai, China). (3) For automated flying, the parallel path plan is used, with 80% longitudinal and 60% lateral overlap. (4) After UAS data collection, photos are sent to the field laptop and analyzed using Photoscan software (now Metashape). In May 2016, UAS flights at 133 m yielded 822 usable photographs. 131 photos were taken above our locations.

2.3. Image Preprocessing and Photogrammetric Processing

Initially, some hazy images are filtered out owing to the swaying of trees caused by wind. The lightness of some shaded pics is enhanced, and the lightness of some pics with overexposure is debased by the set Brightness tool in Photoscan software. To avoid the influence from the difference of color between images in image stitching, the histogram matching was used to adjusted bias images. Sequentially, a series of parameters (medium aligning accuracy, high-quality point cloud, and mild depth-filtering level; refer to Figure 2) were adopted to produce high-quality photogrammetric point cloud data. It is important to emphasize that the point cloud data have 'medium aligning accuracy' (Figure 2). It is because of considering the influence from the high sensibility of aligning algorithm with 'high aligning accuracy', and it may cause an underestimate of canopy height, as no aligning can be made for the tie points at the top of the canopy. Nevertheless, it cannot reflect the real height of nonvegetation if the data with low accuracy. By the way, a point cloud data set sufficient to characterize 3D vegetation communities is constructed. Based on this, according to the orthoimage generation process flow of the software (Photoscan). Finally, a high resolution (0.08 m/pix) orthoimage is produced, and the degree of overlap is bigger than 90%.

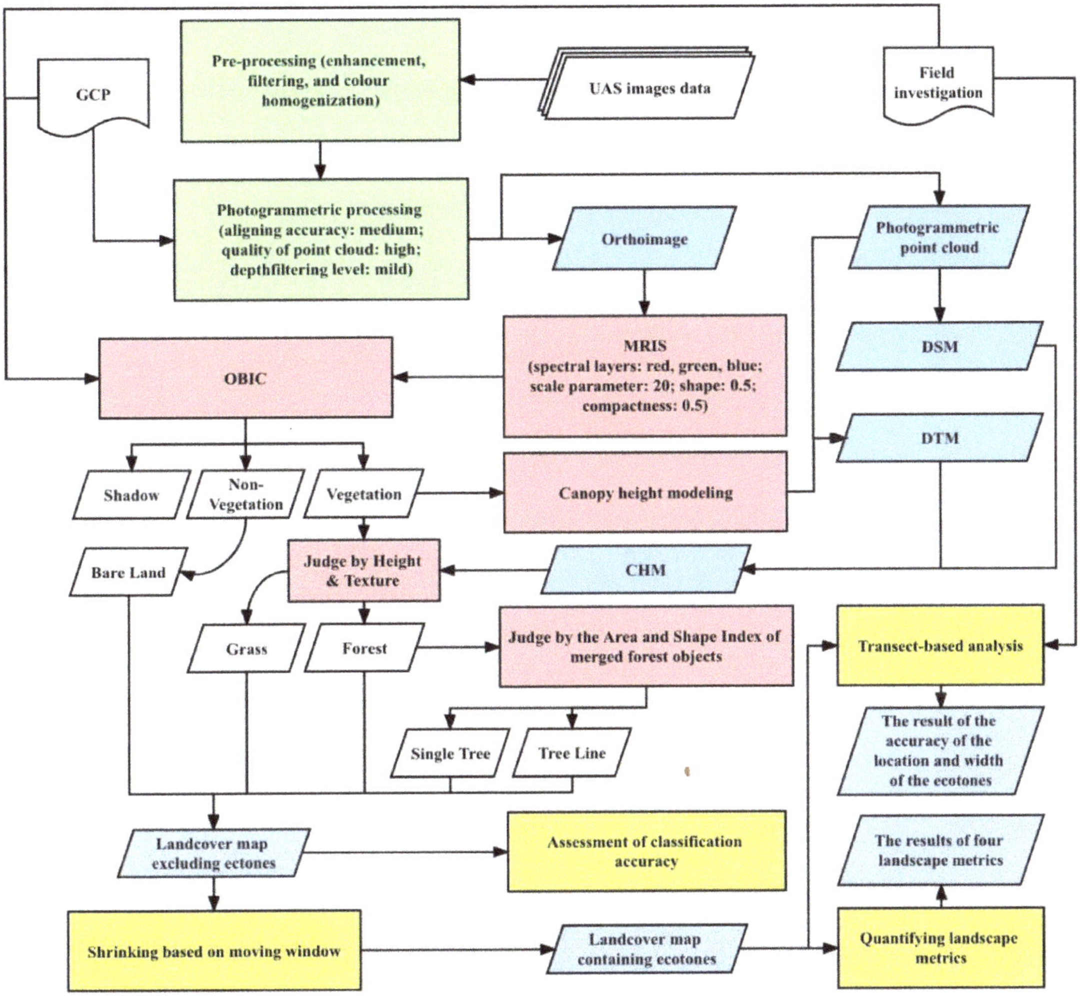

Figure 2. Diagram of detecting and measuring ecotones utilizing UAS images, incorporating a preprocessing flow, a processing workflow of object-based classification, and the CHM extraction workflow. Prepreprocessing is represented by the boxes with a green backdrop. The red-background boxes represent object-based classification processing. Yellow boxes reflect methods for identifying, measuring, and validating ecotones derived from UAS. The parallelograms with a blue background indicate the data or outcomes created by the study framework.

2.4. Object-Based Image Classification (OBIC)

Exploiting the high resolution of UAS imagery, OBIC was chosen to map land cover for the Baijixun study area. In this study, the multiresolution image segmentation (MRIS) algorithm is used [43]. It should be emphasized that in OBIC, the accuracy of segmentation directly affects the accuracy of classification. Hence, the ESP tool was adopted to estimate the optimal value of segmentation parameters. It is an effective method to determine the optimal value of segmentation parameters by controlling variables in the MRIS [44]. In our case, the Shape and Compactness are adjusted to 0.5 and 0.5, respectively (Figures 2 and S3), and the scale parameter related to MRIS is adjusted to 20 to obtain a better result of segmentation (Figure S4)

Afterward, 12581 objects from segmentation are classified into different land cover classes using the random forest (RF) classifier. The RF classifier was chosen mainly because this machine learning algorithm does not need many samples, has fast training speed, and has good antioverfitting ability [45]. In this study, vegetation, nonvegetation, and shadow were distinguished based on orthoimage. The training samples combined the data from the previous situ investigation, the information of GCPs, and some sample points selected based on visual inspection. Combined with field survey data and visual inspection, a total of 138 training sample points were obtained, 30 of which were shaded, 38 of which were nonvegetated, and 70 of which were vegetated (Figure S2). The three bands (R, G, B) were inputted as predictive variables. Out of 113 selected samples, 101 (approximately 90% of the total) sample points were selected as training samples, and the remaining samples were used as validation samples. Then, based on the number of samples and the number of classification targets, the default setting of the RF classifier was accepted. Corresponding objects were determined according to the spatial position of investigation, and the samples of the training classifier were finally involved in the form of objects. In order to detect more small biotopes from the high-resolution orthoimage, characteristic of structure (i.e., CHM), geometrical characteristic (area and shape index), and texture information were crucial. The contrast index and the dissimilarity index were selected for the texture index, generated according to the gray level co-occurrence matrix. Moreover, a rule-based classification scheme was constructed, with different rules and thresholds for each class (Figure 2). Different subclasses have significant differences in different feature dimensions, so every subclass was judged from the parent class by the most sensitive features. We completed the processing of OBIC above in eCognition Developer software.

2.5. Canopy Height Modeling

The nonvegetation area in the land cover classification realized in the above steps was extracted to extract the nonvegetation points from the whole point cloud data. Nonvegetation points were merged with points of GCP on the field survey as modeling samples. As the geostatistical method, the empirical Bayesian kriging (EBK) was adopted after cross-validation with other four common models (S3) to construct a geostatistical model (i.e., digital terrain model (DTM)) for predicting the entire terrain elevation. Before modeling with these nonvegetation points, some points within other classes that could have been misclassified into nonvegetation points needed to be filtered using a slope-based filter method [46]. Furthermore, before modeling by EBK, declustering was performed and the global first-order trend was removed from the data (S3). The point clouds were also used to derive the digital surface model (DSM), a raster-based description of the surface that includes objects on the terrain, such as trees and shrubs. In this work, multilevel B-spline interpolation was used to generate the DSM, which is a high-speed interpolation algorithm for calculating the continuous surface from points based on irregular regional samples [47]. Finally, a superimposed subtraction of the two sets of data produced a canopy height model (CHM) by the grid difference, which is a raster computing tool used to subtract the DTM from the DSM to produce the CHM. The geostatistical analysis was accomplished by the ArcGIS Pro software. The processing of DSM and CHM data was achieved by the SAGAGIS software.

2.6. Extraction and Classification of Forests–Ecotone–Abandoned Landscape

In this study, the ecotones between forest and abandoned agricultural land refer to mixed vegetation above the grass layer at abandoned land but below the forest layer, and they formed height gradient transitional zones. This vegetation height difference of transitions between abandoned agricultural land and forest ecosystems provided a meaningful result in this case and can be used as the characteristic for a three-dimensional landscape [39,48].

For extracting the ecotones between forest and abandoned agricultural land, two prerequisites are used: vegetation height and proportion of ecotones. In this study, the ecotone detection method developed by Hou and Walz [48] was adopted. First, the land cover map obtained using OBIC described above is used as the thematic layer. The forest and agricultural land classes were overlaid with the CHM (DSM-DTM). The height difference is used as a criterion of ecotones. Thus, it was necessary to define the threshold values of height for the distinguish of ecotones. At the Baijixun plot, the average heights of the vegetation in the forest and abandoned agricultural land were computed by the zonal statistic. Second, the proportion of nonecotone pixels (i.e., forest pixels or abandoned land pixels) within a fixed size "window" was assumed to be between 40% and 60% based on the model from Riitters et al. [49]. Unlike the Hou and Walz method, CHM extracted from the photographic point cloud could be completely matched with the land cover map based on the high-resolution classification in our study. Therefore, combined with Hou and Walz's method and the perspective of Riitters, the classification results obtained from the above classification system needed to be shrunken only. In our case, that meant inside a moving window (61 by 61 pixels) surrounding the ecotone pixel, the proportion of both forest and abandoned land pixels should be less than 60%. Combined with the result of classification and CHM data, the average height of forest (H_f) and the average height of abandoned land (Ha) in the landscape were calculated. The detailed step was to form transitions by shrinking grass along the border pixels where CHM $\geq$ Ha m and the relative area of grass pixels in a moving window of (61 by 61) pixels < 60% and shrinkage of the forest where CHM $\leq H_f$ m and the relative area of forest pixels in a moving window of (61 by 61) pixels < 60% (Figure 3). Finally, we assigned ecotone class to the unclassified class and involved eliminating impurities inside ecotones. It should be emphasized that considering for the shrinking from the nongrass land class (i.e., shadow class and bare land class), the shadow class and bare land class were reassigned into the interior of forest class or the interior of grass class according to the proportion of classes in the nearest pixel before do the processing of shrinking. The shrunken work was completed with a tool that is 'pixel-based object resizing' in eCognition Developer software.

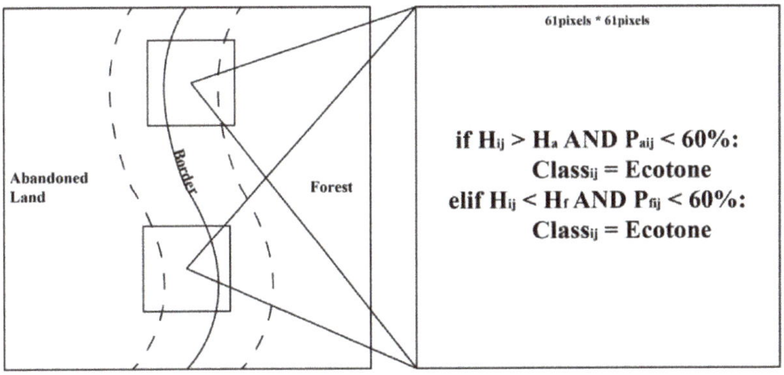

Figure 3. Shrinking process diagram. The moving window along the border between the abandoned land and the forest was used as the judgment basis for the shrinking into the interior of the forest and the interior of the abandoned land where $\boldsymbol{P_a}$ denotes the proportion of pixels in the whole window of the abandoned land. Moreover, the $\boldsymbol{P_f}$ denotes the proportion of pixels in the whole window of the forest. Ha and Hf denote the average height of abandoned land and forest, respectively.

2.7. *Quantifying the Landscape Containing Ecotones*

Compared with the traditional ecotone identification method, this is an obvious advantage of the spatial continuous detection results, that is, the spatial quantization of the ecotone can be realized. Mapping the landscape pattern, including the ecotones, not only shows where the ecotones occur but also makes it possible to quantify the scale of the ecotones. At the CLASS level, a series of landscape metrics are computed in an R package (landscapemetrics [50]) to assess the landscape characteristics of ecotones. The quantitative indicators selected are shown in Table S3. Furthermore, the total area (TA) reflects the size of the corresponding class in the landscape and measures the composition of the landscape. The percentage of landscape (PLAND) is a relative measure that quantifies the landscape weight of each class in the landscape, which is also an important indicator for measuring the landscape composition. The total edge (TE) reflects the total edge length of a class, and longer edge lengths may be associated with larger edge effects. The shape index (SI) is the most straightforward measure of complexity, and the changes in a landscape may reflect the extent of disturbance. In fact, these common indicators are also rich in ecological significance. TA is a direct measure of the scale of landscape elements, and the area of vegetation has an impact on the species in it, that is, the area effect [51]. While PLAND reflects the weight of corresponding classes in the landscape, it also indirectly reflects the contribution of this class to the spatial heterogeneity of landscape elements within the landscape. It is assumed that the larger the PLAND of a class is, the more outstanding the contribution of the corresponding class to the homogeneity of the habitat types within the landscape is, which may influence the biodiversity within the landscape, because it may lead to concentrations of rare species in marginal habitats [52]. TE and SI measure the edge characteristics of land cover types, which may be related to the edge effect of the corresponding land cover class [53].

2.8. Transect-Based Analysis

The ecotones occur between adjacent ecosystems frequently with high heterogeneity. In general, the variability of biodiversity is sensitive to environmental change. The species importance value (IV) is computed in every species per small sample quadrats. In addition, the mean values of the importance values of plants included in each quadrat are calculated to represent the importance values at the level of the quadrat. This work (i.e., the measurement of IV, pH, and OM) has been conducted in previous experiments (Table S1). For analysis of the variation along transects from abandoned land to the interior of the forest, the MSW alone transect was adopted in this study. The MSW is a common approach to reflect the habitat change or indicate how habitats are separated [54]. The discrepancy coefficient was computed between two subwindows (1/2 window) using the squared Euclidean distance (SED) (Formula (1))

$$SED_n = \left(\overline{X}_{iaw} - \overline{X}_{ibw}\right)^2 \quad (1)$$

where n denotes the location of the medium point of two subwindows and a and b respectively represent the two subwindows when the window is n. The w is the size of a window. The m is the number of parameters in a quadrat [55]. The times of removing are determined by subtracting the size of the window from the number of total quadrats and adding 1. In addition, the experiment was carried out several times with a variety of window sizes so that any potential bias caused by the impact of window scale may be eliminated. According to the findings of our earlier study, the position and breadth of the ecotones may be identified more precisely when the size of a window is set to 6. [56]. In this study, eight medium points were determined (The distance along the one-dimensional transect from the starting point is 17.5 m, 25 m, 32.5 m, 40 m, 47.5 m, 55 m, 65 m, and 75 m, respectively) by the window size and the number of all quadrats.

Meanwhile, the obtained results were normalized. The calculated distance coefficient (i.e., SED) was plotted along the transect. The dependent variable was the value of the normalized result, and the independent variable was the position of the starting distance from the transect. Similarly, the width of ecotones occurrence was expressed at the same location of the transects. An interpolation line was set up through the midpoint of each small quadrat, the starting point and ending point of the entire transect. The landscape map containing ecotones was binarized before computing by interpolation line, with one as the ecotones and zero as the nonecotone. Based on the graph, the occurrence of the location of ecotones was detected according to crest and estimate the scale of ecotone according to the width of wave.

2.9. Accuracy Assessment for Ecotone Detection

Two aspects of accuracy need to be verified in this study. One aspect was the classification accuracy of the landcover of biotopes, which will directly affect the ecotones' location of occurrence. Another aspect was the delineation accuracy of the ecotones, which tests the accuracy of the ecotones' width based on our method, reflecting the scale (width) of the transition.

In this study, field surveys are carried out to establish the characteristics and variability of land cover and to acquire reliable field data for training the classifiers of the UAS image data and evaluating the accuracy of the classification results. The ground truth data were combined with the UAS images to assess the accuracy of the classification. The F1-measure was adapted for assessing the accuracy of each class [48]. This indicates the harmonic mean between precision (p_i) and recall (r_i) for each class i, because recall and precision are evenly weighted [57].

$$F_1 = \frac{2p_i r_i}{p_i + r_i} \tag{2}$$

p_i denotes objects correctly classified as class i/(ground reference objects in class i) and r_i denotes objects correctly classified as class i/(number of objects correctly classified as class i + objects falsely classified as class i).

Meanwhile, the F1-measure was also used to test the accuracy of the scale of ecotones occurrence extracted based on proposed method.

$$F_1 = \frac{2p_j r_j}{p_j + r_j} \tag{3}$$

$$p_j = \frac{TP_j}{TP_j + FP_j} \tag{4}$$

$$r_j = \frac{TP_j}{TP_j + FN_j} \tag{5}$$

TP_j denotes the overlap width of the ecotones derived from the UAS data and the ecotones detected from the analysis of the transects based on situ investigation at transect j. FP_j denotes the width of the part not detected by situ data but detected in the UAS data. FN_j denotes the width of the part not detected by UAS data but detected in the situ data. The p_j represents Precision, and the r_j represents Recall.

3. Results

3.1. Canopy Height Estimate

In the Baijixun sample site, the unbiasedness and uncertainty of multiple geostatistical models were compared by cross-validation method, and the EBK was finally selected to model the DTM. A CHM was successfully used for modeling, and it was used to interpolate the height value across the study area (Figure 4). From the results, the predicted slope direction was consistent with the actual slope direction, and the vegetation height changed obviously at the boundary between abandoned land and forest (in the north) (Figures 1 and 4). According to the results of OBIC, we calculated that the average height of the forest class was 3.84 m, and that of the abandoned land class was 0.27 m.

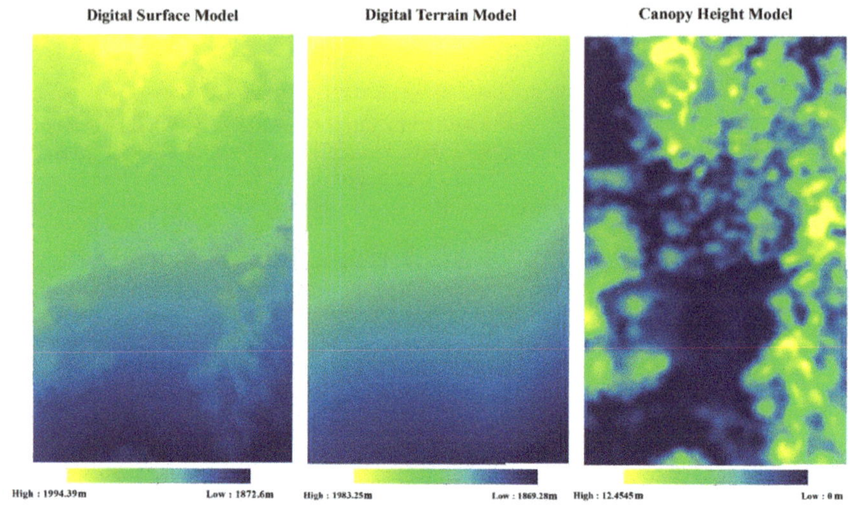

Figure 4. The digital surface model (DSM) (**a**), digital terrain model (DTM) (**b**), and canopy height model (CHM) (**c**) of the Baijixun plot.

3.2. Landscape Pattern with Small Biotopes

In this study, a landscape map was obtained using the object-based image analysis described in the methodology section. The accuracies of the classifications are listed in Table 1, respectively. The overall accuracy of classifying based on UAS photogrammetric orthoimage is 95% (Table 1). This landscape pattern (land cover map) clearly shows a typical landscape of abandoned land surrounded by pine forest communities, as well as a particular pattern of small biotopes (Figure 5b). Based on the high-resolution orthoimage data of UAS, and under the classification framework proposed by us, the detailed object classification was obtained, which is an essential prerequisite for further detecting the ecotones.

Table 1. Accuracy of classification of subtle landscape elements.

Accuracy Index	Forest	Abandoned Land	Bare Land	Single Tree	Tree Line	Shadow
Precision (%)	97.00	87.00	96.00	99.00	97.00	96.00
Recall (%)	94.00	90.00	95.00	98.00	97.00	97.00
F1 (%)	95.00	88.00	95.00	98.00	97.00	96.00
Overall accuracy (%)			95.00			

3.3. Ecotones Detection and Quantification

A window moving along the edge between abandoned land and forest was used to judge pixels under the above two conditions (height restriction and area ratio). A landscape with ecotones between abandoned land and forest boundaries was delineated (Figure 5c). Subsequently, the corresponding landscape metrics calculation results are also calculated based on the fine resolution landscape map (Table 2). In general, the ecotones between abandoned land and forest in the Baijixun plot was irregular to the west and north of the abandoned land (Figure 5c). The ecotones were expanding from forest interior to bare land or abandoned agricultural land. This landscape pattern indicates the gradient of vegetation spontaneous regeneration between abandoned farmland and forest. To the east, the abandoned farmland boundaries left over from the past prior to degradation remained the boundaries between the ecotones and the forests, and their shapes are quite regular (e.g., ecotones elements in this landscape have lower SI values (Table 2)). It may also be disturbed by a small amount of human activity, as shown in Figure 5, where tracklike bare land was identified, and result in a higher landscape contrast of the eastern boundary.

The texture characteristics revealed that the region contained sparse vegetation and large amounts of bare land. From the quantitative results, the ecotone class has the maximum boundary length. It also occupies a large proportion in landscape proportion, almost equal to the type of abandoned land.

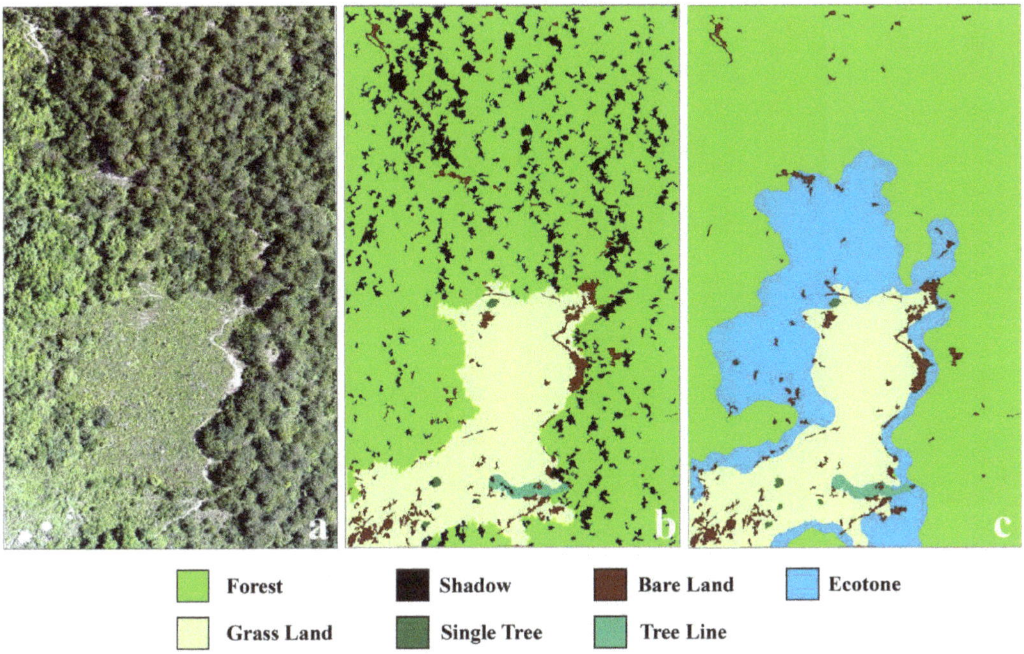

Figure 5. An orthoimage of the study area (**a**), a landscape patter map derived from OBIC (**b**), and (**c**) a landscape including the ecotones extracted (shadows removed).

Table 2. A series of common landscape metrics used to quantify some elements of the landscape.

Landscape Metrics	Forest	Bare Land	Abandoned Land	Ecotones	Tree Line	Single Tree
Total Area (m^2)	1196.90	49.30	288.90	391.90	0.01	2.50
Percentage of Landscape (%)	61.74	2.54	14.90	20.22	0.47	0.13
Total Edge (m)	1100.90	2203.10	1537.70	1631.70	125.00	27.20
Shape Index	1.55	2.39	2.27	1.82	3.27	1.30

3.4. Transect-Based Analysis

Separately, our research group conducted a series of studies on ecotones in the agro-forestry ecosystem of the Baijixun sample plot. These investigations were mainly carried out through the laying of transect lines. Analyses of plant diversity levels within the community and soil properties were used to determine the dynamic changing characteristics and width range of the ecotones. The results showed that the ecotones in the Baijixun plot were 37.5–57.5 m, 30–42.5 m, and 30–57.5 m width in IV, pH, and OM, respectively (Figure 6 and Table S1). However, when the landscape pattern diagram of the sample plot produced in this study was used as the basis, and distances were measured along the same direction as the transect lines laid previously, the widths of the ecotones were found to be 35.43–55.20 m.

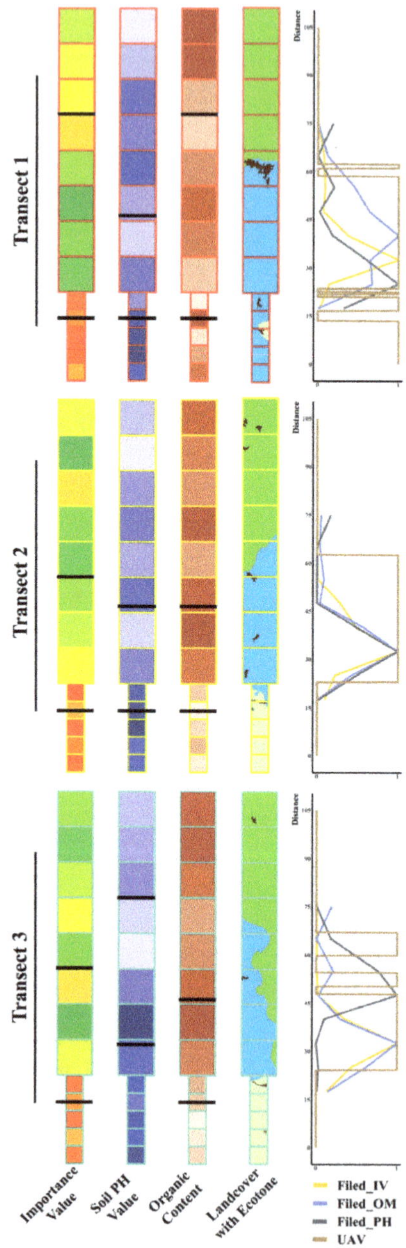

Figure 6. The data obtained in situ were analyzed based on transects. Every transect consists of 8 big quadrats (10 × 10 m) and five small quadrats (5 × 5 m). IV, pH, and OM were surveyed in every quadrat. In this figure, the first column (from left) denotes the distribution of IV along a transect, the second denotes pH, and the third is OM. Three color plates reflect the changes of value in each quadrat, and the higher the value of a quadrat, the deeper the color. The fourth column reflects a land cover with ecotones along a transect. The fifth column shows the changes of different indicators along the direction of the transect. The black lines in the first three columns act as scale to indicate the width of the ecotones detected based on the MSW.

IV, pH, and OM values in each quadrat were measured and presented spatial heterogeneity along the transect (Figure 6). In the three transects, the MSW method was used to calculate the changes of three indicators from field survey along the transects. The changes of field data and the changes of ecotones derived from photogrammetric orthoimage were plotted into a coordinate (Figure 6). The method of F-1 measure was used to test the accuracy of our method on the detection of the ecotones. The overall accuracy was greater than 74% in the ecotones scale (width) detection (Table 3). The location of the crest was the location where the ecotones occurred. From the results, in terms of the accuracy of the occurrence location, the detection accuracy was 100% (Figure 6).

Table 3. Test the accuracies of ecotones width derived from UAS at corresponding transect in different data. For the meaning of TP, FN, and FP, refer to the accuracy assessment section above.

Accuracy Index	Transect 1				Transect 2				Transect 3			
	IV	PH	OM	UAS	IV	PH	OM	UAS	IV	PH	OM	UAS
Width (m)	57.50	30.00	57.50	55.20	37.50	30.00	30.00	39.72	37.50	42.50	30.00	35.43
TP (m)	40.79	28.30	40.79	\	32.05	24.55	24.55	\	28.15	27.08	23.35	\
FN (m)	16.71	1.70	16.71		5.45	5.45	5.45		9.35	15.42	6.65	
FP (m)	14.41	26.89	14.41		7.68	15.18	15.18		7.29	8.35	12.08	
Precision (%)	73.90	51.28	73.90		80.68	61.80	61.80		79.44	76.44	65.89	
Recall (%)	70.94	94.34	70.94		85.46	81.82	81.82		75.05	63.73	77.82	
F1(%)	72.39	66.44	72.39		83.00	70.41	70.41		77.18	69.51	71.36	
Overall accuracy (%)	70.41				74.61				72.68			

4. Discussion

4.1. Mapping Landscape with Small Biotopes

The heterogeneous character of the landscape (mosaic) and the influence of the spatial arrangement of the composing patches of many ecological processes have been recognized [7,9,58]. As shown in Figure 5, this reflects that the ecotones shapes formed a more complicated landscape. However, ecotones are essential components of heterogeneity, which are usually ignored in traditional landscapes. The agricultural ecosystem of the Baijixun sample plot has been abandoned for more than ten years and has been transformed into grassland, with the surrounding forests showing the trend of expansion. Ecotones are growing around the boundary between the abandoned land and forests. It has created the considerable potential for species exchange and materials flux between the forests and abandoned land [56,59,60]. It also indicates that the interaction between the forests and grassland ecosystems is more robust in the ecotones. This would better reflect the process and progress of ecological restoration after the implementation of RFFP. As mentioned above, the RFFP policy has played a significant role in land cover transitions [16,61,62], as well as the ecological processes that are responsible for the spatial distribution of species [18]. However, the conventional landscape pattern without ecotones derived from remote sensing image data generally ignores the gradual transition zones between different landscape units or patches [63,64]. This is also why we conduct a two-step data validation, because the shrunken-based detection method relies on highly accurate land cover mapping, and our validation clearly meets this need (Table 1). The experience based on field investigation provides reliable verification object for verification work. Note that this result is also imaged by OBIC. It also means that the forest and vegetation transition processes caused by the RFFP policy could also be ignored when the landscape or land cover without ecotones maps are used. In this study, the findings revealed the landscape pattern of all ecotones within the sample plot, which facilitated the concurrent measurement of the spatial patterns of ecotones. With continuous investigation and monitoring, it is believed that it will be possible to gain an improved understanding of the entire ecological process of agricultural abandoned land restoration.

4.2. Canopy Height Model

At the aspect of CHM modeling, the point cloud data derived from photogrammetry is used to extract ground points supported by the result of OBIC from the high-resolution orthoimage. The terrain elevation of the unknown region is predicted by using geostatistics (Figure 4). It should be emphasized that this is still a model for prediction based on geostatistics, although this approach has achieved some results in our study. The success of the experiment depends on the topographic heterogeneity of the predicted area and the spatial configuration of the ground points that may be extracted. Moreover, according to the cross-validation results, we found that regardless of the statistical model, the prediction of variability is generally underestimated (Table S2). It reflects the possibility that the fluctuation of topographic heterogeneity on the fine scale is beyond the prediction ability of geostatistical models. Therefore, it may be more effective to restore the changes of terrain elevation at a large scale, as this increases the probability of obtaining effective ground points. It reflects the feasibility of the technology to monitor the landscape with ecotones under the influence of the policy of RFFP.

4.3. Extraction of Ecotones

Data sizes and resolutions can be applied effectively when studying small-scale landscape patterns. It is also noted that transitions between abandoned farmland and forest ecosystems have created height variations among plant communities, and a CHM built by DSM and DTM data could highlight such differences. Hence, this model is used for the extraction of the forest-agriculture ecotones. Hou and Waltz [48,61] have similarly used height variations between different plant communities to segregate and extract small-scale landscape units such as forest–grassland ecotones and tree lines from a 3D perspective. They used airborne LiDAR elevation data and high-resolution satellite image data to classify 3D landscape maps before conducting a comparative analysis between the results and those obtained using the traditional 2D classification method. The results confirmed that 3D height information could reflect gradient changes occurring at forest boundaries and thus be used to extract small-scale landscape units such as ecotones. However, LiDAR scanners are still not considered as standard tools because of the high cost of airborne LiDAR instruments and peripherals (reference targets, tripods, software suites, graphic workstations), as well as data collection [65].

Moreover, some previous study has shown that the point clouds derived from LiDAR and RGB sensors successfully capture the 3D structure [66]. Photogrammetric processing relies on identifiable features to be matched across sequences of images. It should be emphasized that the structural information obtained based on the photogrammetric point cloud is naturally compatible with the orthophoto image, so in this study, the ecotones extraction method is different from that of Hou. As described in the method part, we did not do growing as Hou did. We directly shrink due to the perfect match between the structure information from the photogrammetric point cloud and image data. Image data acquired through UAS photography have higher resolution, which means that the proposed method may be superior to the traditional satellite remote sensing method for regional- and small-scale studies on ecotones. Although the resolution of some satellite images (e.g., Sentinel-2 data [67]) has improved in recent years, and the heights of ground objects can be accurately obtained from airborne or satellite LiDAR (e.g., ICESat-2 data [68]) data, both methods incur substantial data acquisition costs, and the data cycle cannot be prescribed. By contrast, UAS has the advantages of flexibility and controllability. Image data obtained through this method have various characteristics, including being regionalized, customized, and personalized. In addition to satisfying various research requirements, the cost of using this method is also generally acceptable to most research organizations and teams.

4.4. Quantification of Ecotones

Traditionally, studies on ecotones have often relied on transect lines laid along environmental gradients to collect data on plant communities and soil properties and then determined the range of ecotones using moving split-window techniques. Being constrained by human or terrain factors, this method only allows research to be conducted in one direction along the transect lines. By contrast, the remote sensing method based on UAS photographic technology allows comprehensive regional data acquisition. It permits analysis and research from the perspective of landscape patterns, thereby eliminating the necessity for multiple sessions of field investigations and thus making this method obviously advantageous in terms of reduced working hours and intensities. This fully demonstrates the advantages of continuous spatial pattern characterization in landscape ecology research, such as quantified area or edge length (Table 2). The results of this study on the quantification of the landscape containing the ecotones that the ecotones occupy a considerable weight in the whole landscape can be seen from the results of this study. Both TA and PLAND reflect the important spatial landscape status of the ecotones, which is often neglected in previous landscape ecology studies. The neglect of the ecotones may lead to underestimating landscape heterogeneity and the overestimation of landscape contrast. It can be seen from the PLAND index that the dominant patches in the landscape are forest, abandoned land, and ecotones. It also reflects the high landscape heterogeneity. In the TE index, the values of ecotones and abandoned land are close to each other.

4.5. Transect-Based Analysis

Ecotones were mainly determined based on the variations in vegetation height when transitioning between different communities in this study. By contrast, transect surveys use plant diversity levels within the community or soil indicators as the bases. The proliferation and migration of plant species can be determined by surveying sample communities (Figure 6). The herbaceous plants growing within forests can also proliferate over wide extents. However, in this study, we could only determine the height changes at the canopy of the communities but could not assess the specific changes of plants growing in the understory. It might explain the differences in measurement results for the Baijixun sample plot between this study and earlier transect surveys conducted at the traditional community scale. The difference may also be caused by the inaccurate expression of the width in the traditional one-dimensional transect analysis method. Thus, in this study, interpolation lines are used to measure the binarization map with ecotones, and more accurate width measurement results are obtained. It is also worth noting that due to restrictions imposed by the terrains of the sample plots, previous transect surveys could only lay transect lines along one direction: from the base of the mountains up toward the upper edge of the abandoned agricultural land [42,56,60]. Consequently, the data obtained using these transects could only determine the range and dynamic changes of ecotones between the abandoned agricultural land and forests in that particular direction. The survey results revealed that after the abandonment of farmlands and over time, the ecotones gradually became wider in that direction and could eventually be restored as pine forests [42,60]. Additionally, it should be noted that the detection based on MSW technology requires additional consideration of the size of the window, which also brings uncertainty to the ecotones' detection. For example, it can be seen from the results that the initial value of OM in transect 1 and IV in transect 3 is slightly greater than 0, which means that the ecotones that may occur earlier along the transect direction is not detected, which also partially affects the verification of width. However, the precision of the location is irrelevant, as all peaks fall within the range of UAS results. This also reflects the shortcomings of the traditional one-dimensional MSW method.

5. Conclusions

In general, we present a scientifically solid and practical method for extracting 3D ecotones utilizing UAS photographic technology, which can be utilized to enhance the quantification of landscape pattern. The results confirm the advantages of spatial geostatistical techniques in processing photogrammetric derived data, which benefit from high-precision reconstruction of vegetation structure information. In addition, due to the natural matching of structural information and spectral information derived from UAS, the ability of detecting ecotones based on edge 2D moving window technology is further strengthened, which is enough to produce high-resolution land cover mapping with ecotones. In the end, compared to the standard one-dimensional approach of ecotones measurement utilizing transect surveys and the moving split-window methodology, the method described here provides a thorough understanding of the landscape pattern and spatial characteristics of ecotones. This approach has the potential to quantify regional vegetation and forest transition processes. It should be utilized to evaluate and optimize the management zones of existing nature reserves, as well as to establish additional protected areas and corridors for species migration, genetic exchange, and population size. This should coincide with the exhaustive and regular gathering of species diversity census data. Last, the ecotones' extraction method, landscape pattern quantification, and change detection should be linked to the function of ecotones in a landscape and included into biodiversity and ecosystems protection, as well as sustainable planning and management of regional areas. Through constant inquiry and monitoring, it will be possible to get a full understanding of the entire biological process of the restoration of abandoned agricultural land using this ecotones extraction approach.

Supplementary Materials: The following supporting information can be downloaded at: https://www.mdpi.com/article/10.3390/d14050406/s1, Table S1. A summary of field investigation data; Table S2. Summary of cross-validation results; Table S3. Four indices used to quantify the characteristics of the ecotones in the landscape; Figure S1. The pattern of the three transects set; Figure S2. A set of training samples for training classifiers; Figure S3. Based on a control-variable experiment, the optimal shape and the compactness are selected; Figure S4. Based on the optimal shape and compactness selected, an adaptive scale of segmentation is determined through a series of scale changes; Figure S5. Topographic prediction results based on five different geostatistical models.

Author Contributions: B.W., W.W. and Z.Z. designed and conceived this study; field survey and UAS data collection were completed by B.W., H.S., Q.L., Q.X. and Y.D.; B.W., H.S., Y.D. and Q.L. processed all image data and some work about object-based classification; B.W., Y.M., Z.Z. and Q.X. participated in the data extraction, analysis, and validation; Z.Z., A.P.C. and L.L. guided the study and paper revision. The prime draft was completed by B.W. and Z.Z.; W.W., L.L. and A.P.C. finished the revision of the first draft. All authors have read and agreed to the published version of the manuscript.

Funding: This work was supported by grants from the National Natural Science Foundation of China (41761040 and 41361046), the National Key R&D Program of China (No. 2017YFC0505200), the Strategic Priority Research Program of the Chinese Academy of Sciences (Grant No. XDPB0203), and the foundation of Innovation in Culture Adaptation: Fostering Sustainable Community-Based Natural Resource Management in the South-Western Ethnic Minority Region, China (15XSH02). This work was supported by Graduate Research Innovation Fund project of Yunnan University (2020Z58).

Institutional Review Board Statement: Not applicable.

Informed Consent Statement: Not applicable.

Data Availability Statement: All data used in the manuscript are already publicly accessible, and we provided the download address in the manuscript.

Acknowledgments: We are grateful for the kindness and generosity of the people in Forestry Bureau of Weixi County, Yunnan Province who helped us conduct our work. We also thank reviewers and Shiliang Liu for comments that helped us refine our thinking. In addition, we are very grateful to Cameron Proctor from the University of Windsor for his suggestions and amendments to our text.

Conflicts of Interest: The authors declare no conflict of interest.

Abbreviations

Abbreviation	Explanation
CHM	Canopy Height Model
DSM	Digital Surface Model
DTM	Digital Terrain Model
EBK	Empirical Bayesian kriging
ESP	A tool to estimate scale parameter for multiresolution image segmentation of remotely sensed data.
GCP	Ground Control Point
IV	Importance Value
MRIS	Multiresolution image segmentation
MSW	Moving split-window technique
OBIC	Object-based image classification
OM	Soil organic matter
RF	Random Forest
RFFP	Returning Farmland to Forest Program
RGB or R, G, B	Red, Green, Blue band
RTK	Real-time kinematic
SLCP	Sloping Land Conversion Program
UAS	Unmanned Aircraft System

References

1. Clements, F.E. *Research Methods in Ecology*; University Publishing Company: Lincoln, NE, USA, 1905.
2. Delcourt, P.A.; Delcourt, H.R. Ecotone Dynamics in Space and Time. In *Landscape Boundaries: Consequences for Biotic Diversity and Ecological Flows*; Hansen, A.J., di Castri, F., Eds.; Springer: New York, NY, USA, 1992; pp. 19–54. ISBN 978-1-4612-2804-2.
3. Rusek, J. Distribution and Dynamics of Soil Organisms across Ecotones. In *Landscape Boundaries: Consequences for Biotic Diversity and Ecological Flows*; Hansen, A.J., di Castri, F., Eds.; Springer: New York, NY, USA, 1992; pp. 196–214. ISBN 978-1-4612-2804-2.
4. Fortin, M.-J.; Olson, R.J.; Ferson, S.; Iverson, L.; Hunsaker, C.; Edwards, G.; Levine, D.; Butera, K.; Klemas, V. Issues Related to the Detection of Boundaries. *Landsc. Ecol.* **2000**, *15*, 453–466. [CrossRef]
5. Walker, S.; Wilson, J.B.; Steel, J.B.; Rapson, G.L.; Smith, B.; King, W.M.; Cottam, Y.H. Properties of Ecotones: Evidence from Five Ecotones Objectively Determined from a Coastal Vegetation Gradient. *J. Veg. Sci.* **2003**, *14*, 579–590. [CrossRef]
6. Holland, M. SCOPE/MAB Technical Consultations on Landscape Boundaries. Report of a SCOPE/MAB Workshop on Ecotones. A new look at ecotones: Emerging international projects on landscape boundaries. *Biol. Int.* **1988**, *17*, 106.
7. Hansen, A.J.; DiCastri, F. *Landscape Boundaries: Consequences for Biotic Diversity and Ecological Flows*; Springer: New York, NY, USA, 1992.
8. Fortin, M.-J. Edge Detection Algorithms for Two-Dimensional Ecological Data. *Ecology* **1993**, *75*, 956–965. [CrossRef]
9. Farina, A. Scaling Patterns and Processes across Landscapes. In *Principles and Methods in Landscape Ecology*; Farina, A., Ed.; Springer: Dordrecht, The Netherland, 1998; pp. 35–49. ISBN 978-94-015-8984-0.
10. Yarrow, M.M.; Marín, V.H. Toward Conceptual Cohesiveness: A Historical Analysis of the Theory and Utility of Ecological Boundaries and Transition Zones. *Ecosystems* **2007**, *10*, 462–476. [CrossRef]
11. Hufkens, K.; Scheunders, P.; Ceulemans, R. Ecotones in Vegetation Ecology: Methodologies and Definitions Revisited. *Ecol. Res.* **2009**, *24*, 977–986. [CrossRef]
12. Kark, S. Effects of Ecotones on Biodiversity. In *Reference Module in Life Sciences*; Elsevier: Amsterdam, The Netherlands, 2017; p. B9780128096338022000. ISBN 978-0-12-809633-8.
13. Strayer, D.L.; Power, M.E.; Fagan, W.F.; Pickett, S.T.A.; Belnap, J. A Classification of Ecological Boundaries. *BioScience* **2003**, *53*, 723. [CrossRef]
14. Peters, D.P.C.; Gosz, J.R.; Pockman, W.T.; Small, E.E.; Parmenter, R.R.; Collins, S.L.; Muldavin, E. Integrating Patch and Boundary Dynamics to Understand and Predict Biotic Transitions at Multiple Scales. *Landsc. Ecol.* **2006**, *21*, 19–33. [CrossRef]
15. Liu, J.; Li, S.; Ouyang, Z.; Tam, C.; Chen, X. Ecological and Socioeconomic Effects of China's Policies for Ecosystem Services. *Proc. Natl. Acad. Sci. USA* **2008**, *105*, 9477–9482. [CrossRef]
16. Zhang, Z.; Zinda, J.A.; Li, W. Forest Transitions in Chinese Villages: Explaining Community-Level Variation under the Returning Forest to Farmland Program. *Land Use Policy* **2017**, *64*, 245–257. [CrossRef]
17. Zhang, Z.; Sun, C.; Ou, X. Mountain vegetation spatial pattern changes affected by slope land conversation program (SLCP). *J. Mt. Sci.* **2009**, *27*, 513–523.
18. Su, W.; Yue, Y.; Yu, X. Community structure and population spatial pattern of natural Pinus tabulae form. *J. Northeast. For. Univ.* **2009**, *37*, 18–20.
19. Liu, J.; Gao, J.; Lv, S.-H.; Han, Y.; Nie, Y. Shifting Farming–Pastoral Ecotone in China under Climate and Land Use Changes. *J. Arid. Environ.* **2011**, *75*, 298–308. [CrossRef]

20. Zhou, D.; Zhao, S.; Liu, S.; Zhang, L. Modeling the Effects of the Sloping Land Conversion Program on Terrestrial Ecosystem Carbon Dynamics in the Loess Plateau: A Case Study with Ansai County, Shaanxi Province, China. *Ecol. Model.* **2014**, *288*, 47–54. [CrossRef]
21. Li, Y.; Viña, A.; Yang, W.; Chen, X.; Zhang, J.; Ouyang, Z.; Liang, Z.; Liu, J. Effects of Conservation Policies on Forest Cover Change in Giant Panda Habitat Regions, China. *Land Use Policy* **2013**, *33*, 42–53. [CrossRef] [PubMed]
22. Chen, H.; Marter-Kenyon, J.; López-Carr, D.; Liang, X. Land Cover and Landscape Changes in Shaanxi Province during China's Grain for Green Program (2000–2010). *Environ. Monit Assess* **2015**, *187*, 644. [CrossRef]
23. Van Den Hoek, J.; Ozdogan, M.; Burnicki, A.; Zhu, A.-X. Evaluating Forest Policy Implementation Effectiveness with a Cross-Scale Remote Sensing Analysis in a Priority Conservation Area of Southwest China. *Appl. Geogr.* **2014**, *47*, 177–189. [CrossRef]
24. Xiong, D.; Ou, X.; Huang, W.; Wang, T.; Yang, J.; Guo, J.; Zhang, Z. Measurement of ecotone width between agro-forest ecosystems based on soil nutrients. *Ecol. Sci.* **2014**, *33*, 597–602.
25. Liu, X.; Zhang, Z.; Sun, Z.; Ou, X.; Zhang, Y.; Mao, Y. Impacts of Different Disturbances on Vegetation Restoration on the Abandoned Farmland. *Ecol. Environ. Sci.* **2013**, *22*, 983–990.
26. Erdős, L. The Moving Split Window (MSW) Analysis in Vegetation Science—An Overview. *Appl. Ecol. Environ. Res.* **2014**, *12*, 787–805. [CrossRef]
27. Zalatnai, M.; KÖRmÖCzi, L.; Toth, T. Soil-Plant Interrelations and Vegetation Boundaries along an Elevation Gradient in a Hungarian Sodic Grassland. *Cereal Res. Commun.* **2008**, *36*, 231–234.
28. Hill, R.; Granica, K.; Smith, G.; Schardt, M. Representation of an Alpine Treeline Ecotone in SPOT 5 HRG Data. *Remote Sens. Environ.* **2007**, *110*, 458–467. [CrossRef]
29. Ørka, H.O.; Wulder, M.A.; Gobakken, T.; Næsset, E. Subalpine Zone Delineation Using LiDAR and Landsat Imagery. *Remote Sens. Environ.* **2012**, *119*, 11–20. [CrossRef]
30. Aplin, P. Remote Sensing: Ecology. *Prog. Phys. Geogr. Earth Environ.* **2005**, *29*, 104–113. [CrossRef]
31. Rocchini, D.; Boyd, D.S.; Féret, J.; Foody, G.M.; He, K.S.; Lausch, A.; Nagendra, H.; Wegmann, M.; Pettorelli, N. Satellite Remote Sensing to Monitor Species Diversity: Potential and Pitfalls. *Remote Sens. Ecol. Conserv.* **2016**, *2*, 25–36. [CrossRef]
32. Remondino, F.; Barazzetti, L.; Nex, F.; Scaioni, M.; Sarazzi, D. UAV Photogrammetry for Mapping and 3D Modeling: Current Status and Future Perspectives. *Int. Arch. Photogramm. Remote Sens. Spatial Inf. Sci.* **2011**, *38*, C22. [CrossRef]
33. Rosnell, T.; Honkavaara, E. Point Cloud Generation from Aerial Image Data Acquired by a Quadrocopter Type Micro Unmanned Aerial Vehicle and a Digital Still Camera. *Sensors* **2012**, *12*, 453–480. [CrossRef]
34. Zarco-Tejada, P.J.; Diaz-Varela, R.; Angileri, V.; Loudjani, P. Tree Height Quantification Using Very High Resolution Imagery Acquired from an Unmanned Aerial Vehicle (UAV) and Automatic 3D Photo-Reconstruction Methods. *Eur. J. Agron.* **2014**, *55*, 89–99. [CrossRef]
35. Torres-Sánchez, J.; López-Granados, F.; Serrano, N.; Arquero, O.; Peña, J.M. High-Throughput 3-D Monitoring of Agricultural-Tree Plantations with Unmanned Aerial Vehicle (UAV) Technology. *PLoS ONE* **2015**, *10*, e0130479. [CrossRef]
36. Bendig, J.; Bolten, A.; Bennertz, S.; Broscheit, J.; Eichfuss, S.; Bareth, G. Estimating Biomass of Barley Using Crop Surface Models (CSMs) Derived from UAV-Based RGB Imaging. *Remote Sens.* **2014**, *6*, 10395–10412. [CrossRef]
37. Bendig, J.; Yu, K.; Aasen, H.; Bolten, A.; Bennertz, S.; Broscheit, J.; Gnyp, M.L.; Bareth, G. Combining UAV-Based Plant Height from Crop Surface Models, Visible, and near Infrared Vegetation Indices for Biomass Monitoring in Barley. *Int. J. Appl. Earth Obs. Geoinf.* **2015**, *39*, 79–87. [CrossRef]
38. Tilly, N.; Aasen, H.; Bareth, G. Fusion of Plant Height and Vegetation Indices for the Estimation of Barley Biomass. *Remote Sens.* **2015**, *7*, 11449–11480. [CrossRef]
39. Hou, W.; Walz, U. Extraction of Small Biotopes and Ecotones from Multi-Temporal RapidEye Data and a High-Resolution Normalized Digital Surface Model. *Int. J. Remote Sens.* **2014**, *35*, 7245–7262. [CrossRef]
40. Parrott, L.; Proulx, R.; Thibert-Plante, X. Three-Dimensional Metrics for the Analysis of Spatiotemporal Data in Ecology. *Ecol. Inform.* **2008**, *3*, 343–353. [CrossRef]
41. Zhu, F.; An, S.; Guan, B.; Liu, Y.; Zhou, C.; Wang, Z. A Review of Ecotone: Concepts, Attributes, Theories and Research Advances. *Acta Ecol. Sin.* **2007**, *27*, 3032–3042.
42. Nie, W. The Dynamics of the Ecotone between Abandonded Agricultural Land and Forest Caused by the Returing Farmland to Forest Program. Master's Thesis, Yunnan University, Kunming, China, 2017.
43. Bongiovanni, G.; Cinque, L.; Levialdi, S.; Rosenfeld, A. Image Segmentation by a Multiresolution Approach. *Pattern Recognit.* **1993**, *26*, 1845–1854. [CrossRef]
44. Drăguţ, L.; Tiede, D.; Levick, S.R. ESP: A Tool to Estimate Scale Parameter for Multiresolution Image Segmentation of Remotely Sensed Data. *Int. J. Geogr. Inf. Sci.* **2010**, *24*, 859–871. [CrossRef]
45. Naidoo, L.; Cho, M.A.; Mathieu, R.; Asner, G. Classification of Savanna Tree Species, in the Greater Kruger National Park Region, by Integrating Hyperspectral and LiDAR Data in a Random Forest Data Mining Environment. *ISPRS J. Photogramm. Remote Sens.* **2012**, *69*, 167–179. [CrossRef]
46. Vosselman, G. Slope Based Filtering of Laser Altimetry Data. *IAPRS* **2000**, *33*, 935–942.
47. Lee, S.; Wolberg, G.; Shin, S.Y. Scattered Data Interpolation with Multilevel B-Splines. *IEEE Trans. Visual. Comput. Graphics* **1997**, *3*, 228–244. [CrossRef]

48. Hou, W.; Walz, U. Enhanced Analysis of Landscape Structure: Inclusion of Transition Zones and Small-Scale Landscape Elements. *Ecol. Indic.* **2013**, *31*, 15–24. [CrossRef]
49. Riitters, K.; Wickham, J.D.; O'Neill, R.; Jones, K.B.; Smith, E. Global-Scale Patterns of Forest Fragmentation. *Conserv. Ecol.* **2000**, *4*, art3. [CrossRef]
50. Hesselbarth, M.H.K.; Sciaini, M.; With, K.A.; Wiegand, K.; Nowosad, J. *Landscapemetrics*: An Open-source *R* Tool to Calculate Landscape Metrics. *Ecography* **2019**, *42*, 1648–1657. [CrossRef]
51. Loke, L.H.L.; Chisholm, R.A.; Todd, P.A. Effects of Habitat Area and Spatial Configuration on Biodiversity in an Experimental Intertidal Community. *Ecology* **2019**, *100*, e02757. [CrossRef] [PubMed]
52. Umaña, M.N.; Mi, X.; Cao, M.; Enquist, B.J.; Hao, Z.; Howe, R.; Iida, Y.; Johnson, D.; Lin, L.; Liu, X.; et al. The Role of Functional Uniqueness and Spatial Aggregation in Explaining Rarity in Trees. *Glob. Ecol. Biogeogr.* **2017**, *26*, 777–786. [CrossRef]
53. Nams, V.O. Shape of Patch Edges Affects Edge Permeability for Meadow Voles. *Ecol. Appl.* **2012**, *22*, 1827–1837. [CrossRef] [PubMed]
54. Ludwig, J.A.; Cornelius, J.M. Locating Discontinuities along Ecological Gradients. *Ecology* **1987**, *68*, 448–450. [CrossRef]
55. Brunt, J.W.; Conley, W. Behavior of a Multivariate Algorithm for Ecological Edge Detection. *Ecol. Model.* **1990**, *49*, 179–203. [CrossRef]
56. Wang, T.; Ou, X.; Zhang, Z.; Liu, X.; Wang, L.; Sun, Z.; He, B.; Li, F. Measurement of ecotone width between agro-ecosystem and forest ecosystem caused by Grain for Green Program. *J. Yunnan Univ.* **2012**, *34*, 604–612.
57. Halder, A.; Ghosh, A.; Ghosh, S. Aggregation Pheromone Density Based Pattern Classification. *Fundam. Inform.* **2009**, *92*, 345–362. [CrossRef]
58. Pickett, S.T.; White, P.S. *The Ecology of Natural Disturbance and Patch Dynamics*; Academic Press: Orlando, Fla, 1985; ISBN 978-0-12-554520-4.
59. Naiman, R.J.; Décamps, H. *The Ecology and Management of Aquatic-Terrestrial Ecotones*; Man and the biosphere series; Unesco; Parthenon Pub. Group: Park Ridge, NJ, USA, 1990; ISBN 978-0-929858-25-8.
60. Xiong, D. The Characteristics of Plant Community in Abandonded Agricultural Land and the Changes of the Ecotone between Agro-Forest Ecosystems with Temporal Changes. Master's Thesis, Yunnan University, Kunming, China, 2013.
61. Lin, Y.; Yao, S. Impact of the Sloping Land Conversion Program on Rural Household Income: An Integrated Estimation. *Land Use Policy* **2014**, *40*, 56–63. [CrossRef]
62. He, J.; Sikor, T. Notions of Justice in Payments for Ecosystem Services: Insights from China's Sloping Land Conversion Program in Yunnan Province. *Land Use Policy* **2015**, *43*, 207–216. [CrossRef]
63. Arnot, C.; Fisher, P.F.; Wadsworth, R.; Wellens, J. Landscape Metrics with Ecotones: Pattern under Uncertainty. *Landsc. Ecol.* **2004**, *19*, 181–195. [CrossRef]
64. Frazier, A.E.; Wang, L. Modeling Landscape Structure Response across a Gradient of Land Cover Intensity. *Landsc. Ecol.* **2013**, *28*, 233–246. [CrossRef]
65. Dassot, M.; Constant, T.; Fournier, M. The Use of Terrestrial LiDAR Technology in Forest Science: Application Fields, Benefits and Challenges. *Ann. For. Sci.* **2011**, *68*, 959–974. [CrossRef]
66. Wallace, L.; Lucieer, A.; Malenovský, Z.; Turner, D.; Vopěnka, P. Assessment of Forest Structure Using Two UAV Techniques: A Comparison of Airborne Laser Scanning and Structure from Motion (SfM) Point Clouds. *Forests* **2016**, *7*, 62. [CrossRef]
67. Drusch, M.; Del Bello, U.; Carlier, S.; Colin, O.; Fernandez, V.; Gascon, F.; Hoersch, B.; Isola, C.; Laberinti, P.; Martimort, P.; et al. Sentinel-2: ESA's Optical High-Resolution Mission for GMES Operational Services. *Remote Sens. Environ.* **2012**, *120*, 25–36. [CrossRef]
68. Neuenschwander, A.; Pitts, K. The ATL08 Land and Vegetation Product for the ICESat-2 Mission. *Remote Sens. Environ.* **2019**, *221*, 247–259. [CrossRef]

diversity

MDPI

Article

Climatic and Topographical Effects on the Spatiotemporal Variations of Vegetation in Hexi Corridor, Northwestern China

Youyan Jiang [1,2], Wentao Du [1,3,*], Jizu Chen [1], Chunya Wang [4], Jinniu Wang [3,4], Wenxuan Sun [1,3], Xian Chai [5], Lijuan Ma [6] and Zhilong Xu [7]

1 Qilian Shan Station of Glaciology and Eco-Environment, State Key Laboratory of Cryospheric Science, Northwest Institute of Eco-Environment and Resources, Chinese Academy of Sciences (CAS), Lanzhou 730000, China; jiangyouyan1981@163.com (Y.J.); chenjizu@yeah.net (J.C.); swxsyy123@126.com (W.S.)
2 Lanzhou Regional Climate Center, Lanzhou 730000, China
3 University of Chinese Academy of Sciences, Beijing 100049, China; wangjn@cib.ac.cn
4 Chengdu Institute of Biology, Chinese Academy of Sciences, Chengdu 610041, China; wangcy9610@gmail.com
5 School of Foreign Languages and Literatures, Lanzhou University, Lanzhou 730000, China; chaix20@lzu.edu.cn
6 National Climate Center, Beijing 100081, China; malj@cma.gov.cn
7 Zhangye Meteorological Bureau, Zhangye 734000, China; lzxzl008@163.com
* Correspondence: duwentao@lzb.ac.cn

Abstract: Oases, as complex geographical landscapes, are strongly influenced by both natural variation and human activities. However, they have degenerated because of unplanned land use and water resource development. The research of oasis changes has mostly discussed single components, but multiple components, especially spatial changes to oasis vegetation, need further strengthening. Land use and NDVI were extracted based on Landsat 5/8 and Mod13A3, respectively, and a transfer matrix was constructed to analyze changes of land use in the Hexi Corridor during 2000–2020. The significant changes in the area of each land use were also quantified. Combined with regional temperature and precipitation, interpolated from meteorological data, the correlations between regional temperature, precipitation, and vegetation coverage were calculated, especially in the quantized areas with significant associations. The results showed that the area of bare land or desert decreased, while the areas of agricultural and residential land increased. The normalized difference NDVI of the studied oases increased at the rate of 0.021 per decade, which was positively related to precipitation ($p < 0.05$), rather than temperature; of which, farmland and planted grass land were 55.65% and 33.79% in the significantly increased area. In the area of significant positive relation between NDVI and precipitation, the ratio of grassland, farmland, and forest was 79.21%, 12.82%, and 4.06%, respectively. Additionally, changes in oasis vegetation were determined primarily by agricultural activities, which reflected a combination of natural and anthropic influences.

Keywords: oasis; land use; normalized difference NDVI; climate change; human activities

Citation: Jiang, Y.; Du, W.; Chen, J.; Wang, C.; Wang, J.; Sun, W.; Chai, X.; Ma, L.; Xu, Z. Climatic and Topographical Effects on the Spatiotemporal Variations of Vegetation in Hexi Corridor, Northwestern China. *Diversity* **2022**, *14*, 370. https://doi.org/10.3390/d14050370

Academic Editor: Michael Wink

Received: 18 April 2022
Accepted: 5 May 2022
Published: 6 May 2022

1. Introduction

An oasis is a small-to-medium-scale landscape maintained by a water supply in an arid or semi-arid desert area [1]. It is also an unstable ecosystem, with the characteristic of high habitat fragmentation. In China, oases are mainly distributed in arid and semi-arid regions to the west of the Helan Mountians, whose area is only 5% of the total. They support the survival of 90% of people and the economic development of 95% [2]. Therefore, changes in the ecosystem of oases can have a considerable influence on both ecological security and sustainable socioeconomic development [3].

The Hexi Corridor in northwestern China, which is a region affected by human activities, climate change, and tectonic activities [4], is one of the most important, but vulnerable, oasis ecosystems in the arid and semi-arid regions of China [5] It has been reported that the dynamic changing of the oasis area in the Hexi Corridor is closely related to agricultural industry [6]. Moreover, combining vegetation and soil-related endmembers time series data can enhance conventional LUCC analysis, to help fully capture land degradation processes [7]. There is strong negative correlation between the water resource reserves and the NDVI of the area [8]. In particular, the evolution of an artificial oasis is closely correlated with human economic activities and the environmental carrying capacity of oases [9]. Natural factors and human activity controlled the desertification process, but the reversal of desertification mainly resulted from human activities [10]. In respect to oases, the socioeconomic development and ecological/environmental coordination among the different administrative regions within the Hexi Corridor vary greatly [11]. The expansion of cultivated land and unreasonable utilization of groundwater are the main factors associated with the climatic and environmental degradation of the Hexi Corridor [12]. The rapid process of urbanization and coordinated urban–rural development have reduced the ecosystem services of the Hexi Corridor [13]. Remote sensing is an effective way to extract spatial information, to monitor spatiotemporal variation at a regional scale. Of which, land-use maps can describe information about land surface coverage. Thus, this has been mostly used to analyze vegetation change. However, there is still a need for further investigation of both natural and human influences, especially in relation to long-term input data. In addition, the more detailed changes at local level need to be strengthened [14]. Investigation of the ecologic and environmental problems in the Hexi Corridor, which highlight the importance of natural and human activities as the driving forces of oasis change, could help boost the sustainable socioeconomic development of the region.

In this study, Landsat5/TM, Landsat8/OLI, MOD13A3, and PANDA data were used in combination with regional meteorological observations, to discuss the influence of climate change and human activities on oasis vegetation in the Hexi Corridor. The changes in the vegetation of the Hexi Corridor and areas of land use in the fluctuating regions were quantitated. In addition, DEM, population number, livestock number, settlement number, and GDP were collected to analyze their influence on oasis change. The findings of this work improve the understanding of the change mechanisms affecting oasis landscapes and provide a scientific basis to support policy makers.

2. Materials and Methods

2.1. Overview of the Study Area

The Hexi Corridor, which has a length of 900 km and width of 100 km (Figure 1), is located in the arid region of northwestern China. It is bordered by Wushaoling to the east, the Jade Gate to the west, the Qilian and Altun mountains to the south, and the Mazongshan, Heli, and Longshou mountains to the north. Owing to its temperate continental climate, annual precipitation is in the range of 40–300 mm, and annual temperature is in the range of 6.2–9.0 °C. It should be noted that annual evaporation exceeds 1500 mm in most areas. The land use types in this region include farmland, forest, grassland, desert, water bodies, snow or glaciers, and residential land.

2.2. Data Sources

This study used 30-m spatial resolution Landsat5/TM data acquired in 2000, Landsat8/OLI data acquired in 2020, and MOD13A3 data acquired during 2000–2020, which were sourced from both the Aerospace Information Research Institute of the Chinese Academy of Sciences (www.aircas.cas.cn, accessed on 1 February 2021) and the National Aeronautics and Space Administration (www.nasa.gov, accessed on 1 February 2021). Additionally, 1-km spatial resolution PANDA data acquired during 2000–2020 were provided by the National Tibetan Plateau Data Center (http://data.tpdc.ac.cn, accessed on 1 May 2021).

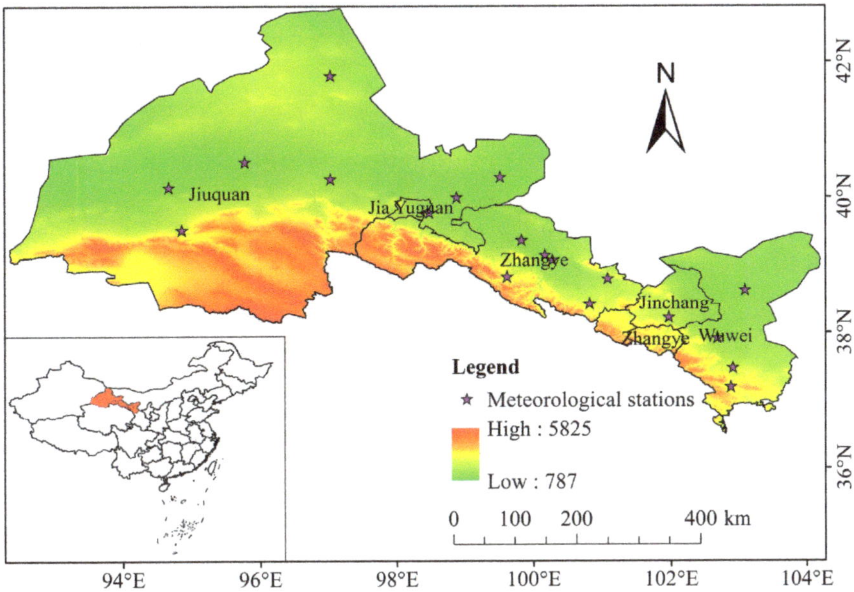

Figure 1. Location of meteorological stations and DEM information in the study area.

To analyze the regional climate change, daily temperature and precipitation data recorded during 2000–2020, at 19 meteorological stations in the Hexi Corridor, were obtained from the Gansu Meteorological Bureau (http://cdc.cma.gov.cn/, accessed on 1 February 2021).

Digital elevation model data with a 30-m spatial resolution (ASTER GDEM V3) were sourced from the Geospatial Data Cloud (http://www.gscloud.cn/, accessed on 1 May 2021), which is a product of jointly developed by Japan's METI and US NASA.

2.3. Research Methods

The data processing and analysis flow are displayed in Figure 2.

2.3.1. Analysis of Land Use Types

Classification of the land use in the Hexi Corridor was performed on the basis of the Landsat5/TM, Landsat8/OLI, and digital elevation model data. The original images were first preprocessed, by means of orthophoto correction, geometric correction, radiometric calibration, atmospheric correction, mosaic and clipping, cloud removal, and shadow processing. Generally, machine-learning approaches, hybrid methods, fuzzy theory, and other methods were the means of obtaining land use information. On the basis of the landscape classification system, supervised classification and manual correction were used to categorize the land use in the Hexi Corridor during 2000–2020 into the following types: farmland, forest, grassland, residential land, bare land or desert, water bodies, and snow or glaciers. Three hundred random samples were extracted to validate the classification of two different images, whose interpretation matched reality, with total accuracies and Kappa coefficients of 87.67% and 0.857 in 2000, and 89.33% and 0.872 in 2020.

A transition matrix can depict in detail the spatial structure and transformation direction of different land use types for different time sequences. This facilitates an accurate understanding of the direction of change of the type, structure, and distribution ratio of land use [15]. The land use classification of the Hexi Corridor during 2000–2020 underwent data fusion, and then an intersection analysis was performed to obtain the transition data for each land use type, to create a land type transition matrix of the overlapping areas.

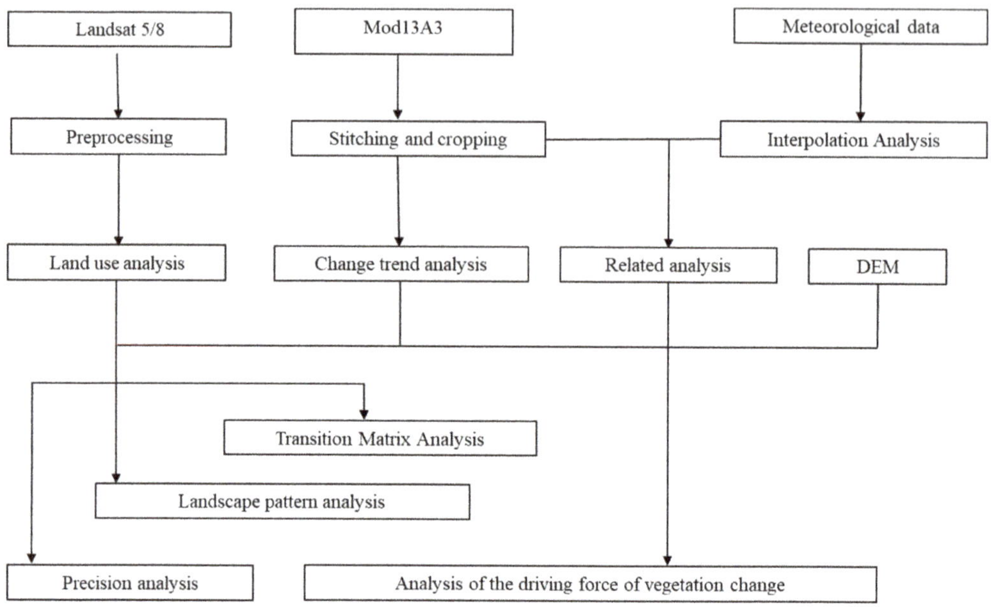

Figure 2. Data processing and analysis flow chart.

2.3.2. Index of Landscape Pattern Indexes

To investigate the changes of landscape pattern over the past 21 years, number of patches (NP), mean patch area (AREA), landscape shape index (LSI), aggregation index (AI), Shannon's diversity index (SHDI), and contagion index (CONTAG) were calculated [16].

2.3.3. NDVI Change Trend Analysis

The maximum value composite method was used to maximize the bimonthly synthetic standard vegetation index, and to obtain annual normalized difference vegetation indexes (NDVIs). Then, following removal of zero and negative values, the annual NDVIs for different vegetation types were obtained by overlaying vegetation maps.

A linear regression analysis was conducted, to simulate the spatial variation of the annual NDVIs for the different vegetation types in the target area [17,18]. The calculation formula can be expressed as follows:

$$\theta_{slope} = \frac{n \times \sum_{j=1}^{n} j \times NDVI_j - \sum_{j=1}^{n} j \sum_{j=1}^{n} NDVI}{n \times \sum_{j=1}^{n} j^2 (\sum_{j=1}^{n} j)^2} \tag{1}$$

where n refers to the cumulative number of monitored years, $NDVI_j$ is the NDVI per year in year j, and θ_{slope} signifies the slope of the trend line. A value of $\theta_{slope} > 0$ ($\theta_{slope} < 0$) indicates that the trend of the NDVI change over n years is increasing (decreasing).

The Theil–Sen Median and Mann–Kendall test were combined to perform a long-term sequence analysis of vegetation change [19–21]. The Mann–Kendall test has been effectively applied in temporal sequence analyses in the fields of hydrology and meteorology. The Theil–Sen median can be used to assess whether temporal sequence data show an upward or downward trend [22,23]. The median of the slope in $n\ (n-1)/2$ data combinations was calculated in the trend analysis. The calculation formula can be expressed as follows:

$$S_{NDVI} = Median \frac{NDVI_j - NDVI_i}{j - i}, \tag{2}$$

where $2000 \leq i < j \leq 2020$. A value of $S_{NDVI} > 0$ ($S_{NDVI} < 0$) indicates that the NDVI is increasing (decreasing).

When an NDVI is tested using the Mann–Kendall test, the NDVI of a certain temporal sequence is regarded as a set of independently distributed sample data, and parameter Z serves as the attenuation index of the NDVI pixel. The calculation formula can be expressed as follows:

$$Z = \begin{cases} \frac{S-1}{\sqrt{s(S)}}, & S > 0 \\ 0, & S = 0 \\ \frac{S+1}{\sqrt{s(S)}}, & S < 0 \end{cases}, \tag{3}$$

$$S = \sum_{j=1}^{n-1} \sum_{i=j+1}^{n} Sgn(NDVI_j - NDVI_i), \tag{4}$$

$$Sgn(NDVI_j - NDVI_i) = \begin{cases} 1, & NDVI_j - NDVI_i > 0 \\ 0, & NDVI_j - NDVI_i = 0, \\ -1, & NDVI_j - NDVI_i < 0 \end{cases} \tag{5}$$

$$S(s) = \frac{n(n-1)(2n+5)}{18}. \tag{6}$$

In Equations (3)–(6), $NDVI_i$ and $NDVI_j$ represent the NDVI for years i and j, respectively, n represents the length of the time sequence, and Sgn is a symbolic function. The value range of statistic Z is $[-\infty, \infty]$. At a given significance level α, values of $|Z| > u_{1-\alpha/2}$ indicate significant change in the studied sequence at the α level. u is a statistic. Here, $Z = \pm 2.58$ is the significance level of $\alpha = 0.01$, $Z = \pm 0.96$ is the significance level of $\alpha = 0.05$, and $Z = \pm 1.65$ is the significance level of $\alpha = 0.10$. In this study, a significance level of $\alpha = 0.05$ was taken to assess the significance of the transformation trend of the NDVI time sequence.

2.3.4. Correlation Analysis

The correlation coefficient between the NDVIs and the climate factors based on a pixel was calculated, as follows:

$$R_{xy} = \frac{\sum_{i=1}^{n} (x_i - \overline{x})(y_i - \overline{y})}{\sqrt{\sum_{i=1}^{n} (x_i - \overline{x})^2} \cdot \sqrt{\sum_{i=1}^{n} (y_i - y)^2}}, \tag{7}$$

where variable i represents the annual serial number, n is assigned the value of 21, x_i represents NDVI data for the i-th year, y_i is the climate factor for the i-th year, $\overline{x}$ and $\overline{y}$ are the mean values of variables x and y, respectively, and R_{xy} is the correlation coefficient between the NDVI and the climate factors. The Mann–Kendall test was used to assess the significance of R_{xy} and to determine its positive and negative correlation thresholds. The analysis was based on the correlation between the annual NDVI, precipitation, and temperature of each pixel.

3. Results and Analysis

3.1. Land Use and Land Cover Change Analysis

As shown in Figure 3, during the study period, snow cover and glaciers were primarily concentrated in high-elevation regions (>3500 m) in the Qilian Mountains. Water bodies included three inland rivers and a number of reservoirs. Forest mainly occupied areas at elevations of 2300–3300 m in the Qilian Mountains. Farmland was mainly distributed in the surroundings of the middle and lower reaches of the rivers. Grassland was divided into natural grassland and artificially planted grassland; the former was located primarily in upstream regions, while the latter was located in middle and lower parts of the river basins. Bare land or desert land covered the largest area in the Hexi Corridor. From 2000 to 2020, the area of land use change was 6.87% of total area.

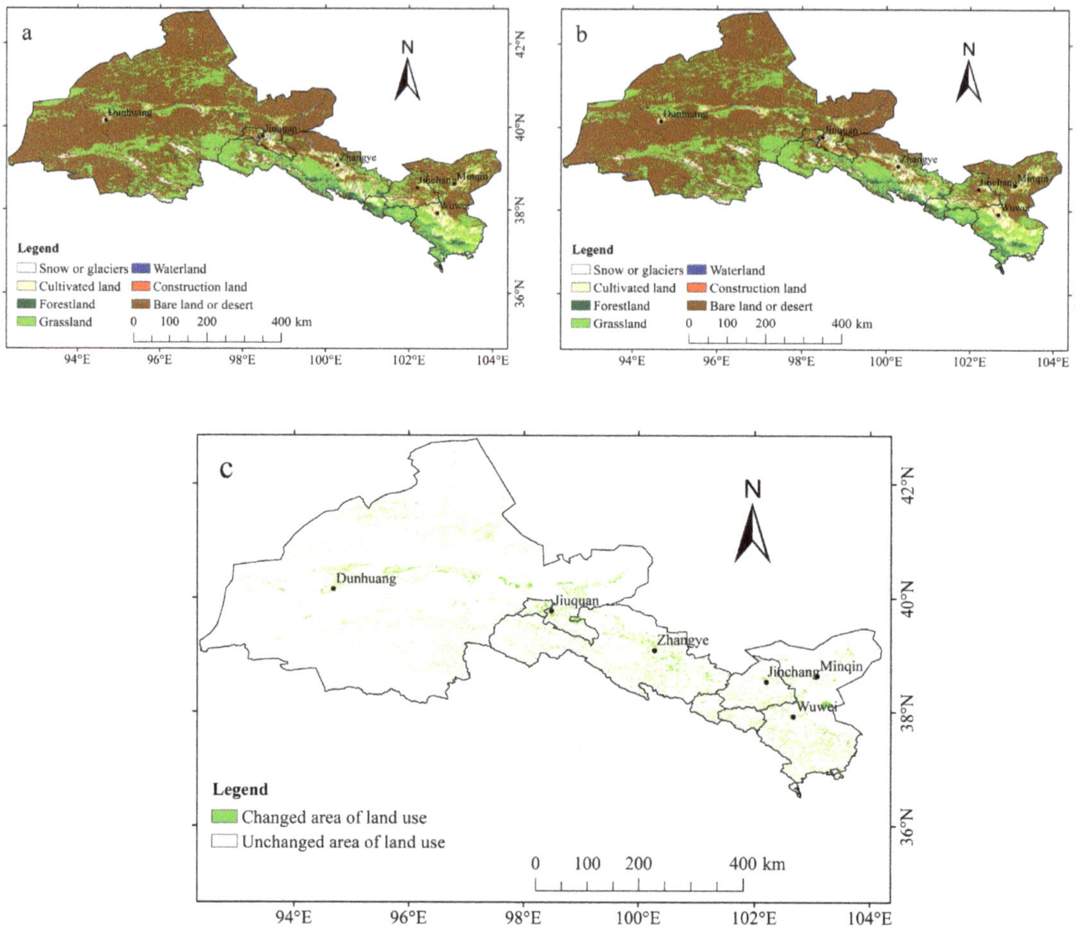

Figure 3. Land use in the Hexi Corridor: (**a**) 2000, (**b**) 2020 and (**c**) changed area.

Table 1 presents a transformation matrix of land use change in the Hexi Corridor from 2000 to 2020, showing the percentage of the area occupied by each land use type. In 2000, the areas of bare land or desert, grassland, farmland, and forest accounted for 68.08%, 21.60%, 5.75%, and 3.05% of the total area, respectively, whereas the area of water bodies, residential land, and snow or glaciers accounted for 0.59%, 0.47%, and 0.46% of the total area, respectively (Table 1). In comparison with 2000, the areas of bare land or desert, grassland, forest, and snow or glaciers in 2020 had decreased to 67.28%, 21.50%, 3.04%, and 0.45% of the total area, respectively; whereas the areas of farmland, residential land, and water bodies had increased to 6.40%, 0.68%, and 0.64% of the total area, respectively. The increases in the area of farmland and residential land represent changes of 0.65% and 0.21%, respectively, whereas the decrease in the area of bare land or desert represents a change of 0.8%. Over the past 21 years, 1.73% of the total area of the Hexi Corridor has been converted from bare land or desert to grassland, while 1.69% of the total area has been converted from grassland to bare land or desert, 0.64% of the total area has been converted from bare land or desert to farmland, and 0.17% of the total area has been converted from farmland to bare land or desert. The interconversion between bare land or desert and forest was equivalent.

Table 1. Transfer matrix of land use types in the oases of the Hexi Corridor from 2000 to 2020.

Land Types	Farm Land	Forest	Grassland	Water Bodies	Snow or Glaciers	Residential Land	Bare Land or Desert	The Total Area
Farm land	5.05	0.06	0.50	0.03	0.00	0.12	0.64	6.40
Forest	0.04	2.53	0.37	0.01	0.00	0.00	0.09	3.04
Grassland	0.30	0.36	19.06	0.05	0.00	0.01	1.73	21.50
Water bodies	0.03	0.01	0.05	0.46	0.00	0.00	0.09	0.64
Snow or glaciers	0.00	0.00	0.00	0.00	0.40	0.00	0.05	0.45
Residential land	0.16	0.00	0.04	0.00	0.00	0.32	0.17	0.68
Bare land or desert	0.17	0.09	1.59	0.04	0.06	0.02	65.32	67.28
The total area	5.75	3.05	21.60	0.59	0.46	0.47	68.08	100

As shown in Table 2, the total NP increased by 15.86, but AREA decreased by 13.69%. Taken together, considering the increase of LSI and SHDI, and the decrease of AI and CONTAG, it can be concluded that the connectivity of different landscape patches decreased, indicating an obvious landscape fragmentation.

Table 2. Landscape pattern index of the entire Hexi Corridor.

Year	NP	AREA	LSI	AI	SHDI	CONTAG
2000	60,765	4.5492	142.6742	98.0313	1.3587	77.6839
2020	70,406	3.9263	145.2059	97.9960	1.3634	77.5832

3.2. *Characteristics of Vegetation Patterns and Analysis of Vegetation Dynamics in the Hexi Corridor*

Given that the NDVI can reflect the overall condition of regional vegetation, the spatiotemporal distributions of annual maximum NDVIs during 2000–2020 were analyzed (Figure 4). The results ranged from 0 to 0.859, and the average value over the 21-year study period was 0.168. Overall, the annual maximum NDVIs increased with time, and the minimum and maximum values appeared in 2001 and 2019, respectively. In 2010, the annual NDVI increased to the mean state during 2000–2020, indicating improved conditions for vegetation growth in the Hexi Corridor. However, clear regional differences were evident. The highest NDVI values were distributed in the southeast, and the lowest values were found in the northwest. The highest annual average and maximum NDVI values were found in the Qilian Mountains in the southeast of the Hexi Corridor, while the lowest annual average and maximum NDVI values were found in the desert of Jiuquan, which might reflect climatic influences, especially that of precipitation.

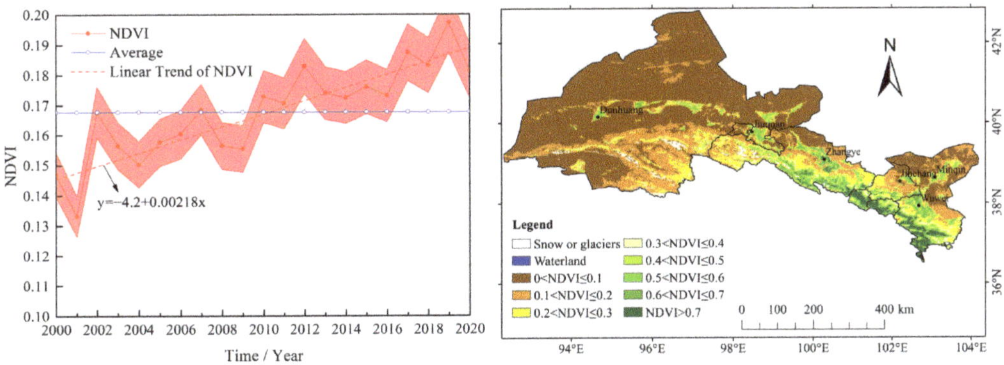

Figure 4. Spatiotemporal distribution of growing season NDVI during 2000–2020.

The land use of regions with higher annual average and maximum NDVI values mainly comprised forest, grassland, and farmland, whereas that of the areas with lower NDVI values was mainly desert vegetation. Over the 21-year study period, the average NDVI of forest, grassland, farmland, bare land, and desert was 0.618, 0.486, 0.515, and 0.112, respectively, indicating that the environment for vegetation growth was better in forest, farmland, and grassland areas than in desert areas.

3.3. Quantitative Analysis of Driving Factors of Interannual Variation of NDVI in the Hexi Corridor

3.3.1. Spatial Patterns of Climatic Effects on Vegetation Changes

The meteorological data were gridded using multivariate regression analysis and Kriging interpolation [24], then used to calculate the correlation with NDVI.

As shown in Figure 5a, except for bare land or desert areas, there was positive correlation between the NDVI and precipitation. Specifically, 41.36% of oasis vegetation areas were correlated positively ($p < 0.05$), of which 13.02% was significant at the 0.01 level. The areas of positive correlation included grassland (79.21%), farmland (12.82%), and forest (4.06%), distributed mainly in the south of Wuwei and Jiuquan counties, to the west of Jinchang, and in the Qilian Mountains. A potential reason for this is that the increasing growth of farmland and grassland vegetation in this area, together with the amount of precipitation, had a substantial impact on the regional NDVI. Only 0.03% of areas had a negative correlation.

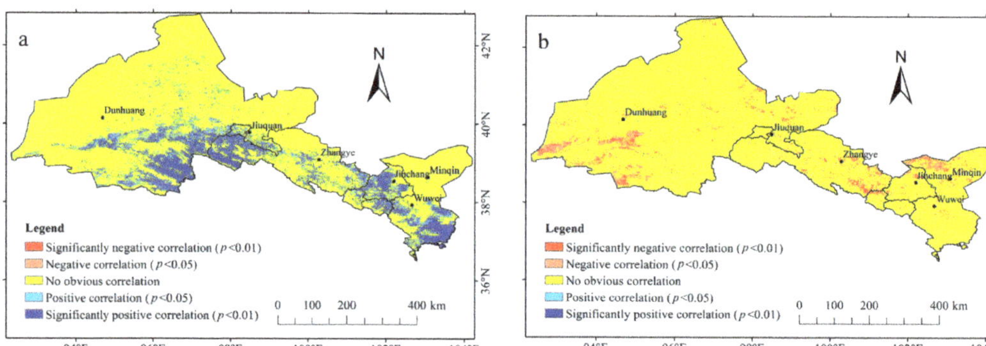

Figure 5. Correlation analysis of NDVI and (**a**) precipitation and (**b**) temperature changes in the Hexi Corridor during 2000–2020.

In terms of the temperature effect (Figure 5b), 5.38% of oasis vegetation areas were correlated negatively ($p < 0.05$), of which 1.65% was significant at the 0.01 level. The areas of negative correlation were bare land and desert (53.23%), grassland (29.2%), and farmland (16.76%); mainly distributed in the northwest of Minqin, the southeast of Zhangye, and the southwest of Jiuquan. This demonstrated that the vegetation change over the entire region was affected more by precipitation than by temperature.

3.3.2. Effects of Topographic Factors on Spatial Patterns of Vegetation Changes

The changes of vegetation at the different elevations during 2000–2020 were discussed based on DEM data. Overall, the NDVI presented an upward trend, but with obvious spatial heterogeneity (Figure 6). For 5.66% of the total area, the NDVI increased significantly ($p < 0.05$), of which 1.85% was significant at the 0.01 level. For 1.1% of the total area, the NDVI decreased significantly ($p < 0.05$), of which 0.21% was significant at the 0.01 level. Generally, the NDVI increased at the edges of the Hexi Corridor oases, while it decreased in southern parts of the city of Dunhuang, central parts of the city of Zhangye, and central parts of Wuwei and Minqin counties. As for the analysis of land use type, 55.65% and 33.79% of the increasing NDVI areas were farmland and planted grassland, while 73.05%

and 13.84% of the decreasing NDVI areas were farmland and building land. The changes to NDVI were in flat areas with >3000 m and <2000 m, but fluctuated greatly in the anthropic area from 2000 to 3000 m. Overall, this was mainly influenced by human activities.

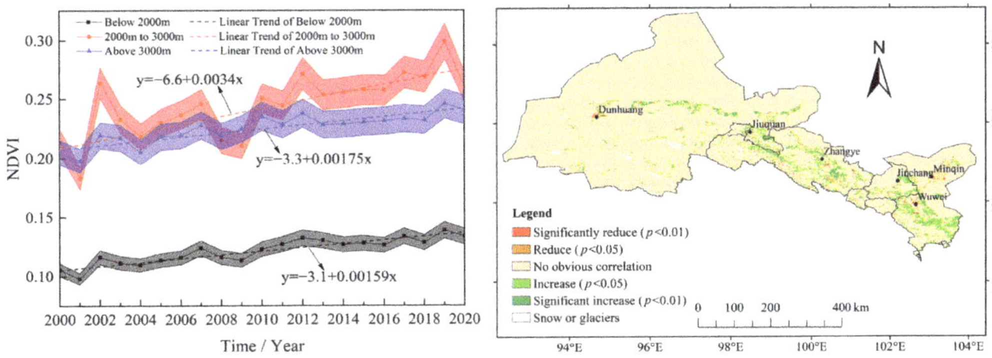

Figure 6. Trend of the NDVI in the Hexi Corridor during 2000–2020.

4. Discussion

4.1. Effects of Climatic Factors on Vegetation Growth

Vegetation growth was found to be strongly correlated with climatic factors, and their interaction was the main focus, regarding global climate change [25–27]. A normalized difference vegetation index (NDVI) could be used to monitor and estimate vegetation activities over different spatiotemporal scales, without damaging or altering vegetation [28]. Considering the differences in the responses of vegetation to climate variability under various eco-environmental conditions [29], in the Hexi Corridor, the response of the NDVI to climate change exhibited a large disparity spatially, especially for different vegetation types. In this study, grassland and farmland showed a more obvious response than other vegetation types to changes in precipitation, consistent with the results of other studies in this area [30,31]. A negative impact of temperature on vegetation growth mainly occurred in relation to bare land and desert, as well as marginal artificial grassland, attributable to the increasing evaporation associated with warming [32].

4.2. Effects of Human Activates on Vegetation Growth

The spatiotemporal change of land use in the Hexi Corridor was found to be closely related to local population growth and economic development. The area of artificial oases was shown to be positively correlated with population number in [33]. According to the data of the Gansu Statistical Yearbook, the figures for the permanent population and settlement changed from 4.66 million with a settlement of 5641 in 2000, to 4.91 million with settlement of 6864 in 2020; while the GDP changed from 25.68 billion to 217.91 billion in the oases of the Hexi Corridor. Meanwhile, the night light index was also increased from 12.62 to 49.5 (Figure 7). However, human factors, such as population changes and increased cultivation of land, appear to have been stronger driving forces, which were directly responsible for the changes in desertification [10]. For example, areas of forest and grassland surrounding farmland were converted to farmland, to meet increasing demand, resulting in serious destruction of the ecological environment, such as soil loosening and increased wind erosion. One mitigation measure adopted to improve the ecological environment was the step-by-step return of farmland to forest, whilst maintaining forest–grassland conservation [34]. The change in forest area was affected by both human and natural factors, of which logging for new housing and overexploitation of groundwater resources were the main factors responsible for the 0.01% reduction in forest area [35]. Meanwhile, the 0.1% decrease in grassland area was primarily ascribed to overgrazing, owing to the annual increase of livestock supporting capacity and economic development needs (according to the Statistical Yearbook of Gansu Province, the livestock numbers in the Hexi Corridor increased from 9.29 million in 2000 to 14.65 million in 2020) [36]. This caused a series of environmental problems and destroyed regional biodiversity. Overgrazing has extensively degraded Chinese grasslands. A reduction in the stocking rate, of 30–50% below the district averages, is required to increase the profitability of livestock production and protect vital ecosystem services. Additionally, short-time exclusions from grazing in a degraded desert have great potential to restore vegetation and soil properties [37].

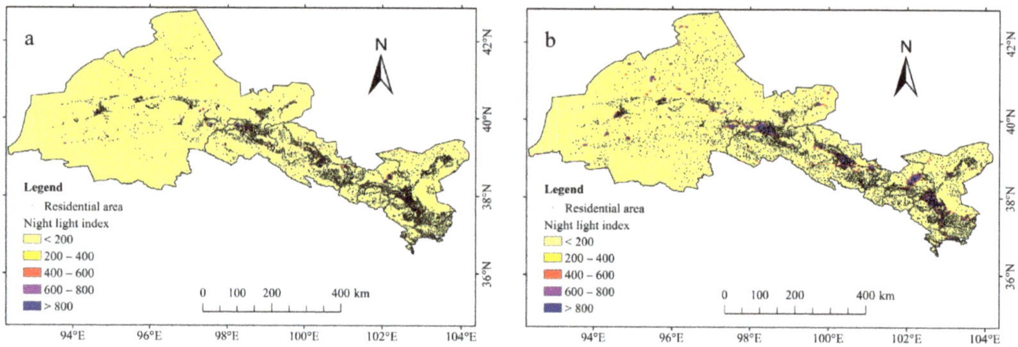

Figure 7. Night light index and residential areas of the Hexi Corridor in (**a**) 2000 and (**b**) 2020.

Further analysis of the spatial distribution of NDVI change associated with the different vegetation types indicated that vegetation (or crop) growth increased in grassland and farmland areas, was relatively stable in forest areas, but decreased slightly in bare land and desert areas. In other regions, vegetation growth exhibited no obvious changes. Under the background of global warming, the vegetation growth in the Hexi Corridor has increased, while the area of farmland, especially that involved in crop growth, has increased since 2007, because the high consumption of water for agricultural production in the middle and lower areas of the river basins has been effectively constrained by governmental control of water resources. Correspondingly, the downstream bare land and desert vegetation presented a slight increase during the study period.

Human activities have had a positive impact on vegetation productivity in the Hexi Corridor, mainly in the oasis areas, which primarily reflects artificial irrigation, fertilization, and other management measures [38]. Irrigation water productivity and food production should be improved through promoting water-saving irrigation technology, maintaining the current use of fertilization, agricultural films and agricultural pesticides, and improving the use efficiencies of agronomic inputs, instead of increasing their amount [39]. On the basis of vegetation change, PANDA night light data, and land use type, the relationship between the changes in oasis vegetation and human activities has been discussed. The results showed that the dynamic changes of oases in the Hexi Corridor were influenced mainly by agricultural activities, supplemented by the effects of natural factors, which is in accord with other similar regions [40–43].

5. Conclusions

Changes in land use and vegetation are the direct outcome of interactions between human activities and the natural environment. The spatial pattern of land use change represents the intensity and mode of the human–land relationships at different region scales. On the basis of the analysis of land use types in the Hexi Corridor, this study found that the area of certain land use types (i.e., bare land or desert, grassland, forest, and snow or glaciers) decreased from 2000 to 2020, while the areas of farmland, residential land, and water bodies increased.

Generally, vegetation growth improved continuously during 2000–2020; however, there were obvious regional differences, i.e., the highest (lowest) values of growth were found in the southeast (northwest) of the region. In the area of marked increasing NDVI, farm land and planted grass land were 55.65% and 33.79%, respectively. In the area of marked increasing NDVI, farm land and construction land were 73.05% and 13.84%, respectively. Vegetation changes at different elevations fluctuated more in the areas with frequent human activities.

Changes in vegetation were more affected by precipitation than by temperature; 41.36% of oasis change had a marked positive correlated with precipitation, while 5.38% of oasis change was negatively correlated with temperature. Based on the analysis of the statistics of the Gansu Statistical Yearbook, settlement numbers, and night light index, it was determined that human activities have gradually come to dominate the NDVI changes in farmland, residential land, and artificial grassland areas.

In China, a number of policies were implemented to improve the ecological environment in oases, whose effect was evaluated by means of remote sensing. The findings of this study highlight the great importance of enhancing the protection and management of existing areas of grassland and forest around bare land or desert areas, by means of both rational use of water resources, and strengthening awareness of ecological protection.

Author Contributions: W.D. and Y.J. conceived the idea and designed the research framework. Y.J. and J.W. discussed the content of the analysis part. J.C., Z.X. and W.S. carried out data collection and preprocessing. Y.J., W.S., C.W. and J.W. undertook data analysis and manuscript preparation. W.D., W.S., L.M. and X.C. contributed to manuscript refinement. All authors have read and agreed to the published version of the manuscript.

Funding: This work was jointly supported by the following research projects: supported by National Natural Science Foundation of China (42071018, 42101139); State Key Laboratory of Cryosphere Science, Northwest Institute of Eco-Environment and Resources, Chinese Academy Sciences (Grant Number: SKLCS-ZZ-2022); Innovation and Development Project of China Meteorological Administration (CXFZ2022J039); Natural Science Foundation of Gansu Province (20JR10RA453); State Key Laboratory of Frozen Soil Engineering (State Key Laboratory of Frozen Soil Engineering, SKLFSE202006), and Opening Foundation of Key Laboratory of Desert and Desertification (KLDD-2020-010).

Institutional Review Board Statement: Not applicable.

Informed Consent Statement: Informed consent was obtained from all subjects involved in the study.

Data Availability Statement: The data that support the findings of this study are available from the corresponding author upon reasonable request.

Conflicts of Interest: The authors declare no conflict of interest.

References

1. Chen, X.; Luo, G.P. Researches and progress of oasis ecology in arid areas. *Arid. Land Geogr.* **2008**, *31*, 487–495.
2. Li, W.D.; Li, Z.Z.; Wang, J.Q. Evaluation of oasis ecosystem risk by reliability theory in an arid area: A case study in the Shiyang River Basin, China. *J. Environ. Sci.* **2007**, *19*, 508–512. [CrossRef]
3. Guo, N.; Wang, X.P.; Cai, D.H.; Qing, J.Z. Analyses on the vegetation index variation and its formation causes in the oases in northwest China recent 22 years. *Arid. Zone Res.* **2010**, *27*, 75–82. [CrossRef]
4. Mu, G.J.; Liu, J.Q. An analysis of the oasis evolution and its control factors. *Quat. Sci.* **2000**, *20*, 539–546.
5. Wang, J.F. Analysis on runoff variation in the Beida River Basin under the influence of climate change and human activities. *J. Arid. Land Resour. Environ.* **2019**, *33*, 86–91.
6. Wang, X.Y.; Liu, S.Z.; Chen, X.S. Dynamic Changes and Driving Factors of Oasis in Hexi Corridor. *J. Desert Res.* **2019**, *39*, 212–219.
7. Sun, Q.Q.; Zhang, P.; Jiao, X.; Han, W.; Sun, Y.; Sun, D. Identifying and understanding alternative states of dryland landscape: A hierarchical analysis of time series of fractional vegetation-soil nexuses in China's Hexi Corridor. *Landsc. Urban Plan.* **2021**, *215*, 104225. [CrossRef]
8. Li, X.J.; Bai, Y.P.; Li, M.; Ma, J.H. Analysis on spatiotemporal correlation between water resource change and ecological Environment in Hexi Corridor. *Bull. Soil Water Conserv.* **2019**, *33*, 86–91.
9. Li, S.; Yan, C. Oasis evolution and human factors analysis in Hexi Corridor over the recent 20 years. *J. Arid. Land Resour. Environ.* **2013**, *27*, 92–98.

10. Han, L.; Zhang, Z.; Qiang, Z.; Xin, W. Desertification assessments in the Hexi corridor of northern China's Gansu Province by remote sensing. *Nat. Hazards* **2015**, *75*, 2715–2731. [CrossRef]
11. Bai, J.; Wang, X.G.; Zhao, C.Z. Evaluation on the coordinated development ability of Eco economic system in Hexi Corridor oases. *Arid. Zone Geogr.* **2010**, *33*, 130–135.
12. Gao, Z.R.; Liu, X.Y.; Yang, Q.H.; Ma, Y.D. Climate-environmental evolution in desert-oasis ecotone in the Hexi corridor. *Arid. Zone Res.* **2010**, *27*, 31–38. [CrossRef]
13. Guan, Q.; Yang, L.; Pan, N.; Lin, J.; Xu, C.; Wang, F.; Liu, Z. Greening and Browning of the Hexi Corridor in Northwest China: Spatial Patterns and Responses to Climatic Variability and Anthropogenic Drivers. *Remote Sens.* **2018**, *10*, 1270. [CrossRef]
14. Sun, P.; Gong, J.; Gao, Y.J.; Xie, Y.C.; Qian, D.W. Spatiotemporal change of oases and their landscape response in arid areas in Northwest China. *Arid. Zone Res.* **2014**, *31*, 355–361.
15. Hu, G.Y.; Dong, Z.B.; Wei, Z. Spatial and temporal change of desertification land of Zoige Basin in recent 30 years and its cause analysis. *Adv. Earth Sci.* **2009**, *24*, 908–916.
16. Wang, Z.; Huang, N.; Luo, L.; Li, X.; Ren, C.; Song, K.; Chen, J. Shrinkage and fragmentation of marshes in the West Songnen Plain, China, from 1954 to 2008 and its possible causes. *Int. J. Appl. Earth Obs. Geoinf.* **2011**, *13*, 477–486. [CrossRef]
17. Stow, D.; Daeschner, S.; Hope, A. Variability of the seasonally integrated normalized difference vegetation index across the north slope of Alaska in the 1990s. *Int. J. Remote Sens.* **2003**, *24*, 1111–1117. [CrossRef]
18. Song, Y.; Ma, M.G. Study on vegetation cover change in northwest china based on SPOT VEGETATION data. *J. Desert Res.* **2007**, *27*, 89–93.
19. Gocic, M.; Trajkovic, S. Analysis of changes in meteorological variables using Mann-Kendall and Sen's slope estimator statistical tests in Serbia. *Glob. Planet. Chang.* **2013**, *100*, 172–182. [CrossRef]
20. Saikia, P.; Konwar, K. Analysis of Changes in Groundwater Levels Using Mann-Kendall and Sen's Slope Estimator in Kamrup (M) District, Assam. In *Geography in the 21st Century Emerging Issues and the Way Forward*; Namya Press: New Delhi, India, 2020.
21. Jiang, W.G.; Yuan, L.H.; Wang, W.J.; Cao, R.; Zhang, Y.F.; Shen, W.M. Spatiotemporal analysis of vegetation variation in the Yellow River Basin. *Ecol. Indic.* **2015**, *51*, 117–126. [CrossRef]
22. Yue, S.; Pilon, P.; Cavadias, G. Power of the Mann Kendall and Spearman's rho tests for detecting monotonic trends in hydrological series. *J. Hydrol.* **2002**, *259*, 254–271. [CrossRef]
23. Fensholt, R.; Langanke, T.; Rasmussen, K.; Reenberg, A.; Prince, S.D.; Tucker, C.; Scholes, R.J.; Le, Q.B.; Bondeau, A.; Eastman, R.; et al. Greenness in semi-arid areas across the globe 1981-2007 an earth observing satellite-based analysis of trends and drivers. *Remote Sens. Environ.* **2012**, *121*, 144158. [CrossRef]
24. Jiang, Y.; Huang, J. The temperature spatial distribution characteristics based on GIS technology in Gansu Province. *J. Arid. Meteorol.* **2013**, *31*, 206–211.
25. Nemani, R.R.; Keeling, C.D.; Hashimoto, H.; Jolly, W.M.; Piper, S.C.; Tucker, C.J.; Myneni, R.B.; Running, S.W. Climate-driven increases in global terrestrial net primary production from 1982 to 1999. *Science* **2003**, *300*, 1560–1563. [CrossRef] [PubMed]
26. Xu, H.M.; Gao, Q.Z.; Huang, Y.M.; Jia, H.K. Simulated the impact of climate change on net primary production in hilly area of Loess Plateau, China. *Acta Ecol. Sin.* **2006**, *26*, 2938–2947.
27. Mu, S.J.; Li, J.L.; Zhou, W.; Yang, H.F.; Zhang, C.B.; Ju, W.M. Spatial-temporal distribution of net primary productivity and its relationship with climate factors in Inner Mongolia from 2001 to 2010. *Acta Ecol. Sin.* **2013**, *33*, 3752–3764.
28. Chen, B.Z.; Xu, G.; Coops, N.C.; Ciais, P.; Innes, J.L.; Wang, G.Y.; Myneni, R.B.; Wang, T.L.; Krzyzanowski, J.; Li, Q.L. Changes in vegetation photosynthetic activity trends across the Asia-Pacific region over the last three decades. *Remote Sens. Environ.* **2014**, *144*, 28–41. [CrossRef]
29. Hou, W.J.; Gao, J.B.; Wu, S.H.; Dai, E. Interannual variations in growing-season NDVI and its correlation with climate variables in the southwestern karst region of China. *Remote Sens.* **2015**, *7*, 11105–11124. [CrossRef]
30. Pan, T.; Wu, S.H.; He, D.M.; Dai, E.F.; Liu, Y.J. Effects of longitudinal range-gorge terrain on the eco-geographical pattern in Southwest China. *J. Geogr. Sci.* **2012**, *22*, 825–842. [CrossRef]
31. Dunkerley, D.L. Intra-event intermittency of rainfall: An analysis of the metrics of rain and no-rain periods. *Hydrol. Processes* **2015**, *29*, 3294–3305. [CrossRef]
32. Li, C.; Zhu, T.; Zhou, M.; Yin, H.; Wang, Y.; Sun, H.; Cao, H.; Han, H. Temporal and spatial change of Net Primary Productivity of vegetation and its determinants in Hexi Corridor. *Acta Ecol. Sin.* **2021**, *41*, 1931–1943.
33. Diao, W.; Zhao, Y.; Zhai, J.; Fan, H.E.; Sui, B.; Zhu, Y. Temporal Spatial Evolution and Driving Force Analysis of Minqin Oasis during 1987–2017. *J. Irrig. Drain.* **2019**, *38*, 106–113.
34. Zhu, Z.Q.; Liu, L.M.; Zhang, J.L. Impact of grain for green project on landscape pattern in hilly loess region in Southern Ningxia: Landscape evolution process assessment of Zhong-zhuangcun small water shed in 1993–2005. *Acta Ecol. Sin.* **2010**, *30*, 146–154.
35. Bai, F.; Li, W.P.; Li, Z.H. Analysis on the main causes resulting in vegetation degeneration in the Heihe River Basin. *Arid. Zone Res.* **2008**, *25*, 219–224. [CrossRef]
36. Lan, Y.C.; Sun, B.M.; Ding, Y.J. Studies on ecological environment changes of Heihe River Basin and its influence factors. *J. Arid. Land Resour. Environ.* **2004**, *18*, 32–39.
37. Zhang, Y.; Zhao, W. Vegetation and soil property response of short-time fencing in temperate desert of the Hexi Corridor, northwestern China. *Catena* **2015**, *133*, 43–51. [CrossRef]

38. Yang, G.H.; Bao, A.M.; Chen, X.; Liu, H.L.; Huang, Y.; Dai, S.Y. Study of the vegetation cover change and its driving factors over Xinjiang during 1998–2007. *Glaciol. Geocryol.* **2009**, *31*, 436–445.
39. Li, X.; Zhang, X.; Niu, J.; Tong, L.; Kang, S.; Du, T.; Li, S.; Ding, R. Irrigation water productivity is more influenced by agronomic practice factors than by climatic factors in Hexi Corridor, Northwest China. *Sci. Rep.* **2016**, *6*, 37971. [CrossRef]
40. Wang, X.Y.; Chen, X.S.; Ding, Q.P.; Zhao, X.Y.; Wang, X.J.; Ma, Z.W.; Lian, J. Vegetation and soil environmental factor characteristics, and their relationship at different desertification stages: A case study in the Minqin desert-oasis ecotone. *Acta Ecol. Sin.* **2018**, *38*, 1569–1580.
41. Wang, X.Y.; Lian, J.; Yang, X.P.; Zhao, X.Y.; Wang, X.J.; Ma, Z.W.; Gong, C.K.; Qu, H.; Wang, B. Variation in vegetation and its response to environmental factors in Maqu County. *Acta Ecol. Sin.* **2019**, *39*, 923–935.
42. Simpson, R.W.; Petroeschevsky, A.; Lowe, I. An ecological footprint analysis for Australia. *Aust. J. Environ. Manag.* **2000**, *7*, 11–18. [CrossRef]
43. Zhang, J.Y.; Li, Y.; Zhao, W.Z.; Shi, X. Tracking analysis on Changes of ecological patterns in Hexi Corridor region. *Water Resour. Prot.* **2015**, *31*, 5–10.

diversity

MDPI

Article

Linking Leaf N:P Stoichiometry to Species Richness and Composition along a Slope Aspect Gradient in the Eastern Tibetan Meadows

Xin'e Li [1,*], Yafei Hu [1], Renyi Zhang [2], Xin Zhao [1] and Cheng Qian [1]

[1] Division of Grassland Science, College of Animal Science and Technology, Yangzhou University, Yangzhou 225009, China; huyafei0816@163.com (Y.H.); zx18990005917@163.com (X.Z.); amanda0599@outlook.com (C.Q.)
[2] Department of Life Science, Lanzhou University, Lanzhou 730000, China; zrenyi@lzu.edu.cn
* Correspondence: lixine@yzu.edu.cn or lixine19@gmail.com

Abstract: As an important topographical factor, slope aspect has an essential influence on plant community structure and leaf traits. Leaf nitrogen (N) and phosphorus (P) stoichiometry is an important leaf trait indicating plant growth. However, it has rarely been studied how leaf N:P stoichiometry correlates with plant community structure along the slope aspect gradient. To understand the variation of leaf N:P stoichiometry and community structure, as well as their correlation with each other, the species composition and leaf N and P in Tibetan meadows were investigated across three slope aspects: the south-, west-, and north-facing slope aspects (i.e., SFS, WFS, and NFS). In our results, leaf N:P ratio was significantly lower on the NFS than on the SFS, indicating N and P limitation on the NFS and SFS, respectively. Richness of forb species and all species was higher on the NFS than on the SFS and was negatively correlated with leaf N concentration, whereas graminoid richness was not statistically different among the slope aspects and showed a negative correlation with leaf P concentration. Thus, our results provide evidence for the functional significance of leaf N:P stoichiometry for species composition along a natural environmental gradient. Our findings could provide applicable guidance in the refinement of grassland management and biodiversity conservation based on topography.

Keywords: topography; community assembly; species richness; leaf functional traits

Citation: Li, X.; Hu, Y.; Zhang, R.; Zhao, X.; Qian, C. Linking Leaf N:P Stoichiometry to Species Richness and Composition along a Slope Aspect Gradient in the Eastern Tibetan Meadows. *Diversity* **2022**, *14*, 245. https://doi.org/10.3390/d14040245

Academic Editors: Lin Zhang, Jinniu Wang and Michael Wink

Received: 5 March 2022
Accepted: 26 March 2022
Published: 27 March 2022

1. Introduction

Leaf nitrogen (N) and phosphorus (P) are the two most limiting elements of terrestrial vegetation relating to plant growth, development, and reproduction [1–3] because N is the key component of proteins that play pivotal roles in plant photosynthesis as enzymes, whereas P is critical to the formation of NADPH, ATP, and ribosomal RNA in the process of protein synthesis [4,5]. Furthermore, the critical leaf N:P ratio has been widely used to diagnose the type of nutrient limitations to plant productivity [6–8]. Leaf N and P concentration and allocation have been found to vary along environment gradients; although numerous studies have focused on their latitudinal and altitudinal patterns [9–16], it has rarely been studied how they change along the slope aspect gradient. Investigations on leaf N:P stoichiometry across slope aspects can provide applicable guidance in the refinement of grassland management and conservation based on topography at the local scale.

Leaf N and P availability also shows important functional significance for community composition [6]. In particular, relationships between plant N:P ratios and species richness are of particular interest in the context of biodiversity conservation and anthropogenic activities. Many studies have found the N:P ratio to be correlated with the richness and composition of species [17–19]. The coexistence of species was suggested to be possibly facilitated by P limitation, since the competition for P with lower mobility is weaker than the competition for resources of high mobility in soil [17,20] and is facilitated by colimitation because differential nutrient limitations may reduce interspecific competition [21,22] but could also be weakened by the P limitation possibly resulting from P deficiency or nutrient imbalance [6] or the disproportionate increase in dominant clonal graminoids [23]. Hence, different correlations of species richness with plant N and P stoichiometry have been reported [17,18,24–26]. Moreover, a high leaf N:P ratio was expected to be associated with more graminoids and fewer forbs in vegetation [23]. However, most of these studies were indirectly demonstrated in experiments with nitrogen fertilization or deposition, and direct evidence is rare along natural environmental gradients. The slope aspect gradient significantly contributed to the heterogeneity of vegetation in mountain areas and thus provided an ideal platform to explore the relationships between leaf N/P stoichiometry and community structure.

Typically, south-facing slopes in the northern hemisphere (i.e., equator-facing slopes) are hot and dry, as the equator-facing orientation is associated with the strongest solar irradiation, whereas north-facing slopes (i.e., polar-facing slopes) are wet and cold, resulting from the lowest solar irradiations; western- or eastern-facing slopes are intermediate in terms of these aspects. Slope aspect substantially contributes to the heterogeneity of vegetation [27–32], possibly owing to the substantial microenvironmental changes, such as solar irradiance, soil moisture and temperature, and soil nutrients [33–36]. However, it is still not clear how vegetation heterogeneity, including species richness and composition, is associated with leaf N:P stoichiometry. Therefore, in the current study, based on a slope aspect gradient in the Tibetan meadow, we mainly aimed to explore the following questions: (1) How do leaf [N] and [P] and the N:P ratio vary in different slope aspects? (2) How do species richness and composition change with slope aspect? (3) How does leaf N:P stoichiometry correlate with species richness and composition along the slope aspect gradient?

2. Materials and Methods

2.1. Study Sites

Our study was conducted in an alpine meadow in the eastern part of the Tibetan Plateau in China. The Research Station of Alpine Meadow and Wetland Ecosystems of Lanzhou University has locations at two elevations: 2960 m and 3650 m (Figure 1A). Data were collected at these two sites: Hezuo (34°44′ N,102°53′ E) and Maqu (33°39′ N,101°53′ E), with the vegetation landscape shown in Figure 1B. We also recorded the precipitation and temperature of these two sites during the period of 1981–2017 according to geographical coordinates using the climate dataset provided by National Tibetan Plateau Data Center (http://data.tpdc.ac.cn (accessed on 10 May 2021)). The mean annual precipitation in Hezuo and Maqu was 570 mm and 690 mm, and the mean annual temperature was around 4 °C and 2 °C, respectively. The monthly mean precipitation and monthly mean, maximum, and minimum temperatures are shown in Figure 1C. Details about the study site can also be found in our previous publications [36,37].

2.2. Leaf N and P Concentration Measurements

In August of 2008, during the peak growing season, a 5 m × 5 m plot was established on each of the south-, west-, and north-facing slopes (i.e., SFS, WFS, and NFS) based on the shape of the hill and the ability to collect leaf samples at each site. At each plot, mature and healthy leaf samples were collected from 3–5 individuals of the dominant species. Across the six plots, 80 observations were collected in total, with 41 observations of 25 species in Hezuo and 39 observations of 20 species in Maqu. The measured species in each site, with their average leaf N and P content per unit of mass (hereafter leaf [N] and [P]), are shown in Table 1. All the species were simply classified into three plant functional groups (PFGs): graminoids (Poaceae and Cyperaceae), non-legume forbs (forbs), and legumes. However, in two plots, dwarf shrub was also found on the NFS.

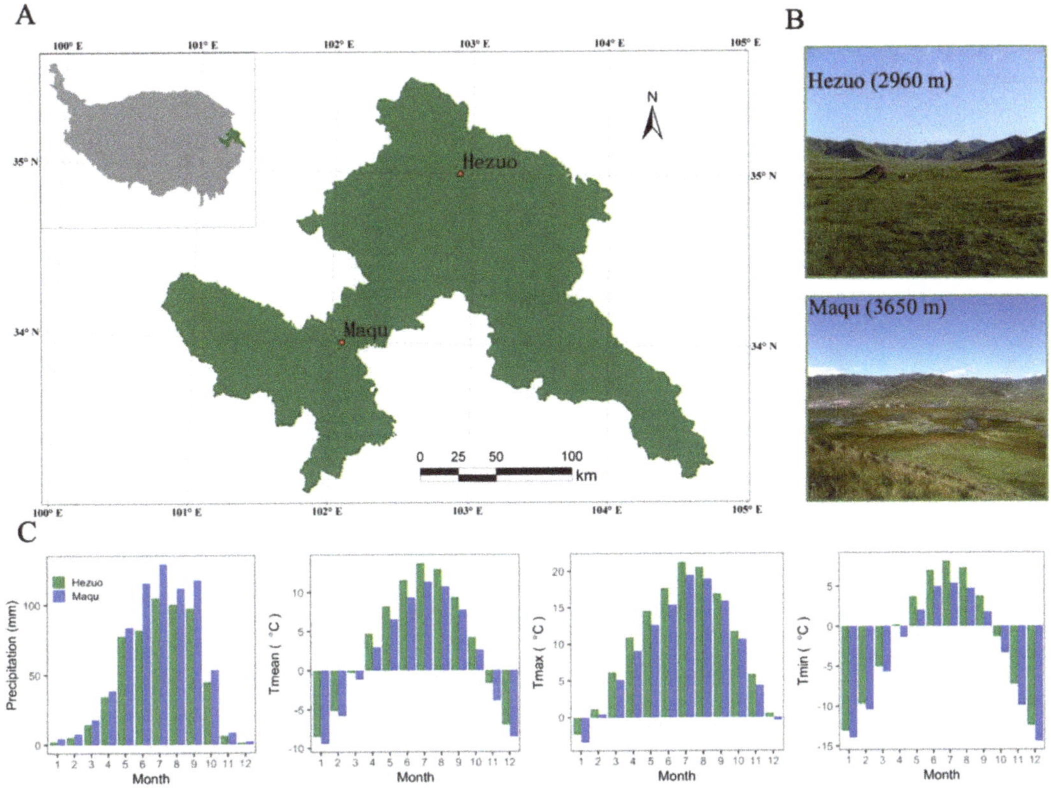

Figure 1. Site location in the Tibetan Plateau (Panel **A**), the vegetation landscape (Panel **B**), and key climate factors (Panel **C**) of the two study sites (Hezuo and Maqu). Tmean, Tmax, and Tmin represent the monthly mean, maximum, and minimum temperature, respectively.

After drying for 48 h at 70 °C, we ground the dry leaves to powder using a mortar. A total of 0.2 g of leaf powder was digested with H_2SO_4-H_2O_2. Digested solution was used to determine leaf [N] with a VAPODEST 40 programmable distillation system (Gerhardt, Germany), and leaf [P] by the vanadium-ammonium molybdate colorimetric method [38].

2.3. Species Composition Measurement

In each plot, we placed three 50 cm × 50 cm quadrates as replicates to survey the species by recording their names, coverage (%), and richness (i.e., the number of species) and then classified all the species into the four plant functional groups: graminoids, forbs, legumes, and shrubs (if any). The coverage was estimated visually, but significantly positive correlation of coverage and species richness confirmed the data reliability ($R^2 = 0.418$, $p < 0.001$).

2.4. Data Analysis

Firstly, the effects of slope aspect on leaf [N], [P], and N:P ratio were detected using a linear mixed model by treating "site" as the random factor (LMM) for the mixed samples, with species-level value as a replicate unit and one common species, *Anaphalis lacteal*, that occurred in each of the six plots. Correlations among leaf [N], [P], and N:P ratio were evaluated in each site using linear models. Comparison of the leaf N:P stoichiometry between different PFGs was also conducted using the LMM. Secondly, the effects of slope aspect on species richness and coverage and their correlations with leaf N:P stoichiometry (using plot-level means) were analyzed using LMM.

All the variables were log10-transformed. All analyses were performed with R version 4.0.3 (R Core Team, 2020) in RStudio version 1.3.1093 (RStudio Team, 2020).

Table 1. Mean value of leaf nitrogen (N) and phosphorus (P) concentrations of the 40 measured species. PFG represents plant functional group.

SPECIES	FAMILY	PFGs	Hezuo		Maqu	
			N (mg g^{-1})	P (mg g^{-1})	N (mg g^{-1})	P (mg g^{-1})
Allium beesianum	Amaryllidaceae	Forbs			16.69	1.43
Allium condensatum	Amaryllidaceae	Forbs			16.28	1.14
Anaphalis hancockii	Asteraceae	Forbs	17.86	1.28		
Anaphalis lactea	Asteraceae	Forbs	26.02	1.66	10.45	0.78
Anemone trullifolia var. linearis	Ranunculaceae	Forbs			13.16	1.04
Aster tataricus	Asteraceae	Forbs			11.02	0.78
Astragalus membranaceus var.membranaceus	Fabaceae	Legumes	24.52	1.73	29.29	1.83
Bupleurum sp	Apiaceae	Forbs			14.01	1.05
Cyperaceae sp	Cyperaceae	Graminoids	17.30	1.24		
Daucus carota	Apiaceae	Forbs			7.80	0.99
Elephantopus scaber	Asteraceae	Forbs	20.89	1.60		
Fragaria ananassa	Rosaceae	Forbs	13.33	1.65		
Gentiana macrophylla	Gentianaceae	Forbs	22.42	1.25		
Gentianopsis barbata	Gentianaceae	Forbs			14.10	1.25
Gueldenstaedtia verna	Fabaceae	Legumes	26.17	1.27	21.44	1.21
Hamamelis mollis	Hamamelidaceae	Shrubs	17.83	1.29		
Kobresia humilis	Cyperaceae	Graminoids	14.61	1.01		
Lancea tibetica	Mazaceae	Forbs	19.10	1.53		
Leontopodium leontopodioides	Asteraceae	Forbs			13.61	0.93
Ligularia virgaurea	Asteraceae	Forbs			11.69	1.10
Medicago falcata	Fabaceae	Legumes			4.36	1.05
Medicago lupulina	Fabaceae	Legumes	32.35	1.90		
Nardostachys jatamansi	Caprifoliaceae	Forbs			14.26	1.07
Nardostachys jatamansi	Caprifoliaceae	Forbs			8.15	0.88
Nepeta cataria	Lamiaceae	Forbs	25.12	1.65		
Oxytropis sp	Fabaceae	Legumes	36.79	2.00		
Pedicularis szetschuanica	Orobanchaceae	Forbs			13.38	1.94
Plantago asiatica	Plantaginaceae	Forbs	16.07	1.56		
Polygonum macrophyllum	Polygonaceae	Forbs			9.63	1.01
Polygonum viviparum	Polygonaceae	Forbs	25.53	1.59	15.41	1.83
Potentilla anserina	Rosaceae	Forbs	13.79	1.57		
Potentilla bifurca	Rosaceae	Forbs	17.76	1.25		
Potentilla fragarioides	Rosaceae	Forbs	12.92	1.03		
Roegneria kamoji	Poaceae	Graminoids	18.00	1.41		
Saussurea graminea	Asteraceae	Forbs			9.40	0.54
Saussurea graminifolia	Asteraceae	Forbs	15.92	1.39		
Saussurea sp	Asteraceae	Forbs	18.62	1.20	8.77	0.84
Scirpus triqueter	Cyperaceae	Graminoids	17.55	0.89		
Stellera chamaejasme	Thymelaeaceae	Forbs	36.87	2.31		
Taraxacum sp	Asteraceae	Forbs	33.27	1.85		

3. Results

3.1. Variations in Leaf [N], [P], and N:P Ratio across the Slope Aspects

The leaf [N], [P], and N:P ratios ranged from 10.09 to 42.96, 0.89 to 3.10, and 8.10 to 23.63 mg g^{-1} in Hezuo and from 4.36 to 33.53, 0.536 to 2.33, and 4.14 to 21.21 mg g^{-1} in Maqu, respectively. Overall, the leaf [N], [P], and N:P ratios were all significantly lower in Maqu, with higher elevations than in Hezuo (Figure 2A,C,E).

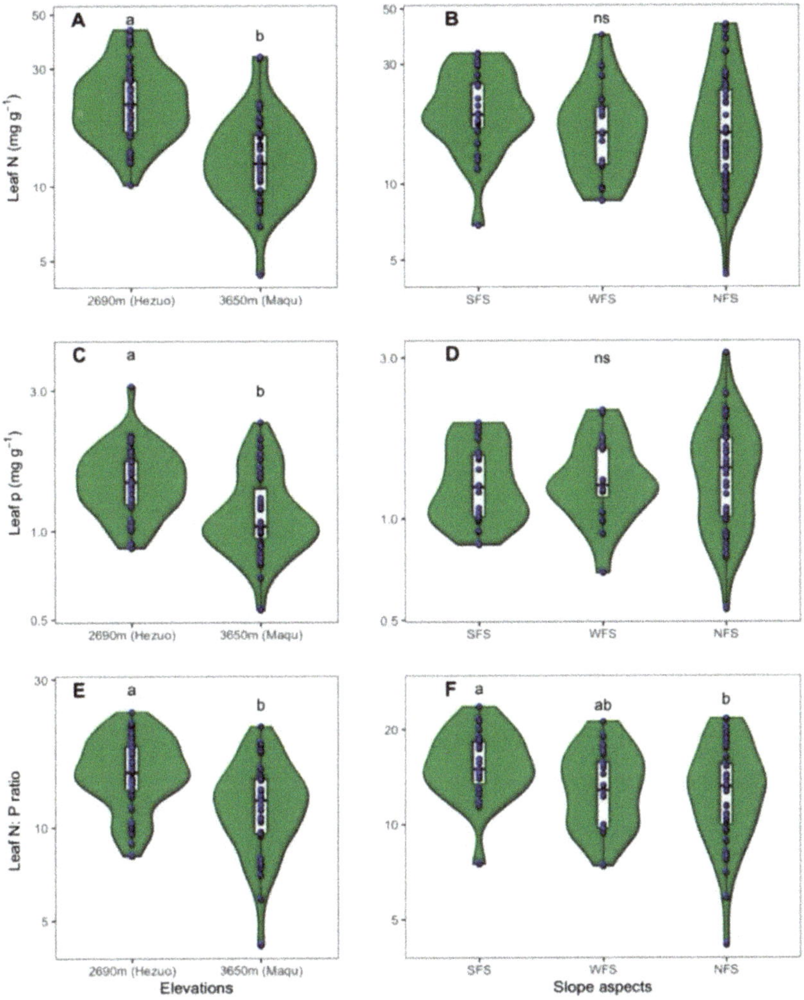

Figure 2. Comparisons of leaf [N] (**A**,**B**), [P] (**C**,**D**), and N:P ratio (**E**,**F**) between sites and among slope aspects. SFS, WFS, and NFS represent south-, west-, and north-facing slope aspects, respectively. Different letters represent significant differences at the level of $p < 0.05$. Variables are in log10 scale.

Although leaf [N] and [P] were not significantly different between slope aspects, leaf N:P ratio was significantly lower on the NFS than on the SFS for the mixed samples (Figure 2). The one common species, *Anaphalis lacteal*, showed the same patterns as those of the mixed samples (Figure S1). Moreover, significant correlations between leaf [N] and [P] and between the leaf N:P ratio and leaf [N] rather than [P] were found in each plot (Figure 3). In addition, legumes showed a significantly higher leaf [N] and [P] than forbs and graminoids and a higher leaf N:P ratio than forbs (Figure S2).

3.2. Correlations of Leaf Traits with Species Richness and Coverage

Significantly higher richness of all species and forbs and lower richness of legumes were found on the NFS than on the SFS and WFS (Figure 4A–D); however, the richness of graminoids and plant coverage for each PFG and for all species did not differ significantly among the slope aspects (Figure 4E–H).

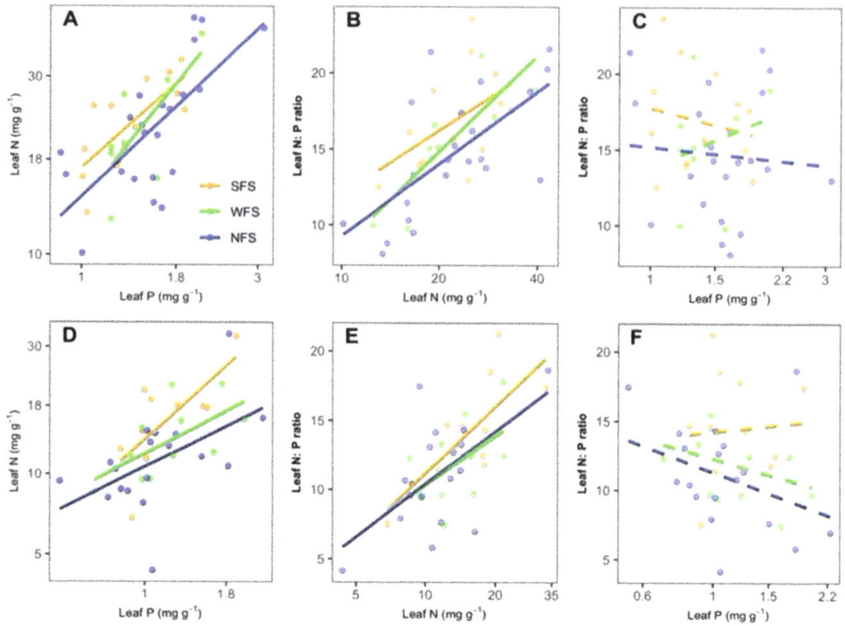

Figure 3. Linear regressions between leaf [N] and [P], between leaf [N] and leaf N:P ratio, and between leaf [P] and leaf N:P ratio in Hezuo (**A–C**) and Maqu (**D–F**). Solid and dashed lines indicate significant and non-significant regressions, respectively. SFS, WFS, and NFS represent south-, west-, and north-facing slope aspects, respectively. Variables are in log10 scale.

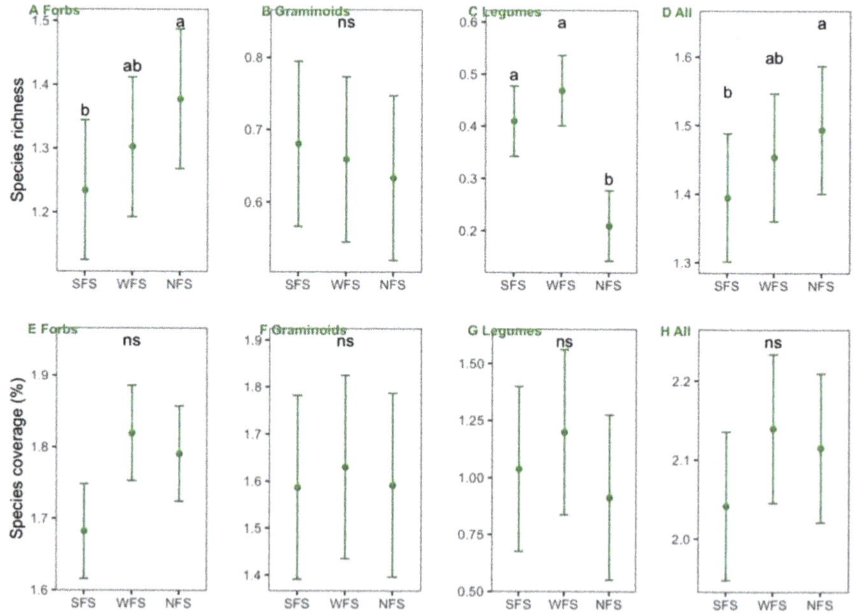

Figure 4. Comparisons of species richness (**A–D**) and coverage (**E–H**) for each plant functional group and the whole community among slope aspects. SFS, WFS, and NFS represent south-, west-, and north-facing slope aspects, respectively. Variables are log10-transformed.

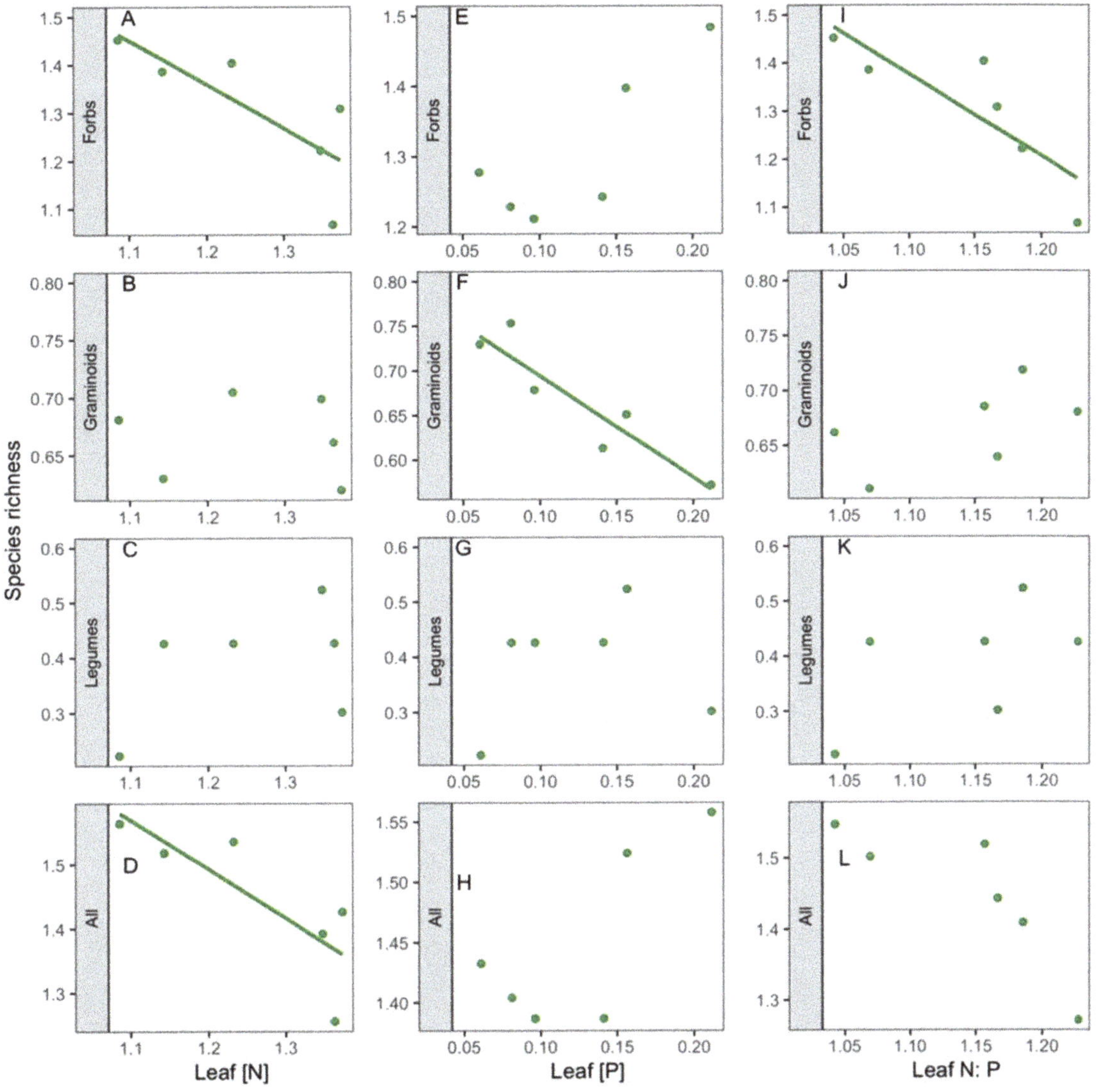

Figure 5. Correlations of species richness of each plant functional group and the whole community with leaf N concentration (**A–D**), with leaf P concentration (**E–H**), and with leaf N:P ratio (**I–L**). Variables are log10-transformed.

The correlations were found to be significantly negative between leaf [N] and richness of all species (R^2 = 0.615, p = 0.047) and forb species (R^2 = 0.570, p = 0.061), between leaf [P] and graminoid richness (R^2 = 0.362, p = 0.093), and between leaf N:P ratios and forb richness (R^2 = 0.675, p = 0.032) (Figure 5). On the other hand, legume coverage was positively correlated with leaf [N] (R^2 = 0.809, p = 0.010), and the coverage of graminoids and all species was negatively and positively correlated with leaf [P] and with marginal significance and significance, respectively (R^2 = 0.051 and 0.704, p = 0.103 and 0.029, respectively) (Figure 6).

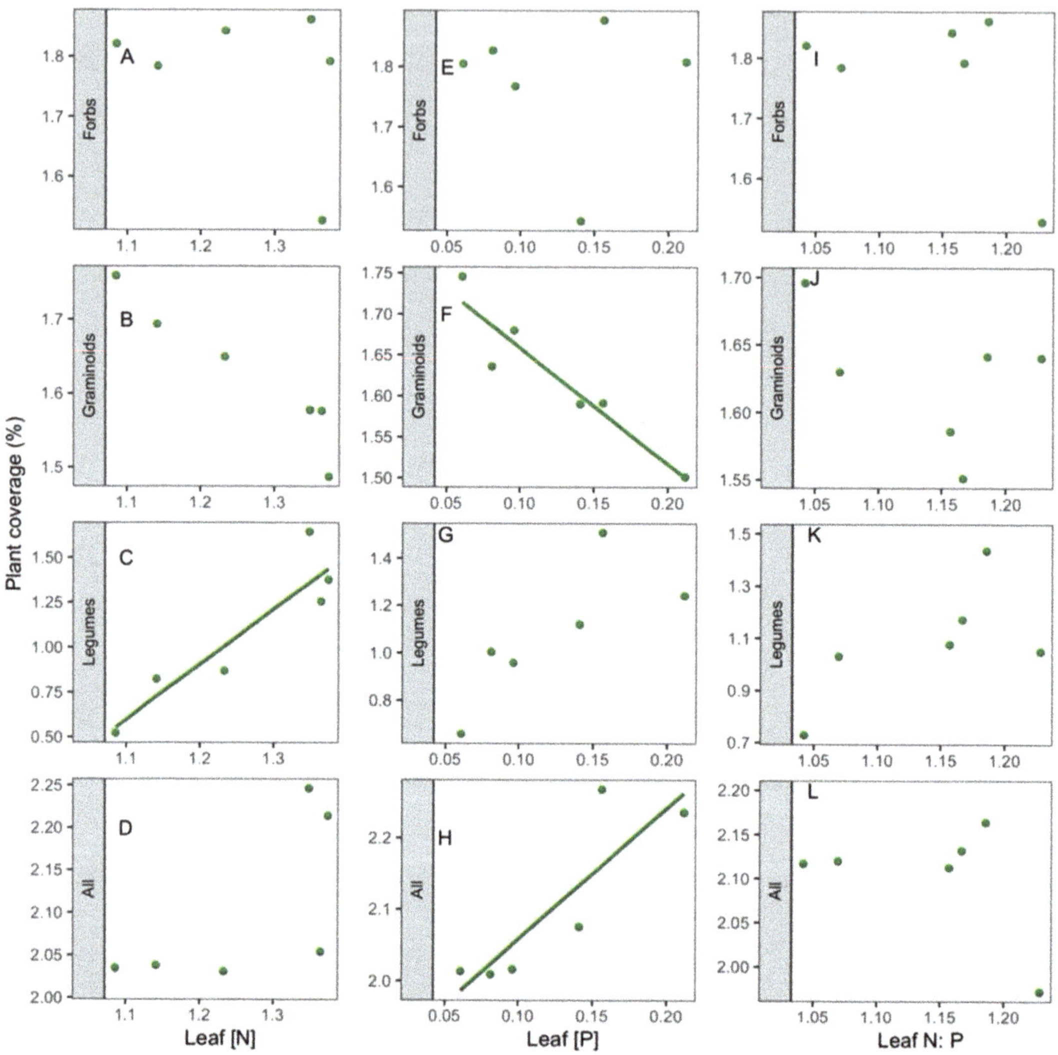

Figure 6. Correlations of species coverage of each plant functional group and the whole community with leaf N concentration (**A–D**), with leaf P concentration (**E–H**), and with leaf N:P ratio (**I–L**). Variables are log10-transformed.

4. Discussion

4.1. Variations in Leaf [N], [P], and N:P Ratio across Different Slope Aspects

Although leaf [N] and [P] did not differ significantly between slope aspects, we found a significantly lower N:P ratio on the NFS, with a mean of 11, and a higher N:P ratio on the SFS, with a mean of 16, which was consistent with a previous study reporting that the lowest N:P ratio occurred on the NFS out of the four slope aspects [39]. Tessier and Raynal [8] pointed out that plants were subject to N-, P-, and co-limitation when the plant N:P ratio was less than 14, higher than 16, and between 14 and 16, respectively. Therefore, our results indicate that plants on the NFS suffered N limitation. The tight correlation of leaf [N] with N:P ratio suggests that leaf N:P is determined by leaf [N] along the slope aspect gradient. A lower soil N availability reported in our previous studies [36] was probably the reason for the lower leaf N:P ratio on the NFS, which supports the biogeochemical hypothesis suggesting that plant N and P are influenced by the availability of soil nutrients [9]. The lower soil N on the NFS was perhaps due to lower temperatures reducing the decomposition and mineralization of organic material.

In addition, the leaf [N], [P], and N:P ratio were lower in Maqu, which has a higher elevation, which is consistent with previous studies of elevational gradients [14,40], indicating an increase in N-limitation along altitudinal gradients.

4.2. Correlations of Leaf N:P Stoichiometry with Species Richness and Coverage across the Slope Aspects

Significantly higher species richness in each quadrate was found on the NFS than on the SFS, which was mainly caused by the variation in forb richness. These results are consistent with previous studies in which species richness was significantly lower on the SFS [31,41], with the possible mechanism of lower soil water content as the limiting factor.

Community species richness was negatively correlated with leaf [N], which is consistent with the negative effects of N-fertilization on species diversity [42–45]. Many N-fertilization experiments showed considerable species loss, with the underlying improved growth of dominating clonal graminoids excluding forbs through competition [45,46]. The possible mechanisms may be that N-fertilization led to a tendency of P limitation, but graminids are often considered stress-tolerant species due to their low requirement for P and K [47]. Therefore, we found negative correlations of leaf [P] with graminoid richness and coverage because higher leaf [P] indicated the rapid growth of P-sensitive species. The significant negative correlations between leaf N:P ratio and forb richness provide direct evidence that on the natural environmental gradient, a higher N:P ratio is associated with fewer forbs and more graminoids, as hypothesized by Güsewell [6].

In addition, the higher legume richness on the SFS and the positive correlation of legume coverage with leaf [N] were consistent with the higher N:P ratio on the SFS due to their N-fixing functions.

5. Conclusions

In this study, we linked species composition and leaf N:P stoichiometry along a slope aspect gradient. Our results reveal different types of nutrition limitation and species compositions in different slope aspects and highlight the functional significance of leaf N:P stoichiometry in community composition along a slope aspect gradient. We found that leaf N:P ratio was significantly lower on the NFS than on the SFS and WFS, although leaf [N] and [P] were not statistically different between slope aspects, implying N limitation and P limitation of plant growth on the NFS and SFS, respectively. Higher richness and coverage of forbs were found on the NFS than on the SFS, whereas those that of graminoids did not differ significantly between slope aspects. The variations in forb and graminoid richness were largely explained by leaf [N] and [P], respectively, with the possible mechanism of different N or P requirements for different PFGs. However, the mechanisms underlying the variations in leaf N:P stoichiometry and community composition across slope aspects are still not clear because we did not explore their correlations with environmental or soil factors.

Supplementary Materials: The following are available online at https://www.mdpi.com/article/10.3390/10.3390/d14040245/s1, Figure S1: Comparisons of leaf [N] (A,B), [P] (C,D), and N:P ratio (E,F) between sites and among slope aspects, respectively, for the species *Anaphalis lacteal*, Figure S2: Comparisons of leaf [N] (Panel A), [P] (Panel B), and N:P ratio (Panel C) between different plant functional groups (PFGs).

Author Contributions: Conceptualization, X.L.; methodology, R.Z.; software, X.L.; formal analysis, X.L.; investigation, X.L.; data curation, Y.H., X.Z. and C.Q.; writing—original draft preparation, X.L.; writing—review and editing, Y.H.; visualization, X.L.; supervision, X.L.; project administration, X.L. and R.Z.; funding acquisition, X.L. All authors have read and agreed to the published version of the manuscript.

Funding: This study was supported by the funding from Yangzhou University.

Data Availability Statement: The data presented in this study are available in article and Supplementary Materials.

Acknowledgments: We thank Zhao Jun, Song Xiaoyu, Nie Yingying, Chen Lingyun, and Zhang Jieqi for their help in the field and lab experiments. We thank Cui Xia for her help in working on Figure 1. This work was supported by the Research Station of Alpine Meadow and Wetland Ecosystems (RSAMWE) of Lanzhou University.

Conflicts of Interest: The authors declare no conflict of interest.

References

1. Vitousek, P.M.; Porder, S.; Houlton, B.Z.; Chadwick, O.A. Terrestrial phosphorus limitation: Mechanisms, implications, and nitrogen–phosphorus interactions. *Ecol. Appl.* **2010**, *20*, 5–15. [CrossRef] [PubMed]
2. LeBauer, D.S.; Treseder, K.K. Nitrogen limitation of net primary productivity in terrestrial ecosystems is globally distributed. *Ecology* **2008**, *89*, 371–379. [CrossRef] [PubMed]
3. Du, E.; Terrer, C.; Pellegrini, A.F.; Ahlström, A.; van Lissa, C.J.; Zhao, X.; Xia, N.; Wu, X.; Jackson, R.B. Global patterns of terrestrial nitrogen and phosphorus limitation. *Nat. Geosci.* **2020**, *13*, 221–226. [CrossRef]
4. Elser, J.; Sterner, R.; Gorokhova, E.a.; Fagan, W.; Markow, T.; Cotner, J.; Harrison, J.; Hobbie, S.; Odell, G.; Weider, L. Biological stoichiometry from genes to ecosystems. *Ecol. Lett.* **2000**, *3*, 540–550. [CrossRef]
5. Taiz, L.; Zeiger, E. Photosynthesis: Physiological and ecological considerations. *Plant Physiol* **2002**, *9*, 172–174.
6. Güsewell, S. N:P ratios in terrestrial plants: Variation and functional significance. *New Phytol.* **2004**, *164*, 243–266. [CrossRef]
7. Schade, J.D.; Espeleta, J.F.; Klausmeier, C.A.; McGroddy, M.E.; Thomas, S.A.; Zhang, L. A conceptual framework for ecosystem stoichiometry: Balancing resource supply and demand. *Oikos* **2005**, *109*, 40–51. [CrossRef]
8. Tessier, J.T.; Raynal, D.J. Use of nitrogen to *Phosphorus* ratios in plant tissue as an indicator of nutrient limitation and nitrogen saturation. *J. Appl. Ecol.* **2003**, *40*, 523–534. [CrossRef]
9. Reich, P.B.; Oleksyn, J. Global patterns of plant leaf N and P in relation to temperature and latitude. *Proc. Natl. Acad. Sci. USA* **2004**, *101*, 11001–11006. [CrossRef]
10. Han, W.; Fang, J.; Guo, D.; Zhang, Y. Leaf nitrogen and phosphorus stoichiometry across 753 terrestrial plant species in China. *New Phytol.* **2005**, *168*, 377–385. [CrossRef]
11. Chen, Y.; Han, W.; Tang, L.; Tang, Z.; Fang, J. Leaf nitrogen and phosphorus concentrations of woody plants differ in responses to climate, soil and plant growth form. *Ecography* **2013**, *36*, 178–184. [CrossRef]
12. He, J.-S.; Wang, L.; Flynn, D.F.; Wang, X.; Ma, W.; Fang, J. Leaf nitrogen: Phosphorus stoichiometry across Chinese grassland biomes. *Oecologia* **2008**, *155*, 301–310. [CrossRef] [PubMed]
13. Zhang, L.; Wang, L.; He, W.; Zhang, X.; An, L.; Xu, S. Patterns of Leaf N:P Stoichiometry along Climatic Gradients in Sandy Region, North of China. *J. Plant Ecol.* **2016**, *11*, 218–225. [CrossRef]
14. Kang, H.; Zhuang, H.; Wu, L.; Liu, Q.; Shen, G.; Berg, B.; Man, R.; Liu, C. Variation in leaf nitrogen and phosphorus stoichiometry in Picea abies across Europe: An analysis based on local observations. *For. Ecol. Manag.* **2011**, *261*, 195–202. [CrossRef]
15. Waigwa, A.N.; Mwangi, B.N.; Wahiti, G.R.; Omengo, F.; Zhou, Y.; Wang, Q.; Schmid, B. Variation of morphological and leaf stoichiometric traits of two endemic species along the elevation gradient of Mount Kenya, East Africa. *J. Plant Ecol.* **2020**, *13*, 785–792. [CrossRef]
16. Zhao, N.; Yu, G.; He, N.; Xia, F.; Wang, Q.; Wang, R.; Xu, Z.; Jia, Y. Invariant allometric scaling of nitrogen and phosphorus in leaves, stems, and fine roots of woody plants along an altitudinal gradient. *J. Plant Res.* **2016**, *129*, 647–657. [CrossRef] [PubMed]
17. Güsewell, S.; Bailey, K.M.; Roem, W.J.; Bedford, B.L. Nutrient limitation and botanical diversity in wetlands: Can fertilisation raise species richness? *Oikos* **2005**, *109*, 71–80. [CrossRef]
18. Blanck, Y.-L.; Gowda, J.; Mårtensson, L.-M.; Sandberg, J.; Fransson, A.-M. Plant species richness in a natural Argentinian matorral shrub-land correlates negatively with levels of plant phosphorus. *Plant Soil* **2011**, *345*, 11–21. [CrossRef]
19. Sasaki, T.; Yoshihara, Y.; Jamsran, U.; Ohkuro, T. Ecological stoichiometry explains larger-scale facilitation processes by shrubs on species coexistence among understory plants. *Ecol. Eng.* **2010**, *36*, 1070–1075. [CrossRef]
20. Huston, M.A.; DeAngelis, D.L. Competition and coexistence: The effects of resource transport and supply rates. *Am. Nat.* **1994**, *144*, 954–977. [CrossRef]
21. Bracken, M.E.; Hillebrand, H.; Borer, E.T.; Seabloom, E.W.; Cebrian, J.; Cleland, E.E.; Elser, J.J.; Gruner, D.S.; Harpole, W.S.; Ngai, J.T. Signatures of nutrient limitation and co-limitation: Responses of autotroph internal nutrient concentrations to nitrogen and phosphorus additions. *Oikos* **2015**, *124*, 113–121. [CrossRef]
22. Braakhekke, W.G.; Hooftman, D.A. The resource balance hypothesis of plant species diversity in grassland. *J. Veg. Sci.* **1999**, *10*, 187–200. [CrossRef]
23. Theodose, T.A.; Roths, J.B. Variation in nutrient availability and plant species diversity across forb and graminoid zones of a Northern New England high salt marsh. *Plant Ecol.* **1999**, *143*, 219–228. [CrossRef]

24. Pekin, B.K.; Boer, M.M.; Wittkuhn, R.S.; Macfarlane, C.; Grierson, P.F. Plant diversity is linked to nutrient limitation of dominant species in a world biodiversity hotspot. *J. Veg. Sci.* **2012**, *23*, 745–754. [CrossRef]
25. Harpole, W.S.; Sullivan, L.L.; Lind, E.M.; Firn, J.; Adler, P.B.; Borer, E.T.; Chase, J.; Fay, P.A.; Hautier, Y.; Hillebrand, H. Addition of multiple limiting resources reduces grassland diversity. *Nature* **2016**, *537*, 93–96. [CrossRef] [PubMed]
26. Olde Venterink, H. Does phosphorus limitation promote species-rich plant communities? *Plant Soil* **2011**, *345*, 1–9. [CrossRef]
27. Armesto, J.J.; Martínez, J.A. Relations between vegetation structure and slope aspect in the mediterranean region of Chile. *J. Ecol.* **1978**, 881–889. [CrossRef]
28. Bennie, J.; Hill, M.O.; Baxter, R.; Huntley, B. Influence of slope and aspect on long-term vegetation change in British chalk grasslands. *J. Ecol.* **2006**, *94*, 355–368. [CrossRef]
29. Gallardo-Cruz, J.A.; Pérez-García, E.A.; Meave, J.A. β-Diversity and vegetation structure as influenced by slope aspect and altitude in a seasonally dry tropical landscape. *Landsc. Ecol.* **2009**, *24*, 473–482. [CrossRef]
30. Kutiel, P. Slope aspect effect on soil and vegetation in a Mediterranean ecosystem. *Isr. J. Plant Sci.* **1992**, *41*, 243–250.
31. Kutiel, P.; Lavee, H. Effect of slope aspect on soil and vegetation properties along an aridity transect. *Isr. J. Plant Sci.* **1999**, *47*, 169–178. [CrossRef]
32. Yang, J.; El-Kassaby, Y.A.; Guan, W. The effect of slope aspect on vegetation attributes in a mountainous dry valley, Southwest China. *Sci. Rep.* **2020**, *10*, 16465. [CrossRef] [PubMed]
33. Ng, E.; Miller, P.C. Soil moisture relations in the southern California chaparral. *Ecology* **1980**, *61*, 98–107. [CrossRef]
34. Sidari, M.; Ronzello, G.; Vecchio, G.; Muscolo, A. Influence of slope aspects on soil chemical and biochemical properties in a Pinus laricio forest ecosystem of Aspromonte (Southern Italy). *Eur. J. Soil Biol.* **2008**, *44*, 364–372. [CrossRef]
35. Singh, S. Understanding the role of slope aspect in shaping the vegetation attributes and soil properties in Montane ecosystems. *Trop. Ecol.* **2018**, *59*, 417–430.
36. Li, X.; Nie, Y.; Song, X.; Zhang, R.; Wang, G. Patterns of species diversity and functional diversity along a south- to north-facing slope in a subalpine meadow. *Community Ecol.* **2011**, *12*, 179–187. [CrossRef]
37. Li, X.; Song, X.; Zhao, J.; Lu, H.; Qian, C.; Zhao, X. Shifts and plasticity of plant leaf mass per area and leaf size among slope aspects in a subalpine meadow. *Ecol. Evol.* **2021**, *11*, 14042–14055. [CrossRef]
38. Olsen, S.R. *Estimation of Available Phosphorus in Soils by Extraction with Sodium Bicarbonate*; US Department of Agriculture: Washington DC, USA, 1954.
39. Qin, Y.; Feng, Q.; Adamowski, J.F.; Zhu, M.; Zhang, X. Community level response of leaf stoichiometry to slope aspect in a montane environment: A case study from the Central Qilian Mountains, China. *Glob. Ecol. Conserv.* **2021**, *28*, e01703. [CrossRef]
40. Zhang, Y.; Li, C.; Wang, M. Linkages of C: N:P stoichiometry between soil and leaf and their response to climatic factors along altitudinal gradients. *J. Soils Sediments* **2019**, *19*, 1820–1829. [CrossRef]
41. Nadal-Romero, E.; Petrlic, K.; Verachtert, E.; Bochet, E.; Poesen, J. Effects of slope angle and aspect on plant cover and species richness in a humid Mediterranean badland. *Earth Surf. Processes Landf.* **2014**, *39*, 1705–1716. [CrossRef]
42. Zhao, Y.; Yang, B.; Li, M.; Xiao, R.; Rao, K.; Wang, J.; Zhang, T.; Guo, J. Community composition, structure and productivity in response to nitrogen and phosphorus additions in a temperate meadow. *Sci. Total Environ.* **2019**, *654*, 863–871. [CrossRef] [PubMed]
43. Midolo, G.; Alkemade, R.; Schipper, A.M.; Benítez-López, A.; Perring, M.P.; De Vries, W. Impacts of nitrogen addition on plant species richness and abundance: A global meta-analysis. *Glob. Ecol. Biogeogr.* **2019**, *28*, 398–413. [CrossRef]
44. Soons, M.B.; Hefting, M.M.; Dorland, E.; Lamers, L.P.; Versteeg, C.; Bobbink, R. Nitrogen effects on plant species richness in herbaceous communities are more widespread and stronger than those of phosphorus. *Biol. Conserv.* **2017**, *212*, 390–397. [CrossRef]
45. De Schrijver, A.; De Frenne, P.; Ampoorter, E.; Van Nevel, L.; Demey, A.; Wuyts, K.; Verheyen, K. Cumulative nitrogen input drives species loss in terrestrial ecosystems. *Glob. Ecol. Biogeogr.* **2011**, *20*, 803–816. [CrossRef]
46. Wilson, S.D.; Tilman, D. Quadratic variation in old-field species richness along gradients of disturbance and nitrogen. *Ecology* **2002**, *83*, 492–504. [CrossRef]
47. De Kroon, H.; Bobbink, R. Clonal plant dominance under elevated nitrogen deposition, with special reference to *Brachypodium pinnatum* in chalk grassland. In *The Ecology and Evolution of Clonal Plants*; Backhuys Publishers: Leiden, The Netherlands, 1997; pp. 359–379.

diversity

Article

Do Mixed *Pinus yunnanensis* Plantations Improve Soil's Physicochemical Properties and Enzyme Activities?

Chen Liang [1], Ling Liu [2], Zhixiao Zhang [2], Sangzi Ze [3], Mei Ji [2], Zongbo Li [4], Jinde Yu [1], Bin Yang [4,*] and Ning Zhao [1,4,*]

1 College of Life Sciences, Southwest Forestry University, Kunming 650224, China; liangchenbb1310@163.com (C.L.); yujinde@swfu.edu.cn (J.Y.)
2 Yunnan Academy of Forestry and Grassland, Kunming 650201, China; liuling_km@126.com (L.L.); zzx_20071988@163.com (Z.Z.); meiji.emma@163.com (M.J.)
3 Yunnan Forestry and Grassland Pest Control and Quarantine Bureau, Kunming 650051, China; zesangzi@163.com
4 Key Laboratory of Forest Disaster Warning and Control of Yunnan Province, Southwest Forestry University, Kunming 650224, China; lizb@swfu.edu.cn
* Correspondence: yangbin48053@swfu.edu.cn (B.Y.); lijiangzhn@swfu.edu.cn (N.Z.)

Abstract: Many survival and ecological problems have emerged in *Pinus yunnanensis* pure pine forest plantations that are usually assumed to be solved by creating mixed plantations. On this basis, we determined the physicochemical properties and enzyme activities of three soil layers in pure and three types of mixed *P. yunnanensis* plantation stands (admixed species: *Alnus nepalensis*, *Celtis tetrandra*, and *Quercus acutissima*) in Southwest China. We used one-way ANOVA with Tukey's test to analyze the effects of plantation type and depth on the soil's properties and variations among different depths. Principal component analysis combined with cluster analysis was used to evaluate the soil quality of different forest types comprehensively. The results showed that the stand with a mixing proportion of 2:1 of *P. yunnanensis* and *A. nepalensis*, *C. tetrandra*, and *Q. acutissima* had higher total porosity, moisture content, total nitrogen, available phosphorus, total phosphorus, sucrase, urease, and catalase enzyme activities than other proportions of mixed forest and *P. yunnanensis* pine pure forest. In general, the mixed *P. yunnanensis* plantation could improve the soil quality, especially its chemical properties and enzymes. This study provides a basis for creating a mixed-mode of *P. yunnanensis* and other tree species that can not only improve the economy of forest land but also enhance the ecological value.

Keywords: *Pinus yunnanensis* pure plantation; *Pinus yunnanensis* mixed plantation; soil; physicochemical properties; enzyme activities; principal component analysis and cluster analysis

Citation: Liang, C.; Liu, L.; Zhang, Z.; Ze, S.; Ji, M.; Li, Z.; Yu, J.; Yang, B.; Zhao, N. Do Mixed *Pinus yunnanensis* Plantations Improve Soil's Physicochemical Properties and Enzyme Activities? *Diversity* **2022**, *14*, 214. https://doi.org/10.3390/d14030214

Academic Editors: Michael Wink, Lin Zhang and Jinniu Wang

Received: 12 January 2022
Accepted: 11 March 2022
Published: 14 March 2022

1. Introduction

Pinus yunnanensis is an ecologically and commercially important tree species distributed in subtropical Southwest China, with over 4 million hectares in plantations and natural forested areas [1]. The *P. yunnanensis* pine forests not only provide a large amount of high-quality timber and industrial raw materials but also play an important role in maintaining species diversity, conserving water resources, and retaining soil [2]. In recent years, because the composition of artificially constructed *P. yunnanensis* pure pine forests has become very simple and the planting area too large, a large number of survival and ecological problems in *P. yunnanensis* pine plantations have emerged. For example, diseases and insect pests are becoming more and more serious, the mountain ecosystem security has been seriously threatened, soil fertility and water retention have declined sharply, and biodiversity has been severely reduced [3,4]. Therefore, the outlook for pure coniferous forests is not optimistic. A change in the species composition and structure of forest stands, the biological characteristics, and the ecology of forest trees is needed to establish a more complex and stable forest ecosystem than pure forests provide in order to increase the ecological and economic benefits of the forest. A large number of studies have proved that mixed forests play an important role in enhancing soil fertility, improving forest nutrient status, increasing forest species diversity, improving the stability of stand structure and productivity, and improving the ecological environment of forest plants [5–7]. Therefore, planting mixed forests is particularly important. Changing the species structure of forest stands leads to more complex and stable mixed forests [8]. Mixed forest is an important form of forest species and tree species allocation in the shelter forest system, and it is also a good way to form a reasonable structure of tree species. By selecting appropriate tree species to mix with *P. yunnanensis* pine, we can make full use of aboveground and underground energy and space so as to increase the biomass of mixed forests and improve the soil' nutrient status of forest land. This provides an important theoretical basis for the growth of forest vegetation and the construction of *P. yunnanensis* pine mixed forests.

Changes in plant diversity are known to affect aboveground and belowground ecosystem functioning, including the diversity of belowground communities of organisms [9–13]. Additionally, higher plant diversity may produce a higher biochemical diversity of root exudates, which further increases the soil's organic matter content [14]. Soil is an important part of the forest ecosystem and plays an important role in maintaining the stability of the terrestrial ecosystem, material circulation, and energy conversion [15]. Soil is also the foundation of plant growth and development. Soil fertility and quality directly determine the biological output and function of the forest. Foresters have always relied on knowledge of the soil's physiochemical properties to assess the value of the forests [16]. Soil fertility is a comprehensive reaction of the soil's physiochemical and biological properties [17]. In terms of chemical properties, the stoichiometric characteristics of soil N and P can reflect not only soil fertility but also the composition of soil's organic matter, soil quality, and the capacity of nutrient supplies [18]. Therefore, the content and distribution pattern of N and P available in the soil is important for plant growth. Soil enzymes can promote chemical reactions in organisms and are the total embodiment of soil's biological activity [19]. In addition, soil enzyme activities regulate the function of ecosystems and play a key role in nutrient cycling, which has been used by scholars as a parameter to evaluate soil quality [20]. The quality of the soil largely depends on the function of the soil, which indicates a combination of its physical, chemical, and biological characteristics [21]. The quality of soil nutrients is the result of the combined effects of soil physics, chemistry, biology, and other factors [22]. It is common to evaluate soil quality and nutrients by measuring the soil's physicochemical and biological indexes.

In recent years, a large number of experimental studies have focused on the innovation of *P. yunnanensis* cultivation technology, but they lack an understanding of the physical and chemical properties of *P. yunnanensis* plantation soil. Therefore, it is important to evaluate the advantages of different mixed forests by studying soil quality. In this study, the *P. yunnanensis* pine pure forest was used as the control group, and different mixed forest groups were set up to analyze and compare the physicochemical properties and enzyme activities of soil. This work provides a theoretical reference for establishing a mixed pattern of *P. yunnanensis* pine and other tree species, which could not only improve the forest economy but also enhance its ecological value.

2. Materials and Methods

2.1. Study Area Overview

The total management area of Dongshan Forest Farm in Dali city, Yunnan Province, China, is 98,300 hm^2, including 58,100 hm^2 of forest land (44,200 hm^2 of natural and 13,900 hm^2 of artificial forest). The topography of the forest farm is high in the east and low in the west. The highest altitude is 2046 m, the lowest altitude is 1416 m, and the average altitude is 1600 m. The annual sunshine in the forest farm is more than 2700 h, the frost-free period is 125–150 days, and the annual average precipitation is 500–550 mm. The soil type in this region is mainly neutral brown soil, with a small amount of cinnamon soil and mountain meadow soil. In 2016, according to the distribution of typical forest types of the study area, six forest stands were constructed and set as study sites; I: *Pinus yunnanensis* × *Alnus nepalensis* (2:1); II: *Pinus yunnanensis* × *Alnus nepalensis* (3:1); III: *P. yunnanensis* × *Quercus acutissima* (2:1); IV: *P. yunnanensis* × *Quercus acutissima* (3:1); V: *P. yunnanensis* × *Celtis tetrandra* (2:1); VI: *P. yunnanensis* pure forest. These forest types belonged to undeveloped forest land before the construction of mixed forests, and the plants on the land were mainly *Heteropogon contortus* and *Dodonaea viscosa*. The six forest types are 100 m apart and do not interfere with each other. The tree species arrangement of the six forest types is shown in Figure 1. The soil's properties before the tree planting were very similar. We assumed that the local soil's properties were largely a consequence of plant growth and soil protection of forest types, and the initial conditions or management for the sites were similar. To reduce the effects of slope and elevation on the soil's properties, all the selected plots were taken at a similar slope (around 4) and elevation (around 1880 m) (Table 1).

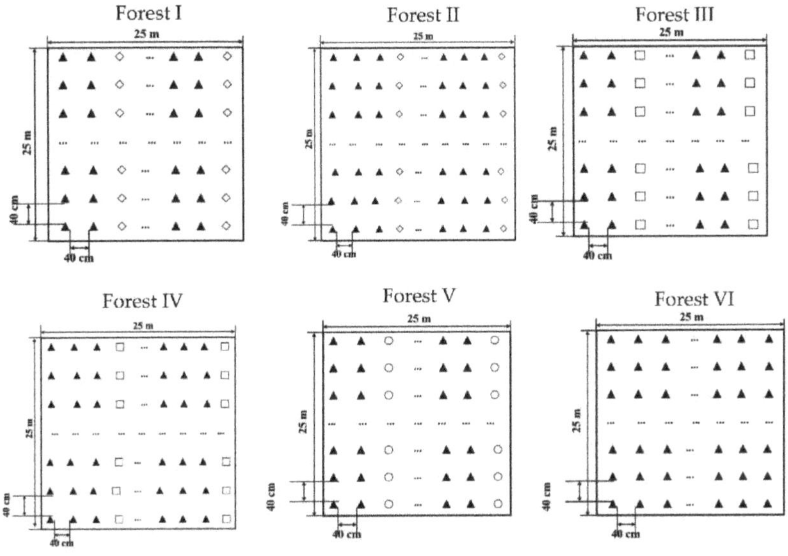

Figure 1. Area settings. Forest I: *Pinus yunnanensis* × *Alnus nepalensis* (2:1); Forest II: *Pinus yunnanensis* × *Alnus nepalensis* (3:1); Forest III: *Pinus yunnanensis* × *Quercus acutissima* (2:1); Forest IV: *Pinus yunnanensis* × *Quercus acutissima* (3:1); Forest V: *Pinus yunnanensis* × *Celtis tetrandra* (2:1); Forest VI: *Pinus yunnanensis* pure forest. ▲: *Pinus yunnanensis*; ◇: *Alnus nepalensis*; □: *Quercus acutissima*; ○: *Celtis tetrandra*, hereinafter the same.

Table 1. Basic information of experimental plots.

Forest Stands	Altitude (m)	Slope (°)	Latitude	Longitude
I	1890	3	N 25°25′13.46″	E 100°29′41.41″
II	1890	5	N 25°25′15.17″	E 100°29′39.50″
III	1879	5	N 25°25′14.17″	E 100°29′34.59″
IV	1895	2	N 25°25′12.13″	E 100°29′52.64″
V	1895	2	N 25°25′12.13″	E 100°29′52.64″
VI	1879	2	N 25°24′56.76″	E 100°29′36.00″

Notes: Forest I: *Pinus yunnanensis* × *Alnus nepalensis* (2:1); Forest II: *Pinus yunnanensis* × *Alnus nepalensis* (3:1); Forest III: *Pinus yunnanensis* × *Quercus acutissima* (2:1); Forest IV: *Pinus yunnanensis* × *Quercus acutissima* (3:1); Forest V: *Pinus yunnanensis* × *Celtis tetrandra* (2:1); Forest VI: *Pinus yunnanensis* pure forest.

2.2. Sample Plot Selection and Soil Sample Collection

In March 2016, we selected *P. yunnanensis* seedlings with consistent and healthy growth cultivated in seedling base to be transplanted to Dongshan Forest Farm in Midu County, Dali, Yunnan Province. The tree species used to set the same kind of mixed forest with different mixed proportions were also seedlings with consistent growth and no diseases and pests in the plantation. The soil samples were collected in mid-June 2016 and mid-June 2018 in this experiment.

Each plot of each forest type sample plot is more than 100 m apart, and the slope and planting management measures of the plantations are basically the same. In each forest type, three plots on the diagonal were taken, and the area of each plot is 625 m^2, i.e., 25 m × 25 m (Figure 2). Thus, a total of 18 standard plots were set up in our study sites. In a 625 m^2 plot, five large sample plots of size 5 m × 5 m were set in the four corners and center of each forest type plot to measure the ground diameter and tree height of *P. yunnanensis*. In each 625 m^2 plot, five small sample plots of size 2 m × 2 m were selected within the five large sample plots, and soil samples were divided into soil layers with a diameter of 4 cm. During sampling, litter on the soil surface was removed, and the sampling depth was 60 cm (divided into three layers of 0–20, 20–40, and 40–60 cm). Soil samples collected from five 2 m × 2 m plots of 625 m^2 were layered and mixed as one replicate, and a total of three 625 m^2 areas were used as three replicates. Samples were sealed in soil bags and taken back to the laboratory for testing. Each layer of soil samples in each repetition was divided into two parts. One part was sieved and stored in a refrigerator at 4 °C to determine the enzyme activities, and the other was dried naturally at room temperature to remove roots, stones, and other impurities in the soil sample and then screened for 2 mm to determine the physicochemical properties of the soil.

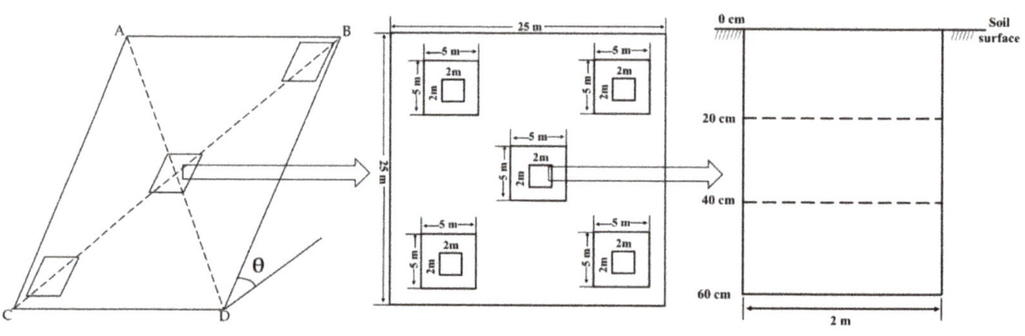

Figure 2. The diagram for the positions to collect soil samples in each plot.

2.3. Measurement Items and Methods

A set of three physical variables of the soil was evaluated: bulk density (BD), soil moisture content (MC), and total porosity (TOP). In the laboratory, undisturbed samples were weighed, dried in a forced-air oven at 105 °C for 48 h, and weighed again. The bulk density (BD, g/cm^3) and total porosity (T porosity, %) were determined by the cutting-ring method [23]. A total of 20 g of fresh soil was dried in an oven at 105 °C to a constant weight, and the soil's moisture content was measured by gravimetric analysis [24].

Four chemical variables of the soil were evaluated, pH, total nitrogen (TN), available phosphorus (AP), and total phosphorus content (TP). Soil pH was determined using a pH meter with a soil/water ratio of 1:2.5 [25]. The total nitrogen (TN) was determined by the automatic Kjeldahl method [26]. The total phosphorus (TP) was determined via the acid solution–molybdenum–antimony anti-colorimetric method [27]. The available phosphorus (AP) was extracted using 0.5 M $NaHCO_3$ (Olsen) and measured by a spectrophotometer [28].

Three enzyme activities of soil samples, which were stored in a refrigerator at 4 °C, were measured: urease activity (Ure), sucrase activity (Suc), and catalase activity (Cat). Soil urease activity (Ure) was determined using the phenol-sodium hypochlorite colorimetric method, sucrase activity (Suc) was determined using the 3,5-dinitrosalicylic acid colorimetric method, and catalase (Cat) activity was determined using the potassium permanganate titration method, respectively [29]. Urease activity and sucrase activity were expressed by the mass of specific substrate (or production of specific substrate) consumed by soil enzymes in 1 g of dried soil per unit time [30]. Catalase activity was expressed by the volume of 0.02 mol/L $KMNO_4$ consumed by soil enzymes in 1 g of dried soil per unit time [31].

We randomly selected ten *P. yunnanensis* pines from five 5 m × 5 m sample plots in each forest type and measured their ground diameter and plant height.

2.4. Statistical Analysis

At the beginning of afforestation (2016) and after two years of afforestation (2018), we selected 10 soil fertility indexes of six forest types for principal component analysis. The PCs receiving high eigenvalues and comprising variables with high factor loading were assumed to be the variables that best represent the system attributes. Therefore, we selected only PCs with eigenvalues $n > 1.0$. The score of each principal component is calculated according to the factor load of each principal component and the standardized mass fraction data. The calculation formula of principal component score (F) is:

$$F = FAC \times \lambda \quad (1)$$

where FAC is the normalized data and λ is the arithmetic square root of the eigenvalue.

The composite principal component score ($F_{\text{overall score}}$) is the sum of the product of each principal component score and its corresponding contribution rate:

$$F_{\text{overall score}} = (F1 \times R1 + F1 \times R1 + + Fn \times Rn)/(R1 + R2 + \cdot Rn) \quad (2)$$

where R is the variance contribution rate.

The comprehensive score of each forest type was taken as a new index to evaluate its comprehensive quality, the difference of soil quality between different forest types was measured by Euclidean distance, and the classification average method and the shortest distance method were used to systematically cluster each treatment.

All statistical data analyses were performed using Microsoft Excel 2010 (Microsoft Redmond, WA, USA), Origin 8.0 software (Origin Lab Corp., Northampton, MA, USA), and SPSS software (Ver. 22.0; SPSS Inc., Chicago, IL, USA) for Windows. One-way analyses of variance (ANOVA) and Duncan's multiple comparison test were used to assess statistically significant differences ($p < 0.05$) among different soil depths under each forest stand. Pearson correlation analysis was conducted to identify relationships among measured the soil's properties.

3. Results

3.1. Effects of Different Forest Types of P. yunnanensis on Soil's Physicochemical Properties

3.1.1. Soil's Physical Properties

The results of three physical properties of soil under six forest stands are shown in Table 2. The soil bulk density showed an upward trend from the top of the soil that decreased with the increase in soil depth both at the beginning of afforestation (2016) and after two years of afforestation (2018). The interesting phenomenon is that the soil bulk density after two years of afforestation of the six forest types was all lower than that at the beginning of afforestation (2016); the soil bulk density of five *P. yunnanensis* mixed forest types was lower than that of the *P. yunnanensis* pure forest. However, total soil porosity and moisture content showed the opposite trend to BD; specifically, the total porosity and moisture content of six forest types showed a decreasing trend from the surface layer as the soil depth increased. Of course, the total porosity and moisture content of the same layer after two years of afforestation (2018) were higher than those at the beginning of afforestation (2016). Besides this, compared with the *P. yunnanensis* pure forest, the *P. yunnanensis* × C. tetrandra (2:1) (Forest V) mixed forest stand in 0–20 cm soil layer had the lowest bulk density and the highest total porosity and water content after two years of afforestation (Figure S1). In addition, the soil's physical properties of different mixed proportions were also different in the unified mixed forest type. After two years of afforestation (2018), the total soil porosity and water content in Forest I and Forest III were all significantly higher than those in Forest II and Forest IV, respectively. However, at the beginning of afforestation (2016), there was no significant difference in the soil's physical properties among different mixed proportions of the same mixed forest type.

Table 2. Description of soil physical properties under six forest stands at 0–20, 20–40 cm, and 40–60 cm (mean ± stand error, $n = 3$).

Year	Depth (cm)	I	II	III	IV	V	VI
				BD (g/cm^3)			
	0–20	1.29 ± 0.04 ^{Aa}	1.26 ± 0.06 ^{Aa}	1.25 ± 0.04 ^{Ba}	1.26 ± 0.02 ^{Ba}	1.29 ± 0.04 ^{Aa}	1.28 ± 0.03 ^{Ba}
2016	20–40	1.37 ± 0.05 ^{ABa}	1.37 ± 0.07 ^{ABa}	1.40 ± 0.08 ^{Aa}	1.39 ± 0.02 ^{Aa}	1.38 ± 0.07 ^{Aa}	1.39 ± 0.04 ^{Aa}
	40–60	1.4 ± 0.05 ^{Aa}	1.41 ± 0.03 ^{Aa}	1.42 ± 0.04 ^{Aa}	1.42 ± 0.04 ^{Aa}	1.42 ± 0.06 ^{Aa}	1.41 ± 0.02 ^{Aa}
	0–20	1.14 ± 0.07 ^{Aab}	1.24 ± 0.04 ^{Aab}	1.10 ± 0.02 ^{Aab}	1.22 ± 0.02 ^{Aab}	1.07 ± 0.15 ^{Ab}	1.27 ± 0.03 ^{Ba}
2018	20–40	1.19 ± 0.07 ^{Aab}	1.28 ± 0.03 ^{Aab}	1.20 ± 0.08 ^{ABab}	1.31 ± 0.07 ^{Aab}	1.13 ± 0.12 ^{Ab}	1.36 ± 0.01 ^{Aa}
	40–60	1.39 ± 0.18 ^{Aa}	1.34 ± 0.05 ^{Aa}	1.30 ± 0.06 ^{Aa}	1.35 ± 0.07 ^{Aa}	1.25 ± 0.05 ^{Aa}	1.39 ± 0.04 ^{Aa}
				TOP (%)			
	0–20	44.53 ± 2.30 ^{Aa}	46.88 ± 5.13 ^{Aa}	46.86 ± 3.01 ^{Aa}	45.88 ± 9.36 ^{Aa}	45.36 ± 6.17 ^{Aa}	44.61 ± 0.46 ^{Ca}
2016	20–40	41.03 ± 1.18 ^{ABa}	41.05 ± 7.02 ^{Aa}	41.97 ± 2.00 ^{Aba}	41.25 ± 4.76 ^{Aa}	41.55 ± 1.89 ^{Aa}	41.69 ± 1.47 ^{Aa}
	40–60	37.51 ± 1.32 ^{Bb}	37.65 ± 4.15 ^{Ab}	37.43 ± 2.47 ^{Bb}	37.21 ± 1.91 ^{Ab}	37.21 ± 5.99 ^{Ab}	37.75 ± 0.73 ^{Ba}
	0–20	56.07 ± 0.57 ^{Aab}	51.75 ± 1.61 ^{Abc}	58.33 ± 4.50 ^{Aab}	53.46 ± 3.08 ^{Abc}	62.96 ± 2.79 ^{Aa}	47.88 ± 1.46 ^{Ac}
2018	20–40	52.89 ± 3.48 ^{ABab}	48.83 ± 2.41 ^{ABbc}	55.59 ± 1.84 ^{Aab}	50.63 ± 1.87 ^{ABbc}	59.2 ± 3.90 ^{Aba}	43.61 ± 1.32 ^{Bc}
	40–60	48.90 ± 0.70 ^{Bab}	46.18 ± 1.04 ^{Bb}	51.16 ± 1.92 ^{Aa}	47.75 ± 1.38 ^{Bab}	52.27 ± 2.71 ^{Ba}	38.28 ± 1.76 ^{Cc}
				MC (%)			
	0–20	27.18 ± 1.01 ^{Aa}	26.63 ± 6.58 ^{Aa}	27.07 ± 11.31 ^{Aa}	27.41 ± 4.49 ^{Aa}	28.14 ± 6.36 ^{Aa}	27.99 ± 1.20 ^{Aa}
2016	20–40	22.32 ± 2.40 ^{Ba}	22.1 ± 6.12 ^{Aa}	23.84 ± 10.45 ^{Aa}	22.07 ± 4.15 ^{Aa}	22.78 ± 8.12 ^{Aa}	22.72 ± 1.65 ^{Ba}
	40–60	18.12 ± 0.24 ^{Ca}	18.5 ± 4.94 ^{Aa}	19.04 ± 4.45 ^{Aa}	18.48 ± 5.35 ^{Aa}	19.55 ± 6.60 ^{Aa}	18.80 ± 1.33 ^{Ca}
	0–20	38.09 ± 2.31 ^{Ba}	34.69 ± 1.02 ^{Cb}	39.44 ± 0.74 ^{Cab}	33.22 ± 0.51 ^{Cb}	42.05 ± 2.56 ^{Bab}	28.90 ± 0.71 ^{Cc}
2018	20–40	34.36 ± 1.19 ^{Aab}	29.34 ± 0.74 ^{Abc}	34.27 ± 2.55 ^{Aa}	28.53 ± 1.04 ^{Ac}	40.64 ± 2.82 ^{Aa}	23.65 ± 0.70 ^{Ad}
	40–60	29.36 ± 1.07 ^{Ab}	25.52 ± 0.32 ^{Bc}	27.56 ± 2.27 ^{Bb}	24.16 ± 1.49 ^{Bc}	27.77 ± 0.82 ^{Aa}	19.65 ± 1.35 ^{Bd}

Notes: BD: bulk density; TOP: total porosity; MC: moisture content. Different capital letters indicate that the indexes of different soil layers of the same forest type are significantly different ($p < 0.05$). Different lowercase letters indicate that there are significant differences in different indexes of different forest types in the same soil layer ($p < 0.05$).

3.1.2. Soil's Chemical Properties

Table 3 shows the results of four of the soil's chemical properties under six forest stands. Although the pH values of the three soil layers were not significantly different ($p > 0.05$), the soil of the six forest types were weakly acid, and its pH value was in the range of 4–7. Compared with the *P. yunnanensis* pure forest, *P. yunnanensis* mixed with *A. nepalensis*, *Q. acutissima*, *C. tetrandra*, and *P. yunnanensis* reduced the soil's acidity. The soil's TN, TP, and AP content from most of the forest stands significantly decreased with the deepening of the soil depth ($p < 0.05$). Interestingly, the TP and AP content after two years of afforestation (2018) were higher than those at the beginning of afforestation (2016). In addition, after two years of afforestation, Forest III had the highest mean TN content of both the soil depths, Forest V had the highest TP content of both the soil depths, and Forest I had the same AP content at both the soil depths (Figure S2). Besides this, the soil chemical properties of different mixed proportions were also different in the unified mixed forest type. After two years of afforestation (2018), the soil's TN, TP, and AP content in Forest I and Forest III were all significantly higher than that in Forest II and Forest IV, respectively.

Table 3. Description of the soil's chemical properties under six forest stands at 0–20, 20–40, and 40–60 cm (mean ± stand error, n = 3).

Year	Depth (cm)	I	II	III	IV	V	VI
				pH			
	0–20	4.71 ± 0.16 ^Ba	4.74 ± 0.26 ^Aa	4.74 ± 0.53 ^Ba	4.74 ± 0.15 ^Aa	4.69 ± 0.53 ^Aa	4.72 ± 0.2 ^Aa
2016	20–40	4.79 ± 0.02 ^ABa	4.82 ± 0.19 ^Aa	4.83 ± 0.52 ^ABa	4.85 ± 0.40 ^Aa	4.78 ± 0.69 ^Aa	4.81 ± 0.18 ^Aa
	40–60	4.92 ± 0.78 ^Aa	4.91 ± 0.22 ^Aa	4.96 ± 0.94 ^Aa	4.92 ± 1.14 ^Aa	4.94 ± 0.43 ^Aa	4.93 ± 0.35 ^Aa
	0–20	4.50 ± 0.72 ^Aa	4.48 ± 0.51 ^Aa	5.08 ± 0.34 ^Aa	4.61 ± 0.27 ^Aa	4.60 ± 0.57 ^Aa	4.96 ± 0.32 ^Aa
2018	20–40	5.30 ± 0.49 ^Aa	4.79 ± 0.77 ^Aa	5.05 ± 0.67 ^Aa	4.68 ± 0.33 ^Aa	4.75 ± 0.36 ^Aa	4.83 ± 0.23 ^Aa
	40–60	5.09 ± 0.72 ^Aa	5.47 ± 0.38 ^Aa	5.24 ± 0.35 ^Aa	5.05 ± 0.16 ^Aa	5.02 ± 0.21 ^Aa	4.78 ± 0.32 ^Aa
				TN (g/kg)			
	0–20	0.46 ± 0.06 ^Aa	0.51 ± 0.16 ^Aa	0.50 ± 0.15 ^Aa	0.47 ± 0.10 ^Aa	0.47 ± 0.28 ^Aa	0.48 ± 0.04 ^Aa
2016	20–40	0.17 ± 0.02 ^Ba	0.22 ± 0.08 ^Ba	0.22 ± 0.09 ^Ba	0.23 ± 0.01 ^Ba	0.21 ± 0.03 ^Ba	0.19 ± 0.04 ^Ba
	40–60	0.15 ± 0.04 ^Ba	0.17 ± 0.05 ^Ba	0.16 ± 0.06 ^Ba	0.18 ± 0.05 ^Ba	0.13 ± 0.02 ^Ba	0.14 ± 0.03 ^Ba
	0–20	0.74 ± 0.03 ^Aab	0.71 ± 0.04 ^Abc	0.81 ± 0.04 ^Aa	0.62 ± 0.02 ^Ac	0.68 ± 0.05 ^Abc	0.51 ± 0.04 ^Ad
2018	20–40	0.49 ± 0.05 ^Ba	0.29 ± 0.04 ^Bb	0.56 ± 0.03 ^Ba	0.26 ± 0.04 ^Bb	0.32 ± 0.03 ^Bb	0.22 ± 0.04 ^Bb
	40–60	0.24 ± 0.03 ^Cab	0.19 ± 0.04 ^Cab	0.28 ± 0.04 ^Ca	0.18 ± 0.03 ^Cb	0.21 ± 0.03 ^Cab	0.17 ± 0.04 ^Cb
				TP (g/kg)			
	0–20	0.24 ± 0.14 ^Aa	0.23 ± 0.08 ^Aa	0.27 ± 0.11 ^Aa	0.23 ± 0.06 ^Aa	0.27 ± 0.02 ^Aa	0.25 ± 0.03 ^Aa
2016	20–40	0.17 ± 0.1 ^Aa	0.18 ± 0.05 ^ABa	0.20 ± 0.07 ^Aa	0.17 ± 0.05 ^ABa	0.18 ± 0.06 ^Ba	0.19 ± 0.04 ^Ba
	40–60	0.15 ± 0.06 ^Aa	0.14 ± 0.04 ^Ba	0.14 ± 0.04 ^Aa	0.14 ± 0.08 ^Ba	0.15 ± 0.08 ^Ba	0.16 ± 0.04 ^Ba
	0–20	0.51 ± 0.34 ^Aab	0.29 ± 0.06 ^Aab	0.37 ± 0.05 ^Aab	0.35 ± 0.04 ^Aab	0.51 ± 0.23 ^Aa	0.27 ± 0.02 ^Ab
2018	20–40	0.29 ± 0.06 ^ABab	0.24 ± 0.04 ^Aab	0.31 ± 0.03 ^Aab	0.27 ± 0.02 ^Bab	0.34 ± 0.04 ^Ba	0.21 ± 0.05 ^Ab
	40–60	0.26 ± 0.03 ^Ba	0.2 ± 0.08 ^Aa	0.25 ± 0.03 ^Ba	0.26 ± 0.03 ^Ba	0.28 ± 0.02 ^Ca	0.19 ± 0.06 ^Ba
				Available P (mg/kg)			
	0–20	2.29 ± 1.00 ^Aa	2.31 ± 0.63 ^Aa	2.32 ± 0.47 ^Aa	2.31 ± 0.47 ^Aa	2.24 ± 0.58 ^Aa	2.27 ± 0.14 ^Aa
2016	20–40	2.10 ± 0.37 ^Aa	2.12 ± 0.24 ^ABa	2.14 ± 0.75 ^Aa	2.13 ± 0.22 ^Aa	2.08 ± 0.54 ^Aa	2.03 ± 0.06 ^Ba
	40–60	1.91 ± 0.82 ^Aa	1.93 ± 0.39 ^Ba	1.94 ± 0.96 ^Aa	1.95 ± 0.48 ^Aa	1.84 ± 0.64 ^Aa	1.87 ± 0.16 ^Ca
	0–20	2.97 ± 0.05 ^Aa	2.68 ± 0.28 ^Aab	2.78 ± 0.24 ^Aab	2.73 ± 0.15 ^Aab	2.94 ± 0.16 ^Aa	2.36 ± 0.11 ^Ab
2018	20–40	2.53 ± 0.12 ^Ba	2.27 ± 0.10 ^Bab	2.45 ± 0.14 ^Bab	2.25 ± 0.09 ^Bab	2.32 ± 0.17 ^Bab	2.14 ± 0.12 ^ABb
	40–60	2.25 ± 0.21 ^Ca	2.11 ± 0.07 ^Ba	2.19 ± 0.14 ^Ba	2.16 ± 0.03 ^Ba	2.13 ± 0.30 ^Ba	2.02 ± 0.21 ^Ba

Notes: Available P: available phosphorus content; TP: total phosphorus content; TN: total nitrogen content. Different capital letters indicate that the indexes of different soil layers of the same forest type are significantly different ($p < 0.05$). Different lowercase letters indicate that there are significant differences in different indexes of different forest types in the same soil layer ($p < 0.05$).

3.2. Effect of Different Forest Types of *P. yunnanensis* on Soil Enzyme Activity

The results of soil enzyme activity in six different *P. yunnanensis* forest stands are shown in Table 4. The soil urease activity levels of *P. yunnanensis* forest types at two years of afforestation (2018) were all higher than those at the beginning of afforestation (2016), among which the urease activity of 0–20 cm surface soil was the highest. After two years of afforestation, the surface soil urease activity of Forest I was the highest, at 563.50 μg/g·24 h, which was significantly higher than that at the beginning of afforestation (2016) ($p < 0.01$) (Figure S3A). The soil sucrase activity of six different *P. yunnanensis* forest types all decreased significantly with the increase in soil depth ($p < 0.05$) both at the beginning of afforestation (2016) and after two years of afforestation (2018). The order of sucrase activity in different soil layers both at the beginning of afforestation (2016) and after two years of afforestation (2018) was 0–20 cm > 20–40 cm > 40–60 cm. As shown in Figure S3B, the sucrase activity of the surface soil after two years of afforestation (2018) of Forest

III was significantly higher than that of the surface soil at the beginning of afforestation (2016) ($p < 0.01$). The soil catalase activities of the six different forest types also showed a decreasing trend with the increase in soil depth both at the beginning of afforestation (2016) and after two years of afforestation (2018). After two years of afforestation (2018), the forest soil surface layer (0–20 cm) had the highest catalase activity, and Forest I had the highest mean catalase activity, which was 8.12 mg/g·24 h, and Forest VI had the lowest the catalase activity, which was 7.11 mg/g·24 h. As shown in Figure S3C, in the 0~20 cm soil layer after two years of afforestation (2018), there was no difference in the catalase activity of the soil under the same forest type. Obviously, the three enzyme activities of different mixed proportions were also different in the unified mixed forest type. After two years of afforestation, the three enzyme activities in Forest I and Forest III were all significantly higher than that in Forest II and Forest IV, respectively.

Table 4. Description of soil enzyme activities under six forest stands at 0–20, 20–40, and 40–60 cm (mean ± stand error, $n = 3$).

Year	Depth (cm)	I	II	III	IV	V	VI
				Ure (μg/g·24 h)			
2016	0–20	268.61 ± 6.89 ^{Ac}	345.85 ± 7.75 ^{Aa}	229.45 ± 2.29 ^{Ae}	226.32 ± 6.09 ^{Ae}	248.72 ± 1.30 ^{Ad}	294.97 ± 6.79 ^{Ab}
	20–40	154.73 ± 4.61 ^{Bcd}	196.69 ± 6.26 ^{Ba}	178.14 ± 5.39 ^{Bb}	199.66 ± 9.20 ^{Ba}	144.16 ± 8.78 ^{Bd}	167.48 ± 3.25 ^{Bbc}
	40–60	127.03 ± 2.14 ^{Ca}	95.00 ± 7.00 ^{Cb}	121.93 ± 4.96 ^{Ca}	116.7 ± 4.89 ^{Ca}	115.06 ± 6.51 ^{Ca}	72.12 ± 9.43 ^{Cc}
2018	0–20	563.4 ± 12.56 ^{Aa}	462.99 ± 5.06 ^{Aab}	481.52 ± 1.27 ^{Aa}	378.52 ± 11.33 ^{Aab}	496.9 ± 5.39 ^{Aab}	378.11 ± 7.04 ^{Ab}
	20–40	447.34 ± 12.74 ^{Ba}	346.08 ± 8.88 ^{Bab}	329.24 ± 7.32 ^{Bab}	263.08 ± 11.81 ^{Bbc}	341.41 ± 5.65 ^{Bb}	214.98 ± 4.18 ^{Bc}
	31.93	125.54 ± 9.72 ^{Ca}	122.84 ± 7.08 ^{Bbc}	181.94 ± 5.26 ^{Ca}	153.38 ± 4.92 ^{Cbc}	154.58 ± 7.34 ^{Cab}	148.88 ± 3.95 ^{Bbc}
				Suc (mg/g·24 h)			
2016	0–20	20.53 ± 1.62 ^{Aa}	20.47 ± 2.78 ^{Aa}	21.11 ± 1.57 ^{Aa}	19.63 ± 2.35 ^{Aa}	20.89 ± 3.57 ^{Aa}	20.68 ± 2.36 ^{Aa}
	20–40	13.15 ± 1.45 ^{Ba}	14.17 ± 4.74 ^{Aa}	12.94 ± 3.60 ^{Ba}	14.04 ± 1.48 ^{Ba}	13.91 ± 2.61 ^{Ba}	13.27 ± 2.17 ^{Ba}
	40–60	6.41 ± 1.40 ^{Ba}	6.75 ± 0.57 ^{Ba}	6.47 ± 1.83 ^{Ca}	6.27 ± 0.61 ^{Ca}	6.16 ± 0.87 ^{Ca}	6.38 ± 0.58 ^{Ca}
2018	0–20	28.58 ± 2.12 ^{Bab}	24.79 ± 2.31 ^{Abc}	33.23 ± 3.81 ^{Aa}	27.91 ± 3.27 ^{Aab}	31.93 ± 3.46 ^{Aa}	22.59 ± 1.37 ^{Ac}
	20–40	19.76 ± 3.58 ^{Ba}	18.56 ± 11.67 ^{ABa}	23.48 ± 2.78 ^{Ba}	18.73 ± 1.99 ^{Ba}	21.13 ± 3.36 ^{Ba}	15.56 ± 1.99 ^{Ba}
	40–60	9.16 ± 0.67 ^{Cbc}	10.46 ± 1.04 ^{Bbc}	13.56 ± 2.54 ^{Ca}	8.97 ± 1.93 ^{Cbc}	11.53 ± 1.92 ^{Cab}	7.63 ± 0.58 ^{Cc}
				Cat (mg/g·24 h)			
2016	0–20	7.09 ± 1.06 ^{Aa}	7.37 ± 0.45 ^{Aa}	7.1 ± 0.51 ^{Aa}	7.28 ± 0.77 ^{Aa}	7.01 ± 0.58 ^{Aa}	7.05 ± 0.14 ^{Aa}
	20–40	6.51 ± 0.72 ^{Aa}	6.87 ± 0.70 ^{Ba}	6.46 ± 0.28 ^{Aa}	6.81 ± 0.17 ^{Aa}	6.45 ± 1.05 ^{Aa}	6.49 ± 0.21 ^{Ba}
	40–60	6.07 ± 0.78 ^{Aa}	6.17 ± 0.48 ^{Ba}	6.02 ± 1.04 ^{Aa}	6.46 ± 0.47 ^{Aa}	6.19 ± 1.55 ^{Aa}	6.06 ± 0.15 ^{Ca}
2018	0–20	8.12 ± 0.21 ^{Aa}	7.48 ± 0.61 ^{Aab}	7.52 ± 0.26 ^{Aab}	7.36 ± 0.2 ^{Aab}	7.74 ± 0.45 ^{Aab}	7.11 ± 0.77 ^{Ab}
	20–40	6.84 ± 0.64 ^{Ba}	6.96 ± 0.77 ^{Aa}	6.93 ± 0.52 ^{ABa}	6.91 ± 0.69 ^{Aa}	6.85 ± 0.51 ^{Ba}	6.63 ± 0.18 ^{Aa}
	40–60	6.65 ± 0.72 ^{Ba}	6.23 ± 0.50 ^{Aa}	6.56 ± 0.32 ^{Ba}	6.54 ± 0.49 ^{Aa}	6.48 ± 0.23 ^{Ba}	6.18 ± 0.48 ^{Ab}

Notes: Ure: urease activity; Suc: sucrase activity; Cat: catalase activity. Different capital letters indicate that the indexes of different soil layers of the same forest type are significantly different ($p < 0.05$). Different lowercase letters indicate that there are significant differences in different indexes of different forest types in the same soil layer ($p < 0.05$).

3.3. *Principal Component Analysis and Cluster Analysis of Soil Fertility in Different Forest Types*

As shown in Table 5, the first two components had eigenvalues of magnitude >1. The variance contribution rate of the principal component 1 (PC1) is 80.755%, which explains most of the variation in the data, the variance contribution rate of the principal component 2 (PC2) is only 11.770%, and the cumulative variance contribution rate of the PC1 and PC2 is 92.525%, which basically explains all the variation in the data. Taking the principal component scores of different forest types as a new index to evaluate the soil fertility quality, it can be seen that the soil fertility quality of different forest types is ranked as follows: 2018-V > 2018-I > 2018-III > 2018-IV > 2018-II > 2018-VI > 2016-II > 2016-III > 2016-IV > 2016-VI > 2016-V > 2016-I.

Table 5. Score and overall score of principal components influencing each index of soil fertility.

Forest Type	FAC1	F1	FCA2	F2	$F_{overall\ score}$	Ranking
2018-I	1.532	4.353	−1.369	−1.485	3.611	2
2018-II	0.411	1.168	−1.130	−1.226	0.863	5
2018-III	1.237	3.515	2.500	2.712	3.413	3
2018-IV	0.384	1.091	−0.359	−0.389	0.903	4
2018-V	1.667	4.736	−0.183	−0.198	4.108	1
2018-VI	−0.513	−1.456	1.149	1.247	−1.113	6
2016-I	−0.888	2.523	−0.187	−0.203	−2.228	12
2016-II	−0.600	−1.704	−0.129	−0.140	−1.505	7
2016-III	−0.725	−2.061	0.097	0.105	−1.785	8
2016-IV	−0.809	−2.298	−0.104	−0.113	−2.020	9
2016-V	−0.864	−2.455	−0.192	−0.208	−2.169	11
2016-VI	−0.832	−2.366	−0.092	−0.100	−2.077	10

Taking Euclidean distance as a measure of the sudden fertility difference of different forest types, the results of the systematic clustering of each forest type by using the shortest distance method are shown in Figure 3. The twelve forest types were grouped into four categories, and 2018-I was separately grouped into one category; the soil quality was first class. 2018-II, 2018-III, and 2018-V were clustered into one class, and the soil quality was second class. 2018-IV, 2018-VI, and 2016-II were clustered into one class; the soil quality was third class. 2016-III, 2016-IV, 2016-V, 2016-I, and 2016-VI were clustered into one class; the soil quality was fourth class.

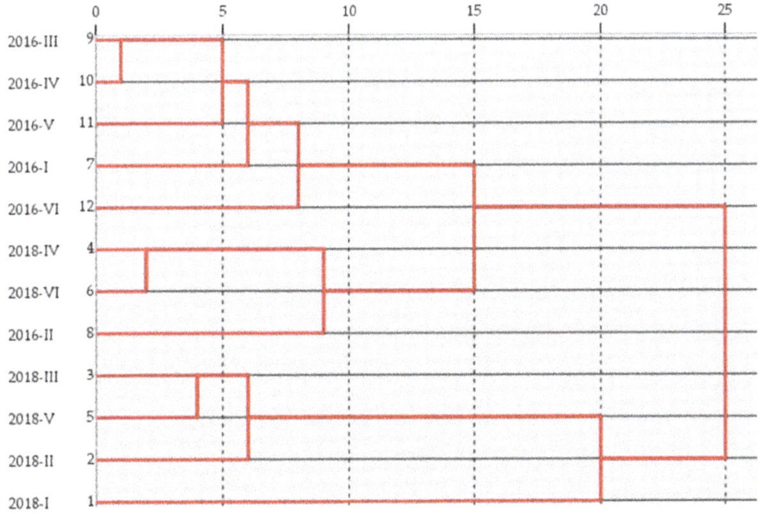

Figure 3. Hierarchical cluster analysis of the soil fertility of different forest types.

Pearson correlation analysis was used to examine the relationships among these properties to reduce redundancy (Table 6). Although the soil bulk density was not positively correlated with pH, it was significantly negatively correlated with other physical and chemical properties and enzyme activities ($p < 0.01$). This indicated that the larger the soil bulk density is, the worse the soil structure is, and the more compact soil is. This phenomenon will also lead to poor soil ventilation performance, which is not conducive to the decomposition and transformation of litter nutrients on the surface of forest land and thus will reduce the mass fraction of forest soil nutrients. There was a significant positive correlation between the soil water content and TN, AP, TP, and total porosity ($p < 0.01$). In addition, the contents of total nitrogen, total phosphorus and available phosphorus, total porosity, and moisture content were significantly positively correlated with the activities of urease, invertase, and catalase ($p < 0.01$).

Table 6. Pearson correlation coefficients among soil properties.

	pH	TN	TP	AP	BD	TOP	MC	Ure	Suc	Cat
pH	1									
TN	−0.176	1								
TP	−0.080	0.658 **	1							
AP	−0.162	0.845 **	0.690 **	1						
BD	0.174	−0.659 **	−0.699 **	−0.597 **	1					
TOP	−0.086	0.624 **	0.723 **	−0.670 **	−0.704 **	1				
MC	−0.094	0.773 **	0.755 **	0.727 **	−0.763 **	0.881 **	1			
Ure	−0.166	0.872 **	0.618 **	0.828 **	−0.579 **	0.625 **	0.768 **	1		
Suc	−0.163	0.859 **	0.690 **	0.745 **	−0.681 **	0.642 **	0.775 **	0.814 **	1	
Cat	−0.011	0.667 **	0.467 **	0.671 **	−0.471 **	0.496 **	0.576 **	0.711 **	0.617 **	1

Notes: **, Significant correlation at $p < 0.01$. BD, soil bulk density; TOP, total porosity; MC: moisture content; TN, total nitrogen; TP, total phosphorus; AP, available phosphorus; Ure, urease; Cat, catalase; Suc, sucrase.

3.4. Growth Status of Various Tree Species after Two Years of Afforestation

By selecting the *P. yunnanensis* of four forest types and measuring its ground diameter and plant height, we found that the *P. yunnanensis* mixed with *A. nepalensis* or *Q. acutissima* has the highest ground diameter and plant height, which indicates that the *A. nepalensis* and *Q. acutissima* are beneficial to the growth of *P. yunnanensis* (Figure 4).

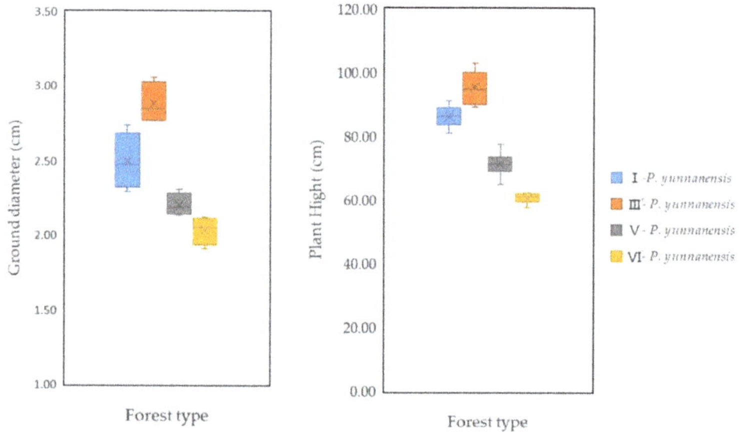

Figure 4. The growth status of *P. yunnanensis* of four forest types after two years of afforestation.

4. Discussion

Soil is the carrier of forest ecological processes and the material base of forest survival. Soil nutrient status directly affects the growth and metabolism of trees [32]. The mixing of different tree species inevitably leads to changes in the forest ecosystem, which directly affects the soil nutrient flow and material circulation, and the soil nutrient content also changes. Mixed forest stands are different, which leads to certain differences in different soil physical and chemical properties of different forest stands [33].

Li [15] and Lai [34] showed that different forest types have a significant impact on soil properties, such as soil's physical and chemical properties. Different forest types lead to certain differences in surface litter reserves and their composition, tree root growth and development, and the litter decomposition rate, resulting in different physical properties of the soil of different forest stands [35]. The soil bulk density affects the circulation, storage, and distribution of water, moisture, and heat in trees. Small bulk density means loose soil texture and good structure. On the contrary, soil that is compact lacks a granular structure. This study found that the soil bulk density of the six forest types increased with increasing soil depth, which is consistent with previous studies [36,37]. Forest III and Forest V had the lowest bulk density, and Forest VI had the highest bulk density on the soil surface. This may be due to the single tree species of the *P. yunnanensis* pine pure forest, the slow decomposition of litter, and the soil becoming compact, which is not conducive to the long-term utilization of forest land. It indicated that *P. yunnanensis* pine mixed with other species had better soil infiltration. *Q. acutissima* is a broad-leaved deciduous tree species, the litter is easy to decompose, and the activity of the microbial community in the decomposition process can reduce the compactness of the soil. Soil moisture is an important component of the forest, which actively participates in the transformation and metabolism of soil substances [38]. In our study, the surface soil water content was higher than that of the deep layer, which may be because the litter layer of the mixed forest is thicker and the evaporation is less, so there is more surface soil moisture storage. The soil water holding capacity was different among different forest types, and the five mixed forests were higher than the *P. yunnanensis* pure pine forest, which was caused by the difference in soil porosity.

The changes in forest type, stand age, and land use can also affect soil's chemical properties significantly [39–42]. Generally, the litter of forest plants is distributed on the soil's surface, and a large number of nutrient elements are released on the surface of the soil. With the increase in soil depth, the litter is less and less [43,44]. Therefore, in this study, the total soil nitrogen, total phosphorus, and available phosphorus of each forest stand show the phenomenon of surface accumulation, which was consistent with the previous research [45]. After two years of afforestation (2018), the soil's TN, TP, and AP content in Forest I and Forest III were all higher than that in other forest types. The reason for this was that the nitrogen content in the middle leaf of *A. nepalensis* is higher than that in the leaves of other plants, and the degradation of leaf litter provides a large amount of nitrogen to the soil and improves soil fertility [46]. Forrester et al. [47] found that nitrogen content and nitrogen assimilation in the soil litter of mixed forests were higher than those of pure forests, which led to the content of available nitrogen and phosphorus in the soil of mixed forests being higher than that of pure forests. However, the soil's TN, TP, and AP content in Forest VI were lowest in six forest types; the reason is that the needles of *P. yunnanensis* have a coarse and hard texture, high cellulose content, waxy epidermis, and poor water permeability, so it is difficult to decompose and transform, which affects the accumulation of organic matter in soil [48]. This study also showed that the difference between TN and TP in mixed forests was basically not significant because soil TP mainly comes from rock weathering and litter decomposition, which is a long process [49]. TN mainly comes from the degradation of litter; the difference in TN among forest types is basically not significant, which may be related to the close litter and slow decomposition. In summary, the *P. yunnanensis* mixed forest had improved the soil quality, which was consistent with the research of Wang [50]. They showed that mixed forests promote increases in soil organic matter, N and P content, improve soil nutrient status, and help to sustain soil fertility.

In this study, after two years of afforestation, three enzymes of five *P. yunnanensis* pine mixed forest types were higher than *P. yunnanensis* pure pine. The results showed that the mixed forest was beneficial to the accumulation and decomposition rate of forest litter, increased the amount of nutrient return, improved nutrient availability, and accelerated soil nutrient mineralization [51]. Pure forest could inhibit the decomposition activities of soil microorganisms due to its allelopathic effect. Interestingly, the three kinds of soil enzyme activities in the surface soil (0–20 cm) of different mixed models were higher than those in the deep soil (40–60 cm). Yang et al. also found that the soil urease, sucrase, and catalase enzyme activities decreased with the increase in soil depth before and after afforestation, which was consistent with the research results of Yang [52]. The reasons for this vertical decrease in soil enzyme activity are that there were more litter and microorganisms in the surface soil of the mixed forest, and the physiological and biochemical reactions of soil microorganisms and various enzymes were intense, while there were less litter and microorganisms in the deep soil, and the physiological and biochemical reactions were relatively stable [53]. Another reason is that the surface soil has a rich root system that penetrates the whole topsoil layer so that the enzyme activity in the surface soil is higher than that in the deep soil [54]. Sucrase comes from plant roots and microorganisms, which can catalyze the hydrolysis of sucrose into glucose and sucrose, which plays an important role in soil carbon and nitrogen cycling [55]. Catalase can promote the oxidation of various compounds by hydrogen peroxide, reduce the toxic effect of hydrogen peroxide in soil, and also reflect the total respiratory intensity in soil. In this study, different mixed modes had no significant effect on catalase activity, which may be related to tree species types and tree species growth stages. Zhou et al. [56] proved that the soil sucrase and catalase activities increased with the increase in forest age of the top Chinese prickly ash green food. In our study, the three enzyme activities in the topsoil of the six forest types after two years of afforestation (2018) were higher than those at the beginning of afforestation (2016). Therefore, long-term monitoring of soil enzyme activity in different mixed forests can be carried out in the future. Urease is a key enzyme in nitrogen cycling, which can catalyze the hydrolysis of urea. NH_3 formed by hydrolysis is one of the nitrogen sources of plants, and its activity can reflect the nitrogen supply capacity and level of soil [57]. In our study, after two years of afforestation, Forest I had the highest urease activities, which is mainly due to the presence of the nitrogen-fixing plant *A. nepalensis*. Nitrogen-fixing plants can often significantly increase soil organic matter and soil nitrogen [58]. For example, Wang et al. showed that the organic matter and nitrogen contents in the surface soil of an artificial forest of nitrogen-fixing trees were 40–50% and 20–50% higher than those of non-nitrogen-fixing trees, respectively [59]. At the beginning of afforestation (2016), there was no significant difference in soil indexes between mixed forest and pure forest. This may be because even though the forest types are different, this is a new mixed forest and their original soil conditions are basically the same.

Correlations were also found between these soil indices. There was a correlation between soil physicochemical properties and enzyme activities [60,61]. In this study, correlation analysis of physicochemical properties and enzyme activities of surface soil of six forest types was carried out. Soil enzymes are very sensitive to environmental changes, such as pH, temperature, and water content [62]. In our study, there was no significant negative correlation between soil pH and three kinds of soil enzyme activities, which indicated that if the soil's acidity is too strong, it will inhibit the enzyme activity. In addition, we selected ten soil fertility indexes of six forest types for principal component analysis to evaluate the soil quality of different forest types. Due to the different types of forest vegetation, there are some differences in the reserves and composition of surface litter, the growth and development of tree roots, and the decomposition rate of litter, resulting in the difference in the soil's nutrient content in different forest types [63]. In our study, through the comprehensive evaluation and ranking of soil quality of different *P. yunnanensis* pine mixed forest types by using the method of principal component analysis and cluster analysis, it was found that the soil quality of Forest V, Forest III, and Forest I were higher after two years of afforestation. The fundamental reason is that the species composition and biological characteristics of different forest types are different, so the quality, quantity, and decomposition rate of the litter of different forests are different, which affects the soil's nutrient content and distribution in different forest types. This indicated that the mixed pattern of *P. yunnanensis* pine growing for a certain number of years could improve the soil's nutrient content compared with the *P. yunnanensis* pure forest. On the other hand, compared with the growth status of *P. yunnanensis* under the *P. yunnanensis* pine pure forest, the growth status of *P. yunnanensis* pine under the three mixed patterns was relatively better, and the growth status of *P. yunnanensis* pine in Forest III was the best. This result was consistent with the result that the soil quality of Forest I, Forest II, and Forest V was better than that of Forest VI.

5. Conclusions

Interplanting different forests under *P. yunnanensis* forests can increase the soil quality, which is significant for improving the ecological environment of *P. yunnanensis* mixed forests. The results of this study showed that, compared with the *P. yunnanensis* pure forest, the *P. yunnanensis* × *A. nepalensis* mixed forest could improve soil physicochemical properties. The ratio of row spacing between *P. yunnanensis* and the selected mixed tree species was 2:1, which could make more rational use of soil nutrients and promote the growth of *P. yunnanensis*. Among them, such two planting patterns as *P. yunnanensis* × *Q. acutissima* mixed forest and *P. yunnanensis* × *A. nepalensis* have outstanding soil improvement effects, and they are worthy of promotion and application in *P. yunnanensis* forests in areas with insect pests.

Although the soil indexes of different *P. yunnanensis* mixed forests were measured in this study, their functions were not fully displayed because they were still in the young forest stage. Therefore, the future soil quality evaluation of the mixed forests needs a lot of research and long-term positioning observation. This study found that the mixed pattern of *P. yunnanensis* pine at the early stage of afforestation was affected by factors such as growth period and external environment, so proving its feasibility is a long-term process. Follow-up studies will continue in the later stage of the project to verify the effect of the *P. yunnanensis* pine mixed forest model and provide excellent companion trees for the sustainable management of *P. yunnanensis* pine mixed forests.

Supplementary Materials: The following supporting information can be downloaded at: https://www.mdpi.com/article/10.3390/10.3390/d14030214/s1, Figure S1: Comparison of the physical properties of surface soil (0–20 cm) at the beginning of afforestation (2016) and after two years of afforestation (2018). Figure S2: Comparison of the chemical properties of topsoil (0–20 cm) in different forest types at the beginning of afforestation (2016) and after two years of afforestation (2018). Figure S3: Comparison of the enzyme activities in the surface soil (0–20 cm) of different forest types at the beginning of afforestation (2016) and after two years of afforestation (2018).

Author Contributions: Conceptualization, funding acquisition, and project administration, N.Z. and B.Y.; data curation. Z.L. and L.L.; formal analysis, C.L., Z.Z. and M.J.; investigation, S.Z. and M.J.; methodology, B.Y. and J.Y.; resources, C.L., S.Z. and N.Z.; software, L.L. and J.Y.; supervision, Z.Z. and Z.L.; validation, L.L. and S.Z.; visualization, C.L., B.Y. and J.Y.; writing—original draft, C.L.; writing—review & editing, C.L. and N.Z. All authors have read and agreed to the published version of the manuscript.

Funding: This work was supported by the National Natural Science Foundation of China (31760210) and the Key Project of Yunnan Applied Basic Research Program (Grant No.202101AS070009; 2018FG001-010).

Institutional Review Board Statement: Not applicable.

Data Availability Statement: Not applicable.

Conflicts of Interest: The authors declare no conflict of interest.

References

1. Huang, X.; Li, S.; Su, J.; Liu, W.; Lang, X. The relationship between species richness and ecosystem multifunctionality in the Pinus yunnanensis natural secondary forest. *Biodivers. Sci.* **2017**, *25*, 1182–1191. [CrossRef]
2. Xu, Y.; Woeste, K.; Cai, N.; Kang, X.; Li, G.; Chen, S.; Duan, A. Variation in needle and cone traits in natural populations of Pinus yunnanensis. *J. For. Res.* **2016**, *27*, 41–49. [CrossRef]
3. Rothe, A.; Binkley, D. Nutritional interactions in mixed species forests: A synthesis. *Can. J. For. Res.* **2001**, *31*, 1855–1870. [CrossRef]
4. Castagneyrol, B.; Jactel, H.; Vacher, C.; Brockerhoff, E.G.; Koricheva, J. Effects of plant phylogenetic diversity on herbivory depend on herbivore specialization. *J. Appl. Ecol.* **2014**, *51*, 134–141. [CrossRef]
5. Zhang, C.; Zhao, X.; Gadow, K.V. Analyzing selective harvest events in three large forest observational studies in North Eastern China. *For. Ecol. Manag.* **2014**, *316*, 100–109. [CrossRef]

6. Huang, X.; Liu, S.; You, Y.; Wen, Y.; Wang, H.; Wang, J. Microbial community and associated enzymes activity influence soil carbon chemical composition in Eucalyptus urophylla plantation with mixing N2-fixing species in subtropical China. *Plant Soil* **2017**, *414*, 199–212. [CrossRef]
7. Zhang, P.; Pang, S.J.; Yang, B.G.; Liu, S.L.; Jia, H.Y.; Chen, J.B.; Guo, D.Q. Effects of different mixed models on Eucalyptus growth, litter and soil nutrients. *J. Northwest A&F Univ.* **2021**, *49*, 31–37. [CrossRef]
8. Kelty, M.J. The role of species mixtures in plantation forestry. *For. Ecol. Manag.* **2006**, *233*, 195–204. [CrossRef]
9. Spehn, E.M.; Joshi, J.; Schmid, B.; Diemer, M.; Körner, C. Aboveground resource use increases with plant species richness in experimental grassland ecosystems. *Funct. Ecol.* **2000**, *14*, 326–337.
10. Kowalchuk, G.A.; Buma, D.S.; de Boer, W.; Klinkhamer, P.G.L.; Van Veen, J.A. Effects of above-ground plant species composition and diversity on the diversity of soil-borne microorganisms. *Antonie Van Leeuwenhoek* **2002**, *81*, 509–520. [CrossRef] [PubMed]
11. Zak, D.R.; Holmes, W.E.; White, D.C.; Peacock, A.D.; Tilman, D. Plant diversity, soil microbial communities, and ecosystem function: Are there any links? *Ecology* **2003**, *84*, 2042–2050. [CrossRef]
12. Heemsbergen, D.A.; Berg, M.P.; Loreau, M.; van Hal, J.R.; Faber, J.H.; Verhoef, H.A. Biodiversity Effects on Soil Processes Explained by Interspecific Functional Dissimilarity. *Science* **2004**, *306*, 1019–1020. [CrossRef] [PubMed]
13. Eisenhauer, N.; Reich, P.B.; Scheu, S. Increasing plant diversity effects on productivity with time due to delayed soil biota effects on plants. *Basic Appl. Ecol.* **2012**, *13*, 571–578. [CrossRef]
14. Lavelle, P.; Lattaud, C.; Trigo, D.; Barois, I. Mutualism and biodiversity in soils. *Plant Soil* **1995**, *170*, 23–33. [CrossRef]
15. Li, Y.; Zeng, C.; Long, M. Variation of soil nutrients and bacterial community diversity of different land utilization types in Yangtze River Basin, Chongqing Municipality. *PeerJ* **2020**, *8*, e9386. [CrossRef] [PubMed]
16. Schoenholtz, S.H.; Miegroet, H.V.; Burger, J.A. A review of chemical and physical properties as indicators of forest soil quality: Challenges and opportunities. *For. Ecol. Manag.* **2000**, *138*, 335–356. [CrossRef]
17. Hayicho, H.; Alemu, M.; Kedir, H. Assessing the Effects of Land-Use and Land Cover Change and Topography on Soil Fertility in Melka Wakena Catchment of Sub-Upper Wabe-Shebelle Watershed, South Eastern Ethiopia. *J. Environ. Prot.* **2019**, *10*, 672–693. [CrossRef]
18. Dise, N.B.; Matzner, E.; Forsius, M. Evaluation of organic horizon C:N ratio as an indicator of nitrate leaching in conifer forests across Europe. *Environ. Pollut.* **1998**, *102*, 453–456. [CrossRef]
19. Salam, A.K.; Katayama, A.; Kimura, M. Activities of some soil enzymes in different land use systems after deforestation in hilly areas of West Lampung, South Sumatra, Indonesia. *Soil Sci. Plant Nutr.* **1998**, *44*, 93–103. [CrossRef]
20. Zhang, Y.J.; Jin, Y.H.; Gu, X.N.; Jia, W.; Tao, Y.; He, H.S.; Wang, A.L.; Liu, Y.X.; Zhu, L.P. Vegetation change in relation to soil microbes, enzyme activity and soil fertility in the tundra of Changbai Mountain. *Chin. J. Ecol.* **2017**, *36*, 3086–3093. [CrossRef]
21. Karlen, D.L.; Ditzler, C.A.; Andrews, S.S. Soil quality: Why and how? *Geoderma* **2003**, *114*, 145–156. [CrossRef]
22. Smaling, E.M.A.; Stoorvogel, J.J.; Windmeijer, P.N. Calculating soil nutrient balances in Africa at different scales. *Fertil. Res.* **1993**, *35*, 237–250. [CrossRef]
23. Ling, Q.; Zhao, X.; Wu, P.; Gao, X.; Sun, W. Effect of the fodder species canola (*Brassica napus* L.) and daylily (*Hemerocallis fulva* L.) on soil physical properties and soil water content in a rainfed orchard on the semiarid Loess Plateau, China. *Plant Soil* **2020**, *453*, 209–228. [CrossRef]
24. Shiau, Y.-J.; Chiu, C.-Y. Changes in Soil Biochemical Properties in a Cedar Plantation Invaded by Moso Bamboo. *Forests* **2017**, *8*, 222. [CrossRef]
25. Li, C.; Yan, L.; Jian, M. Spatial heterogeneity of soil chemical properties at fine scales induced by *Haloxylon ammodendron* (Chenopodiaceae) plants in a sandy desert. *Ecol. Res.* **2011**, *26*, 385–394. [CrossRef]
26. Zhu, T.; Zhang, J.; Meng, T.; Zhang, Y.; Yang, J.; Müller, C.; Cai, Z. Tea plantation destroys soil retention of NO3− and increases N2O emissions in subtropical China. *Soil Biol. Biochem.* **2014**, *73*, 106–114. [CrossRef]
27. Qin, F.; Lirong, S.U.; Zeng, C.; Qin, L.; Tieguang, H.E.; Yuefeng, Y.U.; Nan, W.E.I.; Yuanqing, M.E.N.G.; Aina, W.E.I.; Jinshan, W.E.I.; et al. Improving Soil Fertility with Different Planting Patterns in Rocky Desertification Areas. *Agric. Biotechnol.* **2020**, *9*, 119–124.
28. Ziadi, N.; Whalen, J.K.; Messiga, A.J.; Morel, C. Assessment and Modeling of Soil Available Phosphorus in Sustainable Cropping Systems. *Adv. Agron.* **2013**, *122*, 85–126. [CrossRef]
29. Gopal, M.; Gupta, A.; Arunachalam, V.; Magu, S. Impact of azadirachtin, an insecticidal allelochemical from neem on soil microflora, enzyme and respiratory activities. *Bioresour. Technol.* **2007**, *98*, 3154–3158. [CrossRef] [PubMed]
30. Liu, Y.-M.; Cao, W.-Q.; Chen, X.-X.; Yu, B.-G.; Lang, M.; Chen, X.-P.; Zou, C.-Q. The responses of soil enzyme activities, microbial biomass and microbial community structure to nine years of varied zinc application rates. *Sci. Total Environ.* **2020**, *737*, 140245. [CrossRef] [PubMed]
31. Zhang, C.; Liu, G.; Xue, S.; Song, Z. Rhizosphere soil microbial activity under different vegetation types on the Loess Plateau, China. *Geoderma* **2011**, *161*, 115–125. [CrossRef]
32. Farrer, E.C.; Goldberg, D.E. Litter drives ecosystem and plant community changes in cattail invasion. *Ecol. Appl.* **2009**, *19*, 398–412. [CrossRef] [PubMed]

33. Dong, X.; Gao, P.; Li, P.; Zhang, J.C.; Dong, J.W.; Xu, J.W.; Tun, X.J. Effects of soil microbial community on the litter decomposition in mixed *Quercus acutissima* Carruth. and *Robinia pseudoacacia* L. forest. *Acta Ecol. Sin.* **2021**, *41*, 2315–2325.
34. Zhou, L.; Sun, Y.; Saeed, S.; Zhang, B.; Luo, M. The difference of soil properties between pure and mixed Chinese fir (Cunninghamia lanceolata) plantations depends on tree species. *Glob. Ecol. Conserv.* **2020**, *22*, e01009. [CrossRef]
35. Li, D.S.; Zhang, P.; Zhang, S.L.; Yin, J.D.; Lu, F.D. A Study on Water Conservation Capacity of Forest Soil in Huangqian Reservoir Area. *J. Nanjing For. Univ.* **2004**, *47*, 25.
36. Ming-Liang, L.U.; Chen, Y.F.; Fang, Z.Z.; He, F. Soil physical and chemical properties of different types of ecological forest stands in Kecheng District. *J. Zhejiang For. Sci. Technol.* **2010**, *30*, 70–72. [CrossRef]
37. Chen, X.F.; Lin, G.W.; Hong, T.; Chen, J.Z.; Su, S.C.; Hong, W.; Wu, C.Z.; Lin, H. Soil Physical and Chemical Properties of Cunninghamia lanceolata-Aleurites Montana Mixed Forestand Cunninghamia lanceolata Pure Forest with Different Age. *Chin. J. Trop. Crop.* **2017**, *38*, 1660–1665.
38. Gray, A.N.; Spies, T.A.; Easter, M.J. Microclimatic and soil moisture responses to gap formation in coastal Douglas-fir forests. *Can. J. For. Res.* **2002**, *32*, 332–343. [CrossRef]
39. Scharenbroch, B.; Bockheim, J. Pedodiversity in an old-growth northern hardwood forest in the Huron Mountains, Upper Peninsula, Michigan. *Can. J. For. Res.* **2007**, *37*, 1106–1117. [CrossRef]
40. Zhang, K.; Zheng, H.; Chen, F.L.; Ouyang, Z.Y.; Wang, Y.; Wu, Y.F.; Lan, J.; Fu, M.; Xiang, X.W. Changes in soil quality after converting Pinus to Eucalyptus plantations in southern China. *Solid Earth* **2015**, *6*, 115–123. [CrossRef]
41. Yesilonis, I.; Szlavecz, K.; Pouyat, R.; Whigham, D.; Xia, L. Historical land use and stand age effects on forest soil properties in the Mid-Atlantic US. *For. Ecol. Manag.* **2016**, *370*, 83–92. [CrossRef]
42. Ouyang, W.; Wu, Y.; Hao, Z.; Zhang, Q.; Bu, Q.; Gao, X. Combined impacts of land use and soil property changes on soil erosion in a mollisol area under long-term agricultural development. *Sci. Total Environ.* **2018**, *613–614*, 798–809. [CrossRef] [PubMed]
43. Pan, K.; Xu, Z.; Blumfield, T.; Totua, S.; Lu, M. In situ mineral 15N dynamics and fate of added 15NH4 + in hoop pine plantation and adjacent native forest in subtropical Australia. *J. Soils Sediments* **2008**, *8*, 398–405. [CrossRef]
44. Zhou, Y.D.; Wu, J.S.; Zhao, S.W.; Guo, S.L.; Lu, P. Change of soil organic matter and water holding ability during vegetation succession in Ziwuling region. *Acta Bot. Boreali-Occident. Sin.* **2003**, *06*, 895–900.
45. Dawud, S.M.; Raulund-Rasmussen, K.; Domisch, T.; Finér, L.; Jaroszewicz, B.; Vesterdal, L. Is tree species diversity or species identity the more important driver of soil carbon stocks, C/N ratio, and pH? *Ecosystems* **2016**, *19*, 645–660. [CrossRef]
46. Stewart, W.D.P. A Quantitative Study of Fixation and Transfer of Nitrogen in Alnus. *J. Exp. Bot.* **1962**, *13*, 250–256. [CrossRef]
47. Forrester, D.I.; Bauhus, J.; Cowie, A.L. On the success and failure of mixed-species tree plantations: Lessons learned from a model system of Eucalyptus globulus and Acacia mearnsii. *For. Ecol. Manag.* **2005**, *209*, 147–155. [CrossRef]
48. Liao, Z.Y.; Hui, Y.; Wang, S.J.; Chen, P. Leaf litter decomposition and nutrient return characteristics of *Pinus yunnanensis* at different forest ages. *Ecol. Environ. Sci.* **2018**, *27*, 1981–1986. [CrossRef]
49. Liu, W.D.; Su, J.R.; Li, S.F.; Zhang, Z.J.; Li, Z.J. Stoichiometry study of C,N and P in plant and soil at different successional stages of monsoon evergreen broad-leaved forest in Pu'er, Yunnan Province. *Acta Ecol. Sin.* **2010**, *30*, 6581–6590.
50. Wang, Z.Q.; Wang, Q.C. The spatial heterogeneity of soil physical properties in forests. *Acta Ecol. Sin.* **2000**, *20*, 945–950.
51. Wei, Q.; Zhang, G.; Lei, L.; Chai, C. Water conservation function of litter and soil layer under main forest types in Xinglong Mountain of Gansu. *J. Nanjing For. Univ. Nat. Sci. Ed.* **2013**, *56*, 78.
52. Yang, Y.; Yang, Y.; Geng, Y.; Huang, G.; Cui, X.; Hou, M. Effects of Different Land Types on Soil Enzyme Activity in the Qinghai Lake Region. *Wetlands* **2018**, *38*, 711–721. [CrossRef]
53. Liu, Z.H.; Li, G.; Chen, R.; Wu, M.; Tong, Q.; Tong, F.P. Soil chemical properties of different mixed forests in Hunan province. *J. Cent. South Univ. For. Technol.* **2018**, *38*, 7. [CrossRef]
54. Zou, J.; Li, Y.Y.; Zhang, Y.W.; Yu, L.F. Soil sucrase, phosphatase and catalase enzyme activity change of degraded Karst Vegetation restoration research. *Guangdong Agric. Sci.* **2013**, *40*, 88–91. [CrossRef]
55. Mo, J.; Yan, W.D.; Liu, S.G.; Wu, X.h. Soil enzyme activities and their relations with soil fertility in Camellia oleifera peanut intercropping. *J. Cent. South Univ. For. Technol.* **2017**, *37*, 89–95. [CrossRef]
56. Zhou, Z.Y. Soil enzyme activities under different vegetation types in Beipan River Karst Gorge District. *Sci. Silvae Sin.* **2010**, *46*, 136–141. [CrossRef]
57. Wei, L.; Hao, H.L.; Wu, W.; Wei, Q.K.; Chen, Y.X.; Thies, J.E. Transgenic Bt rice does not affect enzyme activities and microbial composition in the rhizosphere during crop development. *Soil Biol. Biochem.* **2008**, *40*, 475–486.
58. Hoogmoed, M.; Cunningham, S.; Baker, P.; Beringer, J.; Cavagnaro, T. N-fixing trees in restoration plantings: Effects on nitrogen supply and soil microbial communities. *Soil Biol. Biochem.* **2014**, *77*, 203–212. [CrossRef]
59. Wang, F.; Li, Z.; Xia, H.; Zou, B.; Li, N.; Liu, J.; Zhu, W. Effects of nitrogen-fixing and non-nitrogen-fixing tree species on soil properties and nitrogen transformation during forest restoration in southern China. *Soil Sci. Plant Nutr.* **2010**, *56*, 297–306. [CrossRef]
60. Qu, G.H.; Guo, J.X. The relationship between different plant communities and soil characteristics in Songnen grassland. *Acta Prataculturae Sin.* **2003**, *12*, 18–22.
61. Chang, C.; Xie, Z.Q.; Xiong, G.M.; Zhao, C.M.; Shen, G.Z.; Lai, J.S.; Xu, X.W. Characteristics of soil nutrients of different vegetation types in the Three Gorges reservoir area. *Acta Ecol. Sin.* **2009**, *29*, 5978–5985.

62. Trasar-Cepeda, C.; Leirós, C.; Gil-Sotres, F.; Seoane, S. Towards a biochemical quality index for soils: An expression relating several biological and biochemical properties. *Biol. Fertil. Soils* **1998**, *26*, 100–106. [CrossRef]
63. Wang, W.; Linglin, X.U.; Zeng, C.; Tong, C.; Zhang, L. Carbon, nitrogen and phosphorus ecological stoichiometric ratios among live plant-litter-soil systems in estuarine wetland. *Acta Ecol. Sin.* **2011**, *31*, 7119–7124.

diversity

MDPI

Article

Relative Impact of Climate Change and Grazing on NDVI Changes in Grassland in the Mt. Qomolangma Nature Reserve and Adjacent Regions during 2000–2018

Wanglin Zhao [1,2], Tianxiang Luo [1], Haijuan Wei [3], Alamu [1,4] and Lin Zhang [1,2,*]

1 State Key Laboratory of Tibetan Plateau Earth System, Resources and Environment (TPESRE), Institute of Tibetan Plateau Research, Chinese Academy of Sciences, Beijing 100101, China; zhaowanglin@itpcas.ac.cn (W.Z.); luotx@itpcas.ac.cn (T.L.); alamu@itpcas.ac.cn (A.)
2 Motuo Observation and Research Center for Earth Landscape and Earth System, Chinese Academy of Sciences, Linzhi 860712, China
3 Tibet Autonomous Region Remote Sensing Monitoring Center for Ecological Environment, Lhasa 850000, China; weihaijuan_a@163.com
4 University of Chinese Academy of Sciences, Beijing 100049, China
* Correspondence: zhanglin@itpcas.ac.cn; Tel.: +86-010-8409-7055

Abstract: As the roof of the world, the Mt. Qomolangma National Nature Reserve and adjacent regions have a fragile environment and are very sensitive to global climate change. Based on the MODIS and SPOT remote sensing data during 2000–2018, we aimed to explore the change trend and driving factors of grassland in this area under the dual influence of climate change and human activities. Here, temperature and precipitation data were enrolled as the main indicators of climate change, while the number of livestock at the end of the year was regarded as the key indicator of grazing. The results showed that: (1) during 2000–2018, the grassland NDVI reflected an overall increasing trend, and the impact of precipitation was more significant than those of temperature and grazing at both pixel and county levels; (2) probably due to the large population and high grazing intensity, the grassland NDVI in Tingri County was controlled by both precipitation and grazing. In general, precipitation exerts a greater impact on the NDVI changes since this region is characterized by arid and semiarid climates. In some areas, vegetation growth is simultaneously affected by both grazing and climate factors due to the relatively greater pressure of grazing. In the context of future warming, control of the number of tourists for Mt. Qomolangma, as well as that of livestock in Tingri County, will help improve sustainability development and to reduce the adverse effects of grassland degradation.

Keywords: grazing; alpine shrub steppe; climate change; Mt. Qomolangma Nature Preserve

Citation: Zhao, W.; Luo, T.; Wei, H.; Alamu; Zhang, L. Relative Impact of Climate Change and Grazing on NDVI Changes in Grassland in the Mt. Qomolangma Nature Reserve and Adjacent Regions during 2000–2018. *Diversity* **2022**, *14*, 171. https://doi.org/10.3390/d14030171

Academic Editor: Michael Wink

Received: 13 February 2022
Accepted: 25 February 2022
Published: 27 February 2022

1. Introduction

The impact of climate change on terrestrial vegetation growth and its mechanism are important issues in the field of global change research. A large number of studies have proven that global climate change is an important driving factor for changes in terrestrial vegetation in recent decades [1–3]. For the Tibetan Plateau, it is generally believed that simultaneous increases in temperature and precipitation will be beneficial to alpine vegetation growth [4–6]. However, the sensitivity of vegetation productivity to temperature changes at the beginning of this century has declined, and climate warming during nongrowing seasons is considered as one of the main reasons [7]. Climate warming in the nongrowing seasons is not conducive to the preservation of snow in winter, and to a certain extent, it affects vegetation growth indirectly by affecting the water supply in the early growing season [8]. Another study found that temperature increases are likely to affect the plant growth of alpine grassland by changing the forms of precipitation at the beginning of the growing season [9]. In the context of climate change, compared with temperature, although the changes in precipitation are not necessarily the main limiting factor for the variations in the normalized difference vegetation index (NDVI) on the Tibetan Plateau [10], changes in moisture caused by temperature (such as increased evaporation) are likely to affect vegetation growth. In addition, the effects of temperature and precipitation on the NDVI of the Tibetan Plateau vary significantly depending on the seasons and regions [11,12]. Therefore, to explore the impact of climate change on alpine grasslands, we need to fully consider the temporal and spatial changes in temperature and precipitation.

The Tibetan Plateau is one of the main pastoral areas in China, and long-term grazing activities have a dual impact on the grassland of the plateau. On the one hand, grazing affects vegetation biomass and soil compactness through gnawing and trampling [13]. On the other hand, grazing may stimulate plant growth [14] and improve soil fertility through manure [15]. Therefore, it seems that the grazing intensity determines the effect on alpine grassland to a large extent, and the influence of grazing activities on the vegetation in pastoral areas may vary with the grazing density and the length of the grazing period. Exploring a reasonable grazing intensity is a key issue for vegetation protection and sustainable use of forage; however, the related research is still controversial. For example, Zhang et al. (2015) and Lehnert et al. (2016) pointed out that climate is the main controlling factor for changes in alpine grassland ecosystems on the Tibetan Plateau [16,17]. However, Pan et al. (2017) believed that the nonclimatic factors from 1982 to 2013 were responsible for the changes in the alpine grassland on the plateau [18].

As an important national barrier for ecological security and the Asia Tower [19], the Tibetan Plateau plays a unique role in protecting the ecological security of China and even Southeast Asia. However, under the background of global climate change and increasing human activities, the function of the national barrier for ecological security on the Tibetan Plateau is facing serious threats and challenges. The Mt. Qomolangma National Nature Reserve and the seven surrounding counties of Zhongba, Saga, Gyirong, Nyalam, Tingri, Dinggye, and Ngamring, belonging to the Tingri Ecological Observation Station of the Ecology and Environment Bureau in Tibet, are an important ecosystem type on the Tibetan Plateau. The protection of grassland resources is an important part of the construction of ecological security barrier in Tibet [20]. According to the 1:1,000,000 Vegetation Atlas of China [21], the forest, shrub, and grassland in the Mt. Qomolangma National Nature Reserve and adjacent regions accounted for 0.66%, 3.80%, and 68.15%, respectively. According to statistical data, the available grassland area in this region is approximately 82,800 km^2, of which degraded grassland accounts for approximately 42% [22]. Additionally, grassland degradation poses important challenges to the protection and construction of regional ecological security barriers. In this study, based on MODIS (Moderate Resolution Imaging Spectroradiometer) and SPOT (Systeme Probatoire d'Observation de la Terre) remote sensing data, the temperature and precipitation data, and the data of the number of livestock during 2000–2018 in Mt. Qomolangma National Nature Reserve and adjacent regions, we aimed to (1) illustrate the temporal and spatial patterns of grassland NDVI and (2) disclose the relative effects of climate change and grazing on NDVI changes. Here, temperature and precipitation were regarded as the main indicators of climate change and the number of livestock at the end of the year as the key indicators of grazing.

2. Materials and Methods

2.1. Study Area

The Mt. Qomolangma National Nature Reserve and adjacent regions (27.70° N–31.90° N; 82.10° E–88.40° E) are located in the south Tibetan Plateau, including the seven counties of Zhongba, Saga, Gyirong, Nyalam, Tingri, Dinggye, and Ngamring (Figure 1). This area is located between the middle section of the Himalayas and the Gangdise Mountains. The terrain is high in the north and south, with the Yarlung Zangbo River passing between. The middle part is, therefore, characterized by wide valleys and alluvial plains. The landform types are complex and diverse, and the climate exhibits cold and dry conditions. Based on the long-term meteorological observations (1967–2019) at Dingri station (a.s.l., 4300 m), the mean annual, January, and July air temperature and annual rainfall were 3.08, −5.84, 11.69 °C, and 286 mm, respectively [23]. Alpine grassland (mainly comprising alpine meadow and alpine steppe) is the main land-cover type (Figure 1). The common species in alpine steppe contain *Stipa purpurea, Lasiocaryum densiflorum, Artemisia wellbyi, Incarvillea compacta,* and *Androsace tapete,* while those in alpine meadow include *Kobresia pygmaea, Potentilla bifurca, Lancea tibetica, Arenaria bryophylla, Leontopodium pusillum,* and *Gentian* sp. The Mt. Qomolangma National Nature Reserve and adjacent regions have an area of approximately 87,600 km^2, where usable grassland area covers 82,800 km^2. The areas of lightly degraded, moderately degraded, and severely degraded grassland of the study area are 23,400; 10,400; and 33,300 km^2, respectively, accounting for 26.71%, 11.93%, and 3.78% of the total grassland area, respectively [22].

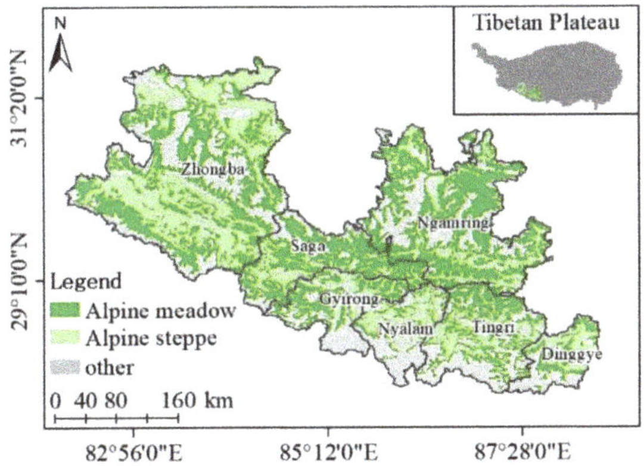

Figure 1. The location and vegetation types of the study area. The map was created using ESRI ArcMap 10.7 software; the topographic base of the map was created with National Catalogue Service For Geographic Information data (https://www.webmap.cn, figure number: GS(2016)2556); the vegetation map source from the CASEarth Data Sharing and Service Portal (http://data.casearth.cn, accessed on 19 December 2018).

2.2. Data Processing

In this study, alpine meadow and steppe in this region were selected as the research object by using the Vegetation Atlas of China (1:1,000,000). Two NDVI datasets were used, including SPOT NDVI from the Resources and Environment Science and Data Center of the Chinese Academy of Sciences, and MODIS NDVI from the NASA website (MOD13A3 datasets). The spatial resolution of the two types of NDVI is 1 km, with the same time interval of 2000–2018. The maximum NDVI during the growing season (May to September) was used in this study. Temperature and precipitation datasets during the early plant-growing season of 2000–2018 were downloaded from the National Tibetan Plateau Data Center (http://data.tpdc.ac.cn, accessed on 20 July 2019). Considering that the maximum NDVI generally occurs at the peak of the plant-growing season (early August), the climatic conditions in the early growing season (May to July) are critical for vegetation growth [24,25]. Therefore, the precipitation (GSP) and average temperature (GST) of the early growing season were selected, and the spatial resolution was 0.1° × 0.1°. The number of livestock (LN) during 2000–2018 in each county of the study area can be found in the National Tibetan Plateau Data Center and Tibet Statistical Yearbook (www.shujuku.org, accessed on 12 February 2020). The absolute numbers of different animals were converted to standard sheep units (SU) [26]. The vector data, such as administrative divisions used in this study, are from the National Catalog Service for Geographic Information. The population data of the study area were obtained from the National Tibetan Plateau Data Center.

2.3. Analysis of NDVI Interannual Change Trend

A linear regression method was used to detect the interannual changes in the vegetation index [27]. The least square method was used to calculate the regression coefficient, namely, the interannual variation trend of NDVI, as follows:

$$\theta_{slope} = \frac{n \times \sum_{i=1}^{n} i \times NDVI_i - \sum_{i=1}^{n} i \sum_{i=1}^{n} NDVI_i}{n \times \sum_{i=1}^{n} i^2 - (\sum_{i=1}^{n} i)^2} \quad (1)$$

where θ_{slope} refers to the interannual trend in NDVI, n is the number of years simulated, and $NDVI_i$ is the NDVI value of the *i*th year. We determined the significance of variation (significant change or no significant change) via T-tests to calculate confidence levels.

2.4. Correlation Analysis between NDVI and Main Influencing Factors

Correlation coefficient calculations and significance tests were used to detect the relationship between grassland growth change and climate change in the study area. The partial correlation method was further used to analyze the relative impact of grazing and climate change on NDVI changes. The mathematical expression of the Pearson correlation coefficient is:

$$R_{X\ NDVI} = \frac{n \times \sum_{i=1}^{n} X_i \times NDVI_i - \sum_{i=1}^{n} X_i \times \sum_{i=1}^{n} NDVI_i}{\sqrt{n \times \sum_{i=1}^{n} X_i^2 - (\sum_{i=1}^{n} X_i)^2} \times \sqrt{n \times \sum_{i=1}^{n} NDVI_i^{\ 2} - (\sum_{i=1}^{n} NDVI_i)^2}} \quad (2)$$

where *R* refers to the correlation coefficient, *n* is the number of years simulated, and X and NDVI are independent and dependent variables, respectively. Based on the NDVI during 2000–2018 and the corresponding temperature and precipitation, this study analyzed the correlation between NDVI and the main climatic factors in the study area, and conducted a significance test. When there are only three variables, a simple correlation coefficient can be used to indirectly calculate the partial correlation coefficient. Setting the three variables as LN, GST, and NDVI, and when LN remains constant, the partial correlation coefficients of GST and NDVI are:

$$R_{NDVI\ GST} = \frac{R_{NDVI\ GST} - R_{GST\ LN} \times R_{NDVI\ LN}}{\sqrt{(1 - R^2_{GST\ LN}) \times \left(1 - R^2_{NDVI\ LN}\right)}} \quad (3)$$

where $R_{NDVI\ GST}$ refers to the partial correlation coefficient of NDVI and GST when LN is fixed, and where $R_{NDVI\ SPEI}$, $R_{NDVI\ LN}$, and $R_{GST\ LN}$ are the correlation coefficients of NDVI and GST, NDVI and LN, and GST and LN, respectively.

2.5. RESTREND Analysis

The residual trend (RESTREND) method is based on the hypothesis that vegetation growth is determined by climate change. After removing the climate influence, the human-induced vegetation change could thus be identified [28,29]. The trends in NDVI residuals ($NDVI_H = NDVI_P - NDVI_A$) between the observed ($NDVI_A$) and the predicted values ($NDVI_P$) were calculated at a 0.1° × 0.1° pixel level. $NDVI_A$ is applied to regress against GST and GSP to generate a statistical model which is then used to compute $NDVI_P$ at each pixel [30–32]. More details can be found in Li et al. (2012) [30] and Cai et al. (2015) [32]. Table 1 shows the relative effects of climate change (*Slope*$NDVI_P$) and human activities (*Slope*$NDVI_H$) on grassland change (*Slope*$NDVI_A$) in six different scenarios [33]. For example, in the first scenario, an increasing trend in $NDVI_A$ and a declining trend in $NDVI_P$ result in a decreasing $NDVI_H$, which means that human activity mainly determines the $NDVI_A$ increase.

Table 1. Evaluation methods of climate change and human activities on grassland dynamics under six possible scenarios.

Scenario	*Slope*NDVI$_A$	*Slope*NDVI$_P$	*Slope*NDVI$_H$	Description
1	+	−	−	Human activity contributes to grassland NDVI$_A$ increase
2	+	+	−	Both climate change and human activity contribute to NDVI$_A$ increase
3	+	+	+	Climate change contributes to grassland NDVI$_A$ increase
4	−	+	+	Human activity contributes to grassland NDVI$_A$ decrease
5	−	−	+	Both climate change and human activity contribute to NDVI$_A$ decrease
6	−	−	−	Climate change contributes to grassland NDVI$_A$ decrease

2.6. Statistical Analysis

The annual variations of NDVI and corresponding climatic factors and the relationships between NDVI and climatic factors were fixed by simple linear correlations. Partial correlations between NDVI and climatic factors of GSP and GST were analyzed at a pixel scale. Since LN data were only available for each county, we applied partial correlation between NDVI and GSP, GST, and LN at a county scale. All statistical analyses were performed using SPSS 19.0 (SPSS Inc., Chicago, IL, USA), and spatial analysis was performed using ArcGIS 10.7 for Desktop. Further, the relative effects of climate change and human activities on NDVI changes were detected by the RESTREND method described above.

3. Results

3.1. Interannual Variations in Grassland NDVI during 2000–2018

Both MODIS NDVI and SPOT NDVI data indicate that the grassland NDVI in the study area increased significantly during 2000–2018 (Figure 2a, MODIS, $R^2 = 0.28$, $p = 0.02$; SPOT, $R^2 = 0.33$, $P = 0.01$). The mean NDVI values were approximately 0.23 during the last 19 years, with their maximum in 2013 and minimum in 2002 or 2009. Figure 2b shows that the MODIS NDVI is highly correlated with the SPOT NDVI in the study area ($R^2 = 0.99$, $P = 0.00$), indicating that the two datasets can be verified by each other. In view of their high correlation, we only present the results based on MODIS NDVI analysis in the following section.

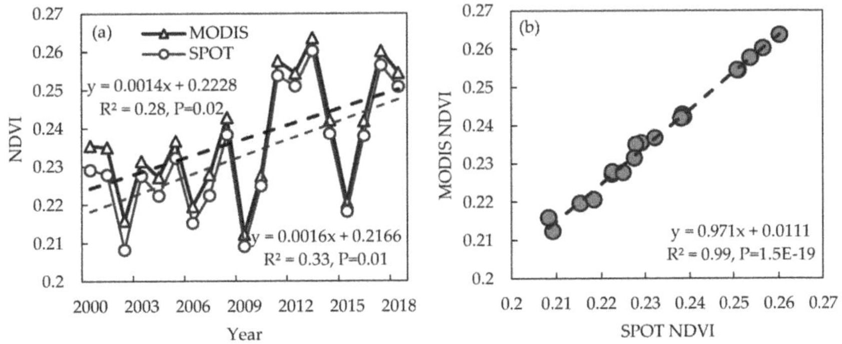

Figure 2. (**a**) Interannual variations in MODIS NDVI and SPOT NDVI and (**b**) their relationships.

3.2. *Variations in Climatic Factors and the Number of Livestock and Their Relationships with NDVI*

In this area, GST (Figure 3a, $R^2 = 0.02$, $P = 0.55$) and GSP (Figure 3b, $R^2 = 0.03$, $P = 0.52$) did not vary during 2000–2018. However, LN reflected a significant decreasing trend (Figure 3c, $R^2 = 0.56$, $P = 0.00$), especially after 2008.

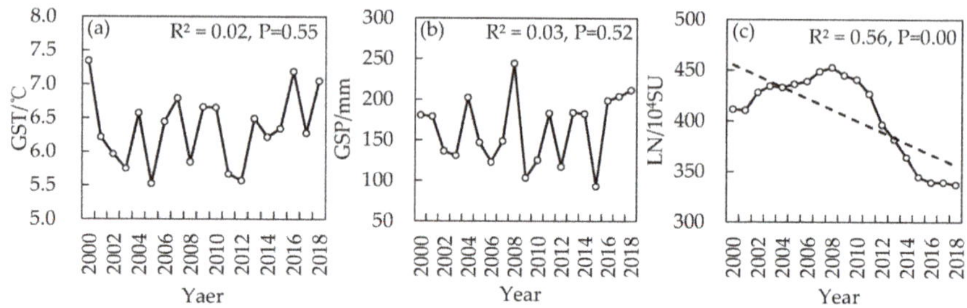

Figure 3. Interannual variations in (**a**) GST, (**b**) GSP, and (**c**) LN.

Pearson correlation analysis showed that NDVI was not correlated with GST (Figure 4a, $R^2 = 0.02$, $P = 0.57$) but was positively related to GSP (Figure 4b, $R^2 = 0.35$, $P = 0.01$) and negatively correlated with LN (Figure 4c, $R^2 = 0.23$, $P = 0.04$).

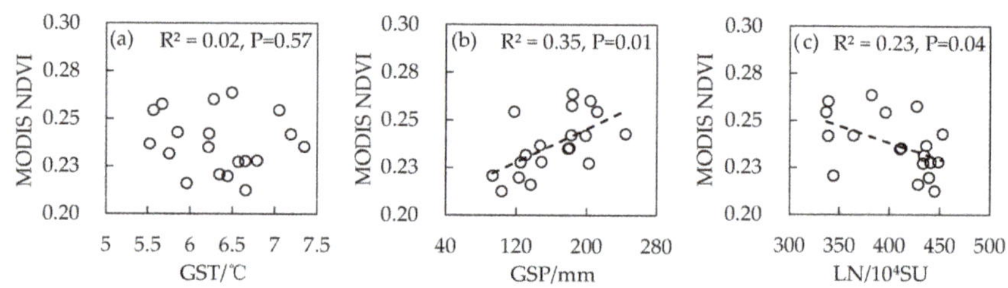

Figure 4. The relationships between NDVI and (**a**) GST, (**b**) GSP, and (**c**) LN.

3.3. *Relative Effects of Climate Factors and Grazing on Grassland NDVI*

At the pixel level, NDVI was more closely related with GSP than with GST. There were 22.31% pixels showing significantly positive partial correlations between NDVI and GSP (Figure 5a), but only 5.42% reflecting significantly positive correlations between NDVI and GST (Figure 5b). Additionally, the partial correlation coefficients between NDVI and GSP showed substantial spatial variations, and those area affected by GSP were mainly distributed in the center part, i.e., Saga and Ngamring.

At the regional level, the partial correlation analysis showed that NDVI variations for most counties in this area were mainly affected by the climatic factor of precipitation, and only that in Tingri County were mediated by both climate and grazing (Figure 6). Among the six counties affected by climate factors, the NDVI for Dinggye County was mainly controlled by GST, while the other five were generally affected by GSP.

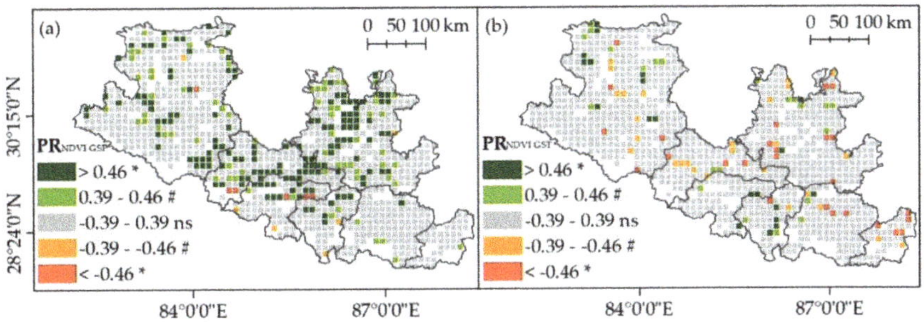

Figure 5. Spatial distribution of the partial correlation coefficients between NDVI and (**a**) GSP, and (**b**) GST. Different colors indicate the classes of partial correlation coefficients. Different symbols denote significant levels: * $p < 0.05$; # $0.05 < p < 0.10$; ns, not significant. The map was created using ESRI ArcMap 10.7 software; the topographic base of the map was created with National Catalogue Service For Geographic Information data (https://www.webmap.cn, figure number: GS(2016)2556).

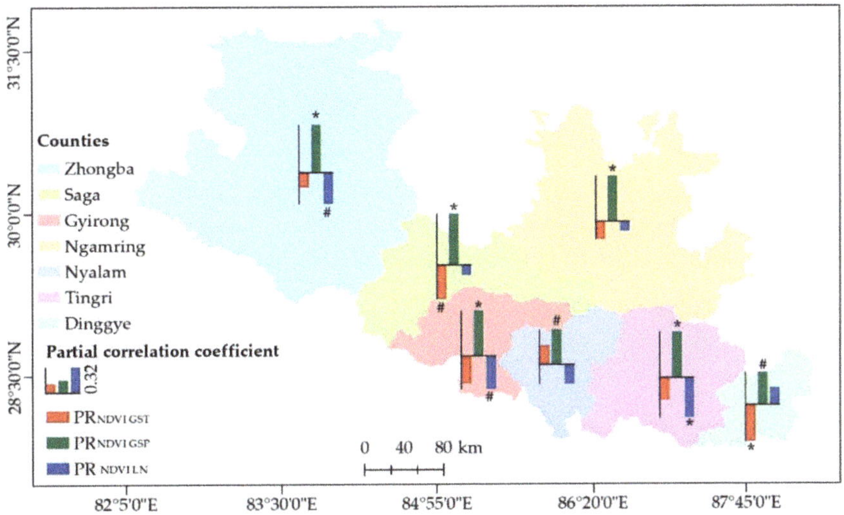

Figure 6. Main drivers of NDVI change in the seven counties of the study area. Different symbols indicate the different significant levels (* $p < 0.05$; # $0.05 < p < 0.10$) according to the partial correlation coefficients. The map was created using ESRI ArcMap 10.7 software; the topographic base of the map was created with National Catalogue Service For Geographic Information data (https://www.webmap.cn, figure number: GS(2016)2556).

The RESTREND analysis, as shown by Figure 7, further indicated that the area showing increasing NDVI accounted for 63.22% of the study area, in which 32.17% was due to climate change, 24.80% was due to human activities, and 6.25% was due to both climate change and human activities. The reason for the NDVI increase by human activities might be mainly due to the implementation of an ecological restoration project (Grazing Withdrawal Program). For the area (36.78%) depicting decreasing NDVI, 21.62% was due to human activities, 12.30% due to climate change, and 2.87% due to both climate change and human activities. The reason for the NDVI decrease by human activities could be mainly due to overgrazing (e.g., Tingri County).

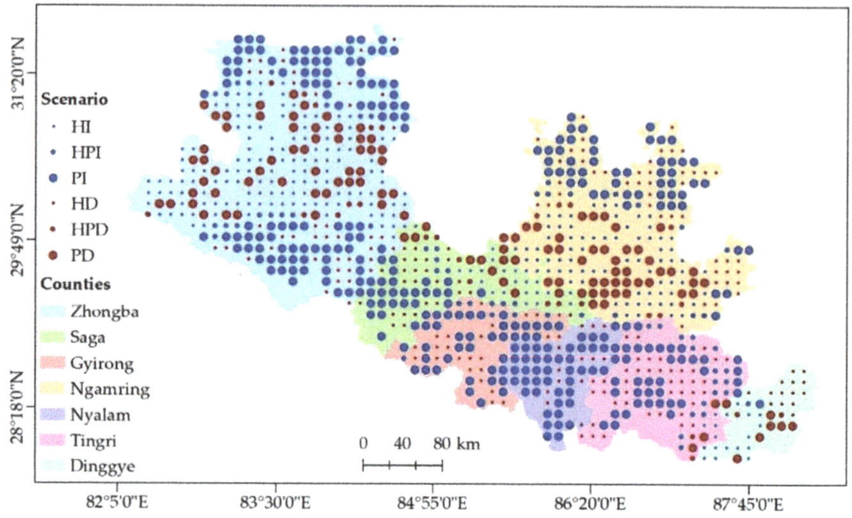

Figure 7. Relative effects of climate change and human activities on NDVI changes in the Mt. Qomolangma Nature Reserve and adjacent regions during 2000–2018. HI denotes human-activity-induced NDVI increase; HPI denotes both climate-change and human-activity-induced NDVI increase; PI denotes climate-change-induced NDVI increase; HD denotes NDVI decrease induced by human activities; HPD denotes NDVI decrease induced by both climate change and human activities; and PD denotes NDVI decrease induced by climate change. The map was created using ESRI ArcMap 10.7 software; the topographic base of the map was created with National Catalogue Service For Geographic Information data (https://www.webmap.cn, figure number: GS(2016)2556).

4. Discussion

The Tibetan Plateau has an average altitude of more than 4000 m and is characterized as a typical alpine ecosystem. Therefore, low temperature is generally considered to be a key factor restricting plant growth in this area [6,10]. However, in arid western Tibet, precipitation is considered to be a more important limiting factor than temperature [34]. Synthesizing the results of Pearson correlation and partial correlation analysis, we found that the variations in grassland NDVI in the Mt. Qomolangma National Nature Reserve and adjacent regions are mainly affected by climatic factors, especially precipitation. This is consistent with existing studies. For example, based on an analysis of the net primary productivity (NPP) of grassland vegetation on the Tibetan Plateau during 2000–2015, He et al. (2019) found that precipitation was the first limiting factor of NPP [35]. This is associated with the effects of rain shadows in the Himalayas. At the same time, the impact of climate change on grassland vegetation is largely related to grazing intensity. In areas with heavy grazing intensity (larger LN), grassland NDVI is more significantly affected by grazing, and when grazing intensity exceeds a certain threshold, the negative effect of grazing on grassland is likely to offset or even exceed the positive effect of climate change on the growth of alpine grassland [36]. In general, human activities have a relatively weak impact on the grassland ecosystems in the study area; therefore, the variation in NDVI is mainly controlled by climate.

However, unlike the other six counties, the variation in grassland NDVI in Tingri County was controlled by both climate (precipitation) and grazing, which indicated that the grazing pressure was heavy in Tingri County to some extent. Figure 8a further illustrated that the population density of Tingri County was the highest among the seven counties, and the grazing intensity was the second highest. Synthesizing the population and grazing intensity data of the seven counties in the study area during the past 20 years, we found that both data were significantly correlated with each other (Figure 8b), i.e., the variation in population density at the county level largely reflected that of the grazing intensity. In addition, the Mt. Qomolangma scenic spot is located in Tingri County. According to the data provided by the Mt. Qomolangma National Nature Reserve Administration, an average of 82,300 tourists was received each year during the past 10 years (2011–2019). This number increased to approximately 119,000 during the past 3 years (2017–2019), which is approximately twice the resident population of Tingri County. A large number of tourists would undoubtedly stimulate the demands for livestock products, which might indirectly cause an increase in grazing intensity and aggravate the negative impact on grassland.

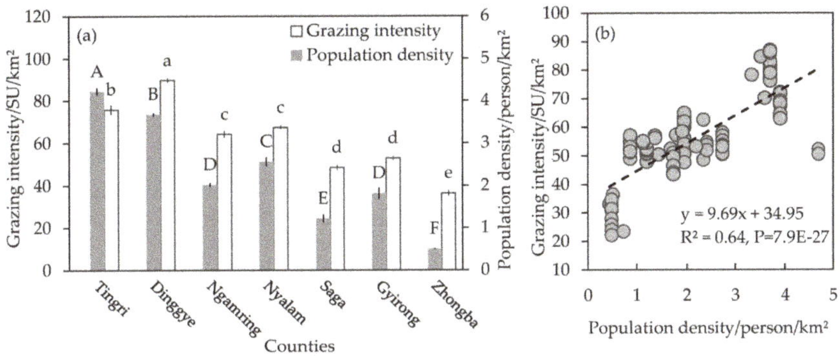

Figure 8. (**a**) Comparisons of the grazing intensity and population density between 7 counties and (**b**) their relationships during 2000–2018. Different lower-case and upper-case letters indicate significant differences in grazing intensity and population density between counties, respectively, by using the Tukey's HSD tests at $\alpha = 0.05$.

It is worth noting that the grazing intensity of Dinggye County ranks first among the seven counties in the study area (Figure 8a). Theoretically, the grazing intensity should be closely related to the changes in NDVI, but partial correlation analysis showed that the grazing effect on the grassland NDVI of the county was not significant. This may be related to the vegetation degradation caused by overgrazing in the county; that is, under heavy grazing conditions, the relationship between NDVI and grazing intensity may be nonlinear [37,38]. In addition, it can be inferred from Figure 9 that NDVI showed a significant negative correlation with grazing intensity after excluding the data in 2008 and 2011, when the precipitation was abundant or the temperature was very low, indicating that the strong fluctuation of the hydrothermal combination factor in different years may play an important role in regulating the relationship between grazing intensity and NDVI. The significantly negative partial correlation between GST and NDVI in Dinggye County further indicates that the future warming climate will not be advantageous for plant growth, which is assumed to associate with increased evapotranspiration and droughts [39,40].

In this study, we found that the NDVI showed an overall increasing trend during 2000–2018, and GSP was the main limiting factor. However, the GSP did not show a significant increasing pattern in the past 19 years, indicating that its interannual fluctuations rather than the changing trend determined the correlation between NDVI and precipitation. This result is consistent with Zhang et al. (2013), who found that the decline in NDVI in the Kosi River Basin in the central Himalayas (involving Tingri, Dinggye, and Nyalam counties) during 1982–1994 and 1994–2000 was related to a sudden drop in precipitation [3]. Therefore, concerning the reasons for vegetation changes, we should pay more attention to the impact of extreme climate events such as extreme precipitation years [41,42].

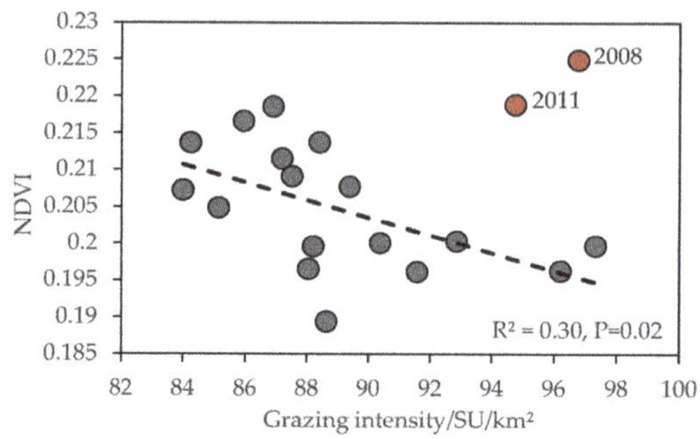

Figure 9. Relationship between NDVI and grazing intensity in Dinggye County.

5. Conclusions

The grassland NDVI in the Mt. Qomolangma National Nature Reserve and adjacent regions showed an overall increasing trend during 2000–2018. This increase is mainly attributed to precipitation rather than temperature or human activities (grazing). However, vegetation growth can also be simultaneously affected by grazing and/or temperature in some areas with relatively heavy grazing pressure, such as in Tingri and Dinggye counties. As a famous scenic spot in the world, the Mt. Qomolangma National Nature Reserve attracted a large number of tourists each year, which would stimulate the demands for livestock products, and cause an increase in grazing intensity and aggravate the negative impact on grassland ecosystems. Therefore, control of the number of tourists for Mt. Qomolangma, as well as that of livestock in Tingri County, will help improve sustainability development and reduce the adverse effects of grassland degradation caused by consistent warming.

Author Contributions: Conceptualization, L.Z., T.L. and W.Z.; methodology, W.Z.; software, W.Z.; validation, A. and H.W.; formal analysis, W.Z.; investigation, L.Z. and H.W.; resources, L.Z.; data curation, W.Z.; writing—original draft preparation, W.Z.; writing—review and editing, W.Z. and L.Z.; visualization, A.; supervision, L.Z.; project administration, L.Z.; funding acquisition, L.Z. and T.L. All authors have read and agreed to the published version of the manuscript.

Funding: The National Natural Science Foundation of China (41830649), the Key technology research and development projects in Xizang Autonomous Regions (XZ202101ZY0005G) and the Second Tibetan Plateau Scientific Expedition and Research (STEP) program (2019QZKK0301-1) provided financial support.

Institutional Review Board Statement: Not applicable.

Data Availability Statement: All data used in the manuscript are already publicly accessible, and we provided the download address in the manuscript.

Acknowledgments: We would like to thank the National Observation and Research Station for Qomolongma Special Atmospheric Processes and Environmental Changes, Tibet, China.

Conflicts of Interest: The authors declare no conflict of interest.

References

1. Zhao, X.; Tan, K.; Zhao, S.Q.; Fang, J. Changing climate affects vegetation growth in the arid region of the Northwestern China. *J. Arid Environ.* **2011**, *75*, 946–952. [CrossRef]
2. Kim, Y.; Kimball, J.S.; Zhang, K.; McDonald, K.C. Satellite detection of increasing northern hemisphere non-frozen seasons from 1979 to 2008: Implications for regional vegetation growth. *Remote Sens. Environ.* **2012**, *121*, 472–487. [CrossRef]
3. Zhang, Y.L.; Gao, J.G.; Liu, L.S.; Wang, Z.F.; Ding, M.J.; Yang, X.C. NDVI-based vegetation changes and their responses to climate change from 1982 to 2011: A case study in the Koshi river basin in the middle Himalayas. *Glob. Planet. Chang.* **2013**, *108*, 139–148. [CrossRef]
4. Wang, G.X.; Hu, H.C.; Wang, Y.B.; Chen, L. Response of alpine cold ecosystem biomass to climate changes in permafrost region of the Tibetan Plateau. *J. Glaciol. Geocryol.* **2007**, *29*, 671–679. (In Chinese)

5. Hu, M.Q.; Mao, F.; Sun, H.; Hou, Y.Y. Study of normalized difference vegetation index variation and its correlation with climate factors in the three-river-source region. *Int. J. Appl. Earth Obs. Geoinf.* **2011**, *13*, 24–33. [CrossRef]
6. Sun, J.; Cheng, G.W.; Li, W.P.; Sha, Y.K.; Yang, Y.C. On the variation of NDVI with the principal climatic elements in the Tibetan Plateau. *Remote Sens.* **2013**, *5*, 1894–1911. [CrossRef]
7. Piao, S.L.; Liu, Z.; Wang, T.; Peng, S.S.; Ciais, P.; Huang, M.T.; Ahlstrom, A.; Burkhart, J.F.; Chevallier, F.; Janssens, I.A.; et al. Weakening temperature control on the interannual variations of spring carbon uptake across northern lands. *Nat. Clim. Chang.* **2017**, *7*, 359–363. [CrossRef]
8. Wang, X.Y.; Wang, T.; Guo, H.; Liu, D.; Zhao, Y.T.; Zhang, T.T.; Liu, Q.; Piao, S.L. Disentangling the mechanisms behind winter snow impact on vegetation activity in northern ecosystems. *Glob. Chang. Biol.* **2018**, *24*, 1651–1662. [CrossRef]
9. Li, R.C.; Luo, T.X.; Mölg, T.; Zhao, J.X.; Li, X.; Cui, X.Y.; Du, M.Y. Leaf unfolding of Tibetan alpine meadows captures the arrival of monsoon rainfall. *Sci. Rep.* **2016**, *6*, 20985. [CrossRef]
10. Zhang, G.L.; Ouyang, H.; Zhang, X.Z.; Zhou, C.P.; Xu, X.L. Vegetation change and its responses to climatic variation based on eco-geographical region of Tibetan Plateau. *Geogr. Res.* **2010**, *29*, 2004–2016. (In Chinese)
11. Sun, J.; Qin, X.J.; Yang, J. Precipitation and temperature regulate the seasonal changes of NDVI across the Tibetan Plateau. *Environ. Earth Sci.* **2016**, *75*, 291. [CrossRef]
12. Zhang, L.; Guo, H.D.; Lei, J.; Lei, L.P.; Wang, C.Z.; Yan, D.M.; Li, B.; Li, J. Vegetation greenness trend (2000 to 2009) and the climate controls in the Qinghai-Tibetan Plateau. *J. Appl. Remote Sens.* **2013**, *7*, 073572. [CrossRef]
13. Owensby, C.E.; Ham, J.M.; Auen, L.M. Fluxes of CO_2 from grazed and ungrazed tallgrass prairie. *Rangel. Ecol. Manag.* **2006**, *59*, 111–127. [CrossRef]
14. Liu, J.S.; Wang, L.; Wang, D.L.; Bonser, S.P.; Sun, F.; Zhou, Y.F.; Gao, Y.; Teng, X. Plants can benefit from herbivory: Stimulatory effects of sheep saliva on growth of *Leymus chinensis*. *PLoS ONE* **2012**, *7*, e29259. [CrossRef] [PubMed]
15. Lin, Y.; Hong, M.; Han, G.D.; Zhao, M.L.; Bai, Y.F.; Chang, S. Grazing intensity affected spatial patterns of vegetation and soil fertility in a desert steppe. *Agric. Ecosyst. Environ.* **2010**, *138*, 282–292. [CrossRef]
16. Zhang, X.Z.; Yang, Y.P.; Piao, S.L.; Bao, W.K.; Wang, S.P.; Sun, H.; Luo, T.X.; Zhang, Y.J.; Shi, P.L.; Liang, E.Y.; et al. Ecological change on the Tibetan Plateau. *Chin. Sci. Bull.* **2015**, *60*, 3048–3056. (In Chinese)
17. Lehnert, L.W.; Wesche, K.; Trachte, K.; Reudenbach, C.; Bendix, J. Climate variability rather than overstocking causes recent large scale cover changes of Tibetan pastures. *Sci. Rep.* **2016**, *6*, 24367. [CrossRef]
18. Pan, T.; Zou, X.T.; Liu, Y.J.; Wu, S.H.; He, G.M. Contributions of climatic and non-climatic drivers to grassland variations on the Tibetan Plateau. *Ecol. Eng.* **2017**, *108*, 307–317. [CrossRef]
19. Zhong, H.X.; Liu, S.Z.; Wang, X.D.; Zhu, W.Z.; Li, X.M.; Yang, L. A Research on the protection and construction of the state ecological safe shelter zone on the Tibet Plateau. *J. Mt. Sci.* **2006**, *24*, 129–136. (In Chinese)
20. Wang, X.D.; Cheng, G.W.; Zhao, T.; Zhang, X.Z.; Zhu, L.P.; Huang, L. Assessment on protection and construction of ecological safety shelter for Tibet. *Bull. Chin. Acad. Sci.* **2017**, *32*, 29–34. (In Chinese)
21. Editorial Board of Vegetation Map of China, Chinese Academy of Sciences. *Vegetation Atlas of China*; Science Press: Beijing, China, 2001. (In Chinese)
22. Tibet Department of Agriculture and Animal Husbandry. *Grassland Resources and Ecological Statistics in Tibet Autonomous Region*; China Agriculture Press: Beijing, China, 2017; pp. 183–200. (In Chinese)
23. Wang, S.J.; Tang, X.Y.; Wang, G.; Li, Y.Q. Study on the climatic characteristics in the mount Qomolangma region during the last 53 Years. *Plateau Mt. Meteorol. Res.* **2021**, *41*, 2–6. (In Chinese)
24. Zhao, W.L.; Luo, T.X.; Zhang, L. Relative impact of climate change and grazing on NDVI variations in typical alpine desert grasslands in Tibet. *Acta Ecol. Sin.* **2019**, *39*, 8494–8503. (In Chinese)
25. Li, X.; Zhang, L.; Luo, T.X. Rainy season onset mainly drives the spatiotemporal variability of spring vegetation green-up across alpine dry ecosystems on the Tibetan Plateau. *Sci. Rep.* **2020**, *10*, 18797. [CrossRef] [PubMed]
26. Yu, C.Q.; Zhang, Y.J.; Claus, H.; Zeng, R.; Zhang, X.Z.; Wang, J.S. Ecological and environmental issues faced by a developing Tibet. *Environ. Sci. Technol.* **2012**, *46*, 1979–1980. [CrossRef]
27. Stow, D.; Daeschner, S.; Hope, A.; Douglas, D.; Petersen, A.; Myneni, R.; Zhou, L.; Oechel, W. Variability of the seasonally integrated normalized difference vegetation index across the north slope of Alaska in the 1990s. *Int. J. Remote Sens.* **2003**, *24*, 1111–1117. [CrossRef]
28. Evans, J.; Geerken, R. Discrimination between climate and human-induced dryland degradation. *J. Arid Environ.* **2004**, *57*, 535–554. [CrossRef]
29. Wessels, K.J.; Prince, S.D.; Malherbe, J.; Small, J.; Frost, P.E.; VanZyl, D. Can human-induced land degradation be distinguished from the effects of rainfall variability? A case study in South Africa. *J. Arid Environ.* **2007**, *68*, 271–297. [CrossRef]
30. Li, A.; Wu, J.G.; Huang, J.H. Distinguishing between human-induced and climate-driven vegetation changes: A critical application of RESTREND in inner Mongolia. *Landsc. Ecol.* **2012**, *27*, 969–982. [CrossRef]
31. Burrell, A.L.; Evans, J.P.; Liu, Y. The addition of temperature to the TSS-RESTREND methodology significantly improves the detection of dryland degradation. *IEEE J. Sel. Top. Appl. Earth Obs. Remote Sens.* **2019**, *12*, 2342–2348. [CrossRef]
32. Cai, H.Y.; Yang, X.H.; Xu, X.L. Human-induced grassland degradation/restoration in the central Tibetan Plateau: The effects of ecological protection and restoration projects. *Ecol. Eng.* **2015**, *83*, 112–119. [CrossRef]

33. Zhang, R.P.; Liang, T.G.; Guo, J.; Xie, H.J.; Feng, Q.S.; Aimaiti, Y. Grassland dynamics in response to climate change and human activities in Xinjiang from 2000 to 2014. *Sci. Rep.* **2018**, *8*, 2888. [CrossRef]
34. Li, P.L.; Hu, Z.M.; Liu, Y.W. Shift in the trend of browning in Southwestern Tibetan Plateau in the past two decades. *Agric. For. Meteorol.* **2020**, *287*, 1–9. [CrossRef]
35. He, K.D.; Sun, J.; Chen, Q.J. Response of climate and soil texture to net primary productivity and precipitation-use efficiency in the Tibetan Plateau. *Pratacultural Sci.* **2019**, *36*, 1053–1065. (In Chinese)
36. Yan, L.; Zhou, G.S.; Zhang, F. Effects of different grazing intensities on grassland production in China: A meta-analysis. *PLoS ONE* **2013**, *8*, e81466. [CrossRef] [PubMed]
37. Li, G.; Sun, W.L.; Zhang, H.; Gao, C.Y. Balance between actual number of livestock and livestock carrying capacity of grassland after added forage of straw based on remote sensing in Tibetan Plateau. *Trans. Chin. Soc. Agric. Eng.* **2014**, *30*, 200–211. (In Chinese)
38. Zhao, J.G.; Liu, C.Q.; Li, H.B.; Luo, N.; Meng, S.J.; Zhang, Z.H.; Wang, J.H. Effects of main climatic factors on plant community characteristics of a meadow steppe under different grazing intensities. *J. Landsc. Res.* **2019**, *11*, 108–110.
39. Wang, Y.J.; Fu, B.J.; Liu, Y.X.; Li, Y.; Feng, X.M.; Wang, S. Response of vegetation to drought in the Tibetan Plateau: Elevation differentiation and the dominant factors. *Agricult. Forest Meteorol.* **2021**, *306*, 108468. [CrossRef]
40. Ding, J.Z.; Yang, T.; Zhao, Y.T.; Liu, D.; Wang, X.Y.; Yao, Y.T.; Peng, S.S.; Wang, T.; Piao, S.L. Increasingly important role of atmospheric aridity on Tibetan alpine grasslands. *Geophys. Res. Lett.* **2018**, *45*, 2852–2859. [CrossRef]
41. Liu, R.; Li, Y.; Liu, Y. Responses of CO_2 flux in a saline desert of Northwest China to two contrasting extreme precipitation years. *Chin. J. Ecol.* **2013**, *32*, 2545–2551. (In Chinese)
42. Liu, D.; Wang, T.; Yang, T.; Yan, Z.J.; Liu, Y.W.; Zhao, Y.T.; Piao, S.L. Deciphering impacts of climate extremes on Tibetan grasslands in the last fifteen years. *Sci. Bull.* **2019**, *64*, 446–454. [CrossRef]

diversity

MDPI

Article

Identification and Characterization of the Detoxification Genes Based on the Transcriptome of *Tomicus yunnanensis*

Wen Li [1], Bin Yang [2], Naiyong Liu [2], Jiaying Zhu [2], Zongbo Li [2], Sangzi Ze [3], Jinde Yu [1] and Ning Zhao [1,2,*]

1 College of Life Sciences, Southwest Forestry University, Kunming 650224, China; liwenaya@163.com (W.L.); yujinde@swfu.edu.cn (J.Y.)
2 Key Laboratory of Forest Disaster Warning and Control of Yunnan Province, Southwest Forestry University, Kunming 650224, China; yangbin48053@swfu.edu.cn (B.Y.); naiyong.liu@swfu.edu.cn (N.L.); jyzhu@swfu.edu.cn (J.Z.); lizb@swfu.edu.cn (Z.L.)
3 Yunnan Forestry and Grassland Pest Control and Quarantine Bureau, Kunming 650051, China; zesangzi@163.com
* Correspondence: lijiangzhn@swfu.edu.cn

Abstract: Bark beetle, as a trunk borer, has caused a large number of tree deaths and seriously damaged the mountain forest ecosystem. Bark beetles oxidize the secondary metabolites of plants, degrade them, and excrete them from the body or convert them into components needed by the body. This process is completed by the cooperation of CYPs, GSTs, and CCEs and occurs in different tissues of the insects, including the gut (i.e., the part where beetle pheromone is produced and accumulated) and antennae (i.e., the olfactory organ used to sense defensive monoterpenes and other plant-related compounds and pheromones in the air). In this study, we identified and characterized three gene superfamilies of CYPs, GSTs, and CCEs involved in the detoxification of endobiotics (e.g., hormones and steroids) and xenobiotics (e.g., insecticides, sex pheromones, and plant allelochemicals) through a combination approach of bioinformatics, phylogenetics, and expression profiles. Transcriptome analyses led to the identification of 113 transcripts encoding 51 P450s, 33 GSTs, and 29 CCEs from *Tomicus yunnanensis* Kirkendall and Faccoli, 2008 (Coleoptera, Scolytinae). The P450s of *T. yunnanensis* were phylogenetically classified into four clades, representing the majority of the genes in the CYP3 clan. The CCEs from *T. yunnanensis* were separately grouped into five clades, and the GST superfamily was assigned to five clades. Expression profiles revealed that the detoxification genes were broadly expressed in various tissues as an implication of functional diversities. Our current study has complemented the resources for the detoxification genes in the family Coleoptera and allows for functional experiments to identify candidate molecular targets involved in degrading plants' secondary metabolites, providing a theoretical basis for insect resistance in mixed forests.

Keywords: cytochrome P450; glutathione S-transferase; carboxylesterases; bark beetles; gene expression

Citation: Li, W.; Yang, B.; Liu, N.; Zhu, J.; Li, Z.; Ze, S.; Yu, J.; Zhao, N. Identification and Characterization of the Detoxification Genes Based on the Transcriptome of *Tomicus yunnanensis*. *Diversity* **2022**, *14*, 23. https://doi.org/10.3390/d14010023

Academic Editor: Luc Legal

Received: 27 November 2021
Accepted: 29 December 2021
Published: 31 December 2021

1. Introduction

To withstand the ingestion of insects, plants have developed a defense mechanism against insects through various methods such as morphology, biochemistry, and molecular regulation in the long-term evolution process [1]. The production of secondary metabolites in plants is a way of defending themselves against insects. Various secondary metabolites in plants can affect the feeding and food utilization of insects, and toxic secondary metabolites can lead directly to insect death or stunted growth [2,3].

However, in the long-term evolutionary process, insects are constantly adapting to plant defense mechanisms. Genetic variation in morphology, including polymorphic wing growth, and differentiation of different types of mouthparts, such as chewing, stabbing, and sucking [4]. Insects can also use the enzymatic detoxification system, which mainly includes cytochrome P450 monooxygenases (CYPs), glutathione S-transferases (GSTs), and carboxylesterases (CCEs), to improve the ability of insects to metabolize secondary substances of plants, and cause insects to adapt to the defense mechanism of the host plant [5].

CYPs are a very large and diverse group of enzymes found in all living things. They constitute a very important system involving endogenous compounds and xenobiotics such as drugs, pesticides, secondary plant metabolites, mutagens, and hormones [6,7]. In addition, CYPs can reflect different levels of phylogenetic information such as subtypes, population, and species. It is a good molecular marker to explore the phylogenetic relationship between populations and population genetic diversity [8,9]. According to the similarity of amino acids, the CYPs of insects can be divided into four clades: CYP2, CYP3, CYP4, and mitochondrial CYP clades [10,11]. When exposed to toxic plant compounds, insects will induce reactions to other exogenous substances. For example, CYP6AE14 in *Helicoverpa armigera* Hübner, 1809 (Lepidoptera, Noctuidae) is involved in the metabolism of gossypol (plant toxins) [12]. GST is a dimeric protein, which plays an important role in intracellular metabolism, chemical cycle, and body defense [13]. Insect GST can be divided into six categories (delta, epsilon, omega, sigma, theta, and zeta). Among them, the delta and epsilon subtypes are unique to insects, and some GSTs play an important role in plant antitoxin [14,15]. In research on *Dendroctonus armandi* Tsai and Li, 1959 (Coleoptera, Scolytinae), it is assumed that DaGSTe1 catalyzes the combination of glutathione with terpenes and phenolic substances from *D. armandi*, so that toxic substances are transported out of the cell, thereby reducing the damage of exogenous toxins from the host to *D. armandi* [16]. In addition to CYPs and GSTs, another detoxification enzyme is CCE. Insect CCEs belong to the carboxyl/cholinesterase family, a branch of the α/β-hydrolase fold superfamily in which enzymes hydrolyze ester bonds, which are present in many plant volatiles, insect pheromones, and hormones, as well as pesticides [17].

Tomicus yunnanensis Kirkendall and Faccoli, 2008 (Coleoptera, Scolytinae) is one of the main pests of *Pinus yunnanensis*, which is widely distributed in Yunnan, Sichuan, and Guizhou. The whole life cycle of *T. yunnanensis* is almost entirely on *P. yunnanensis* [18]. The damage to *P. yunnanensis* by *T. yunnanensis* can be divided into two periods: feeding on treetops and trunks. When feeding on the treetops, the adult worms feed on the pith tissues of the branches and supplement nutrients to complete their sexual development; when feeding on the trunk, the mature adult worms transfer from the treetop to the trunk and feed on phloem, lay eggs, and form mother tunnel and sub tunnel. Then, larvae develop under the bark, pupate, and emerge, and adults will burrow out of the phloem and return to the treetops to feed [19,20]. This concealed lifestyle brings many inconveniences to the study of its habits, prevention, and treatment. *T. yunnanensis* first broke out in central Yunnan Province in the 1980s, and then spread to 65 counties in 15 regions of Yunnan Province, infesting more than 200,000 ha of Yunnan pine forests and blighting more than 93,000 ha of *P. yunnanensis*, and seriously damaged the mountain forest ecosystem [21]. To mitigate the damage of this insect, some common measures can play a certain role for a short time, such as biological control, chemical control and tending, and mixed forest [22]. Among them, the mixed forest has achieved remarkable results in insect resistance and contributed to the protection of mountain forest ecosystem, but its anti-insect mechanism still needs further research. The identification of the detoxification genes of *T. yunnanensis* enriches the detoxification gene family in Coleoptera and provides data support for future research on the function of detoxification genes. We hope that the study can provide a theoretical basis for insect resistance in mixed forests.

2. Materials and Methods

2.1. Insect and Tissue Collection

The adults of *T. yunnanensis* used in this experiment were originally collected from Jiulong Mountain Forest Farm in Zhanyi County, Qujing City, Yunnan Province in June 2018. The antennae, heads, legs, and carcasses (excluding antennae, heads, and legs) of 200 pairs of male and female adults of *T. yunnanensis* were collected separately (three groups of biological replicates). We quickly froze the collected samples with liquid nitrogen and stored them in a refrigerator at −80 °C. qPCR samples were treated the same way.

2.2. Total RNA Extraction and cDNA Synthesis

Total RNA samples of tissues were isolated using TRIzol Reagent according to the manufacturer's protocol (Ambion, Life Technologies, Carlsbad, CA, USA). The quality of RNA was confirmed using a NanoVue UV–vis spectrophotometer (Thermo Fisher Scientific, Waltham, MA, USA), and RNA integrity was verified using a standard 1% agarose gel electrophoresis. Genomic DNA was digested by treatment with DNase I (Fermentas, Thermo Fisher Scientific, USA). First-strand cDNA was synthesized with a first strand cDNA synthesis kit (TaKaRa, Dalian, Liaoning, China). The synthesized cDNA templates were stored at −20 °C.

2.3. Library Construction, Sequencing, and Functional Annotation

Oligosaccharide (DT)-containing magnetic beads were used to enrich the mRNAs in the total RNA of antennae, head, foot, and residue. Then, according to the protocol in NEBNext® Ultra™ RNA Library Prep Kit for llumina® (NEB, Ipswich, MA, USA), we cut them into short fragments with a fragment buffer. The first-strand cDNA was synthesized using this fragment as a template, and then the second-strand cDNA was synthesized by DNA polymerase I and RNaseH. After purification with AMPure XP beads, the second-strand cDNA was repaired, α-end ligated, and ligated with indexed adapters. The products of suitable size were selected, amplified by PCR, and purified with AMPure XP beads to establish a digital gene expression (DEG) library. Raw reads were processed through a rigorous filtering process to eliminate low-quality reads (base calling). The clean reads were then assembled using Trinity (v2.4.0) and then grouped using Corset (v1.0.5) to eliminate redundant data in the assembled transcripts. [23]. Databases used for annotation included the non-redundant nucleotide (Nt), non-redundant database (Nr), Swiss-Prot, Protein family (Pfam), Kyoto Encyclopedia of Genes and Genomes (KEGG) and Clusters of Orthologous Groups (COG/KOG). Gene ontology (GO) analysis was performed using blast2go (b2g4pipe_v2.5) software. Nr, Swiss-Prot, and COG/KOG were analyzed by diamond v0.8.22 software, and Nt was analyzed by NCBI blast v2.2.28+; KEGG was analyzed with KAAS r140224 and Pfam was analyzed with hmmscan HMMER 3. For the DEG analysis, the resulting clean reads were then mapped to the transcriptomic unigenes using RSEM software with default parameters [24]. The differential expression of different conditions/groups (genes and samples) was analyzed by DESeq2 software. The number of read counts normalized by TMM for each mapped gene was used to calculate gene expression levels following the FPKM (fragments per kilobase of transcript sequence per millions base) method [25,26].

2.4. Gene Identification and Sequence Analysis

To identify candidate detoxification genes from *T. yunnanensis*, detoxification gene families from other coleopteran species were selected as queries to search the new stand-alone transcriptome of this beetle. *Anoplophora glabripennis* Motschulsky, 1854 (Coleoptera, Laminae), *Lepeophtheirus salmonis* Kroyer, 1838 (Copepoda, Caligoida), *Dendroctonus ponderosne* Hopkins, 1902 (Coleoptera, Scolytidae), and *Tribolium castaneum* Herbst, 1797 (Coleoptera, Tenebrionidae) were used for CYPs; *D. ponderosae*, *A. glabripennis*, *Rhynchophorus ferrugineus* Oliver, 1790 (Coleoptera: Curculionidae), *D. armandi* and *T. castaneum* for GSTs, and *D. ponderosae*, *D. armandi*, and *T. castaneum* for CCEs (https://www.ncbi.nlm.nih.gov/ (accessed on 7 June 2021)). TBLASTN was used to search and identify candidate detoxification genes against the *T. yunnanensis* transcriptome, with an E-value cutoff of 1×10^{-5}. Further, these identified genes were verified using TBLASTX against the NCBI non-redundant protein sequences database. Open reading frames (ORFs) were identified using the ORF Finder in NCBI (https://www.ncbi.nlm.nih.gov/orffinder/ (accessed on 29 June 2021)). CYP names use the CYP prefix, followed by an Arabic numeral, designates the family (all members nominally >40% identical), a capital letter designates the subfamily (all members nominally >55% identical), and an Arabic numeral designates the individual gene or message and protein [27]. In the set of trees, a multiple sequence alignment was performed using the Muscle method in MEGA7.0 [28]. An ML tree of candidate detoxification genes was constructed by Evoliview [29]. Accession numbers of all protein sequences from other Coleoptera species used in the phylogenetic analysis are listed in Supplementary Materials.

2.5. Quantitative Real-Time PCR

We used qRT-PCR to verify the expression of randomly selected candidate *T. yunnanensis* detoxification genes in different tissues. The cDNA was synthesized with an input of 1 μg of total RNA using the Prime ScriptRT Reagent Kit with gDNA Eraser to remove gDNA (AG, Changsha, China). For quantitative real-time PCR (qPCR), various tissues including antennae, heads (without antennae), legs, and carcasses from female and male adults were collected and immediately immersed in liquid nitrogen. qPCR was performed using SYBR Premix EX Taq™ (AG, Changsha, China) with three technical replicates of each template from three independent biological pools. For the qPCR analysis, the primers (Table S1) were designed by Primer Premier 5.0 [30]. Each reaction contained a total volume of 20 μL, consisting of 10 μL of SYBR Green PCR Master Mix, 0.8 μL of each primer (10 μM), 2 μL (20 ng) of cDNA template, and 6.4 μL of nuclease-free water. The β-actin gene was used as an endogenous control. qPCR cycling parameters were: 94 °C for 4 min followed by 40 cycles at 94 °C for 20 s and 60 °C for 30 s. Relative gene expression level was calculated using the Q-GENE statistical analysis package [31].

3. Results

3.1. Transcriptome Assembly

An Illumina HiSeq platform was used to sequence antennae, head, legs, and carcasses of adult female and male *T. yunnanensis* transcriptomes. We obtained 61.51 million (MA-1), 47.40 million (MA-2), and 52.92 million (MA-3) raw reads from the antennae of a male adult, 58.15 million (MH-1), 50.67 million (MH-2), and 61.09 million (MH-3) raw-reads from the head of a male adult, and there are raw readings from other organizations in Table 1. Filtering resulted in 58.28 million (MA-1), 46.26 million (MA-2), 51.60 million (MA-3), 55.80 million (MH-1), 49.30 million (MH-2), 59.09 million (MH-3), 55.93 million (ML-1), 53.53 million (ML-2), and 56.48 million (ML-3) clean reads (Table 1). The percentages of reads with Q20 and Q30 values for each library were approximately 98% and 93%, respectively. The GC content ranged from 38.55% to 52.60% (Table 1). The final transcript dataset contained 100,455 unigenes with a mean length of 877 bp and an N50 length of 1153 bp (Table 2), indicating the high quality of our assembly.

3.2. Functional Annotation of the Unigenes in T. yunnanensis

A total of 100,455 unigenes were annotated by searching against six databases using BLAST. Specifically, 60,067 (59.79%) unigenes were matched in the non-redundant (NR) database, which accounted for the largest match. Further, 54,782 (54.53%) were annotated in the nucleotide (NT) database, 50,102 (49.87%) were annotated by blasting against the Swiss Port database. Additionally, the euKaryotic Ortholog Groups (KOG) database had the lowest number of annotated unigenes, with 19,808 (19.71%) unigenes (Table 3). Species with the highest proportion of similar genes were *Nephila clavipes* (21.00%), followed by *Dendroctonus ponderosae* (19.2%), *Hyalella azteca* (8.0%), *Toxocara canis* (4.2%) and *Lucilia cuprina* (2.1%) (Figure 1).

Table 1. Summary of the transcriptome sequencing data from *T. yunnanensis*.

Sample	Raw Reads	Clean Reads	Q20 (%)	Q30 (%)	GC (%)
MA_1	615,114,60	582,887,10	97.95	94.15	47.57
MA_2	474,005,30	462,665,14	97.90	93.91	42.76
MA_3	529,246,96	516,082,96	98.06	94.27	42.81
MH_1	581,529,36	558,021,48	97.75	93.46	38.56
MH_2	506,735,06	493,021,74	97.96	94.00	38.74
MH_3	610,987,72	590,914,14	98.04	94.16	40.09
ML_1	582,201,60	559,355,36	98.04	94.29	45.54
ML_2	561,873,52	535,354,48	98.05	94.30	43.28
ML_3	587,793,32	564,878,02	97.88	93.93	42.65
MC_1	640,881,58	628,616,66	97.63	93.21	38.55
MC_2	651,352,06	633,407,70	97.87	93.79	39.15
MC_3	591,512,94	566,883,04	97.10	92.16	41.95
FA_1	523,288,02	510,649,40	98.11	94.44	46.17
FA_2	568,975,58	533,953,70	98.00	94.23	46.38
FA_3	535,300,88	500,120,04	98.00	94.24	46.14
FH_1	513,240,72	502,592,66	97.95	93.94	39.67
FH_2	558,425,54	547,795,96	97.81	93.65	39.48
FH_3	640,915,78	629,613,86	97.62	93.18	39.24
FL_1	586,049,54	572,336,12	97.83	93.72	43.91
FL_2	503,729,98	491,791,08	97.82	93.72	42.40
FL_3	435,955,04	420,544,90	97.94	94.19	52.60
FC_1	572,708,34	560,824,28	97.67	93.30	40.48
FC_2	533,416,24	517,611,84	97.87	93.78	41.24
FC_3	505,727,10	495,473,28	97.77	93.55	39.88

MA: male antennae, MH: male head, ML: male leg, MC: male carcasses, FA: female antennae, FH: female head, FL: female leg, FC: female carcasses.

Table 2. Summary of transcriptome splicing length distribution data from *T. yunnanensis*.

	Min Length	Mean Length	Max Length	N50	Total Nucleotides
Transcripts	154	1005	59,000	1465	184,171,785
Genes	301	877	59,000	1153	880,711,34

Table 3. Statistics of gene annotation success rate.

	Number of Genes	Percentage (%)
Annotated in NR	60,067	59.79
Annotated in NT	54,782	54.53
Annotated in SwissProt	50,102	49.87
Annotated in PFAM	53,205	52.96
Annotated in GO	53,205	52.96
Annotated in KOG	19,808	19.71
Annotated in all databases	1666	1.65
Annotated in at least one database	79,508	79.14
Total unigenes	100,455	100

Using a gene ontology (go) method, annotated genes were divided into three categories (a total of 56 functional groups): biological process, cell composition, and molecular function. Among the biological processes, the subcategories cellular process, metabolic process, and single-organism process contained the most unigenes. In the cellular component class, the subcategories cell and cell part contained the most unigenes. Binding and catalytic activity were the most numerous subcategories in the "molecular function" category (Figure 2a). For the euKaryotic Ortholog Groups (KOG) functional classification, we annotated about 19,808 unigenes and divided them into 25 molecular families (Figure 2b). Among them, the largest category was the general function prediction only, followed by amino acid transport and metabolism and energy production and conversion. Cell motility and nuclear structure were the smallest groups (Figure 2b). A KEGG analysis was used to classify the annotated genes into different KEGG pathway functional categories (Figure 2c). The most representative pathways were amino acid metabolism, carbohydrate metabolism, overview, and signal transduction (Figure 2c).

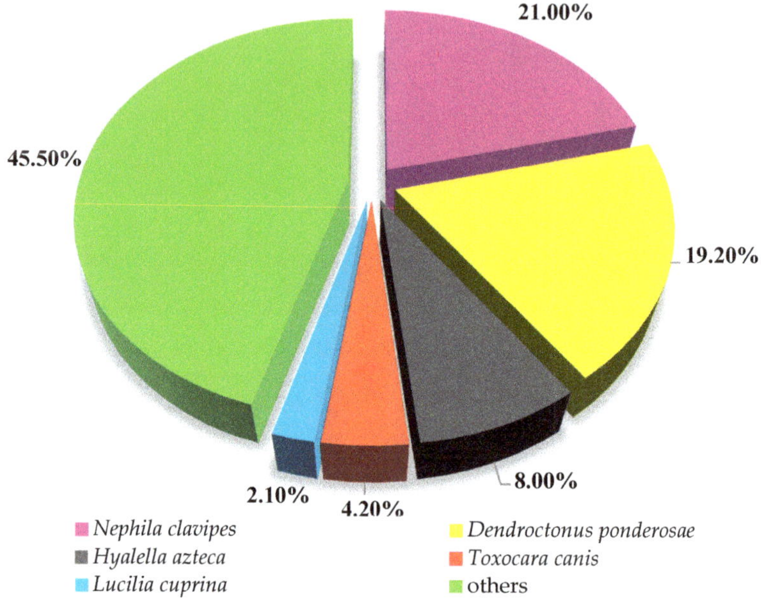

Figure 1. The unigene BLASTx searches against the Nr database for species distribution analysis.

3.3. Identification of Candidate CYPs, GSTs, and CCEs

In this study, a total of 51 predicted CYPs transcripts were identified from the transcriptomes of different issues of *T. yunnanensis* using the BLASTx program. The sequence identities of these candidate CYPs with other Coleoptera insects ranged from 32.69% to 96.54% in the NCBI database (Table 4). According to the CYP nomenclature, we classified 51 CYP sequences into four families (CYP2, CYP3, CYP4, mitochondrial CYP). After removing 25 short sequences (aa < 390) [27], 26 sequences with complete open reading frames were used to construct a phylogenetic tree (Figure 3). The phylogenetic tree showed that TyunCYP had a high homology with DponCYP. The largest family was the CYP3 family, which included 13 members and the CYP3 family contained two subfamilies, CYP6 and CYP9. The results of the phylogenetic tree show that five CYPs (TyunCYP6BW2, TyunCYP6DE1, TyunCYP6DF1, TyunCYP6BX1, TyunCYP6DJ1) belong to the CYP6 subfamily. The second largest family is the CYP4 family; we identified eight genes (TyunCYP393A1, TyunCYP6BK1, TyunCYP411A1, TyunCYP349B2, TyunCYP4CV1, TyunCYP4BQ1, TyunCYP4G2, TyunCYP4BG1) belonging to the CYP4 family.

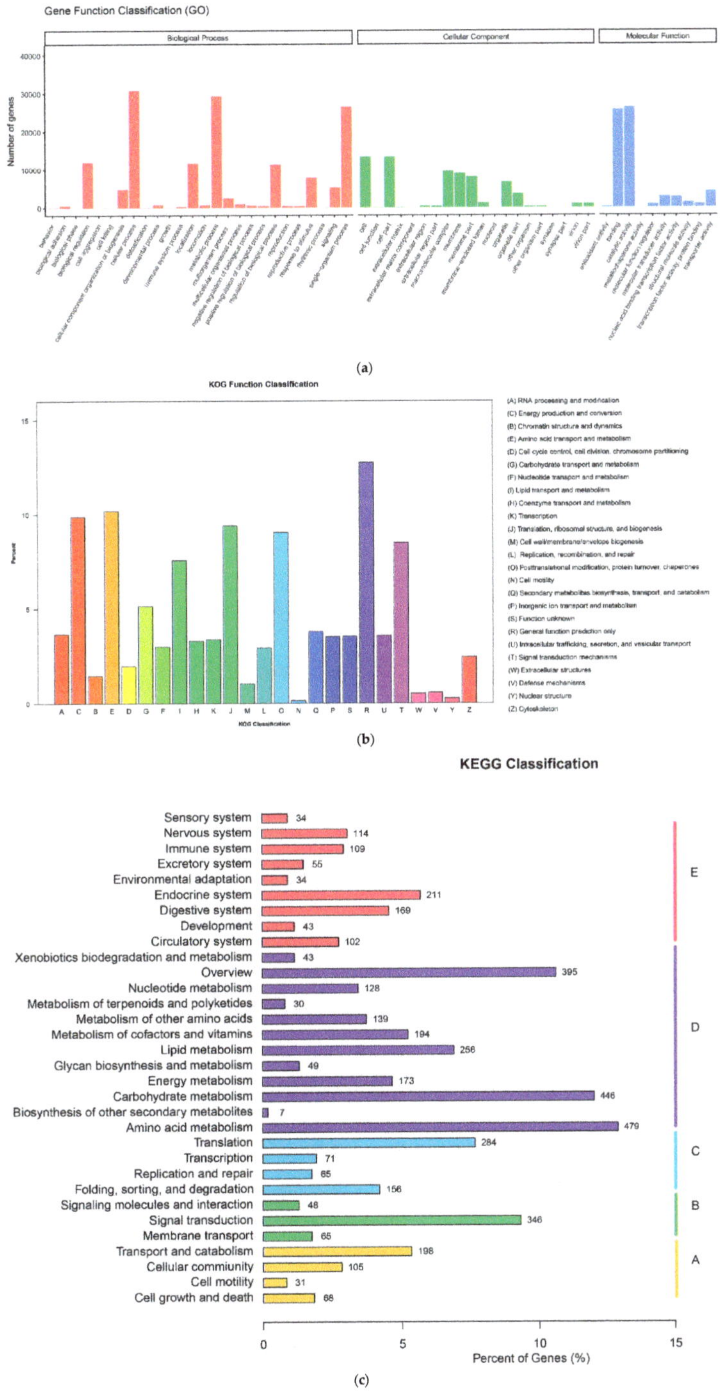

Figure 2. Results of BLASTx matches of *T. yunnanensis* transcriptome unigenes, gene ontology, KOG classification, and KEGG pathway annotation. (**a**) Gene ontology classifications of *T. yunnanensis* unigenes. (**b**) KOG classifications of *T. yunnanensis* unigenes. (**c**) KEGG classification of *T. yunnanensis* unigenes.

Table 4. Best BLASTX matches of *T. yunnanensis* CYPs.

Gene Name	Gene ID	aa	E-Value	Identity	Accession	Species Name
TyunCYP349A1	i2_LQ_TYUN_c39805/f1p0/2021	60	1×10^{-20}	65.00%	AEL88544.1	*Dendroctonus rhizophagus*
TyunCYP9Z1	i2_LQ_TYUN_c22584/f1p0/2717	145	5×10^{-82}	82.64%	AEL88550.1	*Dendroctonus rhizophagus*
TyunCYP314A1	i1_LQ_TYUN_c152368/f1p0/1050	147	6×10^{-78}	83.33%	AFI45004.1	*Dendroctonus ponderosae*
TyunCYP347B1	i4_LQ_TYUN_c2131/f1p0/4666	447	0	81.94%	AFI45010.1	*Dendroctonus ponderosae*
TyunCYP410C1	Cluster-27834.11928	456	1×10^{-107}	39.43%	AFI45013.1	*Dendroctonus ponderosae*
TyunCYP4CV1	Cluster-27834.3883	507	0	77.47%	AFI45020.1	*Dendroctonus ponderosae*
TyunCYP6BW1	i3_LQ_TYUN_c10935/f1p0/4006	283	9×10^{-180}	85.16%	AFI45024.1	*Dendroctonus ponderosae*
TyunCYP6BW2	i1_HQ_TYUN_c416/f4p1/1768	506	0	86.76%	AFI45026.1	*Dendroctonus ponderosae*
TyunCYP6BX1	Cluster-27834.4019	478	0	71.43%	AFI45028.1	*Dendroctonus ponderosae*
TyunCYP6DJ1	i1_LQ_TYUN_c33388/f1p0/1763	507	0	70.71%	AFI45041.1	*Dendroctonus ponderosae*
TyunCYP9Z4	i1_LQ_TYUN_c78340/f1p0/1911	526	0	84.22%	AFI45045.1	*Dendroctonus ponderosae*
TyunCYP6DJ2	i2_LQ_TYUN_c23405/f1p0/2443	66	6×10^{-7}	67.65%	AGF69211.1	*Dendroctonus valens*
TyunCYP4G1	i3_LQ_TYUN_c4999/f1p0/3248	242	5×10^{-158}	93.80%	ALD15896.1	*Dendroctonus armandi*
TyunCYP305F1	Cluster-27834.3827	491	0	87.53%	ALD15904.1	*Dendroctonus armandi*
TyunCYP352B1	i2_LQ_TYUN_c27719/f1p0/2254	225	4×10^{-74}	69.36%	ALD15909.1	*Dendroctonus armandi*
TyunCYP4BD4V	i1_LQ_TYUN_c15852/f1p0/1569	166	8×10^{-68}	67.31%	ALD15912.1	*Dendroctonus ponderosae*
TyunCYP6DF1	Cluster-27834.8537	505	0	72.31%	ALD15922.1	*Dendroctonus armandi*
TyunCYP9Z2	i2_LQ_TYUN_c15205/f1p0/2172	575	0	81.30%	ALD15924.1	*Dendroctonus armandi*
TyunCYP434A1	Cluster-27834.7270	371	7×10^{-147}	71.53%	XP_019754109.1	*Dendroctonus ponderosae*
TyunCYP4BQ1	Cluster-27834.7070	397	0	67.84%	XP_019754731.1	*Dendroctonus ponderosae*
TyunCYP6DE1	i1_HQ_TYUN_c29325/f3p0/1840	509	0	76.23%	XP_019755300.1	*Dendroctonus ponderosae*
TyunCYP315A1	Cluster-27834.2194	292	8×10^{-139}	72.26%	XP_019755328.1	*Dendroctonus ponderosae*
TyunCYP315A2	Cluster-27834.2193	208	2×10^{-20}	70.15%	XP_019755336.1	*Dendroctonus ponderosae*
TyunCYP4G2	i1_HQ_TYUN_c24270/f27p4/1844	563	0	92.11%	XP_019755432.1	*Dendroctonus ponderosae*
TyunCYP4BG1	Cluster-27834.11078	501	0	78.09%	XP_019756327.1	*Dendroctonus ponderosae*
TyunCYP410A1	i1_LQ_TYUN_c6617/f1p1/1851	273	5×10^{-82}	53.52%	XP_019756499.1	*Dendroctonus ponderosae*
TyunCYP411A1	i2_LQ_TYUN_c22094/f1p0/2786	490	0	76.22%	XP_019758316.1	*Dendroctonus ponderosae*
TyunCYP411A2	i2_LQ_TYUN_c11742/f1p1/2726	185	6×10^{-77}	71.43%	XP_019758318.1	*Dendroctonus ponderosae*
TyunCYP345F1	Cluster-26110.0	500	0	82.20%	XP_019759785.1	*Dendroctonus ponderosae*
TyunCYP307B1	Cluster-27834.15214	399	1×10^{-153}	77.06%	XP_019760073.1	*Dendroctonus ponderosae*
TyunCYP307A1	i3_LQ_TYUN_c19449/f1p0/3461	492	0	96.54%	XP_019760634.1	*Dendroctonus ponderosae*
TyunCYP393A2	i0_LQ_TYUN_c4806/f1p0/966	112	5×10^{-17}	71.93%	XP_019760717.1	*Dendroctonus ponderosae*
TyunCYP9AP1	i1_LQ_TYUN_c81629/f1p2/1918	483	0	72.57%	XP_019760793.1	*Dendroctonus ponderosae*
TyunCYP9AN1	i2_LQ_TYUN_c15846/f1p0/2129	523	0	79.31%	XP_019761014.1	*Dendroctonus ponderosae*
TyunCYP433A1	i1_LQ_TYUN_c80202/f1p0/1820	274	2×10^{-128}	70.45%	XP_019761671.1	*Dendroctonus ponderosae*
TyunCYP9Z3	i2_LQ_TYUN_c27749/f1p0/2479	151	2×10^{-44}	73.33%	XP_019764934.1	*Dendroctonus ponderosae*
TyunCYP345E1	i1_HQ_TYUN_c62660/f2p0/1795	304	4×10^{-159}	72.33%	XP_019765603.1	*Dendroctonus ponderosae*
TyunCYP347D1	Cluster-20450.0	498	0	78.93%	XP_019767652.1	*Dendroctonus ponderosae*
TyunCYP303A1	Cluster-27834.21448	484	0	76.80%	XP_019768066.1	*Dendroctonus ponderosae*
TyunCYP18A1	i2_LQ_TYUN_c27030/f1p0/2265	159	8×10^{-82}	79.87%	XP_019768143.1	*Dendroctonus ponderosae*
TyunCYP18A2	i2_HQ_TYUN_c25448/f2p0/2291	525	0	90.50%	XP_019768147.1	*Dendroctonus ponderosae*
TyunCYP393A1	Cluster-27834.3112	461	0	69.78%	XP_019768810.1	*Dendroctonus ponderosae*
TyunCYP334E1	Cluster-27834.2023	583	0	85.74%	XP_019770320.1	*Dendroctonus ponderosae*
TyunCYP347E1	i2_LQ_TYUN_c3851/f1p0/2279	500	0	79.00%	XP_019771168.1	*Dendroctonus armandi*
TyunCYP6DG3	i3_LQ_TYUN_c17722/f1p0/3197	354	0	77.68%	XP_019771861.1	*Dendroctonus ponderosae*
TyunCYP9AZ1	i2_LQ_TYUN_c28708/f1p0/2957	134	2×10^{-70}	79.85%	XP_019773084.1	*Dendroctonus ponderosae*
TyunCYP334E2	i2_LQ_TYUN_c27105/f1p0/2563	498	0	83.77%	XP_019773300.1	*Dendroctonus ponderosae*
TyunCYP349B2	Cluster-40684.0	420	0	82.86%	XP_019773560.1	*Dendroctonus ponderosae*
TyunCYP434A2	i1_HQ_TYUN_c939/f2p0/1483	360	5×10^{-124}	50.00%	XP_019869691.1	*Aethina tumida*
TyunCYP6BW3	i2_LQ_TYUN_c30164/f1p0/2137	382	2×10^{-121}	48.28%	XP_023014703.1	*Leptinotarsa decemlineata*
TyunCYP410C1	i0_LQ_TYUN_c9810/f1p0/524	150	8×10^{-12}	32.69%	XP_030755100.1	*Sitophilus oryzae*

A total of 33 candidate GSTs were identified from the total transcriptome of the different developmental stages of *T. yunnanensis*. The sequence identities of these candidate GSTs with other dipteran insects ranged from 51.96% to 97.14% in the NCBI database (Table 5). After removing short sequences (aa < 150) [32], 18 GSTs were chosen for the phylogenetic analysis. A neighbor-joining tree was subsequently constructed using our identified putative GST proteins and the sequences from four other Coleoptera species, *D. ponderosae*, *T. castaneum*, *D. valens*, and *A. planipennis* (Figure 4). Ten GSTs were identified in the transcriptome of *T. yunnanensis*, including six delta/epsilon class GSTs, four omega class GST, five sigma class GSTs, two theta class GSTs, and one microsomal class GST. The phylogenetic tree showed that TyunGST had a high homology with DarmGST.

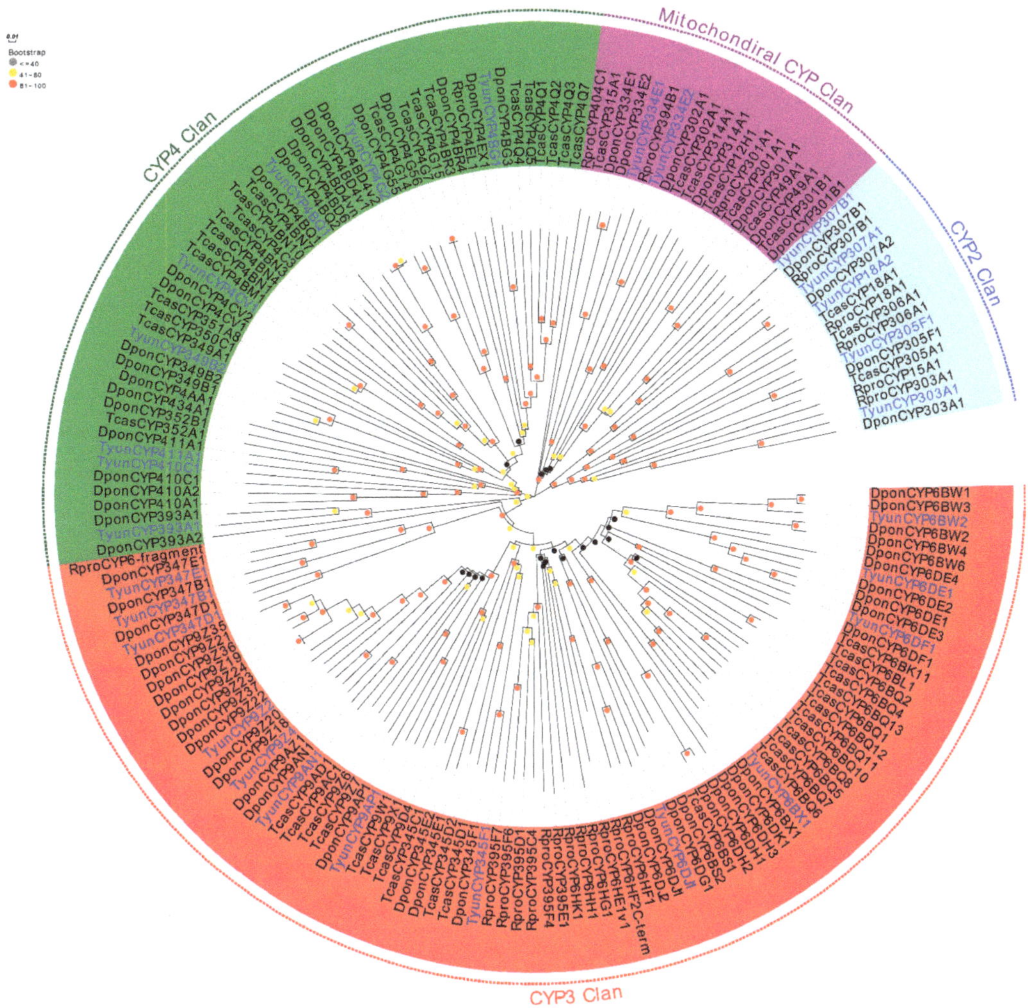

Figure 3. Neighbor-joining tree of candidate CYPs. Bootstrap values after 1000 replications. Dpon, *Dendroctonus ponderosae*; Rpro, *Rhodnius prolixus*; Tcas, *Tribolium castaneum*; Tyun, *Tomicus yunnanensis*.

We identified 29 transcripts encoding CCEs in the *T. yunnanensis* transcriptome by a bioinformatics analysis and all of them were more conserved across other CCEs variants, with 42.86 to 93.03% amino acid identity (Table 6). To guarantee the reliability of the phylogenetic tree, 13 CCEs encoding short sequences wereremoved (aa < 480) [33], and 16 CCEs in our transcriptomes were aligned with CCEs from other coleoptera species (Figure 5). A total of 16 CCE genes were identified in the transcriptome of *T. yunnanensis*, including nine xenobiotic metabolizing enzymes class CCE, three microsomal and alpha-esterases class CCE, two beta- and pheromone esterases class CCE, one JHE class CCE, and one CO class CCE.

Table 5. Best BLASTX matches of *T. yunnanensis* GSTs.

Gene Name	Gene ID	aa	E-Value	Identity	Accession	Species Name
TyunGST1	i2_LQ_TYUN_c36781/f1p0/2077	217	2×10^{-145}	91.71%	AIC76455.1	*Dendroctonus armandi*
TyunGST2	Cluster-27834.9472	70	6×10^{-41}	97.14%	AJE61306.1	*Dendroctonus armandi*
TyunGST3	i0_LQ_TYUN_c12133/f1p0/748	141	3×10^{-48}	57.25%	AJE61307.1	*Dendroctonus armandi*
TyunGST4	i2_LQ_TYUN_c18657/f1p0/2385	112	3×10^{62}	76.74%	AJE61308.1	*Dendroctonus armandi*
TyunGST5	i3_LQ_TYUN_c11135/f1p0/3582	79	5×10^{-33}	82.35%	AJE61309.1	*Dendroctonus armandi*
TyunGST6	Cluster-27834.6624	230	2×10^{-146}	83.48%	AJE61311.1	*Dendroctonus armandi*
TyunGST7	Cluster-27834.8595	244	8×10^{-160}	88.52%	AJE61312.1	*Dendroctonus armandi*
TyunGST8	i1_LQ_TYUN_c149780/f1p1/1152	245	2×10^{-150}	82.04%	AVR54955.1	*Sitophilus oryzae*
TyunGST9	i3_LQ_TYUN_c12086/f1p0/3529	218	2×10^{-142}	88.58%	AVR54957.1	*Sitophilus oryzae*
TyunGST10	i2_LQ_TYUN_c14505/f1p9/2537	52	6×10^{-15}	71.43%	AVR54966.1	*Sitophilus oryzae*
TyunGST11	Cluster-27834.9551	105	9×10^{-41}	64.42%	AVT42177.1	*Lissorhoptrus oryzophilus*
TyunGST12	i0_LQ_TYUN_c7185/f1p0/508	148	6×10^{-48}	62.50%	AVT42178.1	*Lissorhoptrus oryzophilus*
TyunGST13	i2_LQ_TYUN_c7650/f1p0/2527	118	3×10^{-64}	87.85%	AVT42182.1	*Lissorhoptrus oryzophilus*
TyunGST14	Cluster-27834.21190	219	3×10^{-69}	51.96%	AVT42185.1	*Lissorhoptrus oryzophilus*
TyunGST15	i2_LQ_TYUN_c8057/f1p0/2856	294	7×10^{-150}	67.12%	AVT42197.1	*Lissorhoptrus oryzophilus*
TyunGST16	Cluster-27834.9541	219	1×10^{-144}	89.50%	QFU14637.1	*Dendroctonus armandi*
TyunGST17	i4_LQ_TYUN_c8880/f1p0/4497	137	3×10^{-69}	93.69%	QFU14640.1	*Dendroctonus armandi*
TyunGST18	Cluster-1630.0	205	1×10^{-76}	56.78%	QFU14643.1	*Dendroctonus armandi*
TyunGST19	i2_LQ_TYUN_c12200/f1p0/2343	231	3×10^{-137}	77.92%	QFU14646.1	*Dendroctonus armandi*
TyunGST20	i3_LQ_TYUN_c2494/f1p0/3326	218	4×10^{-146}	90.41%	XP_019753833.1	*Dendroctonus ponderosae*
TyunGST21	i2_LQ_TYUN_c34577/f1p0/2723	78	7×10^{-41}	83.54%	XP_019753835.1	*Dendroctonus ponderosae*
TyunGST22	i2_LQ_TYUN_c23261/f1p0/2987	124	2×10^{-73}	93.86%	XP_019755130.1	*Dendroctonus ponderosae*
TyunGST23	i2_LQ_TYUN_c24320/f1p0/2660	150	2×10^{-35}	84.72%	XP_019755704.1	*Dendroctonus ponderosae*
TyunGST24	i7_LQ_TYUN_c267/f1p0/7646	72	7×10^{-22}	67.21%	XP_019755792.1	*Dendroctonus ponderosae*
TyunGST25	i3_LQ_TYUN_c4401/f1p0/3564	111	4×10^{-53}	86.27%	XP_019755793.1	*Dendroctonus ponderosae*
TyunGST26	i3_LQ_TYUN_c26609/f1p0/3037	203	3×10^{107}	72.00%	XP_019755830.1	*Dendroctonus ponderosae*
TyunGST27	i0_LQ_TYUN_c24616/f1p0/906	205	1×10^{-129}	85.37%	XP_019760728.1	*Dendroctonus ponderosae*
TyunGST28	i2_LQ_TYUN_c11257/f1p0/2424	559	0	75.62%	XP_019764595.1	*Dendroctonus ponderosae*
TyunGST29	Cluster-27834.9273	224	4×10^{-124}	73.66%	XP_019766435.1	*Dendroctonus ponderosae*
TyunGST30	i1_LQ_TYUN_c146891/f1p0/1112	237	3×10^{-158}	89.83%	XP_019768505.1	*Dendroctonus ponderosae*
TyunGST31	i0_LQ_TYUN_c13481/f1p0/385	70	4×10^{-32}	82.86%	XP_019770972.1	*Dendroctonus ponderosae*
TyunGST32	i1_LQ_TYUN_c153362/f1p1/1072	206	3×10^{-77}	57.28%	XP_023313068.1	*Anoplophora glabripennis*
TyunGST33	i1_LQ_TYUN_c150961/f1p1/1003	206	2×10^{-78}	58.54%	XP_023313069.1	*Anoplophora glabripennis*

3.4. Tissue Expression Profile of the CYP, GST, and CCE Genes

In order to explore the expression profiles of detoxification genes in different tissues, we screened all 113 genes from the DEG library. In total, 64 genes were found, and 49 genes were missing from the DEG data. The expression profiles based on FPKM values revealed that several detoxification genes (TyunCYP410A1, TyunCYP9AN1, TyunCYP393A1, TyunGST22, TyunGST29, TyunGST31, TyunCCE8, and TyunCCE17) are highly expressed in the antenna (FPKM > 100) and the expression levels of these genes are all higher in male antennae than in female antennae (Figure 6). TyunCYP4G2 is highly expressed in the residue. Tyun315A1, TyunCYP6BK1, TyunCYP6DF1, and TyunGST3 are expressed in all tissues; TyunCYP6DF1 expression in antennae is higher than in other tissues. TyunCYP345E1 and TyunCYP345F1 are little expressed in all tissues (Figure 6a). TyunGST11 and TyunGST33 are expressed in all tissues, the expression of TyunGST11 in foot and antennae is higher than other tissues, and the expression level of male feet is higher than that of female legs. The antenna expression of TyunGST33 is significantly higher than that of other tissues and the male antenna expression is higher than that of the female (Figure 6b). TyunCCE21 is expressed in all tissues. TyunCCE16 is strongly expressed in the head and carcasses, and expression in these two parts of the females is higher than that of the males (Figure 6c).

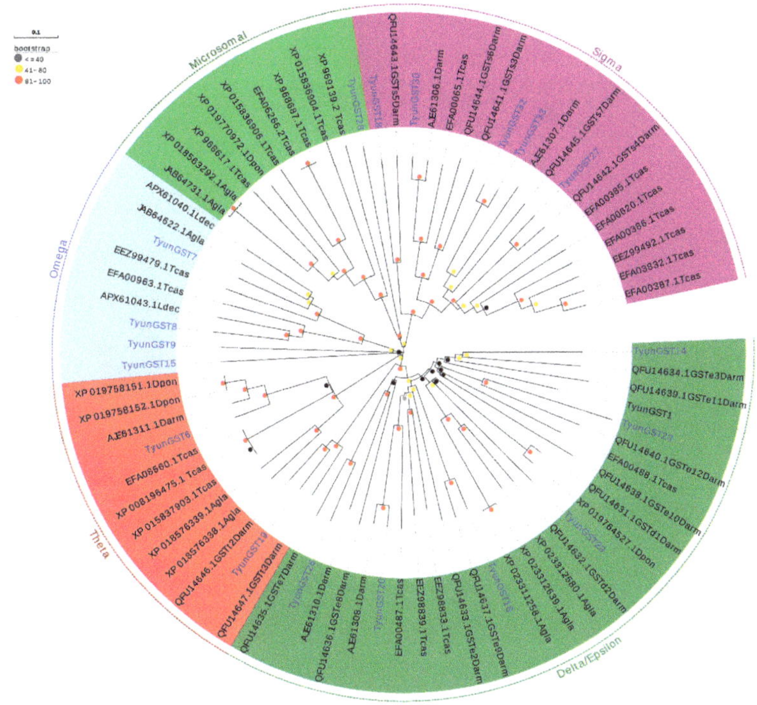

Figure 4. Neighbor-joining tree of candidate GSTs. Bootstrap values after 1000 replications. Dpon, *Dendroctonus ponderosae;* Darm, *Dendroctonus armandi;* Agla, *Anoplophora glabripennis;* Ldec *Leptinotarsa decemlineata*; Tcas, *Tribolium castaneum*; Tyun, *Tomicus yunnanensis*.

Table 6. Best BLASTX matches of *T. yunnanensis* CCEs.

Gene Name		aa	E-Value	Identity	Accession	Species Name
TyunCCE1	i2_LQ_TYUN_c19726/f1p0/2240	580	0	79.41%	AYN64423.1	*Dendroctonus armandi*
TyunCCE2	i1_LQ_TYUN_c11797/f1p0/1951	405	0	93.03%	AYN64424.1	*Dendroctonus armandi*
TyunCCE3	Cluster-27834.6014	133	3×10^{-43}	61.42%	AYN64428.1	*Dendroctonus armandi*
TyunCCE4	Cluster-27834.16668	243	5×10^{-156}	88.48%	AYN64429.1	*Dendroctonus armandi*
TyunCCE5	Cluster-27834.3878	550	5×10^{-164}	45.34%	XP_019754206.1	*Dendroctonus ponderosae*
TyunCCE6	Cluster-27834.11182	570	0	81.05%	XP_019754592.1	*Dendroctonus ponderosae*
TyunCCE7	Cluster-27834.18917	577	0	87.19%	XP_019755320.1	*Dendroctonus ponderosae*
TyunCCE8	i1_LQ_TYUN_c11402/f1p0/1940	563	0	58.70%	XP_019755320.1	*Dendroctonus ponderosae*
TyunCCE9	Cluster-27834.6015	557	0	64.16%	XP_019755963.1	*Dendroctonus ponderosae*
TyunCCE10	i1_HQ_TYUN_c4678/f4p0/1715	558	0	82.08%	XP_019756055.1	*Dendroctonus ponderosae*
TyunCCE11	i1_LQ_TYUN_c2026/f1p0/1764	559	0	85.15%	XP_019756056.1	*Dendroctonus ponderosae*
TyunCCE12	Cluster-27834.3444	122	5×10^{-75}	90.98%	XP_019756397.1	*Dendroctonus ponderosae*
TyunCCE13	Cluster-27834.5714	268	2×10^{-169}	88.01%	XP_019756739.1	*Dendroctonus ponderosae*
TyunCCE14	i1_LQ_TYUN_c29743/f1p0/1703	545	0	50.54%	XP_019757154.1	*Dendroctonus ponderosae*
TyunCCE15	i1_LQ_TYUN_c35851/f1p0/1947	228	8×10^{-114}	75.60%	XP_019758307.1	*Dendroctonus ponderosae*
TyunCCE16	i1_LQ_TYUN_c30292/f1p1/1995	566	0	88.18%	XP_019761960.1	*Dendroctonus ponderosae*
TyunCCE17	i1_LQ_TYUN_c16028/f1p0/1766	347	0	81.23%	XP_019764466.1	*Dendroctonus ponderosae*
TyunCCE18	i1_LQ_TYUN_c9615/f1p0/1928	532	0	76.15%	XP_019765893.1	*Dendroctonus ponderosae*
TyunCCE19	i1_LQ_TYUN_c37855/f1p0/1526	414	0	83.09%	XP_019766062.1	*Dendroctonus ponderosae*
TyunCCE20	Cluster-27834.3692	531	0	85.50%	XP_019766553.1	*Dendroctonus ponderosae*
TyunCCE21	Cluster-27834.10155	569	0	86.12%	XP_019769801.1	*Dendroctonus ponderosae*
TyunCCE22	i3_LQ_TYUN_c17625/f1p0/3579	266	2×10^{-136}	89.52%	XP_019769801.1	*Dendroctonus ponderosae*
TyunCCE23	i1_LQ_TYUN_c34869/f1p0/1995	577	0	69.46%	XP_019769830.1	*Dendroctonus ponderosae*
TyunCCE24	Cluster-27834.16736	569	0	75.66%	XP_019772474.1	*Dendroctonus ponderosae*
TyunCCE25	i2_LQ_TYUN_c30644/f1p0/2041	337	0	78.34%	XP_019773718.1	*Dendroctonus ponderosae*
TyunCCE26	Cluster-27834.18773	65	9×10^{-7}	42.86%	XP_023727217.2	*Cryptotermes secundus*
TyunCCE27	Cluster-27834.13066	308	7×10^{-127}	60.06%	XP_030749177.1	*Sitophilus oryzae*
TyunCCE28	Cluster-27834.7029	572	0	47.51%	XP_030754488.1	*Sitophilus oryzae*
TyunCCE29	i2_LQ_TYUN_c24874/f1p0/2585	308	3×10^{-107}	55.89%	XP_030768230.1	*Sitophilus oryzae*

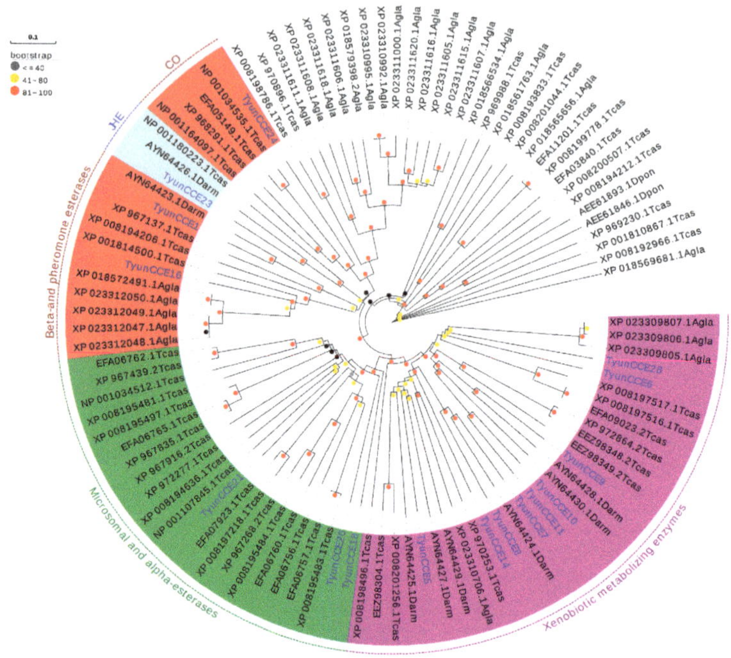

Figure 5. Neighbor-joining tree of candidate CCEs. Bootstrap values after 1000 replications. Dpon, *Dendroctonus ponderosae*; Darm, *Dendroctonus armandi*; Agla, *Anoplophora glabripennis*; Tcas, *Tribolium castaneum*; Tyun, *Tomicus yunnanensis*.

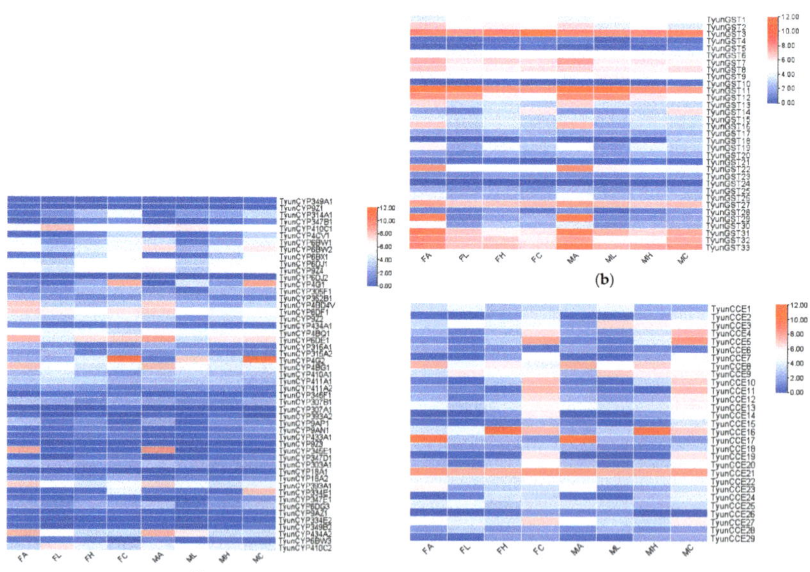

Figure 6. Expression profiles of detoxification genes in *T. yunnanensis*. (**a**) CYP; (**b**) GST; (**c**) CCE. MA: male antennae, MH: male head, ML: male legs, MC: male carcasses, FA: female antennae, FH: female head, FL: female legs, FC: female carcasses.

Further, qPCR was employed to validate the expression of some detoxification genes and to investigate their expression profiles. The qPCR results of TyunCYP4G2, TyunCYP6DF1, TyunGST11, TyunGST33, TyunCCE16, and TyunCCE17 are consistent with the FPKM value analysis results, but the qPCR results of TyunCCE21 differ from the FPKM value analysis, which may be caused by the inconsistency between the qPCR analysis samples and the sequencing samples (Figure 7).

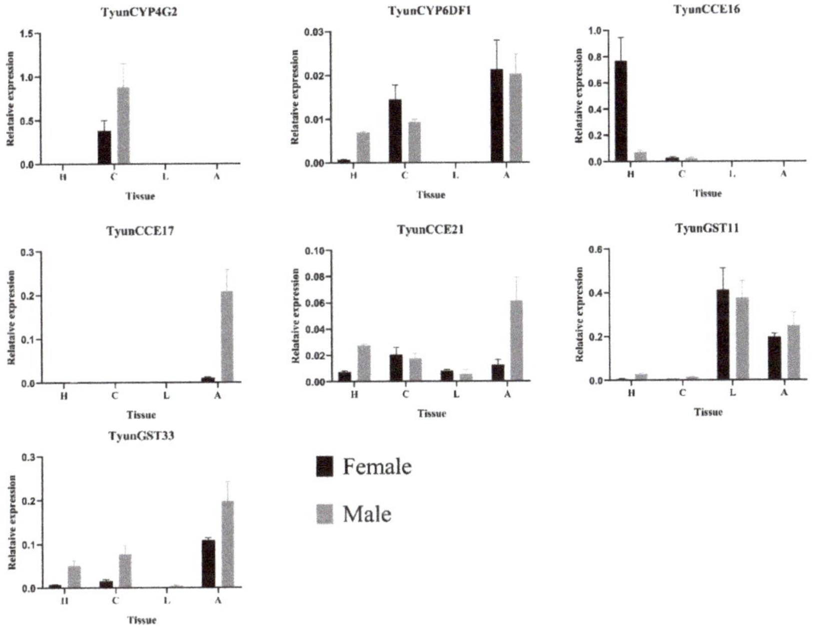

Figure 7. qPCR analysis of *T. yunnanensis* detoxification genes transcript levels in different tissues. H: head; C: carcasses; L: leg; A: antenna.

4. Discussion

Tomicus yunnanensis is one of the most important pests of *Pinus yunnanensis.* In most insect species, detoxification proteins play a key role in the degradation of plant secondary metabolites. In order to better understand the clues of how insects degrade plant secondary metabolites, we first identified candidate detoxification proteins in the transcripts of the antennae, heads, legs, and carcasses of *T. yunnanensis*, and studied some of the detoxification proteins and the expression profiles of four different organizations. Our research results provide new evidence for the molecular basis of the detoxification proteins of *T. yunnanensis* in the metabolic function of detoxification, which may help to develop better methods to control this pest. The identification of at least 51 CYP genes placed *T. yunnanensis* within the middle of the range of the P450 gene family size in insects for which the genome has been sequenced, ranging from a low of 46 in the honeybee (*Apis mellifera* Linnaeus, 1758 (Hymenoptera: Apidae)) to 143 in *T. castaneum* [34], and it was similar to the mountain pine beetle (*D. armandi*), which has 64 CYPs. Previous studies on *T. yunnanensis* were mainly in the direction of olfactory-related proteins, but there are few reports on the detoxification genes of *T. yunnanensis* [35]. In this study, 51 TyunCYPs were analyzed in different adult tissue transcripts, and we found that seven TyunCYPs are predominantly expressed in the antennae of both sexes, belonging to three CYP families, CYP2: TyunCYP305F1; mitochondrial CYP clan: TyunCYP315A1; CYP4: TyunCYP410A1, TyunCYP393A1, and TyunCYP305F1. CYP345E2, a member of CYP3, is an antenna-specific CYP from *D. pondersae* that has been proved to catalyze the oxidation of monoterpene volatile compounds in pine hosts [36]. In our study, we found two CYP345E2 homologues, TyunCYP345E1 and TyunCYP345F1 in *T. yunnanensis*, but the expression levels of TyunCYP345E1 and TyunCYP345F1 in the antennae were both low, which may be caused by differences between species. Some members of the CYP4 family are also involved in odor degradation, and most of them are involved in the synergistic reaction of detoxification and pheromone synthesis [37]. The expression level of TyunCYP393A1 in the antennae of females was higher than that of males, which might be related to the host location of females, while the expression level of TyunCYP410A1 in the antennae of males was higher than that of females, which may be because the male needs to degrade the secondary metabolites produced by the plant defense mechanism. TyunCYP4G2 is specifically and highly expressed in the residue, which may be related to the metabolism of toxic substances in *T. yunnanensis*. In the CYP6 family, few members have been reported to have odorant clearance functions [38]. In many (but not all) studies, genes from the CYP6 subfamily are shown to metabolize xenobiotics and plant natural compounds [27]. Several genes have been reported to have specific expression in the olfactory organs, such as CYP6B48, CYP6B42, and CYP6AE49 expressed in the male and female antennae of *Spodoptera litura* [39]. PxCYP6BG3 and PxCYP6BG6 were found in *Plutella xylostell*, which may be related to larval odor clearance [40]. In this study, we did not find any genes that were specifically and highly expressed in the olfactory organs of the CYP6 family.

The functions of GSTs in many insects' physiological functions, such as insecticide resistance and detoxification (plant secondary metabolites) have been fully demonstrated [40–42]. In this study, we identified 33 TyunGSTs from the *T. yunnanensis* transcriptome. According to some reports, A *Manduca sexta* olfactory specific GST called gstmsolf1 is reported as a degradable plant volatile trans-2-hexanal, belonged to the delta subfamily [40,43], in our study, we identified two new GSTs, TyunGST16 and TyunGST29, that were highly expressed in the antennae, and we found that they belonged to the delta/epsilon GST subfamily, Further family classification is needed for the functional speculation of TyunGST16 and TyunGST29. At present, we speculate that they may be related to odor degradation.. TyunGST30 and DaGSTs1 formed a linage in the sigma subfamily. DaGSTs1 may play a role in reducing the negative effects of terpenoids on beetles [44]. We speculate that TyunGST30 also has this function.

Insect CCE is a superfamily, which participates in many physiological processes and contains a variety of substrates [45]. Some of these are secreted enzymes, which refresh ORs by removing redundant esterase odorants from the surrounding area. Besides Z11-16: Ald, Z11-16: Ac is another major sex pheromone component reported in [46]. We found 29 CCE genes in the transcriptome of *T. yunnanensis*. The total number is higher than that of *D. armandi* (8) [47] and less than that of *T. castaneum* (63) [48]. The phylogenetic tree shows that the CCE genes of *T. yunnanensis* can be divided into five categories. Among them, TyunCCE5 and DarmCCE3 (AYN64425.1) cluster into the same branch, and DarmCCE3 is inferred to play a role in host detoxification [47], which we speculate that TyunCCE5 may also have. Using the FPKM value and a qPCR analysis, we found that three CCEs have a higher expression in the antennae: TyunCCE8, TyunCCE17, and TyunCCE21. TyunCCE8 and DarmCCE2 (AYN64424.1) are clustered into the same clade, whose most enzymes have a dietary detoxification function or ester odor degradation function, such as SlCXE10 [49] and SexCXE10 [50] with antennae dominant expression that can degrade ester plant secondary metabolites. Unlike these three CCEs, TyunCCE8, TyunCCE17, and TyunCCE21 have no obvious gender-biased expression. TyunCCE23, DarmCCE (AYN64426.1), and TcasCCE (NP_001180223.1) are divided into the JHE branch. TcasCCE (NP_001180223.1) has been confirmed to have the function of degrading juvenile hormones. We speculate that TyunCCE23 also has this function; of course, this inference needs to be further verified.

5. Conclusions

Our present work has characterized the detoxification gene families of *T. yunanensis* by transcriptome, together with bioinformatics-based analyses and molecular strategies of qPCR. Our study first lead to the identification of 113 genes associated with detoxification, with as many as 51 CYPs. Further, expression profile studies provided reference data for these genes to explore their potential roles in the detoxification of plant secondary metabolites. Together, this study has complemented the information of detoxification gene families in *T. yunanensis* and will allow for target experiments to screen potential attractants or repellents and to develop novel pest control strategies for controlling this beetle and protect the balance of the forest ecosystem.

Supplementary Materials: The following supporting information can be downloaded at: https://www.mdpi.com/article/10.3390/10.3390/d14010023/s1, Table S1: The primers designed for qPCR analysis; Table S2: Protein sequence of Tomicus yunnanensis.

Author Contributions: Conceptualization, W.L. and N.Z.; methodology, B.Y. and N.L.; software, J.Z. and Z.L.; validation, W.L., S.Z. and N.Z.; formal analysis, N.L.; investigation, J.Z. and Z.L.; resources, W.L. and B.Y.; data curation, J.Y.; writing—original draft preparation, W.L.; writing—review and editing, W.L. and N.Z.; visualization, N.Z.; supervision, N.Z.; project administration, N.Z.; funding acquisition, N.Z. All authors have read and agreed to the published version of the manuscript.

Funding: The research was supported by the National Natural Science Foundation of China (31760210), the Science Research Foundation of Yunnan Provincial Department of Education Project (2021Y269), and Key Project of Yunnan Applied Basic Research Program (grand No.202101AS070009; 2018FG001-010).

Institutional Review Board Statement: Not applicable.

Data Availability Statement: All Data are provided within the paper and supplementary material.

Conflicts of Interest: All the authors declare that they have no conflict of interest.

References

1. Kant, M.R.; Jonckheere, W.; Knegt, B.; Lemos, F.; Liu, J.; Schimmel, B.; Villarroel, C.A.; Ataíde, L.; Dermauw, W.; Glas, J.J.; et al. Mechanisms and ecological consequences of plant defence induction and suppression in herbivore communities. *Ann. Bot.* **2015**, *115*, 1015–1051. [CrossRef]
2. Richard, K.; Anurag, A.A.; Marc, M. The benefits of induced defenses against herbivores. *Ecology* **1997**, *78*, 1351–1355. [CrossRef]
3. Walling, L.L. The Myriad Plant Responses to Herbivores. *J. Plant Growth Regul.* **2000**, *19*, 195–216. [CrossRef] [PubMed]
4. Hermsmeier, D.; Schittko, U.; Baldwin, I.T. Molecular interactions between the specialist herbivore *Manduca sexta* (Lepidoptera.; Sphingidae) and its natural host nicotiana attenuata. I. Large-Scale Changes in the accumulation of growth-and defense-related plant mRNAs. *Plant Physiol.* **2001**, *125*, 683–700. [CrossRef] [PubMed]
5. Chen, C.Y.; Kang, Z.J.; Shi, X.Y.; Gao, X.W. Metabolic adaptation mechanisms of insects to plant secondary metabolites and their implications for insecticide resistance of insects. *Acta Entomol. Sin.* **2015**, *58*, 1126–1139.
6. Ingelman-Sundberg, M. Genetic variability in susceptibility and response to toxicants. *Toxicol. Lett.* **2001**, *120*, 259–268. [CrossRef]
7. Jiang, G.; Zhang, Y.; Chen, F.; Li, J.; Li, X.; Yue, J.; Liu, H.; Li, H.; Ran, C. Differential Analysis of the Cytochrome p450 Acaricide-Resistance Genes in *Panonychus citri* (Trombidiformes: Tetranychidae) Strains. *Fla. Entomol.* **2015**, *98*, 318–329. [CrossRef]
8. Yi, X.Q.; Di, Y.D.; Zhang, Q. Preliminary Study on Genetic Diversity of *Esox reicherti* in Amur River Basin Based on Cytochrome b Gene. *J. Anhui Agric. Sci.* **2009**, *37*, 5403–5405. [CrossRef]
9. Itokawa, K.; Komagata, O.; Kasai, S.; Kawada, H.; Mwatele, C.; Dida, G.O.; Njenga, S.; Mwandawiro, C.; Tomita, T. Global spread and genetic variants of the two CYP9M10 haplotype forms associated with insecticide resistance in *Culex quinquefasciatus* Say. *Heredity* **2013**, *111*, 216–226. [CrossRef] [PubMed]
10. Schuler, M.A.; Berenbaum, M.R. Structure and Function of Cytochrome P450S in Insect Adaptation to Natural and Synthetic Toxins: Insights Gained from Molecular Modeling. *J. Chem. Ecol.* **2013**, *39*, 1232–1245. [CrossRef]
11. Feyereisen, R. Insect CYP Genes and P450 Enzymes ☆. In *Reference Module in Life Sciences*; Elsevier: Amsterdam, The Netherlands, 2019. [CrossRef]
12. Mao, Y.B.; Cai, W.J.; Wang, J.W. Silencing a cotton bollworm P450 monooxygenase gene by plant-mediated RNAi impairs larval tolerance of gossypol. *Nat. Biotechnol.* **2007**, *25*, 1307–1313. [CrossRef]
13. Liu, S.; Shi, X.-X.; Jiang, Y.-D.; Zhu, Z.-J.; Qian, P.; Zhang, M.-J.; Yu, H.; Zhu, Q.-Z.; Gong, Z.-J.; Zhu, Z.-R. De novo analysis of the *Tenebrio molitor* (Coleoptera: Tenebrionidae) transcriptome and identification of putative glutathione S-transferase genes. *Appl. Entomol. Zool.* **2015**, *50*, 63–71. [CrossRef]

14. Vontas, J.G.; Small, G.J.; Nikou, D.C.; Ranson, H.; Hemingway, J. Purification, molecular cloning and heterologous expression of a glutathione S-transferase involved in insecticide resistance from the rice brown planthopper, *Nilaparvata lugens*. *Biochem. J.* **2002**, *362*, 329–337. [CrossRef]
15. Friedman, R. Genomic organization of the glutathione S-transferase family in insects. *Mol. Phylogenetics Evol.* **2011**, *61*, 924–932. [CrossRef] [PubMed]
16. Ma, J.N.; Dai, L.L.; Zhang, R.R.; Chen, H. Cloning and expression of glutathione S-transferase gene DaGSTe1 of *Dendroctonus armandi*. *J. Northwest A F Univ.* **2015**, *43*, 117–122.
17. Durand, N.; Carot-Sans, G.; Chertemps, T.; Bozzolan, F.; Party, V.; Renou, M.; Debernard, S.; Rosell, G.; Maïbèche-Coisné, M. Characterization of an antenna carboxylesterase from the pest moth *Spodoptera littoralis* degrading a host Plant Odorant. *PLoS ONE* **2010**, *5*, e15026. [CrossRef]
18. Yu, L.F.; Huang, H.G.; Ze, S.Z.; Ren, L.L.; Zong, S.X.; Lu, W.J.; Luo, Y.Q. Research on the spatial distribution patterns of *Tomicus* sp. in *Pinus yunnanensis* during the shoot feeding period. *Chin. J. Appl. Entomol.* **2017**, *54*, 940–946.
19. Chen, P.; Lu, J.; Haack, R.A.; Ye, H. Attack pattern and reproductive ecology of *Tomicus brevipilosus* (Coleoptera: Curculionidae) on *Pinus yunnanensis* in Southwestern China. *J. Insect Sci.* **2015**, *15*, 43. [CrossRef] [PubMed]
20. Wang, X.W.; Chen, P.; Wang, Y.X.; Yuan, R.L.; Feng, D.; Li, L.S.; Ye, H.; Pan, Y.; Lv, J.; Zhou, Y.F.; et al. Population Structure and Succession Law of *Tomicus Species* in Yunnan. *For. Res.* **2018**, *31*, 167–172.
21. Lü, J.; Hu, S.J.; Ma, X.Y.; Chen, J.M.; Li, Q.Q.; Ye, H.; Arthofer, W. Origin and Expansion of the Yunnan Shoot Borer, *Tomicus yunnanensis* (Coleoptera: Scolytinae): A Mixture of Historical Natural Expansion and Contemporary Human-Mediated Relocation. *PLoS ONE* **2014**, *9*, e111940. [CrossRef]
22. Wang, J.; Zhang, Z.; Kong, X.; Wang, H.; Zhang, S. Intraspecific and interspecific attraction of three Tomicus beetle species during the shoot-feeding phase. *Bull. Entomol. Res.* **2015**, *105*, 225–233. [CrossRef]
23. Grabherr, M.G.; Haas, B.J.; Yassour, M.; Levin, J.Z.; Thompson, D.A.; Amit, I.; Adiconis, X.; Fan, L.; Raychowdhury, R.; Zeng, Q.; et al. Full-length transcriptome assembly from RNA-Seq data without a reference genome. *Nat. Biotechnol.* **2011**, *29*, 644–652. [CrossRef] [PubMed]
24. Li, B.; Dewey, C. RSEM: Accurate transcript quantification from RNA-Seq data with or without a reference genome. *BMC Bioinform.* **2011**, *12*, 323. [CrossRef] [PubMed]
25. Cock, P.J.A.; Christopher, J.F.; Naohisa, G.; Michael, L.H.; Peter, M.R. The Sanger FASTQ file format for sequences with quality scores, and the Solexa/Illumina FASTQ variants. *Nucleic Acids Res.* **2010**, *38*, 1767–1771. [CrossRef] [PubMed]
26. Trapnell, C.; Williams, B.A.; Pertea, G.; Mortazavi, A.; Kwan, G.; Van Baren, M.J.; Salzberg, S.L.; Wold, B.J.; Pachter, L. Transcript assembly and quantification by RNA-Seq reveals unannotated transcripts and isoform switching during cell differentiation. *Nat. Biotechnol.* **2010**, *28*, 511–515. [CrossRef] [PubMed]
27. Feyereisen, R. 8—Insect CYP Genes and P450 Enzymes. In *Insect Molecular Biology and Biochemistry*; Academic Press: San Diego, CA, USA, 2012; pp. 236–316. [CrossRef]
28. Kumar, S.; Stecher, G.; Tamura, K. MEGA7: Molecular Evolutionary Genetics Analysis version 7.0 for bigger datasets. *Mol. Biol. Evol.* **2016**, *33*, 1870–1874. [CrossRef]
29. Balakrishnan, S.; Gao, S.; Lercher, M.J.; Hu, S.; Chen, W.H. Evolview v3: A webserver for visualization, annotation, and management of phylogenetic trees. *Nucleic Acids Res.* **2019**, *47*, W270–W275. [CrossRef]
30. Singh, V.K.; Mangalam, A.K.; Dwivedi, S.; Naik, S. Primer Premier: Program for Design of Degenerate Primers from a Protein Sequence. *BioTechniques* **1998**, *24*, 318–319. [CrossRef]
31. Simon, P. Q-Gene: Processing quantitative real-time RT–PCR data. *Bioinformatics* **2003**, *19*, 1439–1440. [CrossRef]
32. Shi, H.; Pei, L.; Gu, S.; Zhu, S.; Wang, Y.; Zhang, Y.; Li, B. Glutathione S-transferase (GST) genes in the red flour beetle, *Tribolium castaneum*, and comparative analysis with five additional insects. *Genomics* **2012**, *100*, 327–335. [CrossRef]
33. Schama, R.; Pedrini, N.; Juárez, M.P.; Nelson, D.R.; Torres, A.Q.; Valle, D.; Mesquita, R.D. *Rhodnius prolixus* supergene families of enzymes potentially associated with insecticide resistance. *Insect. Biochem. Mol. Biol.* **2016**, *69*, 91–104. [CrossRef]
34. Feyereisen, R. Evolution of insect P450. *Biochem. Soc. Trans.* **2006**, *34*, 1252–1255. [CrossRef]
35. Liu, N.Y.; Li, Z.B.; Zhao, N.; Song, Q.S.; Zhu, J.Y.; Yang, B. Identification and characterization of chemosensory gene families in the bark beetle, *Tomicus yunnanensis*. *Comp. Biochem. Physiol. Part D Genom. Proteom.* **2018**, *25*, 73–85. [CrossRef]
36. Liu, N.N.; Li, M.; Gong, Y.H.; Liu, F.; Li, T. Cytochrome P450s-Their expression, regulation, and role in insecticide resistance. *Pestic. Biochem. Physiol.* **2015**, *120*, 77–81. [CrossRef]
37. Bozzolan, F.; Siaussat, D.; Maria, A.; Durand, N.; Pottier, M.-A.; Chertemps, T.; Maïbèche-Coisne, M. Antennal uridine diphosphate (UDP)-glycosyltransferases in a pest insect: Diversity and putative function in odorant and xenobiotics clearance. *Insect. Mol. Biol.* **2014**, *23*, 539–549. [CrossRef]
38. Claudia, C.R.; María, F.L.; Ana, K.C.A.; Verónica, P.M.; Brian, T.S.; Gerardo, Z. Isolation and expression of cytochrome P450 genes in the antennae and gut of pine beetle *Dendroctonus rhizophagus* (Curculionidae: Scolytinae) following exposure to host monoterpenes. *Gene* **2013**, *520*, 47–63. [CrossRef]
39. Pottier, M.A.; Bozzolan, F.; Chertemps, T.; Jacquin-Joly, E.; Lalouette, L.; Siaussat, D.; Maïbèche-Coisne, M. Cytochrome P450s and cytochrome P450 reductase in the olfactory organ of the cotton leafworm *Spodoptera littoralis*. *Insect Mol. Biol.* **2012**, *21*, 568–580. [CrossRef]

40. He, P.; Zhang, Y.-F.; Hong, D.-Y.; Wang, J.; Wang, X.-L.; Zuo, L.-H.; Tang, X.-F.; Xu, W.-M.; He, M. A reference gene set for sex pheromone biosynthesis and degradation genes from the diamondback moth, *Plutella xylostella*, based on genome and transcriptome digital gene expression analyses. *BMC Genom.* **2017**, *18*, 219. [CrossRef]
41. You, Y.; Xie, M.; Ren, N.; Cheng, X.; Li, J.; Ma, X.; Zou, M.; Vasseur, L.; Gurr, G.M.; You, M. Characterization and expression profiling of glutathione S-transferasesin the diamondback moth, *Plutella xylostella* (L.). *BMC Genom.* **2015**, *16*, 152. [CrossRef]
42. Zou, X.P.; Xu, Z.B.; Zou, H.W.; Liu, J.S.; Chen, S.N.; Feng, Q.L.; Zheng, S.C. Glutathione S-transferase SlGSTE1 in *Spodoptera litura* may be associated with feeding adaptation of host plants. *Insect. Biochem. Mol. Biol.* **2016**, *70*, 32–43. [CrossRef]
43. Rogers, M.E.; Jani, M.K.; Vogt, R.G. An olfactory-specific glutathione-S-transferase in the sphinx moth *Manduca sexta*. *J. Exp. Biol.* **1999**, *202*, 1625–1637. [CrossRef] [PubMed]
44. Dai, L.L.; Ma, J.N.; Ma, M.Y.; Zhang, H.Q.; Shi, Q.; Zhang, R.; Chen, H. Characterisation of GST genes from the Chinese white pine beetle *Dendroctonus armandi* (Curculionidae: Scolytinae) and their response to host chemical defence. *Pest Manag. Sci.* **2016**, *72*, 816–827. [CrossRef] [PubMed]
45. Oakeshott, J.G.; Claudianos, C.; Campbell, P.M.; Newcomb, R.D.; Russell, R.J. Biochemical Genetics and Genomics of Insect Esterases. *Sci. Ref. Modul. Life Sci.* **2019**, *5*, 309–381. [CrossRef]
46. Tamaki, Y.; Kawasaki, K.; Yamada, H.; Koshihara, T.; Osaki, N.; Ando, T.; Yoshida, S.; Kakinohana, H. Z-11-hexadecenal and Z-11-hexadecenyl acteact: Sex pheromone components of the *Diamonback moth* (Lepideptera: Plutellidae). *Appl. Entomol. Zool.* **1997**, *12*, 208–210. [CrossRef]
47. Dai, L.L.; Gao, H.M.; Ye, J.Q.; Fu, D.Y.; Sun, Y.Y.; Chen, H. Isolation of CarE genes from the Chinese white pine beetle *Dendroctonus armandi* (Curculionidae: Scolytinae) and their response to host chemical defense. *Pest Manag. Sci.* **2019**, *75*, 986–997. [CrossRef] [PubMed]
48. Zhao, Y.J.; Wang, Z.Q.; Zhu, J.Y.; Liu, N.Y. Identification and characterization of detoxification genes in two cerambycid beetles, *Rhaphuma horsfieldi* and *Xylotrechus quadripes* (Coleoptera: Cerambycidae: Clytini). *Comp. Biochem. Physiol. Part B Biochem. Mol. Biol.* **2020**, *243–244*, 110431. [CrossRef]
49. Durand, N.; Carot-Sans, G.; Chertemps, T.; Montagné, N.; Jacquin-Joly, E.; Debernard, S.; Maïbèche-Coisne, M. A diversity of putative carboxylesterases are expressed in the antennae of the noctuid moth *Spodoptera littoralis*. *Insect Mol. Biol.* **2010**, *19*, 87–97. [CrossRef] [PubMed]
50. He, P.; Zhang, Y.N.; Yang, K.; Li, Z.Q.; Dong, S.L. An antenna-biased carboxylesterase is specifically active to plant volatiles in *Spodoptera exigua*. *Pestic. Biochem. Physiol.* **2015**, *123*, 93–100. [CrossRef]

diversity

MDPI

Article

The Influence of Climate Change on Three Dominant Alpine Species under Different Scenarios on the Qinghai–Tibetan Plateau

Huawei Hu [1,2], Yanqiang Wei [2,*], Wenying Wang [3,*] and Chunya Wang [4,5]

1 College of Geosciences, Qinghai Normal University, Xining 810008, China; huhuawei19840618@163.com
2 Key Laboratory of Remote Sensing of Gansu Province, Northwest Institute of Eco-Environment and Resources, Chinese Academy of Sciences, Lanzhou 730000, China
3 College of Life Sciences, Qinghai Normal University, Xining 810008, China
4 Chengdu Institute of Biology, Chinese Academy of Sciences, Chengdu 610041, China; 2019020010@stu.cdut.edu.cn
5 Earth Sciences College, Chengdu University of Technology, Chengdu 610059, China
* Correspondence: weiyq@lzb.ac.cn (Y.W.); wangwy0106@163.com (W.W.)

Abstract: The Qinghai–Tibetan Plateau (QTP) with high altitude and low temperature is one of the most sensitive areas to climate change and has recently experienced continuous warming. The species distribution on the QTP has undergone significant changes especially an upward shift with global warming in the past decades. In this study, two dominant trees (*Picea crassifolia* Kom and *Sabina przewalskii* Kom) and one dominant shrub (*Potentilla parvifolia* Fisch) were selected and their potential distributions using the MaxEnt model during three periods (current, the 2050s and the 2070s) were predicted. The predictions were based on four shared socio-economic pathway (SSPs) scenarios, namely, SSP2.6, SSP4.5, SSP7.0, SSP8.5. The predicted current potential distribution of three species was basically located in the northeastern of QTP, and the distribution of three species was most impacted by aspect, elevation, temperature seasonality, annual precipitation, precipitation of driest month, Subsoil CEC (clay), Subsoil bulk density and Subsoil CEC (soil). There were significant differences in the potential distribution of three species under four climate scenarios in the 2050s and 2070s including expanding, shifting, and shrinking. The total suitable habitat for *Picea crassifolia* shrank under SSP2.6, SSP4.5, SSP7.0 and enlarged under SSP8.5 in the 2070s. On the contrary, the total suitable habitat for *Sabina przewalskii* enlarged under SSP2.6, SSP4.5, SSP7.0 and shrank under SSP8.5 in the 2070s. The total suitable habitat for *Potentilla parvifolia* continued to increase with SSP2.6 to SSP8.5 in the 2070s. The average elevation in potentially suitable habitat for *Potentilla parvifolia* all increased except under SSP8.5 in the 2050s. Our study provides an important reference for the conservation of *Picea crassifolia*, *Sabina przewalskii*, *Potentilla parvifolia* and other dominant plant species on the QTP under future climate change.

Keywords: climate change; potential distribution; MaxEnt model; suitable habitat; average elevation

Citation: Hu, H.; Wei, Y.; Wang, W.; Wang, C. The Influence of Climate Change on Three Dominant Alpine Species under Different Scenarios on the Qinghai–Tibetan Plateau. *Diversity* **2021**, *13*, 682. https://doi.org/10.3390/d13120682

Academic Editor: Anatoliy A. Khapugin

Received: 1 December 2021
Accepted: 14 December 2021
Published: 19 December 2021

Publisher's Note: MDPI stays neutral with regard to jurisdictional claims in published maps and institutional affiliations.

1. Introduction

Climate change is considered to be one of the most important driving factors of species distribution [1–3] According to the report of the sixth Coupled Model Intercomparison Project (CMIP6), the global temperature will continue to increase by the end of the 21st century [4]. The Qinghai–Tibetan Plateau (QTP), famous as the "third pole" in the world with high altitude and low temperature, is one of the most sensitive regions to climate change [5]. With global warming, many species shift their suitable habitats especially upward in altitude in order to adapt to changes in environmental conditions [6,7].

However, it remains unclear what influences climate change will have on alpine species at large regional scales and whether alpine species respond uniformly on the QTP. Two dominant and representative alpine trees (*Picea crassifolia* Kom, *Sabina przewalskii* Kom) and one dominant and representative alpine shrub (*Potentilla parvifolia* Fisch) on the QTP were used in this study, *Picea crassifolia* favors shady slopes, semi-shady slopes and humid valleys in the mountains with an altitude of 1750–3100 m (a.s.l), is endemic to China, and is distributed in the Qilian Mountains, Qinghai, Gansu, Ningxia, Inner Mongolia. *Sabina przewalskii* grows on sunny slopes of 2600–4000 m (a.s.l), is endemic to China, and is distributed in Qinghai, Gansu Hexi Corridor, and the north of Sichuan. *Potentilla parvifolia* favors dry hillside, rock crack, forest edge and forest with an altitude of 900–5000 m (a.s.l), and it is distributed in Heilongjiang, Inner Mongolia, Gansu, Qinghai, Sichuan and Tibet in China. Species on shady slopes are more sensitive to the magnitude of temperature fluctuations, and species on sunny slopes can tolerate larger temperature fluctuations [8]. The previous study was conducted on potential distribution for *Picea crassifolia*, *Sabina przewalskii* and *Potentilla parvifolia*, but they only focused on the potential distribution under different climate scenarios without considering the influence of geographical factors [9].

Species distribution models are popular methods in modeling the potential distributions of species in response to climate change in the past few decades [10]. Many species distribution models are used to predict potential distributions, such as maximum entropy (MaxEnt) [11]), random forests (RFs) [12], CLIMEX, and genetic algorithm for rule set production (GARP) [13]. Among them, MaxEnt is widely selected because it performs excellently with a small number of sample records compared to other models [14]. This research used MaxEnt to predict potential distribution for three species under different shared socio-economic pathways (SSPs) scenarios.

SSPs can be selected to predict greenhouse gas emission scenarios under different climate conditions [15]. SSPs consider the effects of land use and socio-economic with the development of regional climate change and are different from representative concentration pathways (RCPs) [16]. SSPs have a higher beginning point than RCP and the result of prediction is near to the true value [17]). SSP2.6 (Low forced scenario), SSP4.5 (Medium forced scenario), SSP7.0 (Medium-high forced scenario), SSP8.5 (High forced scenario) were selected to predict the potential distribution of three species during the period of the 2050s and 2070s in this study.

The aims of this research are: (1) to predict the potential distribution of three species under different climate scenarios; (2) to assess the key environment variables affecting the distribution of three species; (3) to analyze the area and elevation changes of the suitable habitat of three species in the future climate change. The results of this study will provide an important reference for the conservation of *Picea crassifolia*, *Sabina przewalskii*, *Potentilla parvifolia* and other dominant plant species on the QTP under climate change.

2. Materials and Methods

2.1. Study Area

The Qinghai–Tibetan Plateau (QTP), located in western China, is famous as the "Roof of the World" with the highest and one of the most extensive plateaus on earth [18], It lies between 26° N to 39° N and 73° E to 104° E, and covers a total area of approximately 2.5 million km^2 with an average elevation above 4000 m (a.s.l). Alpine desert ecosystems, alpine meadow, alpine grassland, shrub and forest are distributed from the southwestern to the northeastern of QTP, which is characterized by low annual temperature differences, high daily temperature differences, low air temperature and strong solar radiation [19]). Climate change probably affect species on the QTP more than those in other regions with the same latitude [20,21]).

2.2. Occurrence Data

As the accurate location information of species distribution is the basis of high precision simulation and prediction, the geographical distribution information of *Picea crassifolia*, *Sabina przewalskii* and *Potentilla parvifolia* were obtained from: (1) Chinese Virtual Herbarium (CVH, https://www.cvh.ac.cn/, accessed on 23 September 2021); (2) Global Biodiversity Information Facility (GBIF, http://www.gbif.org/, accessed on 24 September

2021); (3) Relevant literature reports (CNKI, Web of Science, https://www.cnki.net/ https://apps.webofknowledge.com/, accessed on 15 November 2021). Google Earth (http://ditu.google.cn/, accessed on 22 November 2021) was used to proofread specimen distribution information and the duplicate records were removed [22]. Finally, the 172 records of *Picea crassifolia* distribution data, 69 records of *Sabina przewalskii* distribution data and 146 records of *Potentilla parvifolia* distribution data were used (Figure 1). The longitude and latitude of the distribution data and the species name were entered into Excel and converted to csv format for modeling.

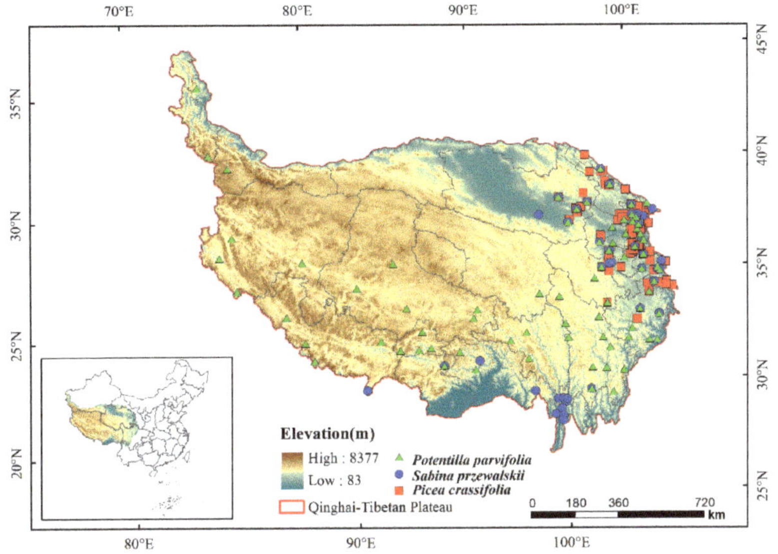

Figure 1. Locations of three species on the Qinghai–Tibetan Plateau.

2.3. *Environment Variables*

2.3.1. Climate Data

The data used for climate assessment were downloaded from the WorldClim global climate database [23] (http://www.worldclim.org, accessed on 15 November 2021).

Current climate data included 19 Bioclimatic variables with 30″ spatial resolution during 1970–2000 [24], which reflect temperature and precipitation.

Future Bioclimatic data were obtained from BCC_CSM1.1 (Beijing Climate Center, China Meteorological Administration, Beijing, China) global circulation model) [8]), which is available for predicting the global climate response to increasing greenhouse gas concentration [22]). Bioclimatic data (at 2.5′ spatial resolution) for four scenarios of Shared Socioeconomic Pathways (SSPs) provided by the sixth Coupled Model Intercomparison Project (CMIP6) were used for modeling [25]). Each SSP includes scenarios of SSP2.6, SSP4.5, SSP7.0, SSP8.5, and analyses the spatial and temporal changes of the annual temperature and precipitation during 2021–2100 [26]). The future climate variables were resampled to the same spatial resolution with current data using ArcGIS 10.7 [27]).

2.3.2. Topographic Data

The DEM data with a cell size of 90 m × 90 m were downloaded from the WIST geodatabase of NASA (http://srtm.csi.cgiar.org/, accessed on 15 November 2021). The variables of slope, aspect and elevation were derived from DEM using ArcGIS 10.7.

2.3.3. Soil Property Data

Soil data used in this study were collected from Harmonized World Soil Database (HWSD, http://www.fao.org/soils-portal, accessed on 15 November 2021).

2.3.4. Processing and Selection of Environment Variables

All environment variables were resampled to 30″ spatial resolution and were processed to the same geographic bounds. In the modeling process, high correlation variables and environment variables that contribute less to the model were removed to improve the accuracy of the model [28,29]. The correlation coefficient was calculated to account for the influence of collinearity on the model accuracy. The variables with r below 0.8 were selected [30]. The 16 variables with low correlation coefficients and high contribution rates were selected for distribution modeling (Table 1).

Table 1. The selection of environmental variables used in this study.

Data Source	Symbol	Variables	Unit	Important Variables for Modelling
WorldClim	Bio1	Annual mean temperature	°C	
	Bio2	Mean diurnal range	°C	
	Bio3	Isothermality (BIO2/BIO7) (×100)	%	
	Bio4	Temperature seasonality (standard deviation ×100)	°C	√
	Bio5	Max temperature of warmest month	°C	
	Bio6	Min temperature of coldest month	°C	
	Bio7	Temperature annual range (BIO5-BIO6)	°C	√
	Bio8	Mean temperature of wettest quarter	°C	
	Bio9	Mean temperature of driest quarter	°C	
	Bio10	Mean temperature of warmest quarter	°C	
	Bio11	Mean temperature of coldest quarter	°C	
	Bio12	Annual precipitation	mm	√
	Bio13	Precipitation of wettest month	mm	
	Bio14	Precipitation of driest month	mm	√
	Bio15	Precipitation seasonality (coefficient of variation)	1	√
	Bio16	Precipitation of wettest quarter	mm	
	Bio17	Precipitation of driest quarter	mm	
	Bio18	Precipitation of warmest quarter	mm	√
	Bio19	Precipitation of coldest quarter	mm	√
DEM	ASL	Elevation	m	√
	SLOP	Slope	°	√
	ASPE	Aspect	°	√

Table 1. *Cont.*

Data Source	Symbol	Variables	Unit	Important Variables for Modelling
HWSD	S_GRAVEL	Subsoil gravel content	%vol	
	S_SAND	Subsoil sand fraction	%wt	
	S_SILT	Subsoil silt fraction	%wt	
	S_CLAY	Subsoil clay fraction	%wt	√
	S_USDA_TEX_CLASS	Subsoil USDA texture classification	name	
	S_REF_BULK_DENSITY	Subsoil reference bulk density	kg/dm^3	
	S_ BULK_DENSITY	Subsoil bulk density	kg/dm^3	√
	S_OC	Subsoil organic carbon	% weight	
	S_PH_H_2O	Subsoil pH (H_2O)	$-\log(H^+)$	
	S_CEC_CLAY	Subsoil CEC (clay)	cmol/kg	√
	S_CEC_SOIL	Subsoil CEC (soil)	cmol/kg	√
	S_BS	Subsoil base saturation	%	√
	S_TEB	Subsoil TEB	cmol/kg	
	S_$CACO_3$	Subsoil calcium carbonate	% weight	
	S_$CASO_4$	Subsoil gypsum	% weight	√
	S_ESP	Subsoil sodicity (ESP)	%	
	S_ECE	Subsoil salinity (Elco)	dS/m	

2.4. Distribution Modeling

MaxEnt with advantages in performance and stability was used to predict the potential distribution of three species (Figure 2) [11,31]. In addition, MaxEnt has the advantage of utilizing continuous and classified data and integrating the interaction between variables [14]. MaxEnt software version 3.4.4k was used to identify the species potential habitat distribution. The MaxEnt was set to run 500 iterations with a maximum of 10,000 background points, a convergence threshold (0.00001), a regularization multiplier of 1, a logistic output grid format, and the algorithm parameters set to "auto feature". The other parameter values were kept in the default settings [32]. A total of 70% of the distribution point data were selected for training, and the rest were used for testing [33]. The Jackknife was used for testing the importance of environmental variables in a model with a small amount of the distribution point records [34].

2.4.1. Accuracy Assessment

The value of the area under the receiver operating characteristic curve (AUC) was selected to assess model accuracy [11]). Model performance can be regarded as fail when it is between 0.5 and 0.6, poor when it is between 0.6 and 0.7, fair when it is between 0.7 and 0.8, good when it is between 0.8 and 0.9; and excellent when it is between 0.9 and 1 [35].

2.4.2. The Area and Elevation Changes of the Habitat Suitability

SDMtoolbox of ArcGIS 10.7 was used to convert the current and future results. The asc format files in the model result were converted to the raster format and reclassified into four suitable habitats, and we calculated the area and average altitude of potential distribution by zonal statistic tool [36]. The area and average altitude changes in the suitable habitat for species distribution were used as an indicator to evaluate the impact of climate change on the distribution of species [8]. The intersection distributions of the three species were obtained through the raster calculator and extracted by attributes tools.

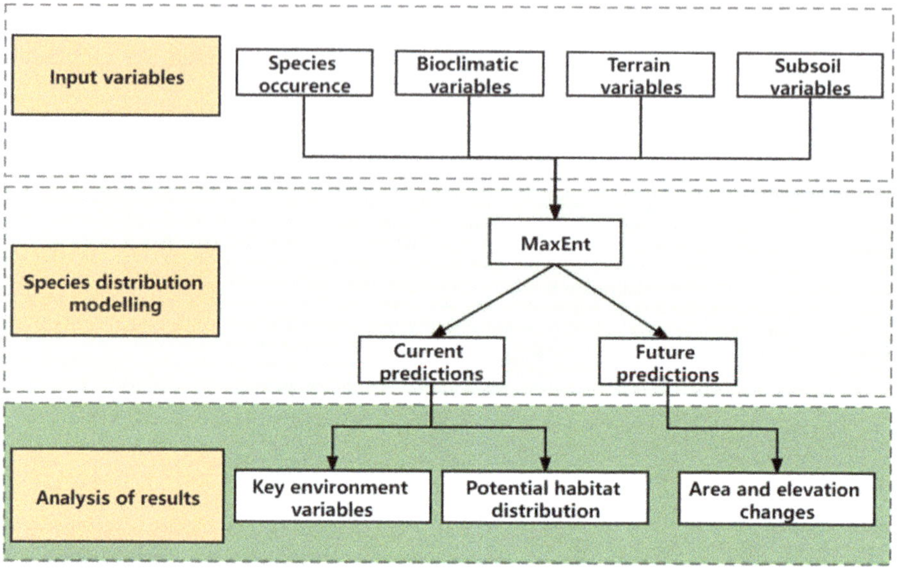

Figure 2. Flow diagram of methodology adopted.

3. Results

3.1. Model Assessment and Key Environmental Variables

The mean AUC of three species in training and testing all exceeded 0.9 in current and future modeling. The model for the geographic distribution prediction performed excellently and had high accuracy.

We selected the variables whose contribution rate for three species is more than 0.1 for analysis. The internal jackknife test of the MaxEnt model for environmental variables' importance showed that aspect was the most critical factor determining the distribution of the three species. Aspect contributed 33.9% to model output for *Picea crassifolia*, 51.7% for *Sabina przewalskii* and 56.1% for *Potentilla parvifolia* (Table 2). In addition, elevation contributed 20.2% to model output for *Picea crassifolia*, 26% for *Sabina przewalskii* and 15.9% for *Potentilla parvifolia*. The following factors were precipitation of driest month (Bio14: 23.1% for *Picea crassifolia*, 3.4% for *Sabina przewalskii* and 3.5% for *Potentilla parvifolia*), annual precipitation (Bio12: 10.7% for *Picea crassifolia*, 4.9% for *Potentilla parvifolia*) and temperature seasonality (Bio4: 6% for *Picea crassifolia*). The total contributions of three subsoil variables (S_CEC_CLAY, S_BULK_DEN, S_CEC_SOIL) did not exceed 3% (Table 2). The results indicated that subsoil conditions had very limited impacts on the potential distribution of *Picea crassifolia*, *Sabina przewalskii* and *Potentilla parvifolia*. The cumulative percentage of aspect, elevation, Bio14, Bio12, Bio4, S_CEC_CLAY, S_BULK_DEN, and S_CEC_SOIL was 95.6% for *Picea crassifolia*, 82.5% for *Sabina przewalskii* and 83.5% for *Potentilla parvifolia*, respectively.

3.2. Potential Distribution of Three Species at Current Climate Scenarios

The total suitable habitat area for *Picea crassifolia* was 99,203.04 km^2 (account for 3.86% of QTP), and was mainly concentrated in the northeastern of QTP. The total suitable habitat area for *Potentilla parvifolia* was 102,179.35 km^2 (account for 3.98% of QTP) and was mainly distributed in the southern and northeastern of QTP. While the total suitable habitat area (21,283.4 km^2) for *Sabina przewalskii* was much lower than *Picea crassifolia* and *Potentilla parvifolia* (only account for 0.82% of QTP) (Table 3, Figure 3).

Table 2. The contribution and cumulative percentage of key environmental variables.

Symbol	Variables	*Picea crassifolia*		*Sabina przewalskii*		*Potentilla parvifolia*	
		Contribution (%)	Cumulative Percentage (%)	Contribution (%)	Cumulative Percentage (%)	Contribution (%)	Cumulative Percentage (%)
ASPE	Aspect	33.9	33.9	51.7	51.7	56.1	56.1
Bio14	Precipitation of driest month	23.1	57.0	3.4	55.1	3.5	59.6
ASL	Elevation	20.2	77.2	26.0	81.1	15.9	75.5
Bio12	Annual precipitation	10.7	87.9	0.0	81.1	4.9	80.4
Bio4	Temperature seasonality	6.0	93.9	0.0	81.1	0.2	80.6
S_CEC_CLAY	Subsoil CEC (clay)	1.2	95.1	0.4	81.5	0.5	81.1
S_BULK_DEN	Subsoil bulk density	0.3	95.4	0.9	82.4	2.3	83.4
S_CEC_SOIL	Subsoil CEC (soil)	0.2	95.6	0.1	82.5	0.1	83.5

Table 3. The area of suitable habitat for three species under different climate scenarios.

	Suitable Habitat	The Area of Potential Suitable Habitat for Three Species (km^2)								The Area at Current (km^2)
		2050s				2070s				
		SSP2.6	SSP4.5	SSP7.0	SSP8.5	SSP2.6	SSP4.5	SSP7.0	SSP8.5	
P. crassifolia	High	10,271.02	10,171.20	8483.41	10,813.65	6294.73	4882.79	9386.29	4767.55	8830.95
	Moderately	33,347.47	34,940.89	42,712.89	30,644.28	4143.71	34,927.27	26,032.80	54,663.53	32,876.52
	Low	55,668.94	47,247.23	57,771.42	53,073.74	43,502.34	58,147.09	48,217.26	48,614.71	57,495.57
	Total	99,287.43	92,359.32	108,967.72	94,531.67	53,940.78	97,957.15	83,636.35	108,045.79	99,203.04
S. przewalskii	High	625.21	1883.79	361.15	0	774.02	1046.25	232.30	120.69	194.19
	Moderately	7243.89	4968.09	3235.84	1260.40	3468.14	4416.38	6272.95	1838.42	3327.49
	Low	32,359.29	18,678.21	16,681.90	13,747.32	26,963.81	22,868.65	31,810.31	18,537.56	17,761.72
	Total	40,228.39	25,530.09	20,278.89	15,007.72	31,205.97	28,331.28	38,315.56	20,496.67	21,283.40
P. parvifolia	High	3047.10	991.80	2301.20	1096.16	1153.32	3744.90	2755.82	915.58	3167.78
	Moderately	20,949.47	13,109.41	18,451.36	9,434.38	14,813.53	15,636.56	13,672.01	19,009.42	21,045.65
	Low	86,588.17	83,692.61	92,607.04	59,658.84	81,386.87	80,101.97	99,566.01	99,144.06	77,965.92
	Total	110,584.74	97,793.82	113,359.60	70,189.38	97,353.72	99,483.43	115,993.84	119,069.06	102,179.35

3.3. Potential Distribution of Three Species under Future Climate Scenarios

There were significant differences in the distribution area of low suitable habitat, moderately suitable habitat and high suitable habitat under future climate scenarios as compared to current. The high suitable habitat for *Picea crassifolia* decreased under SSP2.6, SSP4.5 and SSP8.5 in the 2070s. Especially, the high suitable habitat for *Sabina przewalskii* declined to zero under SSP8.5 in the 2050s. The total suitable habitat for three species *all* decreased under SSP8.5 in the 2050s (Figure 4). The total suitable habitat for *Picea crassifolia* shrank under SSP2.6, SSP4.5 and SSP7.0 and enlarged under SSP8.5 in the 2070s. On the contrary, the total suitable habitat for *Sabina przewalskii* enlarged under SSP2.6, SSP4.5 and SSP7.0 and shrank under SSP8.5 in the 2070s. The total suitable habitat for *Potentilla parvifolia* continued increasing from SSP2.6 to SSP8.5 in the 2070s (Table 3, Figure 5)).

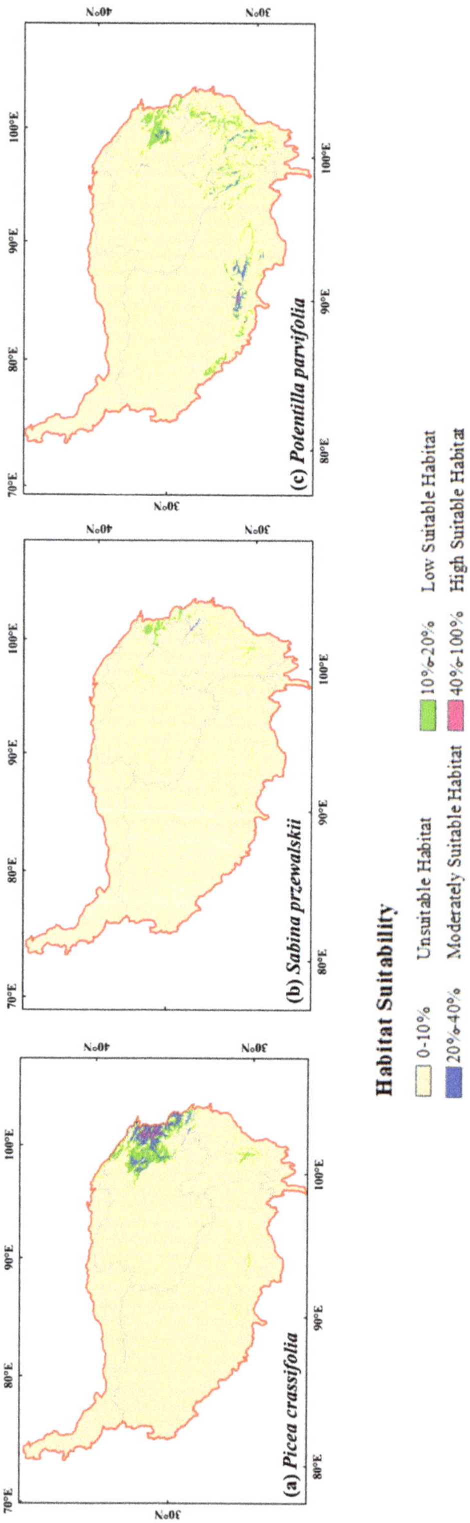

Figure 3. The current potential geographical distribution of the three species.

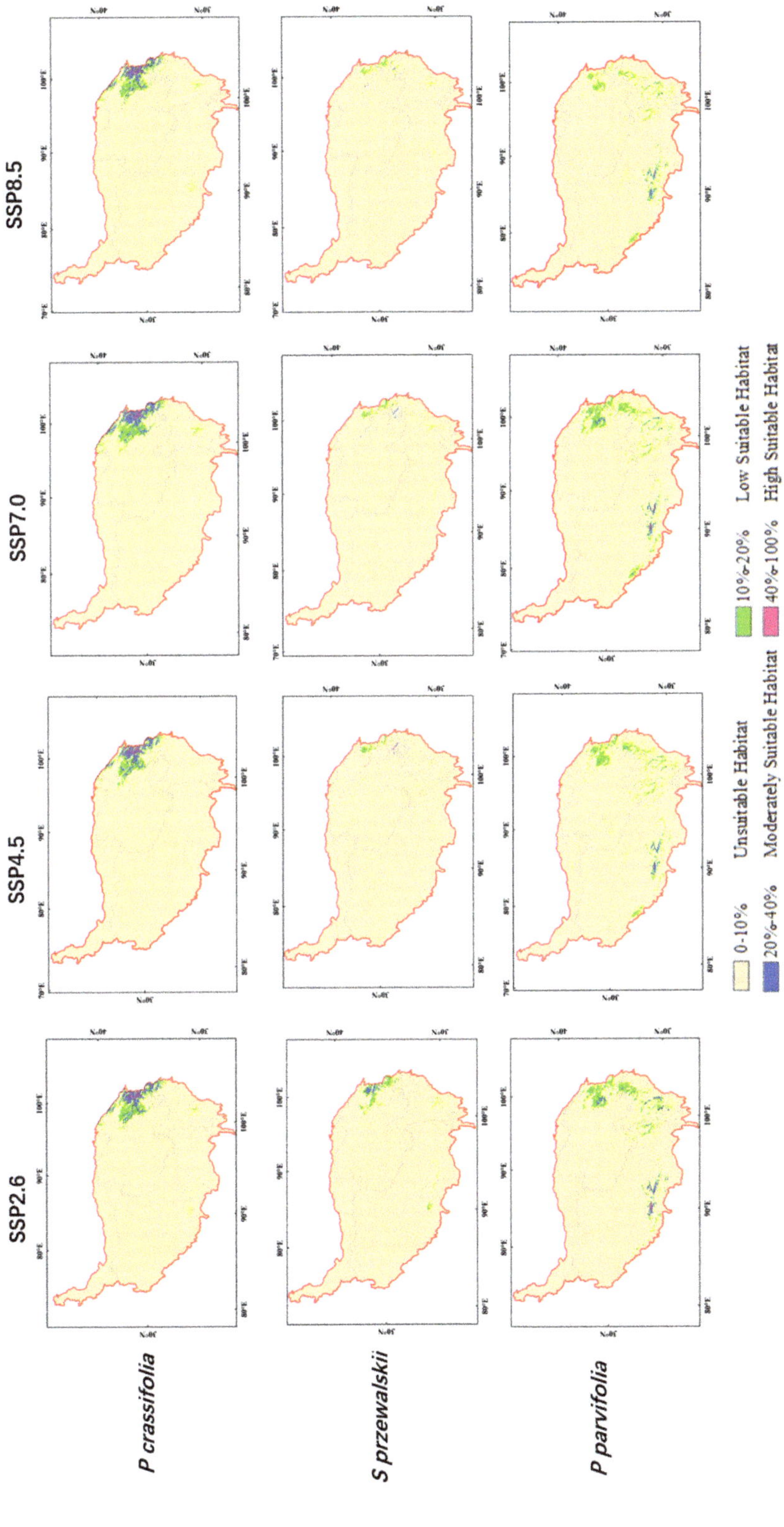

Figure 4. Prediction results of potential geographical distribution of *three species* in 2050s.

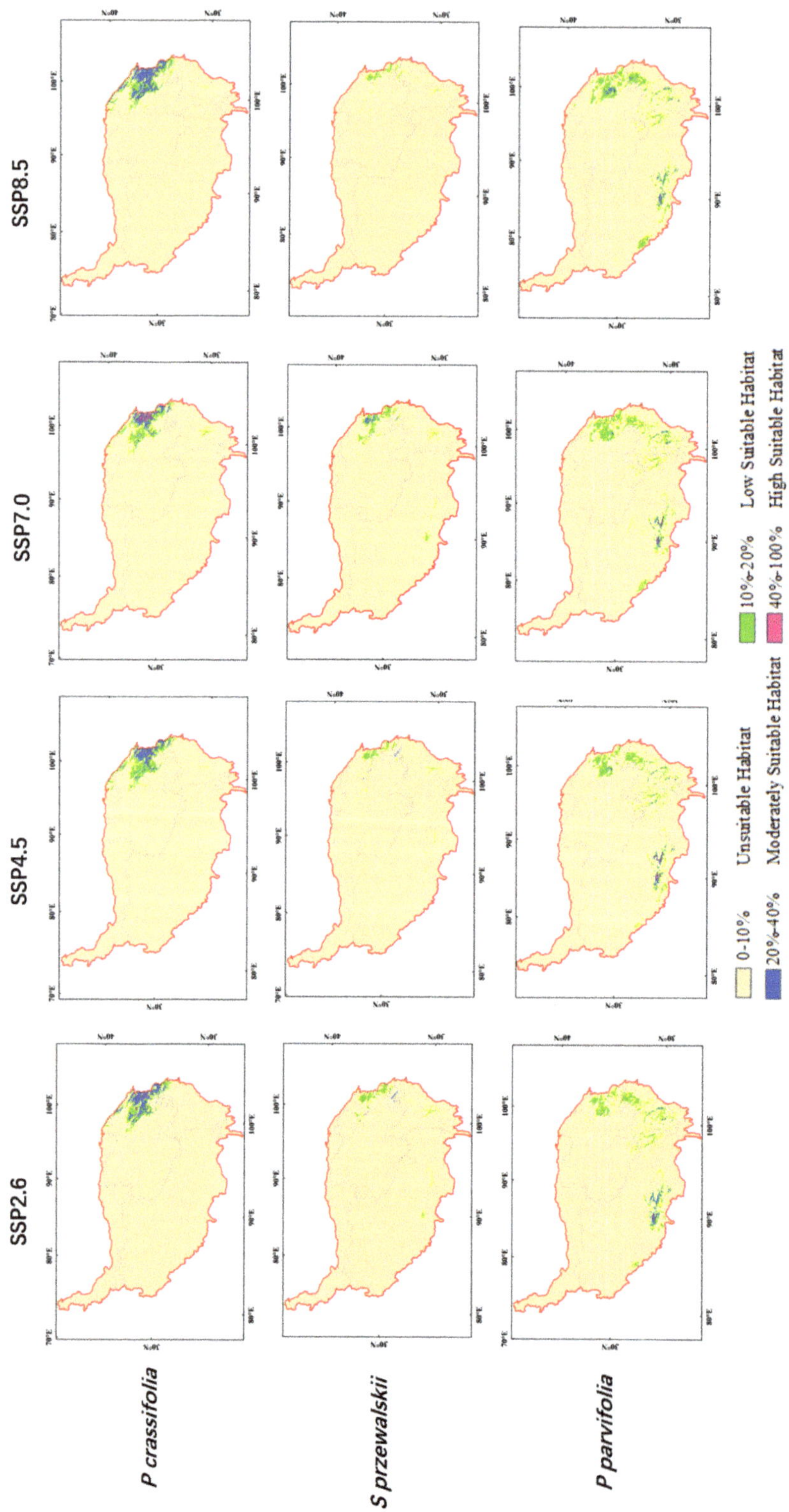

Figure 5. Prediction results of potential geographical distribution of *three species* in 2070s.

The average elevation in potentially suitable habitat for three species showed a slight upward shift under many climate scenarios in the 2050s and 2070s as compared to the average elevation of their potential distribution at current. For instance, the average elevation in potentially suitable habitat for *Potentilla parvifolia* all increased except under SSP8.5 (3552 m) in the 2050s. The average elevation of high suitable habitat for *Picea crassifolia* was 2758 m at current. However, the average elevation of highly suitable habitat for *Picea crassifolia* varied from 2773 m (SSP2.6) to 2849 m (SSP8.5) in the 2050s and increased to 2961 m (SSP2.6), 2891 m (SSP4.5) and 3043 m (SSP8.5) in the 2070s. The average elevation of highly suitable habitat, medium suitable habitat and low suitable habitat for *Sabina przewalskii* was 2951 m, 3175 m and 3083 m, respectively, at current. However, the average elevation of potentially suitable habitat for *Sabina przewalskii* varied from 2597 m to 3487 m in future climate scenarios. The changes of the average elevation in potentially suitable habitat for *Picea crassifolia* and *Sabina przewalskii* under different climate scenarios were not obvious, while the mean elevation in potentially suitable habitat for *Potentilla parvifolia* basically rose from the current period to the 2070s (Table 4).

Table 4. The average elevation of suitable habitat for three species under different climate scenarios.

	Suitable Habitat	The Average Elevation under Different Climate Scenario (m)								Average Elevation at Current (m)
		2050s				2070s				
		SSP2.6	SSP4.5	SSP7.0	SSP8.5	SSP2.6	SSP4.5	SSP7.0	SSP8.5	
P. crassifolia	High	2773	2834	2790	2849	2961	2891	2691	3043	2758
	Moderately	3033	3000	3124	3022	2972	2964	2987	3077	3031
	Low	3353	3333	3374	3383	3298	3335	3077	3375	3359
S. przewalskii	High	2597	3093	3103	——	3090	2894	2764	2649	2951
	Moderately	2765	3487	3305	3149	3334	3224	2775	2878	3175
	Low	3079	3312	3130	3049	3066	3081	3090	2996	3083
P. parvifolia	High	3712	3987	3795	3552	3831	3728	3640	4220	3617
	Moderately	3578	3597	3537	3717	3822	3605	3625	3586	2636
	Low	3499	3539	3502	3711	3587	3482	3520	3629	3475

3.4. The Changes of the Intersection Distribution of Three Species

Due to the large distribution area in the northeastern of QTP for three species, we analyzed the changes of intersection distributions for three species under future climate scenarios. Figure 6 showed the modeled vegetation fractional cover and spatial distribution of suitable habitat under future climate scenarios in the 2050s and 2070s in the northeastern of QTP. There was a decreasing trend for the intersection distribution area of three species (34,745 km^2 for SSP2.6, 15,441 km^2 for SSP4.5 and 7822 km^2 for SSP8.5 in the 2050s; 18,584 km^2 for SSP2.6, 17,060 km^2 for SSP4.5 and 15,440 km^2 for SSP8.5 in 2070s), which expanded their distribution area to the northeast. Under SSP8.5, the distribution for *Potentilla parvifolia* enlarged but the distribution for *Picea crassifolia* and *Sabina przewalskii* contracted. The total suitable habitat area for *Potentilla parvifolia* in the QTP would increase from 97,353.72 km^2 under SSP2.6 to 119,069.06 km^2 under SSP8.5 in the 2070s. However, the total suitable habitat area for *Sabina przewalskii* in the QTP would shrink from 31,205.97 km^2 under SSP2.6 to 20,496.67 km^2 under SSP8.5 in the 2070s.

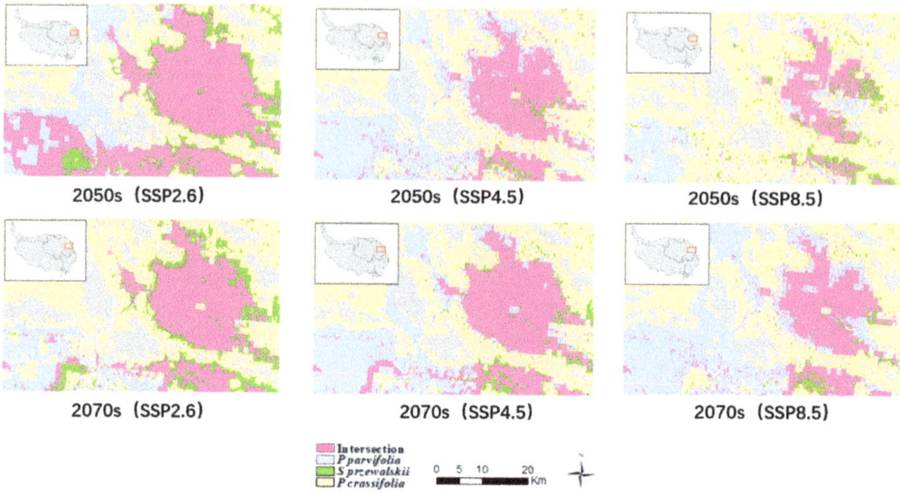

Figure 6. Intersection for *Picea crassifolia*, *Sabina przewalskii* and *Potentilla parvifolia* under SSP2.6, SSP4.5 and SSP8.5 scenarios.

4. Discussion

4.1. Influence of Environmental Variables on the Potential Distribution of Three Species

It is widely known that the species distributions are not only determined by climatic factors but also impacted by local topography, human activities and species interactions [37]. Terrain characteristics, i.e., slope, altitude, and aspect are key environmental variables for shaping the vegetation distribution by changing moisture and heat especially for alpine trees [38,39]. In this study, analyses of environmental variables showed that aspect and elevation are critical factors restricting the distribution of the three species. The three species live in different aspects and their suitable habitats so they have their own ecological characteristics. Species distribution is primarily affected by elevation and aspect in alpine forest ecosystems [40]. According to the results of this study, *Picea crassifolia* is distributed between 2691 and 3375 m. The mean elevation of the highest habitat Suitability of *Picea crassifolia* under SSP8.5 in the 2050s is about 2849 m, which is similar to the previous study [41].

Temperature and precipitation are two major climate factors affecting the species distribution, especially growth-season temperatures, cold tolerance and the available water supply for alpine trees [42]. The results showed that precipitation of driest month (Bio14), annual precipitation (Bio12), temperature seasonality (Bio4) are major climatic factors that influence the distribution of *Picea crassifolia*, *Sabina przewalskii* and *Potentilla parvifolia*. Higher precipitation of the driest month and annual precipitation have a positive impact on species distribution. Temperature and precipitation are the key factors influencing species distribution in the drier upper sites. However, species distribution is more restricted by precipitation than the temperature in the wetter upper sites [37]). Temperature seasonality is positively related to elevation and strong seasonal variation in temperature may inhibit the growth of trees [43].

Soil provides the necessary space and nutrients for species to survive and limits their distributions [44]. The soil thickness at different sites is the reason for the spatial difference of species distribution [45]. Soil thickness ≥40 cm can store enough available water to allow trees to survive during drought periods [40]. In this study, we used physical and chemical characteristics of subsoil (30–100 cm) variables to further evaluate the suitable distribution of three species. We found subsoil CEC (clay), subsoil bulk density and subsoil CEC (soil) have a little influence on species distributions the QTP.

4.2. *Average Elevation Changes of Potential Suitable Habitat for Three Model Species*

The average elevation in potentially suitable habitat for three species showed a slight upward shift under many climate scenarios in the 2050s and 2070s as compared to the average elevation of their potential distribution at current. Especially the average elevation in potentially suitable habitat for *Potentilla parvifolia* all increased except under SSP8.5 (3552 m) in the 2050s. These results are similar to other studies, which show plant species shift to higher elevation and cooler habitats responding to climate warming [46,47]. In order to adapt to climate change at local, regional, and global scales, alpine species shape the mechanism of shifting suitable climatic niches to relatively cooler habitats [48]. For three species in this study, the modeled predictions indicate species would shift to a higher elevation to occupy the current climate niche by the 2070s.

The changes of the average elevation in potentially suitable habitat for trees (*Picea crassifolia* and *Sabina przewalskii*) under different climate scenarios were not obvious, while the mean elevation in potentially suitable habitat for shrub (*Potentilla parvifolia*) had a basically rising from current to 2070s. The modeling results suggested there would be a competitive relationship between shrubs and trees. The existence of shrubs restricts the growth of trees to higher altitudes, so the average altitude of *Picea crassifolia* and *Sabina przewalskii* did not increase further with climate warming. The previous study found the changes in the mean elevation could be influenced by other factors rather than climate alone [49].

4.3. *Influence of Other Factors on the Potential Distribution of Three Species*

The vegetation is currently growing to a higher altitude [50,51], and this expansion will probably go on in the future. It is mainly responsible for climate change due to temperature or water availability [52,53]. In this study, we found that the positive interaction between shrubs and trees can promote the upward movement of vegetation. These interactions occur at slightly higher altitudes [54,55]. Shrubs are expected to expand to a higher elevation than trees with the same critical survival temperature. The snow cover is the protection of shrubs because it alleviates the influence of the temperature on shrubs [56]. The interaction between shrubs and trees may become more and more important to explain changes in the species composition and structure on the QTP. The expansion of shrubs could be discontinuous spatially and temporally which actually covers up tree expansion.

In addition, the two Ips species (*Ips nitidus* Eggers and *Ips shangrila* Cognato and Sun) are the most destructive secondary bark beetles on *Picea crassifolia* and always cause mortality of trees by their cooperation [57]. Increasing human interventions, such as harvesting, grazing and mining, may also result in distribution changes of the three species. The human population on the QTP has expanded dramatically in the past decades. Some suitable habitats for alpine species were converted to other land uses, such as pastures or settlements [40].

5. Conclusions

In this study, we explored the influence of climate change on two dominant alpine trees (*Picea crassifolia* Kom and *Sabina przewalskii* Kom) and one dominant alpine shrub (*Potentilla parvifolia* Fisch) under different climate scenarios on the Qinghai–Tibetan Plateau. The predicted current potential distribution of three species was basically located in the northeastern of Qinghai–Tibetan Plateau, and the distribution of three species was most impacted by aspect, elevation, temperature seasonality, annual precipitation, precipitation of driest month, Subsoil CEC (clay), Subsoil bulk density and Subsoil CEC (soil). There were significant differences in the potential distribution of the three species under four climate scenarios in the 2050s and 2070s including expanding, shifting, and shrinking. The mean elevation in potentially suitable habitat for *Potentilla parvifolia* basically rose from the current period to the 2070s. Our study provides an important reference for the conservation of *Picea crassifolia*, *Sabina przewalskii*, *Potentilla parvifolia* and other dominant plant species under climate change. However, our research only used the friendly MaxEnt model without considering other models. In future studies, we will select the ensemble model which can improve the reliability and accuracy of forecast results to further predict species distribution.

Author Contributions: H.H. carried out the main data search, processing, and paper writing work; Y.W. proposed the paper ideas and carried out the paper revision work; W.W. gave valuable comments in the paper writing and helped to collect and check the data; C.W. carried out the paper revision work. All authors have read and agreed to the published version of the manuscript.

Funding: This research was funded by [the Strategic Priority Research Program of Chinese Academy of Sciences] grant number [XDA19040500]; [the Second Tibetan Plateau Scientific Expedition and Research (STEP) Program] grant number [2019QZKK0502]; [the 2020 Joint Research Project of Three-River National Park of the Chinese Academy of Sciences and the People's Government of Qinghai Province] grant number [LHZX-2020-08]; [the Qinghai Province Research Project] grant number [2019-ZJ-913].

Institutional Review Board Statement: Not applicable.

Data Availability Statement: All environmental variables used in the manuscript are already publicly accessible, and we provided the download address in the manuscript; relevant sampling site information can be found in the online version.

Acknowledgments: Our deepest gratitude goes to the anonymous reviewers for their careful work and thoughtful suggestions that have helped improve this paper substantially. This research was jointly supported by the Strategic Priority Research Program of the Chinese Academy of Sciences (grant no. XDA19040500), the Second Tibetan Plateau Scientific Expedition and Research (STEP) Program (2019QZKK0502); the 2020 Joint Research Project of Three-River National Park of the Chinese Academy of Sciences and the People's Government of Qinghai Province (LHZX-2020-08) and the Qinghai Province Research Project (2019-ZJ-913). We are very grateful for their generous funding.

Conflicts of Interest: The authors declare no conflict of interest.

References

1. Araújo, M.B.; Rahbek, C. How Does Climate Change Affect Biodiversity? *Science* **2006**, *313*, 1396–1397. [CrossRef]
2. Beaumont, L.J.; Pitm, A.J.; Poulsen, M.; Hughes, L. Where will species go? Incorporating new advances in climate modelling into projections of species distributions. *Glob. Chang. Biol.* **2007**, *13*, 1368–1385. [CrossRef]
3. Lenoir, J.; Gegout, J.C.; Marquet, P.A.; de Ruffray, P.; Brisse, H. A significant upward shift in plant species optimum elevation during the 20th century. *Science* **2008**, *320*, 1768–1771. [CrossRef] [PubMed]
4. Xin, X.G.; Wu, T.W.; Zhang, J.; Yao, J.C.; Fang, Y.J. Comparison of CMIP6 and CMIP5 simulations of precipitation in China and the East Asian summer monsoon. *Int. J. Climatol.* **2020**, *40*, 6423–6440. [CrossRef]
5. Liu, X.D.; Chen, B.D. Climatic warming in the Tibetan Plateau during recent decades. *Int. J. Climatol.* **2000**, *20*, 1729–1742. [CrossRef]
6. Hickling, R.; Roy, D.B.; Hill, J.K.; Fox, R.; Thomas, C.D. The distributions of a wide range of taxonomic groups are expanding polewards. *Glob. Chang. Biol.* **2006**, *12*, 450–455. [CrossRef]
7. Pecl, G.T.; Araújo, M.B.; Bell, J.D.; Blanchard, J.; Bonebrake, T.C.; Chen, I.-C.; Clark, T.D.; Colwell, R.K.; Danielsen, F.; Evengård, B.; et al. Biodiversity redistribution under climate change: Impacts on ecosystems and human well-being. *Science* **2017**, *355*. [CrossRef] [PubMed]
8. Naudiyal, N.; Wang, J.; Ning, W.; Gaire, N.P.; Peili, S.; Yanqiang, W.; Jiali, H.; Ning, S. Potential distribution of Abies, Picea, and Juniperus species in the sub-alpine forest of Minjiang headwater region under current and future climate scenarios and its implications on ecosystem services supply. *Ecol. Indic.* **2021**, *121*, 107131. [CrossRef]
9. Xu, Z.; Zhao, C.; Feng, Z.; Peng, H.; Chao, W. The impact of climate change on potential distribution of species in semi-arid region: A case study of Qinghai spruce (Picea crassifolia) in Qilian Mountain, Gansu province, China. In Proceedings of the 2009 IEEE International Geoscience and Remote Sensing Symposium, Cape Town, South Africa, 12–17 July 2009.
10. Kearney, M.R.; Wintle, B.A.; Porter, W.P. Correlative and mechanistic models of species distribution provide congruent forecasts under climate change. *Conserv. Lett.* **2010**, *3*, 203–213. [CrossRef]
11. Phillips, S.J.; Anderson, R.P.; Schapire, R.E. Maximum entropy modeling of species geographic distributions. *Ecol. Model.* **2006**, *190*, 231–259. [CrossRef]
12. Bradter, U.; Kunin, W.E.; Altringham, J.D.; Thom, T.J.; Benton, T.G. Identifying appropriate spatial scales of predictors in species distribution models with the random forest algorithm. *Methods Ecol. Evol.* **2013**, *4*, 167–174. [CrossRef]
13. Stockwell, D.; Peters, D. The GARP modelling system: Problems and solutions to automated spatial prediction. *Int. J. Geogr. Inf. Sci.* **1999**, *13*, 143–158. [CrossRef]

14. Phillips, S.J.; Dudik, M. Modeling of species distributions with Maxent: New extensions and a comprehensive evaluation. *Ecography* **2008**, *31*, 161–175. [CrossRef]
15. Eyring, V.; Bony, S.; Meehl, G.A.; Senior, C.A.; Stevens, B.; Stouffer, R.J.; Taylor, K.E. Overview of the Coupled Model Intercomparison Project Phase 6 (CMIP6) experimental design and organization. *Geosci. Model Dev.* **2016**, *9*, 1937–1958. [CrossRef]
16. Zhang, L.; Chen, X.; Xin, X. Short commentary on CMIP6 Scenario Model Intercomparison Project (ScenarioMIP). *Progress. Inquisitiones De Mutat. Clim.* **2019**, *15*, 519–525. [CrossRef]
17. Riahi, K.; van Vuuren, D.P.; Kriegler, E.; Edmonds, J.; O'Neill, B.C.; Fujimori, S.; Bauer, N.; Calvin, K.; Dellink, R.; Fricko, O.; et al. The Shared Socioeconomic Pathways and their energy, land use, and greenhouse gas emissions implications: An overview. *Glob. Environ. Chang. Hum. Policy Dimens.* **2017**, *42*, 153–168. [CrossRef]
18. Liu, J.; Milne, R.I.; Cadotte, M.W.; Wu, Z.-Y.; Provan, J.; Zhu, G.-F.; Gao, L.-M.; Li, D.-Z. Protect Third Pole's fragile ecosystem. *Science* **2018**, *362*, 1368. [CrossRef] [PubMed]
19. Xia, M.; Jia, K.; Zhao, W.W.; Liu, S.L.; Wei, X.Q.; Wang, B. Spatio-temporal changes of ecological vulnerability across the Qinghai-Tibetan Plateau. *Ecol. Indic.* **2021**, *123*, 107274. [CrossRef]
20. Liu, J.Q.; Sun, Y.S.; Ge, X.J.; Gao, L.M.; Qiu, Y.X. Phylogeographic studies of plants in China: Advances in the past and directions in the future. *J. Syst. Evol.* **2012**, *50*, 267–275. [CrossRef]
21. Qiu, Y.X.; Fu, C.X.; Comes, H.P. Plant molecular phylogeography in China and adjacent regions: Tracing the genetic imprints of Quaternary climate and environmental change in the world's most diverse temperate flora. *Mol. Phylogenetics Evol.* **2011**, *59*, 225–244. [CrossRef]
22. Zhang, K.L.; Yao, L.J.; Meng, J.S.; Tao, J. Maxent modeling for predicting the potential geographical distribution of two peony species under climate change. *Sci. Total Environ.* **2018**, *634*, 1326–1334. [CrossRef] [PubMed]
23. Hijmans, R.J.; Cameron, S.E.; Parra, J.L.; Jones, P.G.; Jarvis, A. Very high resolution interpolated climate surfaces for global land areas. *Int. J. Climatol.* **2005**, *25*, 1965–1978. [CrossRef]
24. Fick, S.E.; Hijmans, R.J. WorldClim 2: New 1-km spatial resolution climate surfaces for global land areas. *Int. J. Climatol.* **2017**, *37*, 4302–4315. [CrossRef]
25. Kebede, A.S.; Nicholls, R.; Allan, A.; Arto, I.; Cazcarro, I.; Fernandes, J.A.; Hill, C.T.; Hutton, C.; Kay, S.; Lázár, A.N.; et al. Applying the global RCP-SSP-SPA scenario framework at sub-national scale: A multi-scale and participatory scenario approach. *Sci. Total Environ.* **2018**, *635*, 659–672. [CrossRef]
26. Wu, T.; Lu, Y.; Fang, Y.; Xin, X.; Li, L.; Li, W.; Jie, W.; Zhang, J.; Liu, Y.; Zhang, L.; et al. The Beijing Climate Center Climate System Model (BCC-CSM): The main progress from CMIP5 to CMIP6. *Geosci. Model Dev.* **2019**, *12*, 1573–1600. [CrossRef]
27. Booth, T.H. Why understanding the pioneering and continuing contributions of BIOCLIM to species distribution modelling is important. *Austral Ecol.* **2018**, *43*, 852–860. [CrossRef]
28. Jiang, F.; Li, G.; Qin, W.; Zhang, J.; Lin, G.; Cai, Z.; Gao, H.; Zhang, T. Setting priority conservation areas of wild Tibetan gazelle (Procapra picticaudata) in China's first national park. *Glob. Ecol. Conserv.* **2019**, *20*, e00725. [CrossRef]
29. Zeng, Y.; Low, B.W.; Yeo, D.C.J. Novel methods to select environmental variables in MaxEnt: A case study using invasive crayfish. *Ecol. Model.* **2016**, *341*, 5–13. [CrossRef]
30. Yan, H.; Feng, L.; Zhao, Y.; Feng, L.; Zhu, C.; Qu, Y.; Wang, H. Predicting the potential distribution of an invasive species, Erigeron canadensis L., in China with a maximum entropy model. *Glob. Ecol. Conserv.* **2020**, *21*, e00822. [CrossRef]
31. Wisz, M.S.; Hijmans, R.J.; Li, J.; Peterson, A.T.; Graham, C.H.; Guisan, A. Effects of sample size on the performance of species distribution models. *Divers. Distrib.* **2008**, *14*, 763–773. [CrossRef]
32. Elith, J.; Phillips, S.J.; Hastie, T.; Dudík, M.; Chee, Y.E.; Yates, C.J. A statistical explanation of MaxEnt for ecologists. *Divers. Distrib.* **2011**, *17*, 43–57. [CrossRef]
33. Jose, S.; Nameer, P.O. The expanding distribution of the Indian Peafowl (Pavo cristatus) as an indicator of changing climate in Kerala, southern India: A modelling study using MaxEnt. *Ecol. Indic.* **2020**, *110*, 105930. [CrossRef]
34. Pearson, R.G.; Raxworthy, C.J.; Nakamura, M.; Peterson, A.T. Predicting species distributions from small numbers of occurrence records: A test case using cryptic geckos in Madagascar. *J. Biogeogr.* **2007**, *34*, 102–117. [CrossRef]
35. Fourcade, Y.; Engler, J.O.; Rodder, D.; Secondi, J. Mapping Species Distributions with MAXENT Using a Geographically Biased Sample of Presence Data: A Performance Assessment of Methods for Correcting Sampling Bias. *PLoS ONE* **2014**, *9*, e97122. [CrossRef]
36. Brown, J.L. SDMtoolbox: A python-based GIS toolkit for landscape genetic, biogeographic and species distribution model analyses. *Methods Ecol. Evol.* **2014**, *5*, 694–700. [CrossRef]
37. Peng, J.; Gou, X.; Chen, F.; Li, J.; Liu, P.; Zhang, Y.; Fang, K. Difference in tree growth responses to climate at the upper treeline: Qilian Juniper in the Anyemaqen Mountains. *J. Integr. Plant Biol.* **2008**, *50*, 982–990. [CrossRef] [PubMed]
38. Wang, J.; Wang, Y.; Feng, J.; Chen, C.; Chen, J.; Long, T.; Li, J.; Zang, R.; Li, J. Differential Responses to Climate and Land-Use Changes in Threatened Chinese Taxus Species. *Forests* **2019**, *10*, 766. [CrossRef]
39. Zhao, Z.; Guo, Y.; Wei, H.; Ran, Q.; Liu, J.; Zhang, Q.; Gu, W. Potential distribution of Notopterygium incisum Ting ex H. T. Chang and its predicted responses to climate change based on a comprehensive habitat suitability model. *Ecol. Evol.* **2020**, *10*, 3004–3016. [CrossRef] [PubMed]

40. Yang, W.J.; Wang, Y.H.; Webb, A.A.; Li, Z.Y.; Tian, X.; Han, Z.T.; Wang, S.; Yu, P. Influence of climatic and geographic factors on the spatial distribution of Qinghai spruce forests in the dryland Qilian Mountains of Northwest China. *Sci. Total Environ.* **2018**, *612*, 1007–1017. [CrossRef]
41. Zhang, L.; Yu, P.; Wang, Y.; Wang, S.; Liu, X. Biomass Change of Middle Aged Forest of Qinghai Spruce along an Altitudinal Gradient on the North Slope of Qilian Mountains. *Sci. Silvae Sin.* **2015**, *51*, 4–10. [CrossRef]
42. Zhong, L.; Ma, Y.; Salama, M.S.; Su, Z. Assessment of vegetation dynamics and their response to variations in precipitation and temperature in the Tibetan Plateau. *Clim. Chang.* **2010**, *103*, 519–535. [CrossRef]
43. Jianmeng, F. Spatial patterns of species diversity of seed plants in China and their climatic explanation. *Biodivers. Sci.* **2008**, *16*, 470–476. [CrossRef]
44. Liu, J.; Yang, Y.; Wei, H.; Zhang, Q.; Zhang, X.; Gu, W. Assessing Habitat Suitability of Parasitic Plant Cistanche deserticola in Northwest China under Future Climate Scenarios. *Forests* **2019**, *10*, 823. [CrossRef]
45. Tromp-van Meerveld, H.J.; McDonnell, J.J. On the interrelations between topography, soil depth, soil moisture, transpiration rates and species distribution at the hillslope scale. *Adv. Water Resour.* **2006**, *29*, 293–310. [CrossRef]
46. Feeley, K.J. Distributional migrations, expansions, and contractions of tropical plant species as revealed in dated herbarium records. *Glob. Chang. Biol.* **2012**, *18*, 1335–1341. [CrossRef]
47. Wolf, A.; Zimmerman, N.B.; Anderegg, W.R.L.; Busby, P.E.; Christensen, J. Altitudinal shifts of the native and introduced flora of California in the context of 20th-century warming. *Glob. Ecol. Biogeogr.* **2016**, *25*, 418–429. [CrossRef]
48. He, X.; Burgess, K.S.; Yang, X.F.; Ahrends, A.; Gao, L.M.; Li, D.Z. Upward elevation and northwest range shifts for alpine Meconopsis species in the Himalaya-Hengduan Mountains region. *Ecol. Evol.* **2019**, *9*, 4055–4064. [CrossRef]
49. Turnbull, L.A.; Crawley, M.J.; Rees, M. Are plant populations seed-limited? A review of seed sowing experiments. *Oikos* **2000**, *88*, 225–238. [CrossRef]
50. Frost, G.V.; Epstein, H.E. Tall shrub and tree expansion in Siberian tundra ecotones since the 1960s. *Glob. Chang. Biol.* **2014**, *20*, 1264–1277. [CrossRef] [PubMed]
51. Zhu, Z.C.; Piao, S.L.; Myneni, R.B.; Huang, M.T.; Zeng, Z.Z.; Canadell, J.G.; Ciais, P.; Sitch, S.; Friedlingstein, P.; Arneth, A.; et al. Greening of the Earth and its drivers. *Nat. Clim. Chang.* **2016**, *6*, 791. [CrossRef]
52. Jiang, Y.; Zhuang, Q.; Schaphoff, S.; Sitch, S.; Sokolov, A.; Kicklighter, D.; Melillo, J. Uncertainty analysis of vegetation distribution in the northern high latitudes during the 21st century with a dynamic vegetation model. *Ecol. Evol.* **2012**, *2*, 593–614. [CrossRef] [PubMed]
53. Zhang, W.X.; Miller, P.A.; Smith, B.; Wania, R.; Koenigk, T.; Doscher, R. Tundra shrubification and tree-line advance amplify arctic climate warming: Results from an individual-based dynamic vegetation model. *Environ. Res. Lett.* **2013**, *8*, 034023. [CrossRef]
54. Druel, A.; Ciais, P.; Krinner, G.; Peylin, P. Modeling the Vegetation Dynamics of Northern Shrubs and Mosses in the ORCHIDEE Land Surface Model. *J. Adv. Model. Earth Syst.* **2019**, *11*, 2020–2035. [CrossRef]
55. Holmgren, M.; Lin, C.-Y.; Murillo, J.E.; Nieuwenhuis, A.; Penninkhof, J.; Sanders, N.; Van Bart, T.; Van Veen, H.; Vasander, H.; Vollebregt, M.E.; et al. Positive shrub-tree interactions facilitate woody encroachment in boreal peatlands. *J. Ecol.* **2015**, *103*, 58–66. [CrossRef]
56. Druel, A.; Peylin, P.; Krinner, G.; Ciais, P.; Viovy, N.; Peregon, A.; Bastrikov, V.; Kosykh, N.; Mironycheva-Tokareva, N. Towards a more detailed representation of high-latitude vegetation in the global land surface model ORCHIDEE (ORC-HL-VEGv1.0). *Geosci. Model Dev.* **2017**, *10*, 4693–4722. [CrossRef]
57. Liu, L.; Wu, J.; Luo, Y.-Q.; Li, Z.-Y.; Wang, G.-C.; Han, F.-Z. Morphological and biological investigation of two pioneer Ips bark beetles in natural spruce forests in Qinghai Province, northwest China. *For. Stud. China* **2008**, *10*, 19–22. [CrossRef]

diversity

MDPI

Review

Progress on Geographical Distribution, Driving Factors and Ecological Functions of Nepalese Alder

Chenxi Xia [1,5], Wanglin Zhao [1,2], Jinniu Wang [3], Jian Sun [1], Guangshuai Cui [1] and Lin Zhang [1,2,4,*]

1 State Key Laboratory of Tibetan Plateau Earth System, Resources and Environment, Institute of Tibetan Plateau Research, Chinese Academy of Sciences, Beijing 100101, China
2 Muoto Observation and Research Center for Earth Landscape and Earth System, Institute of Tibetan Plateau Research, Chinese Academy of Sciences, Beijing 100101, China
3 Chengdu Institute of Biology, Chinese Academy of Sciences, Chengdu 610041, China
4 Tibet Autonomous Region Institute of Science and Technology Information, Lhasa 850000, China
5 University of Chinese Academy of Sciences, Beijing 100049, China
* Correspondence: zhanglin@itpcas.ac.cn; Tel.: +86-010-8409-7055

Abstract: As the oldest species of Betulaceae, Nepalese alder (*Alnus nepalensis*) shows a high capacity for nitrogen fixation, rapid growth rate, and strong adaptability to stress environments, and it plays an important role in maintaining the structure and function of forest and agroforestry ecosystems. We explored its geographic distribution and the corresponding environmental drivers through collecting specimen records and published literature for Nepalese alder over the world during the past 40 years. The research trends, the growth limiting factors, the physiological characteristics, and ecological functions were all summarized as well. In terms of geographical distribution and limiting factors, Nepalese alder is mainly distributed in southern mountainous areas of the Himalayas and southwest China. Since it presented a clear northern limit of distribution and an upper limit of elevation, temperature is assumed to be the main environmental limiting factor. According to historical development, the research history of Nepalese alder could be divided into three main periods: the initial development (before 2001), the fast development (2002–2015), and the high-quality development (2016–2022), with the two key points in 2002 and 2015 relating to the conversion of cropland to a forest project that the government conducted and the application from theory to practice, respectively. As can be seen from the ecological functions, Nepalese alder could form symbiotic nodules with Frankia, which plays an important role in improving soil physical and chemical properties and facilitating vegetation secondary succession. Overall, the present review provides a reference for further studies on ecological adaptability and sustainable utilization of Nepalese alder under climate change, and also for regional ecosystem service, forestry production practice, and vegetation restoration.

Keywords: biological nitrogen fixation; bibliometrics; climate change; Frankia; vegetation restoration

Citation: Xia, C.; Zhao, W.; Wang, J.; Sun, J.; Cui, G.; Zhang, L. Progress on Geographical Distribution, Driving Factors and Ecological Functions of Nepalese Alder. *Diversity* **2023**, *15*, 59. https://doi.org/10.3390/d15010059

Academic Editors: Thomas Fickert and Michael Wink

Received: 30 November 2022
Revised: 24 December 2022
Accepted: 28 December 2022
Published: 4 January 2023

1. Introduction

Nepalese alder (*Alnus nepalensis*), a plant of the alder genus *Alnus* in the family of Betulaceae, is described in the Flora of China as growing in river beach wetlands or gully terrace forests (Figure 1) at a wide range of altitudes between about 700 and 3600 m. It grows fast in warm and humid environments, with a height exceeding 13 m over 5 years [1], and it has a strong carbon sequestration capacity [2]. This species is resistant to barrenness, and is often used for riverside berms and barren mountain beautification. Thus, it has been selected as an ideal tree species for ecological shelter forest and mixed afforestation [3,4]. At the same time, as a high-quality papermaking raw material, the Nepalese alder is an important tree species used to create short-cycle industrial wood raw material forests. Therefore, this species is listed in China's "Forest Resources Development and Protection Project" (FRDPP), and it is the main broad-leaved tree species for the construction of papermaking industries [5]. The characteristics of Nepalese alder as a pioneering species, its rapid growth, and its high-quality fuel carbon are inseparable from its nitrogen fixation performance by the symbiotic nodules where the Frankia is attached [6]. This nitrogen fixation performance increases soil organic matter and nitrogen and phosphorus nutrient content [7,8], promotes soil microbial reproduction, and increases soil enzyme activity [9]. To some extent, the improvement of soil properties also promotes the growth and development of other plants within the same area, so Nepalese alder is often introduced into agroforestry to enhance the biological production of target species. In the eastern Himalayas, agroforestry ecosystems dominated by cardamom (*Amomum subulatum*) and Nepalese alder are specially created [10], and Sharma et al. (2002b) found that the energy conversion efficiency and net energy increase of alder–cardamom were significantly higher than those of an ordinary cardamom system, and the production potential of the ecosystem was also optimized [11]. In addition, studies have pointed out that the new branches and leaves of Nepalese alder reflected high moisture content and are not prone to canopy fires. Therefore, it can be cultivated in large quantities as a biological fireproof tree species [12]. In all, the Nepalese alder plays an important role in maintaining the stability of forest ecosystems, agroforestry ecosystems, and the sustainable development of forestry.

Figure 1. Community structure and organ characters of Nepalese alder (Motuo, Tibet).

According to the Flora of China, the distribution of Nepalese alder spans a large altitudinal gradient near 3000 m, which indicates that this species is well-adapted to different environments. In addition, the clear altitude limit (especially the upper limit) indicates that there is an obvious hydrothermal constraint. However, at present, research on the correlation between alder and environmental factors mainly focused on the effects of nitrogen- and phosphorus-addition treatments on the growth of alder seedlings in Nepal [13,14], whereas research concentrating on the relationships between alder and hydrothermal factors such as temperature and precipitation is still lacking on a large scale [15]. In addition, Frankia attached to the roots of Nepalese alder are closely related to the nitrogen fixation ability of the tree itself, and as an actinomycete in microorganisms, its activity is also affected by environmental factors [16]. Thus, the response of Frankia to environmental factors will also affect the growth of Nepalese alder. Accordingly, this paper first discusses the global geographical distribution and limiting factors of Nepalese alder by querying and collecting the relevant literature of Nepalese alder from both domestic and foreign specimen databases, and then it analyzes the relevant research hotspots and development trends based on Citespace software and literature synthesis, respectively. We further reviewed the main physio-ecological characteristics and ecological functions of Nepalese alder, which might provide a reference for in-depth exploration of the ecological adaptability and sustainable use of alder under the background of climate change. At the same time, this paper may offer a theoretical reference for regional forestry production practice and vegetation restoration.

2. Data and Methods

2.1. Literature Collection

In the core database of the Web of Science (WOS) Science Citation Index (SCIE), a search for literature between 1981 and 2022 with a subject term of "Alnus nepalensis*" yielded a total of 140 records. In the China National Knowledge Infrastructure (CNKI) database, with the subject terms "Nepalese alder", "Handonggua", or "Mengzi alder", we found a total of 366 records. After manually removing impurities from the search data such as patents, reports, notices, and other documents, as well as some of the "Handonggua" literature which documented other alder species rather than *Alnus nepalensis*, 128 English papers and 334 Chinese documents were finally obtained.

2.2. Geographic Data Analysis

Based on the geographic information (longitude and latitude) of study sites in the existing literature described above (66 sites, the detailed geographic information is not available in all the papers), as well as that of Nepalese alder specimens from the Global Biodiversity Information Facility (GBIF, https://www.gbif.org, 184 sites, accessed on 30 September 2022) and the iPlant (http://www.iplant.cn, 71 sites, accessed on 30 October 2022), we analyzed the geographical distribution pattern about this species by using ArcGIS 10.7 for Desktop. The Shuttle Radar Topographic Mission Digital Elevation Model (SRTM DEM), the most comprehensive high-resolution digital topographic database, was applied to obtain the elevation for the above 321 sites. The SRTM DEM data with a spatial resolution of 90 m were obtained from the USGS (https://earthexplorer.usgs.gov/, accessed on 10 October 2022). Slope aspect of the study site was obtained by using visual interpretation of high-resolution remote sensing imagery (World Imagery, https://www.arcgis.com/, accessed on 2 November 2022).

2.3. Bibliometrics Analysis

With the help of CiteSpace visualization software (version number: 6.1.R3), keyword clustering analysis was carried out for the English and Chinese literature, respectively. The keyword clustering map of Nepalese alder researches were obtained, and the relevant research hotspots were revealed according to the analysis. Since the literature related to the Nepalese alder was mainly in Chinese (accounting for about 72%), we summarized the development processes about *A. nepalensis* researches based on the published papers in Chinese.

3. Geographic Distribution of Nepalese Alder and Corresponding Limiting Factors

As shown in Figure 2, the Nepalese alder is widely distributed, with a latitude ranging between 38°47′ S and 34°26′ N, and a longitude from 156°55′ W to 177°43′ E. It is mainly concentrated in the southern foothills of the Himalayas and the southwestern part of China (especially Yunnan province) with a middle and low latitude of 20°01′–30°32′ N, and a longitude of 79°06′–104°20′ E. Additionally, a few of the records were found in North America, Oceania, and Ecuador in South America. Perhaps due to the synthetic influence of climatic, geological, and topographical factors, such as the southwest monsoon, the east Asia monsoon, and the uplift of the Himalayas, coupled with the shade-intolerant and adaptive characteristics of Nepalese alder, the Himalayas, Hengduan Mountain, and their southern mountains have become the main distribution area of Nepalese alder. Compared with other alder species of the congener (*A. formosana, A. trabeculosa*), Nepalese alder has a wider range at similar latitudes and a larger altitude span (more than 3000 m); for example, *A. formosana* is only distributed at altitudes of no more than 2500 m [17], whereas *A. trabeculosa* is only distributed between 200 and 1000 m [18]. Therefore, Nepalese alder is more adaptable to complex and varied habitats.

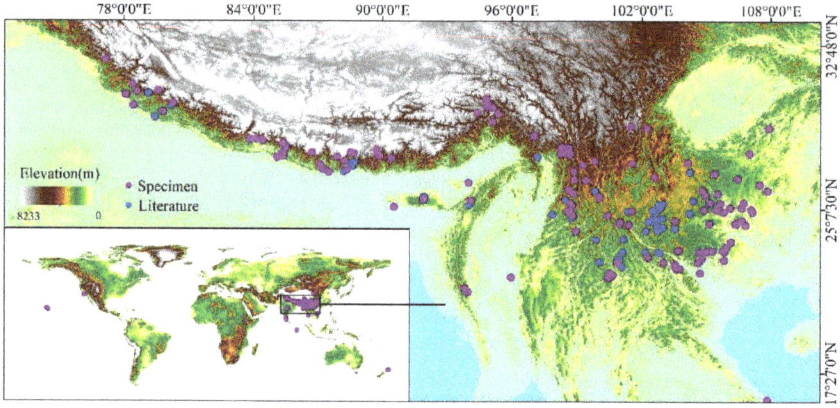

Figure 2. Geographic distribution of Nepalese alder based on information from specimen and literature.

From the respect of its northern boundary, the westernmost is in the district of Korlu in Himachal Pradesh in northeastern India, with a latitude of 32°60′ N. The northern boundary of the central region is mainly located in the Dongjiu area of Bayi District, Nyingchi City, with a latitude of 30°00′ N, whereas those of the eastern region were in Shimian County of Ya'an City, Sichuan Province and Wuchuan Gelao and Miao Autonomous County of Zunyi City, Guizhou Province, with latitudes of 28°56′ N and 28°55′ N, respectively. The meteorological data of the above four sites of northern limits were obtained from the WorldClim website (https://worldclim.org/, accessed on 11 October 2022). After justification according to the differences between gridded and exact altitudes, the annual mean temperatures were calculated as 15.4 °C, 9.6 °C, 10.8 °C, and 14.7 °C, respectively. Therefore, it seemed that the annual mean temperature of about 10 °C might be the thermal factor shaping its geographic distribution, which is consistent with Yang et al. who reported a low temperature limit of about 10 °C for Nepalese alder [4].

In terms of altitudinal distribution, nearly three-quarters of the research sites and specimen collection sites of Nepalese alder were concentrated between 1000 and 2500 m (Figure 3a), and only 12.79% of the sites sat below 1000 m, and a very small number (less than 1.3%) were distributed in alpine regions above 3500 m (mainly on Hengduan Mountain). The highest altitude record, 3860 m, happened in the northwest of Huili County in southern Sichuan Province, at a latitude of 26°39′ N. In addition, several sites with an altitude of more than 3600 m are distributed in Weixi Lisu Autonomous County, Yunnan, with a latitude of about 27°12′ N. Accordingly, it can be seen that the Nepalese alder is mainly distributed in typical mountainous terrain environments, and it is rarely distributed in low mountain plains and alpine areas above 3500 m.

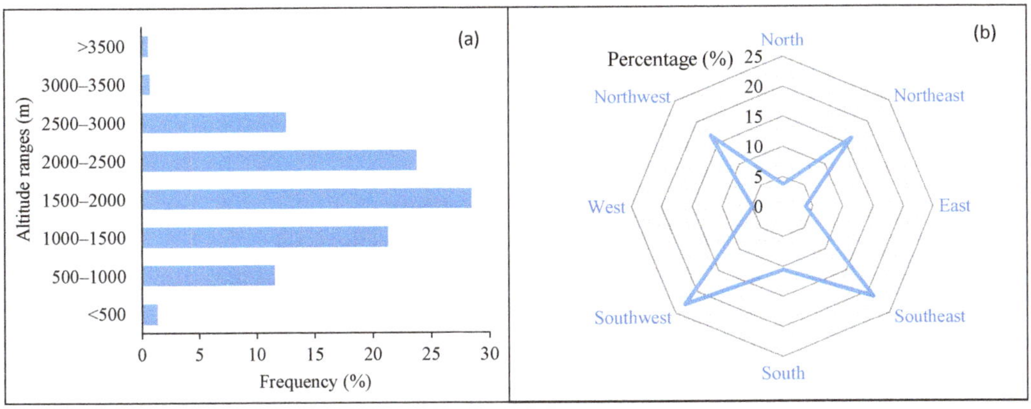

Figure 3. (**a**) Altitude and (**b**) slope aspect distribution patterns of Nepalese alder.

Slope aspect may also exert impacts on the distribution of Nepalese alder. Nearly 60% of the sites were distributed in south-facing slopes (Figure 3b), and only 37% of the samples were distributed in north-facing slopes, indicating that the Nepalese alder prefers sunny slope habitats, consistent with its sunward and light-loving characteristics [4]. This is also associated with its distribution in the southern foothills of the Himalayas and the mountainous areas of southwest China, both of which showed the same topographic characteristic of high in the north and low in the south.

4. Development History of Nepalese Alder

The whole development history can be roughly summarized into three periods: the initial development (before 2001), the fast development (2002–2015), and the high-quality development (2016–2022) (Figure 4). Among them, the number of papers published in the early period was small, with a total of only 30 papers, and the research topics mainly focused on the growth habit and forest community structure of Nepalese alder (Figure 5a). For example, the Yunnan Institute of Forestry [19] reported its growth characteristics and corresponding habit, which is a relatively systematic study of Nepalese alder in China in this stage. Guo et al. (1999) found that the alder tree layer exerted a significant effect on precipitation interception in the central Yunnan plateau [20]. In 2002, research development entered into an accelerated period, during which the number of publications increased significantly and reached a peak in 2015 (39 articles), which was about 10 times that of 2005 (the least number of articles published in this period). This might be related to the national policy to fully launch the project of returning farmland to forest in 2002, as well as the implementation of related projects which greatly promoted the research on the fast-growing alder tree species. The focus at this stage gradually shifted to vegetation restoration and seedling growth (Figure 5b). For example, Yang et al. (2004) found that the community coverage and species composition were relatively stable in phosphate mining areas [21], so it had important application value in the restoration of natural vegetation in phosphate mining wasteland. Zhou et al. (2012) found that the presence of this species significantly increased the available nitrogen and potassium content in the soil, promoted the formation of soil agglomerate structure, and effectively accelerated the restoration process of vegetation in phosphate collection areas [22]. After 2015, the number of publications showed a downward trend, which might be related to the fact that the research has focused more on practical applications. This stage focused on soil nutrients and vegetation restoration, and related research gradually shifted to practical applications (Figure 5c); for example, the Yuanyang Rice Terraces had become a research hotspot [23,24].

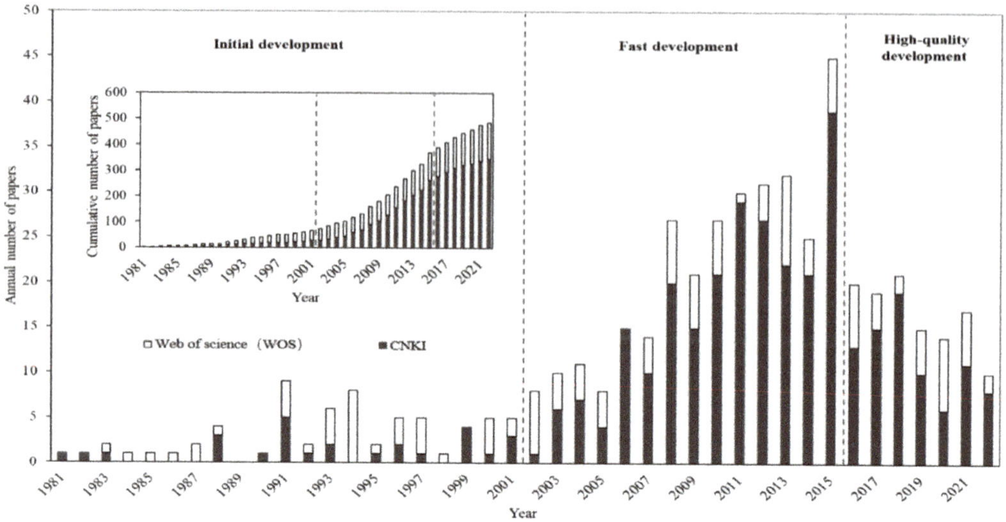

Figure 4. Changes in the number of publications of Nepalese alder since 1981.

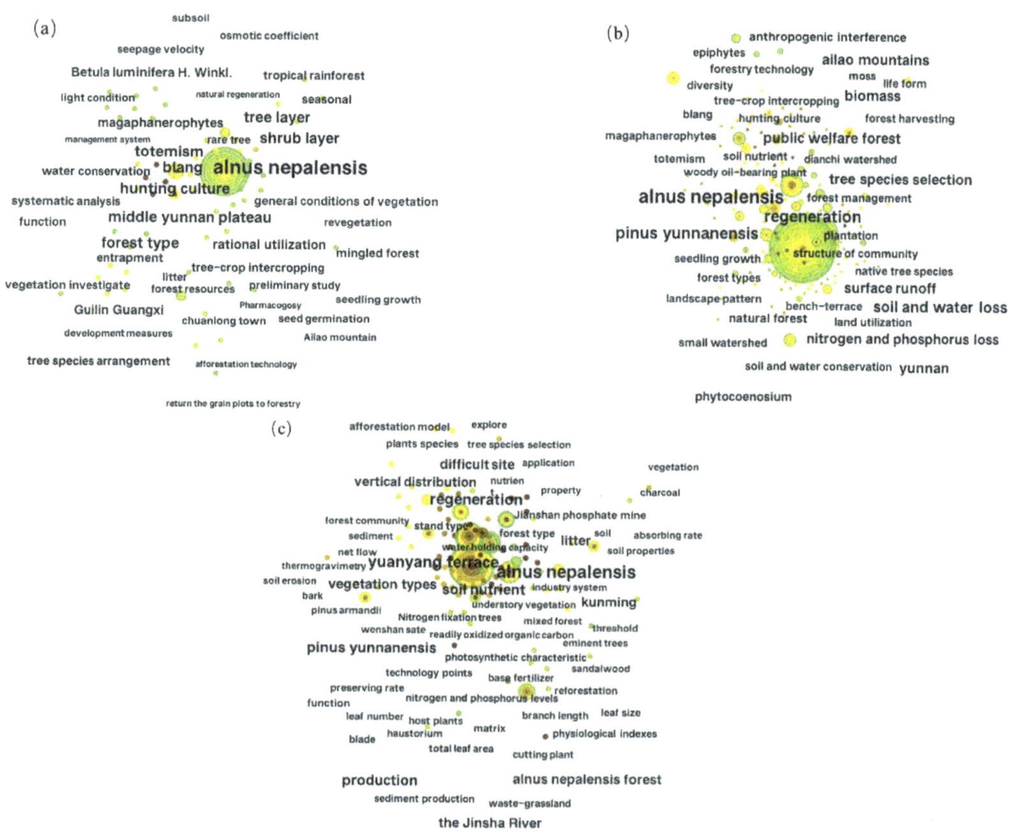

Figure 5. Keywords network map for Nepalese alder in different development stages. (**a**) Initial development, (**b**) fast development, (**c**) high-quality development.

5. Growth Limiting Factors and Sustainable Utilization of Nepalese Alder

5.1. Growth Limiting Factors of Nepalese Alder

The evident northern boundary and upper limit of Nepalese alder suggested that the thermal condition should be the dominant limiting factor. Noshiro et al. studied the effects of tree height, diameter at breast height, and altitude on the anatomical characteristics of Nepalese alder wood in the eastern Himalayas and found that the pore diameter, vessel length, and fiber pipe length of wood were negatively correlated with altitude, indicating that temperature was the main factor affecting its anatomical structure [25]. However, further research by Noshiro et al. suggested that at altitudes above 1800 m, the vessel and fiber lengths of Nepalese alder exhibited a decreasing trend with altitude that may be related to moisture constraints [15]. Quantifying the relative contributions of different hydrothermal factors to its growth and distribution will help to understand the biogeographic distribution pattern of this species; moreover, the selection of suitable ecological sources for plantation can provide important genetic resources reserved for mountain vegetation restoration.

In addition to hydrothermal factors, nitrogen and phosphorus are also important factors affecting the growth of Nepalese alder. Simulated nitrogen deposition experiments showed that the nitrogen addition at lower levels significantly promoted the growth of alder seedlings, whereas that at higher levels inhibited seedling biomass and reduced the investment to allocation of photosynthetic organs [13]. Phosphorus deficiency also decreased chlorophyll content in Nepalese alder seedlings [14]. In addition, environmental factors could indirectly influence the growth of Nepalese alder by affecting the activity or diversity of Frankia. For example, Jha et al. found that phosphate fertilizer treatment reduced mycorrhizal infection in alder seedlings, but it significantly stimulated the nodulation of alder seedlings [26]. It can be seen that phosphorus plays a very important role in the growth of Nepalese alder itself and Frankia, especially in humid tropical and subtropical areas where forest growth is largely affected by phosphorus restriction [27]. In all, the influence of climate, soil, and other factors on the growth and distribution of Nepalese alder should be comprehensively considered in the future.

5.2. Physio–Ecological Characteristics and Ecological Functions of Nepalese Alder

As a pioneer species, Nepalese alder often occupies barren hillsides due to their strong adaptability, rapid growth, and strong tolerance to barren soil, and the nitrogen fixation effect of the symbiotic root with Frankia improves soil chemical status to a certain extent. Therefore, by improving the growth environment, Nepalese alder further promotes the growth and development of other plants. At present, most scholars have discussed the physiological and ecological characteristics of Nepalese alder (Figure 6). For example, some studies have explored the growth characteristics of the Nepalese alder and cardamom agroforestry system [11], and the productivity of mixed Nepalese alder and tea plants in gardens [9]. In addition, studies have also found that chemicals in Nepalese alder might affect the growth and development of other plants through allelopathy [28]. The study of Frankia has been a major focus of alder-related research (Figure 6). At this stage, most research has focused on genetic diversity. Studies have shown that Frankia are rich in genetic diversity, which is influenced by various factors such as climate, topography, and altitude. Through the study of the diversity of Frankia, it not only helps people to understand the origin and evolution process of Frankia, so as to build an efficient symbiotic system with strong nitrogen fixation ability and stress resistance, but also provides a scientific basis for revealing the mechanisms of the nitrogen fixation and soil improvement [29].

5.2.1. Soil Improvement

The roots of Nepalese alder attached with Frankia generally form symbiosis nodulation with nitrogen fixation functions [6], thereby changing the soil physicochemical properties and increasing the soil organic matter and nitrogen and phosphorus content (Figure 6). Mishra et al. (2018) analyzed the soil fertility in different forests in the eastern Himalayas, and found that the soil organic matter content of Nepalese alder forests was significantly higher than that of other forest types, and the presence of Nepalese alder accelerated soil nutrient cycling [8]. Other studies also found that the soil total nitrogen and phosphorus, alkali-hydrolyzed nitrogen, available phosphorus, and soil microbial biomass in Nepalese alder forests were significantly higher than in other vegetation types (e.g., forests dominated by *Pseudognaphalium affine*, *Cunninghamia lanceolata*, *Michelia oblonga*, *Parkia roxburghii*, and *Pinus kesiya*) in the Yuanyang Terraces of Yunnan Province [30] and the hilly ecosystems of Northeast India [31]. Studies have also pointed out that Frankia in the roots of Nepalese alder can not only improve soil physical–chemical properties and increase soil-nutrient content, but also facilitate the reproduction of soil microorganisms and improve the activity of soil enzymes [9]. Li et al. further found that the presence of Nepalese alder increased the number of soil nitrogen-fixing bacteria in seven forest communities (*Clerodendrum bangei*, *Cunninghamia lanceolata*, *Camellia sinensis+Alnus nepalensis*, *Alnus nepalensis*, *Gnaphalium affine*, *Choerospondias axillaris*, and *Neolitzea chui+Schima superba*) in Yuanyang Terrace [32]. Among six different vegetation types in Yuanyang terraces, it was also found that Nepalese alder significantly increased the activity of soil proteases [30].

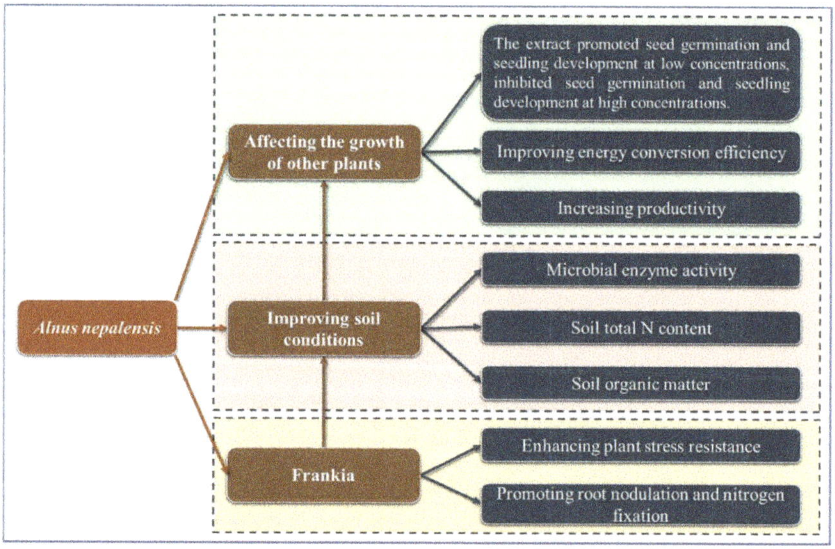

Figure 6. A framework for physio-ecological characteristics and ecological functions of Nepalese alder.

5.2.2. Effect of Nepalese Alder on Other Plants

The presence of Nepalese alder can improve soil physical and chemical properties, which in turn promotes the growth of surrounding plants (Figure 6). For example, Mortimer et al. transplanted Nepalese alder to tea (*Camellia sinensis* var. *assamica*) garden, and found that the existence of Nepalese alder increased the variety and number of fungi and bacteria in the soil, which in turn improved the productivity of tea [9]. Sharma et al. found that Nepalese alder significantly increased the soil nitrogen content in the mixed agroforestry system, thereby improving the productivity and energy conversion efficiency of cardamom communities [10]. In addition, the allelopathy of Nepalese alder can also exert a positive or negative impact on the growth of other plants to some extent. Wang et al. studied the effect of fresh leaf aqueous extract of Nepalese alder on the growth of Yunnan pine seedlings and found that the lower concentration of aqueous extract (<5 g/L) can promote the growth of seedlings, enhance root vitality, and increase chlorophyll content [28]. However, a higher concentration above 5 g/L will inhibit the growth of seedlings. Wang et al. further treated the seeds of *P. Yunnanesis* with different concentrations of extracts (800, 400, 150 mg/kg) from different organs of Nepalese alder, and found that the high concentration extracts significantly inhibited the germination and seedling growth of *P. yunnanensis* [33]. Besides, this inhibition was weakened with the decrease of extract concentration, and eventually turned into a promoting effect. In summary, the effect of Nepalese alder on the growth of other plants is mainly manifested as a promoting effect, while the allelopathy on other plants is a bit more complicated. Generally speaking, a lower concentration of aqueous extract yields a promotion effect, and a higher one may lead to inhibition. That is, when the allelopathic intensity of Nepalese alder itself is high, it will have a negative impact on the growth of other plants, which may explain why Nepalese alder forms a pure forest in the juvenile stage with few other coexisting tree species.

5.2.3. Frankia Infection

Frankia infection is the most representative physiological and ecological characteristic of Nepalese alder, and Frankia can form nitrogen-fixing nodules in symbiosis with Nepalese alder roots, which improves the soil conditions in the alder forest to a certain extent (Figure 6). At present, research on Frankia with Nepalese alder has focused on its genetic diversity, nutrient absorption, and morphological variation. Among them, the genetic diversity of Frankia has attracted a lot of attention. Studies have shown that the genetic diversity of Frankia nodules is influenced by a variety of factors, including topography, climate, and altitude [29]. On one hand, Frankia is widely distributed in multiple species, indicating that it can survive in diverse habitats [34]. On the other hand, in different habitats, especially at high altitudes where the environment is harsh, environmental factors such as stronger ultraviolet radiation and drought may cause genetic instability, resulting in more replication errors and higher genetic diversity to meet survival needs [35,36]. Xiong et al. studied the genetic diversity of Frankia in five regions of Yunnan, and found that the distribution and genetic structure of Frankia were closely related to the environment, and there were dominant genotypes in different regions [37]. Tang et al. used rep-PCR to study the genetic diversity of Frankia in Nepalese alder nodules under different habitats in Yunnan, and found the genetic diversity was positively correlated with the degree of environmental stress [29]. Similarly, Dai et al. [34] found that the genetic diversity of the Frankia strain in samples of Nepalese alder nodules in the Hengduan Mountains is related to climate and glacial history. In addition to topographical and climatic factors, altitude is also an important factor affecting the genetic diversity of Frankia, and it is assumed that the greater the altitude gradient, the richer the genetic diversity of Frankia [38,39].

6. Perspectives

Although there exist uncertainties about the geographical origin of *Alnus* and the role of climatic factors in shaping its distribution, the above statement indicates that low temperature should be one of the key limiting factors for its distribution. Therefore, in the context of future climate change, Nepalese alder is assumed to dominate in higher latitudes and/or altitudes, which is consistent with the hypothesis that the distribution range of Nepalese alder will expand significantly under the background of climate change, as proposed by Ranjitkar et al. [40]. Based on the maximum entropy model and related climate, soil, and topographic data, the future distribution trend of Nepalese alder under different scenarios can be further simulated.

In addition to the influence of climatic factors, the loss of phosphorus will have an impact on the coupling of Nepalese alder and Frankia to a certain extent [26]. However, there are relatively few studies concentrated on the impact of multi-environmental factors on the growth of Nepalese alder at this stage, which limits our in-depth understanding of the change trend and distribution range of Nepalese alder in future situations. In addition, the Himalayan mountains are very young and still in an active stage. Affected by geology, topography, and climatic factors, as well as the uplift of the Himalayas, the mountains are prone to geological disasters such as landslides and debris flows, which have recently caused a series of ecological and environmental problems such as forest degradation and species habitat destruction [41,42]. Therefore, strengthening the study of environmental impacts on the growth of Nepalese alder is of great necessity to make full use of Nepalese alder which may improve soil chemical and physical status in early succession, and to accelerate the positive evolution of ecosystem functions.

Previous studies have mentioned the effects of temperature and precipitation on the anatomical structure (i.e., stem traits) of Nepalese alder. Whereas, there is still a lack of research on plant functional traits, especially root and leaf traits, which limits our understanding of the response of Nepalese alder to the changes of hydrothermal factors from different sources. In particular, this species is mainly distributed in tropical and subtropical regions, and the phenomenon of leaf feeding by insects is obvious, and the analysis of leaf morphology and related traits is significant for understanding the plant–insect interaction and its mechanism in light of climate change. This work needs to be carried out urgently.

At this stage, most of the research on Frankia focused on its genetic diversity, especially the environmental impacts (terrain, climate, and altitude) on the genetic diversity, while research on the growth and reproduction of Frankia itself has rarely been reported. Relevant research is important to explore the interconnection between Frankia and Nepalese alder, and to reveal the mechanism affecting the distribution and growth of Nepalese alder. The relationship between Frankia and alder is the key factor in using Nepalese alder for ecological restoration, and it has important theoretical and practical significance for practicing General Secretary Xi Jinping's thesis that lucid waters and lush mountains are golden mountains, and the secret to establishing an ecological civilization highland on the Qinghai-Tibet Plateau.

Author Contributions: Conceptualization, L.Z. and C.X.; methodology, L.Z. and C.X.; software, C.X. and W.Z.; validation, L.Z.; formal analysis, C.X.; investigation, L.Z. and C.X.; resources, L.Z.; data curation, C.X. and W.Z.; writing—original draft preparation, C.X.; writing—review and editing, C.X., L.Z., J.W., J.S. and G.C.; visualization, L.Z.; supervision, L.Z.; project administration, L.Z.; funding acquisition, L.Z. All authors have read and agreed to the published version of the manuscript.

Funding: The Second Tibetan Plateau Scientific Expedition and Research (STEP) Program (2019QZKK0301-1) and the Key technology research and development projects in Xizang Autonomous Regions (XZ202101ZY0005G) provided financial support.

Institutional Review Board Statement: Not applicable.

Data Availability Statement: All data used in the manuscript are already publicly accessible, and we provided the download address in the manuscript.

Conflicts of Interest: The authors declare no conflict of interest.

References

1. Forestry and Grassland Bureau of Lanping County. Series of Cultivation Techniques of Main Silvicultural Tree Species in Yunnan Forest Industry (*Alnus nepalensis*). Available online: https://www.lanping.gov.cn/xxgk/015280551/info/2016-96381.html (accessed on 30 October 2022).
2. Zhang, P.C.; Zhang, Y.P.; Yang, G.P.; Zheng, Z.; Liu, Y.H.; Tan, Z.H. Carbon storage and sequestration of tree layer in subtropical evergreen broadleaf forests in Ailao Mountain of Yunnan. *Chin. J. Ecol.* **2010**, *29*, 1047–1053.
3. Zhou, G.L.; Yang, C.H. Alnus nepalensis. *Guizhou For. Sci. Technol.* **1988**, *1*, 93–95.
4. Yang, C.H. Alnus nepalensis. *Soil Water Conserv. China* **1990**, *6*, 41–42.
5. Wang, J.H.; Gu, W.C.; Li, B.; Guo, W.Y.; Xia, L.F. Study on selection of *Alnus cremastogyne* provenance/ family—Analysis of growth adaptation and genetic atability. *Sci. Silvae Sin.* **2000**, *36*, 59–66.
6. Varghese, R.; Chauhan, V.S.; Misra, A.K. Evolutionary implications of nucleotide sequence relatedness between Alnus nepalensis and Alnus glutinosa and also between corresponding Frankia micro symbionts. *Plant Soil* **2003**, *254*, 219–227. [CrossRef]
7. Chaudhry, S.; Singh, S.P.; Singh, J.S. Performance of Seedlings of Various Life Forms on Landslide-Damaged Forest Sites in Central Himalaya. *J. Appl. Ecol.* **1996**, *33*, 109–117. [CrossRef]
8. Mishra, G.; Giri, K.; Pandey, S. Role of Alnus nepalensis in Restoring Soil Fertility: A Case Study in Mokokchung, Nagaland. *Natl. Acad. Sci. Lett.* **2018**, *41*, 265–268. [CrossRef]
9. Mortimer, P.E.; Gui, H.; Xu, J.; Zhang, C.; Barrios, E.; Hyde, K.D. Alder trees enhance crop productivity and soil microbial biomass in tea plantations. *Appl. Soil Ecol.* **2015**, *96*, 25–32. [CrossRef]

10. Sharma, G.; Sharma, R.; Sharma, E.; Singh, K.K. Performance of an age series of Alnus-cardamom plantations in the Sikkim Himalaya: Nutrient dynamics. *Ann. Bot.* **2002**, *89*, 273–282. [CrossRef]
11. Sharma, G.; Sharma, E.; Sharma, R.; Singh, K.K. Performance of an age series of alnus-cardamom plantations in the Sikkim Himalaya: Productivity, energetics and efficiencies. *Ann Bot.* **2002**, *89*, 261–272. [CrossRef]
12. Li, S.Y.; Guan, X.Y.; Chang, N.N.; Wang, Q.H.; Shu, L.F. Comparative study on combustion characteristics of different size live ranches of *Alnus nepalensis*. *J. Cent. South Univ. For. Technol.* **2011**, *31*, 61–64.
13. Tang, H.Y.; Xu, L.P.; Li, S.F.; Huang, X.B.; Yang, L.H. Effects of simulated nitrogen deposition on the growth, twig and leaf traits of *Alnus nepalensis* seedlings in the southern subtropical region. *J. Northwest For. Univ.* **2018**, *33*, 162–166.
14. Liu, Z.N. Effects of nitrogen and phosphorus stress on physiological indexes response and changes of mixed planting seedlings of *pinus yunnanensis* and *Alnus nepalensis*. *For. Inventory Plan.* **2019**, *44*, 225–230.
15. Noshiro, S.; Suzuki, M.; Joshi, L.; Ikeda, H.; Ohba, H. Ecological wood anatomy of *Alnus nepalensis* (Betulaceae) throughout Nepal. *Iawa J.* **2020**, *41*, 261–277. [CrossRef]
16. Sharma, G.; Sharma, R.; Sharma, E. Impact of altitudinal gradients on energetics and efficiencies of N_2 -fixation in alder–cardamom agroforestry systems of the eastern Himalayas. *Ecol. Res.* **2010**, *25*, 1–12. [CrossRef]
17. Zhao, Z.H. *Effect of Fertilization in Late-Season on Physiological Growth of Alnus formosana (Burkill) Makino*; Guangxi University: Guangxi, China, 2020.
18. Zhang, Y.N. *Effect of Environmental Factors on Seed Germination and Seedling Growth of the Wetland Plant Alnus trabeculosa*; Central China National University: Wuhan, China, 2011.
19. Yunnan Institute of Forestry Science. *Alnus nepalensis. J. West China For. Sci.* **1975**, 18–19. Available online: https://www.oriprobe.com/journals/caod_2440.html (accessed on 30 October 2022).
20. Guo, L.Q.; Wang, Q.H.; Zhou, H.C.; Yang, W. Rainfall interception of forest plants of main forest types in central Yunnan plateau. *Yunnan For. Sci. Tech.* **1999**, *1*, 13–22.
21. Yang, L.P.; Ye, Q.Y.; Yang, S.H.; Wang, B.R. The features of *Alnus nepalensis* community in phosphate mining wasteland and its role in vegetation restoration. *J. Yunnan Univ.* **2004**, *26*, 234–237.
22. Zhou, Q.H.; Duan, X.M.; Yue, F. Effects of *Alnus nepalensis* plantation on soil physical and chemical properties in the cutting area of Kunming phosphate mine. *Jiangsu Agric. Sci.* **2012**, *40*, 324–325.
23. Li, M.P.; Liao, N.; Liu, S.R. Effect of nitrogen-fixing tree species Alnus nepalensis on the degraded soils and understory restoration in the upper reaches of the Jinsha River, China. *Acta Ecol. Sin.* **2022**, *42*, 2321–2330.
24. Zhao, C.; He, L.P.; Li, G.X.; Shao, J.P.; Chai, Y. Impacts of vegetation restoration on the soil organic carbon storage in Kunyang phosphorite mine. *Res. Soil Water Conserv.* **2017**, *24*, 168–171.
25. Noshiro, S.; Joshi, L.; Suzuki, M. Ecological wood anatomy of *Alnus nepalensis* (Betulaceae) in East Nepal. *J. Plant Res.* **1994**, *107*, 399–408. [CrossRef]
26. Jha, D.K.; Sharma, G.D.; Mishra, R.R. Mineral nutrition in the tripartite interaction between Frankia, Glomus and Alnus at different soil phosphorus regimes. *New Phytol.* **1993**, *123*, 307–311. [CrossRef]
27. Hou, E.Q.; Luo, Y.Q.; Kuang, Y.W.; Chen, C.R.; Liu, X.K.; Jiang, L.F.; Luo, X.Z.; Wen, D.Z. Global meta-analysis shows pervasive phosphorus limitation of aboveground plant production in natural terrestrial ecosystems. *Nat. Commun.* **2020**, *11*, 637. [CrossRef] [PubMed]
28. Wang, X.L.; Cao, Z.L.; Zhu, X. Allelopathy of water extract of *Alnus nepalensis* on the seedling growth of *Pinus yunnanensis*. *J. Gansu Agric. Univ.* **2012**, *47*, 76–79.
29. Tang, X.M.; Dai, Y.M.; Xiong, Z.; Zhang, Z.Z.; Zhang, C.G. Effect of nature stress on genetic diversity of *Frankia* in *Alnus* nodules. *Chin. J. Appl. Ecol.* **2003**, *14*, 1743–1746.
30. Wu, L.T.; Zhang, Z.H.; Xie, G.L.; Di, X.Y.; Shi, H. Contents of soil nutrients and characteristics of enzyme activities in different vegetation types of the Yuanyang terraces. *Guangdong Agric. Sci.* **2022**, *49*, 86–95.
31. Ramesh, T.; Manjaiah, K.M.; Tomar, J.M.S.; Ngachan, S.V. Effect of multipurpose tree species on soil fertility and CO_2 efflux under hilly ecosystems of Northeast India. *Agrofor. Syst.* **2013**, *87*, 1377–1388. [CrossRef]
32. Li, M.R.; Li, T.G.; Li, B.; Qin, L.; Jiang, M. Soil microorganisms quantities and nutrients contents of different forest communication at core zone on Yuanyang terrace. *J. West China For. Sci.* **2019**, *48*, 49–55.
33. Wang, X.L.; Cao, Z.L.; Zhu, X. Allelopathic effect of ether ethano extraction from Alnus nepalensis organs on pinus yunnanensis seed germination. *J. Southwest For. Univ.* **2010**, *30*, 21–23.
34. Dai, Y.M.; He, X.Y.; Zhang, C.G.; Zhang, Z.Z. Characterization of genetic diversity of Frankia strains in nodules of Alnus nepalensis (D. Don) from the Hengduan Mountains on the basis of PCR-RFLP analysis of thenif D-nif KIGS. *Plant Soil* **2004**, *267*, 207–212. [CrossRef]
35. Cai, M.J.; Feng, H.Y.; An, L.Z.; Wang, X.L. Present status and prospects in research on effect of enhanced UV-B radiation on plants. *Chin. J. Appl. Ecol.* **2002**, *13*, 359–364.
36. Nevo, E. Evolution of genome–phenome diversity under environmental stress. *Proc. Natl. Acad. Sci. USA* **2001**, *98*, 6233–6240. [CrossRef] [PubMed]
37. Xiong, Z.; Tang, X.M.; Dai, Y.M.; Zhang, C.G.; Zhang, Z.Z.; Xu, L.H. Genetic diversity of Frankia strains in *Alnus nepalensis* nodules in Yunnan revealed by rep-PCR. *Chin. J. Appl. Environ.* **2006**, *12*, 623–627.

38. Khan, A.; Myrold, D.D.; Misa, A.K. Distribution of Frankia genotypes occupying Alnus nepalensis nodules with respect to altitude and soil characteristics in the Sikkim Himalayas. *Physiol. Plant.* **2007**, *130*, 364–371. [CrossRef]
39. Xiong, Z.; Li, W.J.; Zhang, Z.Z.; Jiang, C.L. The influence of altitude on the genetic diversity of microsymbionts of *Alnus nepalensis*—Frankia. *J. Southwest For. Univ.* **2001**, *21*, 205–209.
40. Ranjitkar, S.; Sujakhu, N.M.; Lu, Y.; Wang, Q.; Wang, M.C.; He, J.; Mortimer, P.E.; Xu, J.C.; Kindt, R.; Zomer, R.J. Climate modelling for agroforestry species selection in Yunnan Province, China. *Environ. Model. Softw.* **2016**, *75*, 263–272. [CrossRef]
41. Dand, C.; Sun, X.F.; Cheng, Z.L. Vertical zonality characteristics of debris flow along ZhaMo highway in Tibet. *Bull. Soil Water Conserv.* **2018**, *38*, 328–333.
42. Yang, T.J. Characteristics and control measures of debris flow along the ZhaMo highway. *China Sci.* **2018**, *13*, 1055–1059.

diversity

MDPI

Review

Incorporating Effect Factors into the Relationship between Biodiversity and Ecosystem Functioning (BEF)

Jian Hou *, Haobo Feng and Menghan Wu

Jixian National Forest Ecosystem Observation and Research Station, National Ecosystem Research Network of China, School of Soil and Water Conservation, Beijing Forestry University, Beijing 100083, China; cfw12356@126.com (H.F.); 15032957056@163.com (M.W.)
* Correspondence: houjian@bjfu.edu.cn

Abstract: Generally, the high levels of biodiversity found in natural ecosystems have positive effects on ecosystem functions (EFs), though the intensity and direction of such effects can vary. This is associated with the impacts of other EF-driving factors. In this study, the factors that affect biodiversity-ecosystem functioning (BEF) are reviewed and summarized, and current gaps in the research on the effects of these factors on BEF are discussed. Moreover, a new conceptual model, the generating-presentation model, accounting for links between effect factors and EFs, is built to provide a systematic means of understanding how different factors affect BEF. The model shows that the correlation between biodiversity and EFs can be described as involving a cascade process, while the separation of biodiversity and EFs from ecosystems without considering integrated features is not appropriate for BEF-related research. The generating-presentation model can comprehensively reflect the effects of different factors on EFs and thus has major theoretical and applied implications.

Keywords: community structures; conceptual model; multifunctionality; trophic level

Citation: Hou, J.; Feng, H.; Wu, M. Incorporating Effect Factors into the Relationship between Biodiversity and Ecosystem Functioning (BEF). *Diversity* **2022**, *14*, 274. https://doi.org/10.3390/d14040274

Academic Editors: Lin Zhang, Jinniu Wang and Michael Wink

Received: 14 March 2022
Accepted: 3 April 2022
Published: 5 April 2022

1. Introduction

Biodiversity decline over the past two decades has drawn more attention to the effects of biodiversity loss on ecosystem functions (EFs) [1,2]. This has led to the development of an ecological field examining the relationship between biodiversity and ecosystem functioning (BEF) [3]. Generally, high levels of biodiversity have positive effects on ecosystem functions. However, the intensity and direction of such effects can vary [4]. This is associated with the impacts of other EF-driving factors. For example, van der Plas [3] compared the effects of biodiversity, community composition, and abiotic conditions on EFs based on a review of BEF-related studies on terrestrial and aquatic ecosystems. The author's results show that the effects of biodiversity on EFs are dependent not only on community factors but also on the environment and the scale of the given community [3]. Therefore, to maintain a sustainable level of EF, the protection of abiotic conditions that maintain species and their functional traits and sustainable biodiversity is required. How can various factors of BEF be considered in a methodical way? Which effect factors should be considered in BEF research? These questions urgently need to be addressed.

The purpose of this study is to integrate the different factors that affect BEF and to propose a systematic way of understanding how different factors affect BEF. We first review recent BEF research, particularly that published over the past four years, and summarize the factors that affect BEF. Then, we describe current gaps in the research on the effects of these factors on BEF. Moreover, from conceptual models of links between effect factors and EFs provided in the literature, a new conceptual model is built to summarize the factors that affect BEF and to comprehensively describe ways of measuring the effects of different factors on EFs. This new conceptual model can clearly express the relationships between various effect factors and BEF, which has important theoretical and applied implications.

2. Community-Related Factors Affecting BEF

2.1. Community Structure

Community structures have a significant effect on BEF [5,6]. As variations in biodiversity characterize the dynamics of community composition, the integration of community-related factors (e.g., species abundance and distribution) into BEF research based on studies on biodiversity variation can further our understanding of the effects of community changes on EFs. For example, Cao et al. [7] quantified the species abundance and EFs of benthic invertebrates in Dian Lake (Yunnan, China) to evaluate the effects of the former on the latter. Their results indicate that the relative abundance of Annelida animals shows a significant linear correlation with ecosystem multifunctionality (MEF) [7]. Moreover, Huang et al. [8] performed a study on large-scale forest ecosystems and found high levels of tree species richness to be positively correlated with forest yields while this positive correlation is affected by tree growth rates. Bannar-Martin et al. [9] partitioned the relative contributions of species richness and community composition into EFs using the Pice equation. The study showed that considering biodiversity and community composition in temporal dynamics can predict EF variations more effectively than considering biodiversity alone [9]. However, species richness can be determined by the number of species in a community and the functional traits related to a given species. Therefore, the partitioning of the relative contributions of species richness and community compositions to EF can prove challenging [9]. This may account for the lack of case studies conducted in this area [10], which must be remedied in future BEF research.

2.2. Trophic Level

Despite the significant effects of community trophic levels on BEF, few studies have been conducted in this area [11]. The complementarity effect is often employed to measure EF enhancement resulting from biodiversity [12]. In particular, the complementarity effect is related to numerous community-related features, including the niche difference of species within a community and interspecific competition [13]. However, trophic structures can affect the compensation effect by limiting interspecific competition, which is linked to the community compositions of trophic levels [14]. Hence, the effects of community trophic level structures on BEF are evident. However, research on the effects of EFs of biodiversity at different trophic levels and especially at high trophic levels remains limited [11,15]. Schuldt et al. [16] investigated the effects of biodiversity on MEF at multiple trophic levels for a forest ecosystem with rich levels of diversity in the sub-tropical zone. The study showed that the EF effects of species richness at high trophic levels can be transferred via the food web [16]. This indicates that biodiversity at high trophic levels can indirectly affect EFs via its effects on biodiversity at adjacent lower trophic levels. This cascade effect of biodiversity at high trophic levels on EFs highlights the importance of understanding the effects of biodiversity at multiple trophic levels on EFs. Moreover, Schuldt et al. [16] asserted that BEF-related studies focused on producers alone may underestimate the overall effect of biodiversity on EFs. Research on the effects of biodiversity or species composition at high trophic levels on EF may thus facilitate the development of theories and applications of ecology.

At high trophic levels, species are exposed to high levels of extinction risk, which may have severely negative impacts on EFs [17]. Therefore, research on the mechanisms dominating BEF under complex food web conditions is of great significance. Biodiversity at high trophic levels is not only key for MEF predictions but is also an important management objective [15]. The diversity of heterotrophic species is often regarded in studies as a component of MEF rather than as a factor driving EF [16]. However, these studies neglect the effects of biodiversity at high trophic levels on EFs, limiting a complete understanding of BEF [18,19]. The diversity of heterotrophic species cannot be replaced with plant diversity. However, relationships between MEF and biodiversity at high trophic levels have not yet been revealed. At the same time, for most ecosystems, biodiversity mechanisms at high trophic levels impacting MEF via the food web are not fully understood [20].

While few BEF studies have considered the food web, research on food chain energy fluxes is relatively mature [21]. The application of energy flux theories to biological chains can serve as a novel means of researching the formation of BEF in complex ecosystems [21]. For example, intraguild predation, which refers to predatory behavior between two species with the same prey, is often observed in food chain energy flows [17]. Intraguild predation can extend the vertical niche space and reduce energy flows from a low trophic level to a high trophic level, affecting biodiversity food web dynamics. Wang et al. [17] simulated a food web involving five species to show that the correlation between biodiversity and EFs is enhanced under intraguild predation. The study demonstrates that plant BEF model improvements based on studies considering multiple trophic levels are effective in clarifying the effects of biodiversity on EF [17]. Hence, future studies should focus more heavily on tests involving multiple trophic levels.

2.3. Parasite-Host Relations

Species interactions have been widely investigated due to their significant effects on species functions, species diversity, and ecological processes [22]. However, the parasite-host relation, a key factor shaping species interactions, is rarely discussed in BEF studies. Few studies have explored the effects of parasite communities on EFs and ecosystem productivity [23]. Parasitic species impact the quantity and traits of hosts and can consequently further impact EFs. Thus, BEF-related studies must further consider parasite-host relations [23]. In particular, microbes, a key parasite species, rapidly respond to environmental changes. Based on this, the effects of the microbe community on plant community yields under global climate change and significant human disturbance must be better understood to maintain and optimize primary productivity.

Several studies have investigated the mechanisms underlying the effects of parasites on BEF. For example, Ferlian et al. [24] explored interactions between plants and mycorrhiza and demonstrated that the effects of plant diversity on EF tend to be more pronounced when intraspecific competition exceeds interspecific competition. Mycorrhiza can affect plant competition by providing water and nutrients to plants in exchange for photosynthetic products [25]. Moreover, mycorrhiza traits vary across plants [26]. Thus, mycorrhiza plays a key role in maintaining plant diversity and sustainable BEF. Laforest-Lapointe et al. [27] found a positive correlation between leaf bacteria diversity and ecosystem productivity. In particular, the authors demonstrate that both the functional characteristics of the host species and functional diversity have great impacts on the structure and diversity of leaf bacteria communities [27]. In other words, microbes can affect EFs through changes in the adaptability and phenotype of the host. Leaf bacteria diversity is proportional to the host yield, which can be attributed to the following three mechanisms. First, leaf bacteria can improve the host's resistance to viruses by increasing resource competitiveness, reducing the nutrient base and enhancing antibiotic production. Second, leaf bacteria can affect the production of plant hormones such as auxin and cytokinin. Third, nitrogen-fixing bacteria can facilitate the collection of nitrogen available to plants [27].

3. External Factors Affecting BEF

3.1. Effects of Environmental Factors on BEF

Several studies have discussed the impacts of environmental changes on BEF, demonstrating the importance of environmental factors to BEF-related research [28–30]. For example, Nunes et al. [31] investigated environmental factors that affect beetles and revealed that community attributes and environmental variables can simultaneously affect EFs. In reference to grassland ecosystems, Zirbel et al. [32] demonstrated that multifunctionality variation shapes landscape compositions four times more than functional diversity. Moreover, the correlation between EF and environmental factors was found to be stronger than that between EF and biodiversity. Smeti et al. [33] investigated correlations between species diversity along the bottom muddy layers of rivers and several EFs (e.g., decomposition rate of organic substances) under environmental stresses of river basin pollutants. The study demonstrated that BEF is significantly affected by environmental stress [33]. Thus, BEF-related analyses must cover both community-related attributes and environmental factors.

Much attention has been paid to the effects of environmental changes on biodiversity, and mechanisms underlying environmental factors affecting biodiversity and EF are currently being explored at length [34]. These mechanisms can be grouped into four categories: resource availability efficiency, the spatial differentiation of resources, community habitat space, and the functional differences between species [35]. Due to differences in the methods and scales used, the reported mechanisms of environmental factors affecting EF under different experimental conditions show variations even for similar EFs [36]. For example, Spaak et al. [37] applied a theoretical analysis model to demonstrate that minor environmental impacts can lead to the collapse of EFs at a constant level of species richness. Moreover, the effects of environmental factors on EF can be expressed via community attributes such as population density and community composition rather than based on biodiversity [37]. For example, Liu et al. [38] investigated the effects of habitat fragmentation on EFs and revealed that non-biological environmental factors in fragmented landscapes can affect BEF under constant biodiversity levels. Habitat fragmentation can affect EF by shaping community composition (directly) or other environmental conditions (indirectly). The underlying mechanism associated with this process involves four factors [38]. First, habitat fragmentation can drive the nonrandom selection of species, while changes in community structure caused by the nonrandom selection of species may further influence EF. Second, habitat fragmentation results in the nonrandom decline of key functional traits, also affecting EF. Third, habitat fragmentation has a negative impact on species interactions and particularly on those of symbiotic species. This results in a reduced ecological complementarity effect, thus affecting EF [39]. Fourth, habitat fragmentation may affect species at different trophic levels (e.g., negative impacts on top predators), which consequently affects plant species via the top-down cascade effect [40]. Furthermore, the boundary effect induced by habitat fragmentation can affect environmental conditions such as temperature, humidity, wind, and light, subsequently affecting EF. The above works highlight the utility of considering various environmental factors in BEF-related research to investigate the direct and indirect effects of environmental factors on EF.

3.2. *Effects of Climate Change on BEF*

While the positive correlation between biodiversity and EFs is widely recognized, the majority of evidence of this relationship is based on small-scale observations [41]. As the effects of environmental factors (particularly of climate change) on BEF are still not fully understood, large-scale BEF-related studies face great challenges [35]. While both climate change and a decline in biodiversity will have a synergistic effect on ecosystems, the research on this topic remains limited [42]. On one hand, biodiversity may limit the effects of climate change on EF; on the other hand, climate change may alter the underlying mechanisms of BEF [43]. Thus, due to interactions between climate change and biodiversity loss, the effects of these two factors on EF cannot be separated.

BEF-related studies generally refer to climate change as the variation in precipitation and temperature. Numerous studies have used large-scale data platforms while others are based on controlled tests [44]. For example, Ratcliffe et al. [35] investigated the large-scale effects of potential driving factors on BEF based on the European Forest Data Centre platform. Their results suggest that the correlation between biodiversity and EF is strengthened under arid climates. In particular, due to the increasing levels of water stress induced by changes in climate, the importance of the effects of biodiversity on EFs in European forests may be enhanced [35]. Pires et al. [43] investigated the synergistic effects of precipitation variation and litter diversity on EF using controlled tests. Their results reveal that the effects of precipitation variation on scavenger communities are negatively related to litter diversity [43]. Thus, litter diversity can resist the effects of climate change. Variations in precipitation affect the underlying mechanisms that shape the effects of litter diversity on decomposition by hindering the complementarity effects of ecosystems while strengthening the selection effect. In reference to temperature variations, Wieczynski et al. [45] established a coupling relation for the trait-climate factors shaping forest ecosystems based on a global forest database. Functional diversity was found to decline with increases in latitude and altitude. Moreover, the study demonstrated that temperature and vapor pressure have strong impacts on the functional composition of the community and the ecological strategies of species [45]. Based on controlled tests, Garcia et al. [46] revealed a synergetic relation between global warming and biodiversity decline, which was in turn observed to have a negative impact on the functions of microbe communities. In terms of variations in the atmospheric composition, Eisenhauer et al. [47] investigated the effects of grassland plant diversity on MEF under four environmental conditions (natural ambient conditions, increased CO_2 concentrations, addition of nitrogen, and simultaneous increases in carbon dioxide and nitrogen). Their results showed a positive correlation between plant diversity and MEF that weakened with nitrogen enrichment [47].

According to the above studies, climate change has a significant impact on BEF. Therefore, the effects of climate change, as a natural disturbance, on EF can be adjusted by biodiversity [48]. Hisano et al. [49] reviewed BEF-related studies to reveal that biodiversity can mitigate (exacerbate) the negative (positive) impacts of climate change on EF. Similarly, numerous BEF-related studies have claimed that biodiversity can improve the productivity, stability, and ecological elasticity of an ecosystem, which consequently counteracts the impacts of climate change [50]. Hisano et al. [49] suggest that climate change can be regarded as a disturbance and hence can be incorporated into BEF research on disturbances. In particular, each form of climate change corresponds with different types of disturbances. For example, extreme and long-term climate events correspond to pulse and press disturbances, respectively. Biodiversity can promote the positive effects of disturbances on resource enrichment and productivity [48]. Furthermore, biodiversity can relieve the negative effects that disturbances may have on productivity, improving ecosystem stability [51]. For example, biodiversity can reduce the negative impacts of climate events on tree growth by regulating the demographic patterns, stability, and ecological elasticity of natural forest ecosystems.

4. Spatial and Temporal Scale Effects on BEF

4.1. Spatial Scale Effects on BEF

BEF-related research conducted over different spatial scales and particularly of larger scales can facilitate a thorough and accurate understanding of BEF [52]. At present, biodiversity decline is widely observed with much research demonstrating that declines in the density of local plant species can damage both EF and ecosystem services [1]. However, most BEF-related studies focus on a specific region, and it thus remains unclear how biodiversity interactions across different spatial scales affect EF [53]. This limits our understanding of BEF variations as a function of spatial scale [54,55]. Recent studies have attempted to correlate the effects of biodiversity across different scales from metapopulation dynamics or new scale-up methods. For example, Bracken et al. [55] evaluated the scale dependence of the effects of species decline on the biomass of intertidal algae based on data from the coast of Maine in the USA. The study reveals weak (strong) effects of biodiversity on EF at small (large) scales. Hautier et al. [56] analyzed eight EFs over 65 globally distributed grassland ecosystems and found that plant diversity helps maintain the ecosystem services provided by grasslands at both regional and landscape scales. The study suggested that diversity at a single spatial scale can affect the EFs of other spatial scales [56]. Moreover, Lefcheck et al. [54] correlated the β diversity of tropical fish in Dominica with their herbivorous rate over varying spatial scales. Their results reveal that the herbivorous nature of fish is positively correlated with β diversity. This result is consistent with the spatial insurance hypothesis, i.e., due to spatial heterogeneity, biodiversity is particularly important at regional scales with different landscape units. However, research on local EFs that considers regional background information is lacking [54]. As natural resource management is usually performed at regional scales, integration at this scale is of great significance. Moreover, an understanding of BEF across different spatial scales can help thoroughly and accurately reveal the underlying mechanisms of BEF and is thus of great significance to both theoretical studies and management practices used in the industry.

4.2. Effects of Temporal Scale on BEF

Understanding of the effects of biodiversity on EF as a function of time is essential to predict the effects of biodiversity change and management measures on EF and ecosystem services [57]. Research has indicated that the positive effects of biodiversity on EF can strengthen with time as a result of the enhancement of the interspecific complementarity effect with time [58]. For example, Guerrero-Ramírez et al. [59] used long-term biodiversity-controlled tests to demonstrate the positive correlation between biodiversity effects on EF for grasslands with time. This can be attributed to variations in species distribution patterns as a function of time. However, Guerrero-Ramírez et al. [59] did not observe any regular BEF trend as a function of time for forest ecosystems. To date, factors driving BEF trajectories with time have yet to be determined, and whether these factors are caused by changes in the environment must be clarified. It is also unclear whether these factors are consistent across environmental conditions. In particular, while soil characteristics may affect both transient BEF and its trend as a function of time, the exact mechanisms involved remain unclear [60]. For this reason, long-term BEF experiments in natural environments must be pursued.

Numerous controlled tests reported in the BEF-related literature do not consider temporal scales. In these tests, the natural reproduction and succession of species are usually terminated by the removal of non-sown species [61–63]. This not only enhances ecosystem selection and complementarity effects but may also reduce the effects of competition exclusion, resulting in indeterminate results [64]. For example, BEF studies conducted under non-weeding environmental conditions remain relatively limited compared to those of weeding environmental conditions [62]. The establishment of such non-weeding communities echoes the ecological restoration of abandoned farmland. Among these communities, the biodiversity of the initial community is controlled by a sowing event while the diversity and yield of subsequent communities is dependent on the development of the initial community. However, the effects of community development on BEF remain to be clarified. Veen et al. [65] observed that during the natural succession of a community, initial species (selected manually) affect the composition and biodiversity of the community at an early stage while biodiversity and EF are independently or negatively correlated during secondary community successions. The mechanisms behind these processes, however, require further investigation.

5. Generating-Presentation Model

5.1. Model Significance and Structure

The above review of the different effect factors discussed in BEF-related research shows that numerous research gaps on effect factors in BEF-related research remain and must be explored further. Thus, it is urgently necessary to construct a way to systematically consider factors that affect BEF and to provide suggestions for future research. We thus build a conceptual model with a circle-like structure called the generating-presentation model (Figure 1). The conceptual model summarizes the effect factors of BEF and comprehensively explores the effects of different factors on EFs. Because biodiversity can be regarded as a community-related factor that affects EF [9], a conceptual model accounting for the relationships among various effect factors and EFs can be used to express the links between effect factors and BEF.

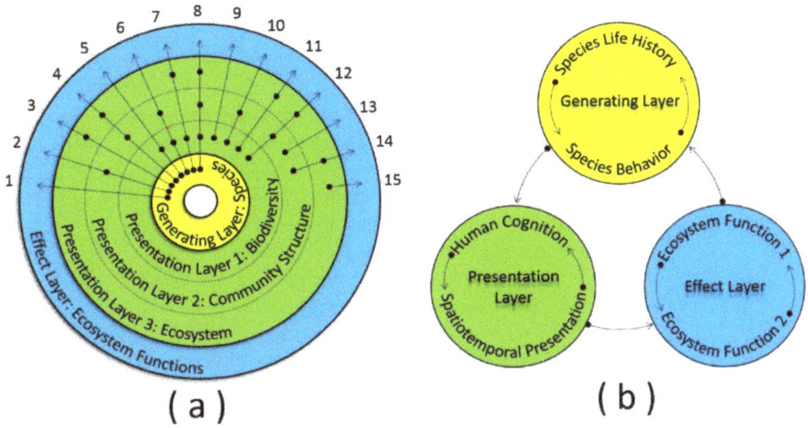

Figure 1. The generating-presentation model. (**a**) The main section of the generating-presentation model includes three layers: the generating layer, presentation layer, and effect layer from the inner layer to the outer layer. The presentation layer includes three sublayers: presentation layers 1–3 from the inner layer to the outer layer. The focuses of each layer and sublayer are illustrated in the figure. "A•→B" denotes the effect of A on B. Ecosystem functions (EFs) can be affected according to 15 approaches. For example, approach 1 indicates that EF is directly affected by species life history or species behavior. Approach 2 states that EF can be affected by community structures and species life history or species behavior. (**b**) There is a feedback loop between the generating, presentation, and effect layers: variations in the presentation layer generated from variations in the generating layer can affect the effect layer, and the effect layer can also have a feedback effect on the generating layer. The generating layer shows an interaction effect between species life history and species behavior. The presentation layer illustrates interactions between human cognition and spatiotemporal scale. The effect layer reveals tradeoffs and synergies among different EFs.

The main section of the generating-presentation model is composed of three layers: a generating layer, presentation layer, and effect layer from the inner layer to the outer layer. The presentation layer includes three sublayers: presentation layers 1–3 from the inner layer to the outer layer. Each layer and sublayer has its own structure and implications. The outer layer or sublayer includes an inner layer or sublayer except for the effect layer.

5.2. Generating Layer

The generating layer represents the effect factors related to a species' life history and behavior. Due to biodiversity, community structures and biological parts of ecosystems can be considered as the existential state of species at a higher scale, and this layer can then be defined as the generating layer. Variations in biodiversity, community structures or biological parts of ecosystem can result from variations in the generating layer similar to how bodily variations can result from gene variations. The generating layer covers two features: species life history (e.g., the birth, senility, illness, and death of a species) and species behavior (e.g., competitiveness, mutualism, predation, etc.). There is an interaction effect between these two sections. For example, Maynard [22] quantified the links among a competitive network structure, biodiversity, and community functioning, and found that a competitive network can determine the direction of BEF relationships.

5.3. Presentation Layer

The presentation layer represents the effect factors related to biodiversity, community structures, and ecosystems. The layer represents the existential state of species and is thus regarded as a presentation layer because variations in the generating layer can be represented by this layer. The presentation layer includes three sublayers. The first sublayer (presentation layer one) is related to biodiversity and can be considered part of the community structure. Biodiversity often refers to functional diversity, phylogenetic diversity, species classification diversity, and so on. However, according to Oehri et al. [52], landscape diversity is also positively linked to EFs. The second sublayer (presentation layer two) represents the community structure and can be viewed as a part of an ecosystem. According to the works reviewed above, many effect factors of community structure are related to different aspects of community structure aside from biodiversity, including species' spatial patterns and trophic level structures. However, Delgado-Baquerizo et al. [66] found that the BEF relationship can also be affected by the biodiversity of other biomes. The authors show that the positive relationship between plant diversity and multifunctionality can be affected by soil biodiversity. The last sublayer (presentation layer three) is related to ecosystems and covers not only biological features but also abiotic environments. Due to the strong effects of environmental factors such as temperature and geology on EFs [67–69], presentation layer three reflects abiotic factors in most cases.

Moreover, the presentation layer reflects the interactions between human cognition and spatiotemporal scales. First, the presentation of species in this layer depends on levels of human cognition. The development of human cognition can enhance the form of presentation used. For example, according to human cognitive development, numerous biodiversity indexes with different ecological implications have been built to reflect certain aspects of biodiversity [53,70]. According to cognitive development, different methods for expressing community structures such as pattern indexes have also been proposed [71]. Second, as it is meaningless to study biodiversity, communities, and ecosystems without considering a spatiotemporal scale, human cognition about biodiversity, communities, and ecosystems is limited by the spatiotemporal scale. Moreover, the development of human cognition can also improve our understanding of scale effects and can further enhance the accuracy of species presentation in the presentation layer. Over the last three decades, numerous studies have focused on BEF at fine scales and over short time periods [72]. Such studies reveal the mechanisms of BEF at small scales but do not test the mechanisms involved at larger scales. Further understanding the role of scale in BEF research can help guide policies on BEF management [53].

5.4. Effect Layer

The effect layer focuses on EFs and forms the outermost layer of the generating-presentation model. Because EF is related to the effects of ecosystems, the layer is termed the effect layer. Within this layer, there are tradeoffs and synergies among different EFs. In addition, variations in the presentation layer generated from variations in the generating layer can influence the effect layer. The effect layer can also have feedback effects on the generating layer. A feedback loop thus exists between these three layers.

6. Discussion

6.1. Measuring Factors Affecting EFs

From the generating-presentation model we can conclude that EFs can be affected by different factors corresponding to different layers or sublayers through 15 approaches. For example, according to approach 1, EFs are directly affected by species life history or species behavior. According to approach 2, EFs can be affected by community structures and species life history or species behavior. Among the 15 approaches, 8 (approaches 5–12) are related to biodiversity. Based on the factors identified to affect BEF in recent research, various effect factors are incorporated into the generating-presentation model (Table 1). It should be noted that the generating-presentation model can incorporate not only factors affecting BEF but also those affecting EFs. In other words, biodiversity is identified as a common factor that affects EFs in the model.

Table 1. The approaches to express the effects of factors on ecosystem functions (EFs) in generating-presentation model.

Approach	Meaning
1	EF is affected by species life history or species behavior directly.
2	EF is affected by the community structure, and species life history or species behavior.
3	EF is affected by the ecosystem, and species life history or species behavior.
4	EF is affected by the ecosystem, community structure, and species life history or species behavior.
5	EF is affected by biodiversity, and species life history or species behavior. Examples can be found in studies [23,24,57,65].
6	EF is affected by the community structure, biodiversity, and species life history or species behavior. Examples can be found in studies [14].
7	EF is affected by the ecosystem, biodiversity, and species life history or species behavior. Examples can be found in studies [43].
8	EF is affected by the ecosystem, community structure, biodiversity, and species life history or behavior. Examples can be found in studies [29,34,37,45].
9	EF is affected by biodiversity directly. Examples can be found in studies [53,55,66].
10	EF is affected by the community structure and biodiversity. Examples can be found in studies [11,16,21].
11	EF is affected by the ecosystem and biodiversity. Examples can be found in studies [32,35,49,54,59].
12	EF is affected by the ecosystem, community structure, and biodiversity. Examples can be found in studies [7,28,31,33,38,47].
13	EF is affected by the community structure directly.
14	EF is affected by the ecosystem and community structure.
15	EF is affected by the ecosystem directly.

The generating-presentation model provides a useful reference system for EFs or ecosystem services research. In future BEF research, the approaches of effects of biodiversity on EFs or ecosystem services can be clearly presented by this model. For instance, some recent BEF studies cited in this article can be grouped in approaches 5–12 (Table 1).

6.2. Integrating Factors Affecting BEF with the Generating-Presentation Model

In general, the ultimate goal of BEF research is to evaluate the state of EF. In some cases, the correlation between biodiversity and EF has been described as involving a cascade process [72]. On one hand, as expressed by the approaches illustrated in the generating-presentation model, this cascade process indicates that biodiversity, which is affected by species, can influence the community structures and compositions of ecosystems. On the other hand, as Isbell et al. [72] showed in their research, because of cascade processes, the impacts of human activities on BEF may increase at larger spatial and longer temporal scales. Furthermore, community structures and compositions can affect ecosystem processes, which can in turn further lead to a variety of EFs. However, as a dynamic and open system, the natural ecosystem has a typical dissipative structure whose factors have complex sources. As a result, the separation of biodiversity and EFs from the ecosystem without considering integrated features is not appropriate for BEF-related research.

Biodiversity decline has drawn attention to the effects of biodiversity loss on EFs [2,73]. However, biodiversity is only a factor that affects EFs. Thus, when the research object is to evaluate the state of EF, not only biodiversity but other factors must be studied—especially when the main effect factors do not include biodiversity (approaches 1–4 and 13–15 of the generating-presentation model). However, when the research objective is to evaluate the state of biodiversity, factors related to biodiversity, such as land-use change, may be considered [74]. These two research objectives are relatively independent of each other, though relationships between biodiversity and EFs have been found in numerous studies. Therefore, research on BEF may not be relevant in reference to certain areas, as relationships between biodiversity and EFs are not always significant in certain regions.

7. Conclusions

As a new avenue for research in ecology, BEF research generally focuses on the effects of biodiversity loss on EF. However, other factors, with the exception of biodiversity, also affect EF and play an important role in BEF-related studies. In this work, the different factors that affect BEF have been reviewed and summarized, and a new conceptual model, the generating-presentation model, accounting for the links between effect factors and EF has been built to provide a systematic means of understanding how different factors affect BEF. The model shows that correlations between biodiversity and EF can be described as involving a cascade process and that the separation of biodiversity and EFs from the ecosystem without considering integrated features is not appropriate for BEF-related research. The generating-presentation model thus comprehensively reveals the effects of different factors on EFs and thus has key theoretical and applied implications. It is suggested that the integration of effect factors in BEF could allow us to greatly advance our knowledge of BEF.

Author Contributions: Conceptualization, writing—original draft, J.H.; writing—review and editing, H.F.; investigation, M.W. All authors have read and agreed to the published version of the manuscript.

Funding: This research was supported by the National Key Research and Development Program of China [no. 2016YFC0500802].

Institutional Review Board Statement: Not applicable.

Data Availability Statement: Not applicable.

Acknowledgments: The authors would like to thank anonymous referees for their constructive suggestions.

Conflicts of Interest: The authors declare no conflict of interest.

References

1. Paul, C.; Hanley, N.; Meyer, S.T.; Fürst, C.; Weisser, W.W.; Knoke, T. On the functional relationship between biodiversity and economic value. *Sci. Adv.* **2020**, *6*, eaax7712. [CrossRef] [PubMed]
2. Loreau, M.; Oteng-Yeboah, A.; Arroyo, M.T.K.; Babin, D.; Barbault, R.; Donoghue, M.; Gadgil, M.; Hauser, C.; Heip, C.; Larigauderie, A.; et al. Diversity without representation. *Nature* **2006**, *442*, 245–246. [CrossRef] [PubMed]
3. van der Plas, F. Biodiversity and ecosystem functioning in naturally assembled communities. *Biol. Rev. Camb. Philos. Soc.* **2019**, *94*, 1220–1245. [CrossRef] [PubMed]
4. Isbell, F.; Calcagno, V.; Hector, A.; Connolly, J.; Harpole, W.S.; Reich, P.B.; Scherer-Lorenzen, M.; Schmid, B.; Tilman, D.; van Ruijven, J.; et al. High plant diversity is needed to maintain ecosystem services. *Nature* **2011**, *477*, 199–202. [CrossRef]
5. Leibold, M.A.; Chase, J.M.; Ernest, S.K.M. Community assembly and the functioning of ecosystems: How metacommunity processes alter ecosystems attributes. *Ecology* **2017**, *98*, 909–919. [CrossRef]
6. Jaillard, B.; Rapaport, A.; Harmand, J.; Brauman, A.; Nunan, N. Community assembly effects shape the biodiversity-ecosystem functioning relationships. *Funct. Ecol.* **2014**, *28*, 1523–1533. [CrossRef]
7. Cao, X.; Chai, L.; Jiang, D.; Wang, J.; Liu, Y.; Huang, Y. Loss of biodiversityalters ecosystem function in freshwater streams: Potential evidence from benthic macroinvertebrates. *Ecosphere* **2018**, *9*, e02445. [CrossRef]
8. Huang, Y.; Chen, Y.; Castro-Izaguirre, N.; Baruffol, M.; Brezzi, M.; Lang, A.; Li, Y.; Hardtle, W.; von Oheimb, G.; Yang, X.; et al. Impacts of species richness on productivity in a large-scale subtropical forest experiment. *Science* **2018**, *362*, 80–83. [CrossRef]

9. Bannar-Martin, K.H.; Kremer, C.T.; Ernest, S.K.M.; Leibold, M.A.; Auge, H.; Chase, J.; Declerck, S.A.J.; Eisenhauer, N.; Harpole, S.; Hillebrand, H.; et al. Integrating community assembly and biodiversity to better understand ecosystem function: The Community Assembly and the Functioning of Ecosystems (CAFE) approach. *Ecol. Lett.* **2018**, *21*, 167–180. [CrossRef]
10. Thompson, P.L.; Gonzalez, A. Ecosystem multifunctionality in metacommunities. *Ecology* **2016**, *97*, 2867–2879. [CrossRef]
11. Wang, M.-Q.; Li, Y.; Chesters, D.; Anttonen, P.; Bruelheide, H.; Chen, J.-T.; Durka, W.; Guo, P.-F.; Haerdtle, W.; Ma, K.; et al. Multiple components of plant diversity loss determine herbivore phylogenetic diversity in a subtropical forest experiment. *J. Ecol.* **2019**, *107*, 2697–2712. [CrossRef]
12. Wright, A.J.; Wardle, D.A.; Callaway, R.; Gaxiola, A. The Overlooked Role of Facilitation in Biodiversity Experiments. *Trends Ecol. Evol.* **2017**, *32*, 383–390. [CrossRef]
13. Barry, K.E.; Mommer, L.; van Ruijven, J.; Wirth, C.; Wright, A.J.; Bai, Y.; Connolly, J.; De Deyn, G.B.; de Kroon, H.; Isbell, F.; et al. The Future of Complementarity: Disentangling Causes from Consequences. *Trends Ecol. Evol.* **2019**, *34*, 167–180. [CrossRef]
14. Heilpern, S.A.; Weeks, B.C.; Naeem, S. Predicting ecosystem vulnerability to biodiversity loss from community composition. *Ecology* **2018**, *99*, 1099–1107. [CrossRef]
15. Soliveres, S.; van der Plas, F.; Manning, P.; Prati, D.; Gossner, M.M.; Renner, S.C.; Alt, F.; Arndt, H.; Baumgartner, V.; Binkenstein, J.; et al. Biodiversity at multiple trophic levels is needed for ecosystem multifunctionality. *Nature* **2016**, *536*, 456–459. [CrossRef]
16. Schuldt, A.; Assmann, T.; Brezzi, M.; Buscot, F.; Eichenberg, D.; Gutknecht, J.; Hardtle, W.; He, J.S.; Klein, A.M.; Kuhn, P.; et al. Biodiversity across trophic levels drives multifunctionality in highly diverse forests. *Nat. Commun.* **2018**, *9*, 2989. [CrossRef]
17. Wang, S.; Brose, U.; Gravel, D. Intraguild predation enhances biodiversity and functioning in complex food webs. *Ecology* **2019**, *100*, e02616. [CrossRef]
18. Enquist, B.J.; Abraham, A.J.; Harfoot, M.B.J.; Malhi, Y.; Doughty, C.E. The megabiota are disproportionately important for biosphere functioning. *Nat. Commun.* **2020**, *11*, 699. [CrossRef]
19. Zavaleta, E.S.; Pasari, J.R.; Hulvey, K.B.; Tilman, G.D. Sustaining multiple ecosystem functions in grassland communities requires higher biodiversity. *Proc. Natl. Acad. Sci. USA* **2010**, *107*, 1443–1446. [CrossRef]
20. Lefcheck, J.S.; Byrnes, J.E.K.; Isbell, F.; Gamfeldt, L.; Griffin, J.N.; Eisenhauer, N.; Hensel, M.J.S.; Hector, A.; Cardinale, B.J.; Duffy, J.E. Biodiversity enhances ecosystem multifunctionality across trophic levels and habitats. *Nat. Commun.* **2015**, *6*, 6936. [CrossRef]
21. Barnes, A.D.; Jochum, M.; Lefcheck, J.S.; Eisenhauer, N.; Scherber, C.; O'Connor, M.I.; de Ruiter, P.; Brose, U. Energy Flux: The Link between Multitrophic Biodiversity and Ecosystem Functioning. *Trends Ecol. Evol.* **2018**, *33*, 186–197. [CrossRef]
22. Maynard, D.S.; Crowther, T.W.; Bradford, M.A. Competitive network determines the direction of the diversity–function relationship. *Proc. Natl. Acad. Sci. USA* **2017**, *114*, 11464–11469. [CrossRef]
23. Frainer, A.; McKie, B.G.; Amundsen, P.A.; Knudsen, R.; Lafferty, K.D. Parasitism and the Biodiversity-Functioning Relationship. *Trends Ecol. Evol.* **2018**, *33*, 260–268. [CrossRef]
24. Ferlian, O.; Cesarz, S.; Craven, D.; Hines, J.; Barry, K.E.; Bruelheide, H.; Buscot, F.; Haider, S.; Heklau, H.; Herrmann, S.; et al. Mycorrhiza in tree diversity–ecosystem function relationships: Conceptual framework andexperimental implementation. *Ecosphere* **2018**, *9*, e02226. [CrossRef]
25. Merrild, M.P.; Ambus, P.; Rosendahl, S.; Jakobsen, I. Common arbuscular mycorrhizal networks amplify competition for phosphorus between seedlings and established plants. *New Phytol.* **2013**, *200*, 229–240. [CrossRef]
26. Scheublin, T.R.; Van Logtestijn, R.S.P.; Van Der Heijden, M.G.A. Presence and identity of arbuscular mycorrhizal fungi influence competitive interactions between plant species. *J. Ecol.* **2007**, *95*, 631–638. [CrossRef]
27. Laforest-Lapointe, I.; Paquette, A.; Messier, C.; Kembel, S.W. Leaf bacterial diversity mediates plant diversity and ecosystem function relationships. *Nature* **2017**, *546*, 145–147. [CrossRef]
28. Hu, A.; Wang, J.; Sun, H.; Niu, B.; Si, G.; Yeh, C.F.; Zhu, X.; Lu, X.; Zhou, J.; Yang, Y.; et al. Mountain biodiversity and ecosystem functions: Interplay between geology and contemporary environments. *ISME J.* **2020**, *14*, 931–944. [CrossRef]
29. Lewington-Pearce, L.; Parker, B.; Narwani, A.; Nielsen, J.M.; Kratina, P. Diversity and temperature indirectly reduce CO_2 concentrations in experimental freshwater communities. *Oecologia* **2020**, *192*, 515–527. [CrossRef]
30. Theis, S.; Ruppert, J.L.W.; Roberts, K.N.; Minns, C.K.; Koops, M.; Poesch, M.S. Compliance with and ecosystem function of biodiversity offsets in North American and European freshwaters. *Conserv. Biol. J. Soc. Conserv. Biol.* **2020**, *34*, 41–53. [CrossRef]
31. Nunes, C.A.; Braga, R.F.; de Moura Resende, F.; de Siqueira Neves, F.; Figueira, J.E.C.; Fernandes, G.W. Linking Biodiversity, the Environment and Ecosystem Functioning: Ecological Functions of Dung Beetles Along a Tropical Elevational Gradient. *Ecosystems* **2018**, *21*, 1244–1254. [CrossRef]
32. Zirbel, C.R.; Grman, E.; Bassett, T.; Brudvig, L.A. Landscape context explains ecosystem multifunctionality in restored grasslands better than plant diversity. *Ecology* **2019**, *100*, e02634. [CrossRef] [PubMed]
33. Smeti, E.; von Schiller, D.; Karaouzas, I.; Laschou, S.; Vardakas, L.; Sabater, S.; Tornes, E.; Monllor-Alcaraz, L.S.; Guillem-Argiles, N.; Martinez, E.; et al. Multiple stressor effects on biodiversity and ecosystem functioning in a Mediterranean temporary river. *Sci. Total Environ.* **2019**, *647*, 1179–1187. [CrossRef] [PubMed]
34. Dib, V.; Pires, A.P.F.; Casa Nova, C.; Bozelli, R.L.; Farjalla, V.F. Biodiversity-mediated effects on ecosystem functioning depend on the type and intensity of environmental disturbances. *Oikos* **2020**, *129*, 433–443. [CrossRef]
35. Ratcliffe, S.; Wirth, C.; Jucker, T.; van der Plas, F.; Scherer-Lorenzen, M.; Verheyen, K.; Allan, E.; Benavides, R.; Bruelheide, H.; Ohse, B.; et al. Biodiversity and ecosystem functioning relations in European forests depend on environmental context. *Ecol. Lett.* **2017**, *20*, 1414–1426. [CrossRef]

36. Gammal, J.; Järnström, M.; Bernard, G.; Norkko, J.; Norkko, A. Environmental Context Mediates Biodiversity–Ecosystem Functioning Relationships in Coastal Soft-sediment Habitats. *Ecosystems* **2018**, *22*, 137–151. [CrossRef]
37. Spaak, J.W.; Baert, J.M.; Baird, D.J.; Eisenhauer, N.; Maltby, L.; Pomati, F.; Radchuk, V.; Rohr, J.R.; Van den Brink, P.J.; De Laender, F. Shifts of community composition and population density substantially affect ecosystem function despite invariant richness. *Ecol. Lett.* **2017**, *20*, 1315–1324. [CrossRef]
38. Liu, J.; Wilson, M.; Hu, G.; Liu, J.; Wu, J.; Yu, M. How does habitat fragmentation affect the biodiversity and ecosystem functioning relationship? *Landsc. Ecol.* **2018**, *33*, 341–352. [CrossRef]
39. Smith, M.D.; Knapp, A.K. Dominant species maintain ecosystem function with non-random species loss. *Ecol. Lett.* **2003**, *6*, 509–517. [CrossRef]
40. Poisot, T.; Mouquet, N.; Gravel, D. Trophic complementarity drives the biodiversity-ecosystem functioning relationship in food webs. *Ecol. Lett.* **2013**, *16*, 853–861. [CrossRef]
41. Frank, D.M. Science and values in the biodiversity-ecosystem function debate. *Biol. Philos.* **2022**, *37*, 7. [CrossRef]
42. Rogers, A.D.; Frinault, B.A.V.; Barnes, D.K.A.; Bindoff, N.L.; Downie, R.; Ducklow, H.W.; Friedlaender, A.S.; Hart, T.; Hill, S.L.; Hofmann, E.E.; et al. Antarctic Futures: An Assessment of Climate-Driven Changes in Ecosystem Structure, Function and Service Provisioning in the Southern Ocean. *Annu. Rev. Mar. Sci.* **2020**, *12*, 7.1–7.34. [CrossRef]
43. Pires, A.P.F.; Srivastava, D.S.; Marino, N.A.C.; MacDonald, A.A.M.; Figueiredo-Barros, M.P.; Farjalla, V.F. Interactive effects of climate change and biodiversity loss on ecosystem functioning. *Ecology* **2018**, *99*, 1203–1213. [CrossRef]
44. Gazol, A.; Julio Camarero, J. Functional diversity enhances silver fir growth resilience to an extreme drought. *J. Ecol.* **2016**, *104*, 1063–1075. [CrossRef]
45. Wieczynski, D.J.; Boyle, B.; Buzzard, V.; Duran, S.M.; Henderson, A.N.; Hulshof, C.M.; Kerkhoff, A.J.; McCarthy, M.C.; Michaletz, S.T.; Swenson, N.G.; et al. Climate shapes and shifts functional biodiversity in forests worldwide. *Proc. Natl. Acad. Sci. USA* **2019**, *116*, 587–592. [CrossRef]
46. Garcia, F.C.; Bestion, E.; Warfield, R.; Yvon-Durocher, G. Changes in temperature alter the relationship between biodiversity and ecosystem functioning. *Proc. Natl. Acad. Sci. USA* **2018**, *115*, 10989–10994. [CrossRef]
47. Eisenhauer, N.; Hines, J.; Isbell, F.; van der Plas, F.; Kazanski, C.E.; Lehmann, A.; Liu, M.; Hobbie, S.E.; Lochner, A.; Rillig, M.C.; et al. Plant diversity maintains multiple soil functions in future environments. *eLife* **2018**, *7*, e41228. [CrossRef]
48. Wright, A.J.; Ebeling, A.; de Kroon, H.; Roscher, C.; Weigelt, A.; Buchmann, N.; Buchmann, T.; Fischer, C.; Hacker, N.; Hildebrandt, A.; et al. Flooding disturbances increase resource availability and productivity but reduce stability in diverse plant communities. *Nat. Commun.* **2015**, *6*, 6092. [CrossRef]
49. Hisano, M.; Searle, E.B.; Chen, H.Y.H. Biodiversity as a solution to mitigate climate change impacts on the functioning of forest ecosystems. *Biol. Rev. Camb. Philos. Soc.* **2018**, *93*, 439–456. [CrossRef]
50. Isbell, F.; Craven, D.; Connolly, J.; Loreau, M.; Schmid, B.; Beierkuhnlein, C.; Bezemer, T.M.; Bonin, C.; Bruelheide, H.; de Luca, E.; et al. Biodiversity increases the resistance of ecosystem productivity to climate extremes. *Nature* **2015**, *526*, 574–577. [CrossRef]
51. Poorter, L.; Bongers, F.; Aide, T.M.; Almeyda Zambrano, A.M.; Balvanera, P.; Becknell, J.M.; Boukili, V.; Brancalion, P.H.S.; Broadbent, E.N.; Chazdon, R.L.; et al. Biomass resilience of Neotropical secondary forests. *Nature* **2016**, *530*, 211–214. [CrossRef]
52. Oehri, J.; Schmid, B.; Schaepman-Strub, G.; Niklaus, P.A. Terrestrial land-cover type richness is positively linked to landscape-level functioning. *Nat. Commun.* **2020**, *11*, 154. [CrossRef]
53. Gonzalez, A.; Germain, R.M.; Srivastava, D.S.; Filotas, E.; Dee, L.E.; Gravel, D.; Thompson, P.L.; Isbell, F.; Wang, S.; Kefi, S.; et al. Scaling-up biodiversity-ecosystem functioning research. *Ecol. Lett.* **2020**, *23*, 757–776. [CrossRef]
54. Lefcheck, J.S.; Innes-Gold, A.A.; Brandl, S.J.; Steneck, R.S.; Torres, R.E.; Rasher, D.B. Tropical fish diversity enhances coral reef functioning across multiple scales. *Sci. Adv.* **2019**, *5*, eaav6420. [CrossRef]
55. Bracken, M.E.S.; Douglass, J.G.; Perini, V.; Trussell, G.C. Spatial scale mediates the effects of biodiversity on marine primary producers. *Ecology* **2017**, *98*, 1434–1443. [CrossRef]
56. Hautier, Y.; Isbell, F.; Borer, E.T.; Seabloom, E.W.; Harpole, W.S.; Lind, E.M.; MacDougall, A.S.; Stevens, C.J.; Adler, P.B.; Alberti, J.; et al. Local loss and spatial homogenization of plant diversity reduce ecosystem multifunctionality. *Nat. Ecol. Evol.* **2018**, *2*, 50–56. [CrossRef]
57. van Moorsel, S.J.; Schmid, M.W.; Hahl, T.; Zuppinger-Dingley, D.; Schmid, B. Selection in response to community diversity alters plant performance and functional traits. *Perspect. Plant Ecol. Evol. Syst.* **2018**, *33*, 51–61. [CrossRef]
58. Fargione, J.; Tilman, D.; Dybzinski, R.; Hille Ris Lambers, J.; Clark, C.; Harpole, W.S.; Knops, J.M.H.; Reich, P.B.; Loreau, M. From selection to complementarity: Shifts in the causes of biodiversity-productivity relationships in a long-term biodiversity experiment. *Proc. R. Soc. B Biol. Sci.* **2007**, *274*, 871–876. [CrossRef]
59. Guerrero-Ramírez, N.R.; Craven, D.; Reich, P.B.; Ewel, J.J.; Isbell, F.; Koricheva, J.; Parrotta, J.A.; Auge, H.; Erickson, H.E.; Forrester, D.I.; et al. Diversity-dependent temporal divergence of ecosystem functioning in experimental ecosystems. *Nat. Ecol. Evol.* **2017**, *1*, 1639–1642. [CrossRef]
60. Hautier, Y.; Seabloom, E.W.; Borer, E.T.; Adler, P.B.; Harpole, W.S.; Hillebrand, H.; Lind, E.M.; MacDougall, A.S.; Stevens, C.J.; Bakker, J.D.; et al. Eutrophication weakens stabilizing effects of diversity in natural grasslands. *Nature* **2014**, *508*, 521–525. [CrossRef]
61. Roscher, C.; Temperton, V.M.; Buchmann, N.; Schulze, E.-D. Community assembly and biomass production in regularly and never weeded experimental grasslands. *Acta Oecol. Int. J. Ecol.* **2009**, *35*, 206–217. [CrossRef]

62. Doherty, J.M.; Callaway, J.C.; Zedler, J.B. Diversity-function relationships changed in a long-term restoration experiment. *Ecol. Appl.* **2011**, *21*, 2143–2155. [CrossRef] [PubMed]
63. Bezemer, T.M.; van der Putten, W.H. Ecology—Diversity and stability in plant communities. *Nature* **2007**, *446*, E6–E7. [CrossRef] [PubMed]
64. Grace, J.B.; Anderson, T.M.; Seabloom, E.W.; Borer, E.T.; Adler, P.B.; Harpole, W.S.; Hautier, Y.; Hillebrand, H.; Lind, E.M.; Paertel, M.; et al. Integrative modelling reveals mechanisms linking productivity and plant species richness. *Nature* **2016**, *529*, 390–393. [CrossRef]
65. Veen, G.F.; Putten, W.H.; Bezemer, T.M. Biodiversity-ecosystem functioning relationships in a long-term non-weeded field experiment. *Ecology* **2018**, *99*, 1836–1846. [CrossRef]
66. Delgado-Baquerizo, M.; Reich, P.B.; Trivedi, C.; Eldridge, D.J.; Abades, S.; Alfaro, F.D.; Bastida, F.; Berhe, A.A.; Cutler, N.A.; Gallardo, A.; et al. Multiple elements of soil biodiversity drive ecosystem functions across biomes. *Nat. Ecol. Evol.* **2020**, *4*, 210–220. [CrossRef]
67. Bestion, E.; Barton, S.; Garcia, F.C.; Warfield, R.; Yvon-Durocher, G. Abrupt declines in marine phytoplankton production driven by warming and biodiversity loss in a microcosm experiment. *Ecol. Lett.* **2020**, *23*, 457–466. [CrossRef]
68. Smith, M.D.; Koerner, S.E.; Knapp, A.K.; Avolio, M.L.; Chaves, F.A.; Denton, E.M.; Dietrich, J.; Gibson, D.J.; Gray, J.; Hoffman, A.M.; et al. Mass ratio effects underlie ecosystem responses to environmental change. *J. Ecol.* **2020**, *108*, 855–864. [CrossRef]
69. Perkins, D.M.; Bailey, R.A.; Dossena, M.; Gamfeldt, L.; Reiss, J.; Trimmer, M.; Woodward, G. Higher biodiversity is required to sustain multiple ecosystem processes across temperature regimes. *Glob. Chang. Biol.* **2015**, *21*, 396–406. [CrossRef]
70. Mason, N.W.H.; Mouillot, D.; Lee, W.G.; Wilson, J.B. Functional richness, functional evenness and functional divergence: The primary components of functional diversity. *Oikos* **2005**, *111*, 112–118. [CrossRef]
71. Hou, J.; Wang, H.; Fu, B.; Zhu, L.; Wang, Y.; Li, Z. Effects of plant diversity on soil erosion for different vegetation patterns. *Catena* **2016**, *147*, 632–637. [CrossRef]
72. Isbell, F.; Gonzalez, A.; Loreau, M.; Cowles, J.; Diaz, S.; Hector, A.; Mace, G.M.; Wardle, D.A.; O'Connor, M.I.; Duffy, J.E.; et al. Linking the influence and dependence of people on biodiversity across scales. *Nature* **2017**, *546*, 65–72. [CrossRef]
73. Cardinale, B.J.; Duffy, J.E.; Gonzalez, A.; Hooper, D.U.; Perrings, C.; Venail, P.; Narwani, A.; Mace, G.M.; Tilman, D.; Wardle, D.A.; et al. Biodiversity loss and its impact on humanity. *Nature* **2012**, *486*, 59–67. [CrossRef]
74. Titeux, N.; Henle, K.; Mihoub, J.-B.; Regos, A.; Geijzendorffer, I.R.; Cramer, W.; Verburg, P.H.; Brotons, L. Biodiversity scenarios neglect future land-use changes. *Glob. Chang. Biol.* **2016**, *22*, 2505–2515. [CrossRef]

MDPI
St. Alban-Anlage 66
4052 Basel
Switzerland
Tel. +41 61 683 77 34
Fax +41 61 302 89 18
www.mdpi.com

Diversity Editorial Office
E-mail: diversity@mdpi.com
www.mdpi.com/journal/diversity

www.ingramcontent.com/pod-product-compliance
Lightning Source LLC
Chambersburg PA
CBHW061834210926
51850CB00006B/883
* 9 7 8 3 0 3 6 5 8 3 0 6 8 *